This book is to be returned on or before
the last date stamped below.

New technologies for the exploration and exploitation of oil and gas resources

Volume 2

New technologies for the exploration and exploitation of oil and gas resources

Proceedings of the 2nd E.C. Symposium held in Luxembourg, 5-7 December 1984

Volume 2

Edited by

R. De Bouw
E. Millich
J.P. Joulia
D. Van Asselt

Commission of the European Communities,
Directorate-General Energy, Brussels

Published by
Graham & Trotman
for the Commission of the European Communities

6/22/04

First published in 1985 by

Graham & Trotman Limited
Sterling House
66 Wilton Road
London SW1V 1DE

Graham & Trotman Inc.
13 Park Avenue
Gaithersburg
MD 20877, USA

ISBN 0 86010 675 6 (Volume 1)
    0 86010 676 4 (Volume 2)

EUR 10 168

Organisation of the conference:
Commission of the European Communities
Directorate-General Energy, Brussels

Publication arrangements:
D. NICOLAY
Commission of the European Communities
Directorate-General Information Market and Innovation, Luxembourg

D
622.338
NEW

Printed in Belgium by Ceuterick

C O N T E N T S

VOLUME I

DRILLING - PILING - SOIL INVESTIGATION

PLATFORMS AND OFFSHORE STRUCTURES

PRODUCTION OPERATIONS – EQUIPMENT – SUBSEA COMPLETIONS

VOLUME II

ENHANCED OIL RECOVERY - HEAVY OIL

PIPELINES

ENHANCED OIL RECOVERY – HEAVY OIL

Enhanced oil recovery processes for the Cortemaggiore field, Italy (05.08/77)

Piropo : a heavy-oil deposit in the Adriatic Sea – Studies for possible exploitation (05.10/78)

Development of an electrical system to prevent water coning on producing oil wells (05.12/78)

Electro-dispersion : feasibility (05.41/82)

Downhole steam generator for enhanced oil recovery (05.16/80)

Steam injection pilot project into Emeraude offshore reservoir (05.33/81; 05.42/82)

Injection of steam in the Marienbronn reservoir (05.19/80)

Exploitation of heavy oil : the FBH process (03.67/78)

Heavy oil pilot project Nordhorn (05.20/80)

Pilot plant for enhanced oil recovery in a heavy oil – Ponte Dirillo field – Italy (05.14/79)

Upgrading of heavy oil and bitumen (05.22/81)

Pretreatment of heavy oils at the oilfield (03.94/80; 05.30/81; 05.43/82)

Method pilot project and industrial pilot project for injection of microemulsion and polymer into Chateaurenard field (05.02/76; 05.21/80; 05.28/81)

Development of brine soluble polymers and associated chemicals for the enhanced recovery of petroleum (05.06/77)

Enhanced oil recovery from the Egmanton oilfield by carbon dioxide miscible flooding (05.15/80)

Pilot project for injection of miscible gas into Pecorade reservoir (05.29/81; 05.44/82; 05.48/83)

Nitrogen injection in North Sea Reservoirs (05.34/82)

Search for drainage methods for heavy and viscous oil reservoirs involving the risk of water inflows - application to Rospo Mare (05.09/78)

Production test on Rospo Mare heavy oil reservoir (Italy) (05.24/81)

Horizontal drilling in Rospo Mare oil reservoir (05.36/82)

Method of enhanced petroleum recovery by injection of CO_2 into Coulommes-Vaucourtois field (05.23/81)

Design and development of a homing-in device for blow-out control (03.112/81)

Research on improved hydrocarbon recovery from chalk deposits (05.04/76)

Efficient exploitation and utilization of Schandelah oil shale (05.26/81)

Treatment fluids to improve sea water injection (05.01/76)

Conditions of exploitation of tar shales in the Toarcian (05.05/76)

Pilot project for recovery of oil by steam injection (upper Lacq field) (05.07/77)

Tar shales : the Tranqueville in situ combustion pilot project (05.31/81)

Improving recovery from very heavy oil reservoirs (05.11/78)

Enhancement of distillation effects during steamflooding of heavy oil reservoirs (05.38/82)

(05.08/77)
ENHANCED OIL RECOVERY PROCESSES
FOR THE CORTEMAGGIORE FIELD, ITALY

E. CAUSIN, G.L. CHIERICI, M. ERBA, G. MIRABELLI and C. TURRIANI
Agip S.p.A.
Petroleum Engineering

Summary

The Cortemaggiore oilfield, discovered in 1949, consists of several pools, the deepest ones containing light oil. Oil production began in 1951 and was practically exhausted by 1978.

A revision of the reservoir geology indicates that the two most important oil pools, corresponding to levels E and F_1+F_2, are separated into blocks by transcurrent faults which, however, maintain the hydraulic continuity among the blocks through a common aquifer. This condition is ideal for testing different EOR techniques within the same field.

The possible application of the following EOR processes, to the reservoir under consideration, has been investigated in the laboratory: miscible displacement by CO_2; immiscible displacement by nitrogen; micellar/polymer flooding; vaporization of the residual oil into dry gas cycled through the oil reservoir. Numerical model studies have been performed. By drilling a new well and by working over some old ones, the primary oil production has been resumed. No pilot plant for testing the most promising EOR processes will be undertaken before the current primary production becomes exhausted.

1. RESERVOIR GEOLOGY

The Cortemaggiore oilfield is made up of several sand layers at different depths. The uppermost levels are of Pliocene age and contain condensate gas, whereas the lowermost ones, which are of Tortonian age, contain light oil. Levels E and F_1+F_2 are the most important ones and have been the object of detailed investigations. The reservoir-rock at these levels is made up of turbidites of the Marnoso-arenacea formation, showing remarkable heterogeneity both laterally and vertically.

Due to post-depositional tectonics the layers have been folded and faulted. Particularly important are the extensive post-Miocene transcurrent faults caused by tangential stresses striking SW-NE, which divided levels E and F_1+F_2 into six blocks (Fig. 1). Such a subdivision of the two levels into blocks is proven by the behavior of the wells and by their pressure history in particular.

2. PRODUCTION BEHAVIOR

The initial reservoir oil volume amounted to $598 \times 10^3 m^3$ STO for le-

vel E and 1492 x $10^3 m^3$ STO for level $F_1 + F_2$. The primary production mecha
nism for all blocks of the two levels was mainly edge water drive with lo
cal dissolved gas drive due to pressure decrease below the bubble point.
For each level the aquifer is common to all blocks.
At the time the reservoir was abandoned (1978) the average oil recovery
was 25.9% and 28.4% for levels E and $F_1 + F_2$ respectively. However, because
of the irregular well distribution, these recovery percentages varied wi
dely from one block to the other within the same level. At level E percen
tages varied from 56.7 for block 5 down to 1.5 for block 3, whereas at le
vel $F_1 + F_2$ they varied from 56.6 to 3.9 for blocks 4 and 6 respectively.
A combination of radioactive logs (CNL+TDT-K) recorded in seven wells re
vealed the presence of 'isles' of reservoir-rock not displaced by water
and thus containing oil with a saturation very close to the initial value.

3. PVT STUDIES ON RESERVOIR OIL

The whole thermodynamic study has been carried out on reservoir oil
of the $F_1 + F_2$ level, which is the most important of the field. Its initial
pressure of 21.6 MPa dropped to a minimum of 13.7 MPa during exploitation,
to rise back to 20.1 MPa at the time of oil sampling. The reservoir tempe
rature is 54°C. The stock tank oil has an API gravity of 39.5° and a visco
sity of 2.7 mPa·s.
The oil bubble-point pressure is 15.8 MPa and its viscosity at reservoir conditions has a value of 0.55 mPa·s.
In order to evaluate the effect of gas injection on oil thermodynamic
properties, the PVT behavior of the reservoir oil saturated with natural
gas (99.5% C_1) in one case, and with CO_2 in a second case, at a saturation
pressure equal to the present field pressure (20.1 MPa) has been studied.
It has been shown that natural gas has a minor effect on the thermodynamic
characteristics of the oil, whereas saturation with CO_2 lowers the viscosi
ty to 0.28 mPa·s and increases the volume factor (B_o) from 1.245 for the
oil as such, to 2.831 for CO_2-saturated oil. To investigate the possibility of vaporizing part of the residual oil into gas cycled in the reservoir
(1), multiple-contact tests between reservoir oil and natural gas (99,5%
C_1) in PVT cells have been carried out (2). The results are shown in Fig.
2. It can be seen that after 10 cycles (equivalent to contacting 2,200
sm^3 of gas with each m^3 of stock-tank oil) the residual oil has vaporized
to the extent of 22% into the cycled gas. Practically the whole C_2-C_{10}
fraction of the oil has vaporized into the gas, with a marked increase of
its heating value.

4. SELECTION OF SURFACTANTS AND POLYMERS FOR EOR

The reservoir brine in the oilbearing pools of Cortemaggiore contains
153 g/l of salts, of which 22 g/l are calcium and magnesium chlorides;
this makes it very difficult to identify polymers and surface-active

agents compatible with the reservoir environment. Moreover, preflooding
the reservoir-rock with fresh water does not assure the displacement of
reservoir brine and would certainly cause a severe reduction in permeability due to swelling of the clay present in the rock.

A large number of polymers and surface-active agents had to be scree
ned before those compatible with the reservoir environment could be identified. Among the polymers the xanthanic gums, after proper treatment for
reduction of their insoluble fractions and macro-aggregates, are the most
promising. By using one of these biopolymers, at a shear rate of 10 s^{-1}
apparent viscosities of 40, 60 and 120 cp have been obtained with biopolymer concentrations in reservoir water of 500, 1,000 and 1,500 ppm respectively, at reservoir temperature.

As for the surfactants, a synthetic compound resulting from a mixture of alkyl etoxy-carboxylate and alkyl etoxylate was finally selected.
The ternary diagram for the system surfactant/reservoir oil (with dissolved gas)/reservoir brine, at reservoir pressure and temperature, is shown
in Fig. 3. It is a typical Winsor III system. At the triple point the interfacial tension between the microemulsion and the aqueous phase in equi
librium has a value of 5.5 x 10^{-4} mN/m; between microemulsion and oil pha
se in equilibrium the interfacial tension is $2.5 \cdot 10^{-3}$ mN/m.

5. TESTS ON POROUS MEDIUM AT RESERVOIR CONDITIONS

In order to check the efficiency of the various EOR processes which
in principle could be employed in the reservoir, many flooding tests were
carried out in a column packed with sand saturated with reservoir oil at
reservoir temperature (54°C) and present reservoir pressure (20.1 MPa).
The column used was three meters long, had a diameter of 0.015 m and was
packed with sand from cores from the F_1+F_2 pool.

All during these displacement tests the column was held in a vertical position, and flow rates were kept below the critical value (3) in
order to avoid fingering of the displacing fluid.

Three runs were carried out using CO_2 as the displacing fluid. Two
runs were made with the column at oil saturation conditions equal to those initially present in the reservoir (S_o = 86%), whereas a third run was
carried out after flooding the column with water up to the residual oil
saturation (S_o = 33.4%). For all three runs, oil recovery at breakthrough
was between 95% and 99.6% of the residual oil and this indicates that the
displacement occurred in miscible conditions. At the outlet of the column
a miscibility buffer between reservoir oil and CO_2 was noted, thus indica
ting that miscibility was reached by multi-contact mechanism.

Two runs were made using nitrogen as displacing fluid; the breakthrough recovery of 75% indicates a displaceemnt under conditions of non-
miscibility.

For the micellar/polymer flooding tests, the microemulsion was prepa
red with surfactant (16%), cosurfactant (7.8%), formation brine (6%) and
Cortemaggiore stock-tank oil (70.2%). The microemulsion cushion was fol-

lowed by a solution of 1,500 ppm of biopolymer in formation brine. For material balance calculations the following chemical tracers were added to the fluids: Ca^{++} to reservoir brine, Li^+ to the water contained in the microemulsion, Cd^{++} to the thickened water, and Zn^{++} to the oil in the microemulsion.

Three runs were carried out in the column under the following conditions of initial water saturation: 14% ($=S_{iw}$); 54% and 83% ($=1-S_{or}$). In all three cases the microemulsion cushion was equivalent to 10% of the pore volume. Recoveries, at the end of each run, were equal to 97.6; 86.3 and 69.8% of the reservoir oil.

The behavior during the three runs is shown by the diagrams in Fig. 4 which indicate, in each case, the presence of a stable water-oil bank, thus suggesting favorable recovery possibilities for the reservoir.

For checking the possibility of successfully vaporizing the residual oil in dry gas cycled within the reservoir, four runs were made with a horizontal column after pre-flooding with water up to the point of residual oil saturation. The column was flooded with dry gas (99.5% C_1) and the outcoming gas analyzed by chromatography. The results of one of these runs are are presented in Fig. 5 which shows a residual oil recovery approximately equal to 45% of its C_{3+} fraction (or 40% when referred to the C_{7+} fraction) after running a quantity of dry gas, equal to 15 times the volume of the pores, through the reservoir. This residual oil is vaporized into the gas and it can be recovered as a liquid if the gas is made to go through a suitable gas-separation unit.

6. NUMERICAL MODEL STUDIES

The past behavior (1951-1978) of the F_1+F_2 zone has been matched on a numerical model so as to check the validity of the geological model and to predict the behavior of the reservoir under various production schedules. A Beta-type three-dimensional three-phase model (3D-3P) was used, made up of 17 x 14 x 2 blocks for a total of 302 active blocks.

The model investigation has made evident the existence of reservoir areas which had been neither drained nor displaced by the water encroaching from the lateral aquifer. For this reason the improvement in oil recovery attainable by the use of infill wells has been examined. Four cases have been investigated with basic data listed in Table I, and one case with natural gas injection into block 5 (Fig. 1) through well No. 60 with oil production from well No. 58.

For all four cases the drilling of infill wells brings a substantial increase in oil recovery (Table I). For the case of gas injection, recovery increases from 30% to about 48% with an additional production of 35,000 m^3 of stock-tank oil.

This makes clear that the oil which can be produced by a primary process from the field areas where the oil saturation is still high should be recovered by means of infill wells prior to using EOR processes.

7. INFILLING OF PRODUCTION WELLS

Because of the results obtained through the model study, well No.103 within block 1 (Fig. 1) was drilled and six wells were worked over, four of which were producing from level E and the other two from the upper levels.

Well 103 has been found to be oil-bearing at both levels E and F_1+F_2. In the latter a clay lens separates zones F_1 and F_2; the upper part of zone F_1 is made up of a conglomerate with clay cement and has a low permeability. As a result of the oil production, the current water/oil contact has risen by 12. 5 m above the original one. The well has been put into production with gravel packing from level F_1+F_2 and is now producing oil with 70% water cut. All six old worked-over wells have proven to be oil producers with variable percentages of water. From September 1980 through June 1984 oil production from the four worked-over wells at level E has been : 8,950 m^3 STO with a 1.5% increase of the oil recovery factor.

8. CONCLUSIONS

The investigations described in this paper bring the following conclusions:

1. Pools E and F_1+F_2 of the Cortemaggiore field are of particular interest for comparing various EOR processes as applied to the same reservoir, since these pools are subdivided into blocks by transversal faults.

2. Laboratory tests have brought to the conclusion that the EOR processes best suited to the Cortemaggiore oil field are: miscible displacement with CO_2; displacement by micellar solutions followed by thickened water; vaporization into dry gas cycled in the reservoir.

3. A polymer (xanthanic gum) and a family of surfactants (alkyl etoxylates and etoxy-carboxylates) have been identified which are capable of withstanding the severe reservoir conditions at Cortemaggiore (salt content of 153 g/l and calcium and magnesium chlorides to the extent of 22 g/l).

4. The results obtained from numerical model studies, and those from well 103 and from those wells which were worked-over, show the presence, in the reservoir, of rock "isles" which have not as yet been displaced by water and, in any case, contain high oil saturation. These are giving a new primary production which has to be completed before any EOR pilot can be undertaken.

REFERENCES

1. COOK, A.B., JOHNSON, F.S., SPENCER, G.B. and BAYAZEED, A.F.: "The Role of Vaporization in High Percentage Oil Recovery by Pressure Maintenance", J. Pet. Tech. (February 1967) 245-250.

2. FUSSELL, D.D., SHELTON, J.L. and GRIFFITH, J.D.: "Effect of Rich Gas

Composition on Multiple-Contact Miscible Displacement - A Cell-to-Cell Flash Model Study", Soc. Pet. Eng. J. (December 1976) 310-316.

3. DIETZ, D.N.: "A Theoretical Approach to the Problem of Encroaching and By-Passing Edge Water", Proc. Akad. van Wetenschappen, 56-B, 83-84 (1953).

Table I - PRODUCTION FORECASTS WITH INFILL WELLS
POOL $F_1 + F_2$

Case	Block	Additional wells	Initial rate (m^3/d)	Years of production	Oil recovery (%)	
					present	final
1	1	1	10		13.5	30.0
	2	1	10	10	19.9	48.6
	5	1	10		30.2	49.5
2	1	1	20		13.5	30.0
	2	1	20	5	19.9	48.7
	5	1	20		30.2	49.5
3	5	1	40	5	30.2	68.1
4	5	2	40	5	30.2	68.2

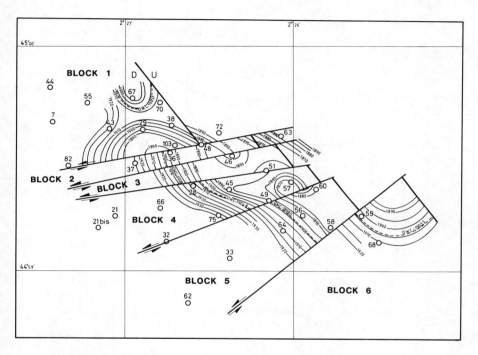

Fig. 1 – Map of top of zone F_1 + F_2

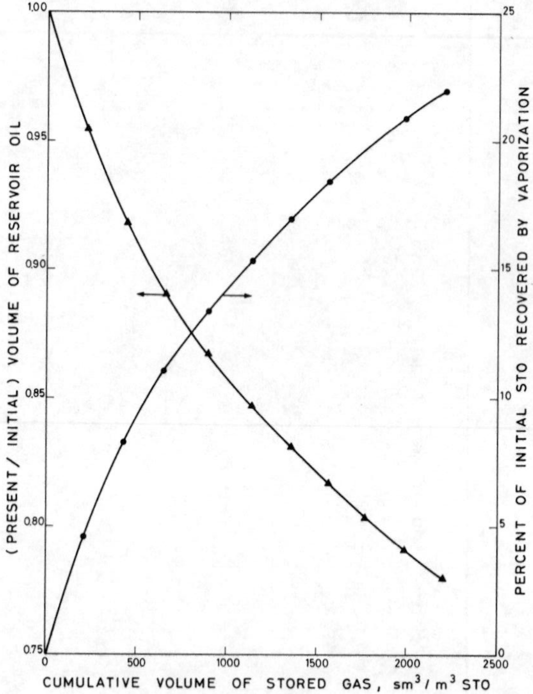

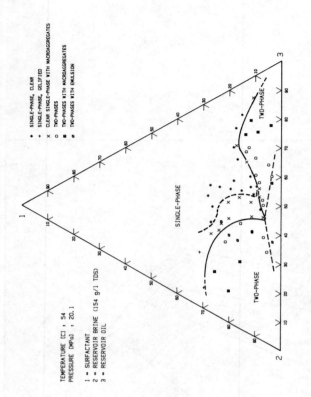

Fig.3 – Pseudo-ternary diagram at reservoir conditions:
Cortemaggiore reservoir oil/reservoir brine/sur-
factant (ethoxycarboxylate).

Fig.2 – Vaporization of residual oil by
multiple contacts with natural
gas at reservoir conditions.

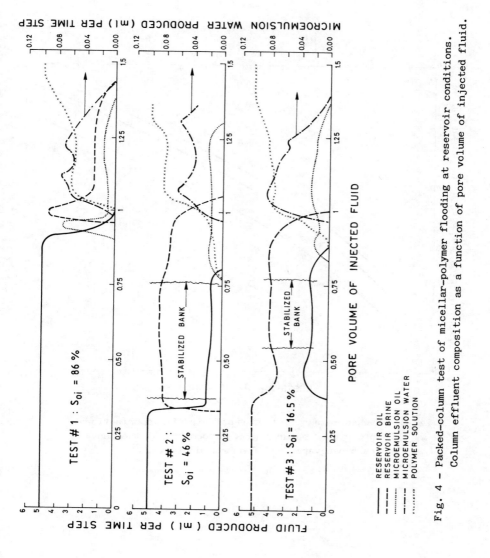

Fig. 4 - Packed-column test of micellar-polymer flooding at reservoir conditions. Column effluent composition as a function of pore volume of injected fluid.

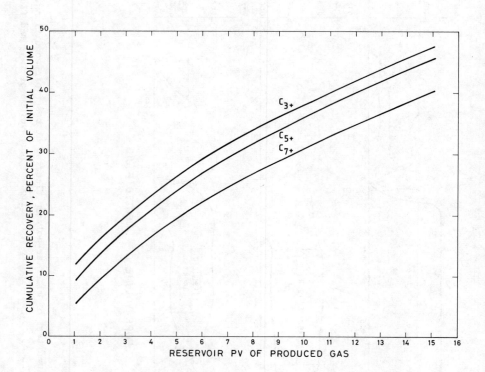

Fig. 5 – Packed–column test of reservoir oil vaporization
into cycled natural gas. Percentage of the heavy
fractions vaporized from residual oil.

(05.10/78)

PIROPO: A HEAVY-OIL DEPOSIT IN THE ADRIATIC SEA
STUDIES FOR POSSIBLE EXPLOITATION

E. BELLA, D. BILGERI, E. CAUSIN, G.L. CHIERICI, V. GILI, G. MIRABELLI,
I. SOZZI
Agip S.p.A.
Exploration and Production

Summary

The heavy-oil deposit of Piropo was discovered in 1975 in a fractured carbonate. The stock tank oil has a 21.3° API gravity. The productivity of the discovery well was too low for exploitation. For evaluating the feasibility of commercial production, a research project has been set up, phased as follows:
- application of acoustic impedance (Sismolog) to the seismic survey to pin-point the most fractured (hence, the most productive) area within the reservoir;
- drilling of a well in the area just mentioned, for gathering all possible geological and fluid dynamics data and performing a massive hydraulic fracturing (MHF) to obtain an estimate of the maximum oil rate attainable;
- theoretical and experimental evaluation of procedures potentially capable of reducing reservoir oil viscosity so as to achieve a higher production rate. It turned out that carbon dioxide injection alternated with production is, from a technical point of view, the most promising process for exploiting this heavy-oil deposit.

1. GEOLOGICAL CONDITIONS

One of the most important targets for oil exploration in the central part of the Adriatic sea is represented by levels with primary and secondary porosity interbedded in the Eocene-Mesozoic carbonate series. These series, from Eocene to Upper Cretaceous, are characterized by a pelagic type deposition with occurrences of carbonatic turbidites (Scaglia Fm.). This deposition mechanism has produced lenticular bodies of porous and permeable rock which become most frequent close to the continental slope.

Tectonic episodes, reaching their peak during Pliocene, have caused the formation of a system of faults and fractures through which hydrocarbons have migrated from the source rock underlying the porous and permeable areas in the Eocene-Cretaceous carbonate rock. These hydrocarbon accumulations make up the existing oil deposits.

Because of this situation the well productivity is strictly related to the intensity of the rock fracturing.

In order to locate the area of maximum fracturing, where to drill the Piropo 2 well, the seismic surveys have been processed through the Sismolog technique. Seismic data were converted into lithologic data, as a re-

sult of the interpretation of changes in the acoustic impedance of the
rock layers.

A comparison with the information from cores and with those obtained
by sonic and density logs on Piropo 1 well, while taking into account the
tectonic history of the area, makes it possible to pin-point those areas
where the presence of fractures or, in any case, of secondary porosity, is
highly probable.

Fig. 1 shows an overlay of the structural map and the expected porosi
ty map. The latter was obtained by matching the reflecting surface of the
top of Scaglia Fm. with an acoustic velocity field consistent with the for
mation interval velocities for Piropo 1 well. The geological model worked
out with all of the seismic sections was also taken into account.

2. PIROPO 2 WELL

Piropo 2 well was drilled in an area where the reservoir rock shows
a maximum of fracturing (Fig. 1). Water depth was 51 m.

The drilling was carried out by the Scarabeo 2 semi-submersible platform. The final depth (3,695 m) was reached after 111 days of drilling,
followed by 98 days of testing.

The Scaglia Fm. was continuously cored for a total length of 335 m.
Standard and special logs for fracture location, i.e. WFs/VDL, HDT/FIL and
CMS Schlumberger, were recorded in the reservoir section of the well.

Through core analyses it was possible to establish the deepwater natu
re of the deposition environment. Sediments are mostly pelagic limestones
(Mudstone-Wackestone) interbedded with carbonate turbidites made of Wack-
stone-Packstone and breccia of platform contribution (Fig. 2).

Only for turbidites the matrix characteristics are such as to suggest
the presence of a reservoir (Fig. 3).

Heavy oil is present both in the matrix of high-porosity Packstone in
terbeddings, and in fractures and stylolites which are randomly distributed over the entire well length. The fracture system appears to be made up
of an interconnected network of macro- and microfractures, which is respon
sible for the hydraulic continuity throughout the entire deposit.

The oil-bearing zone has a thickness of 160 m and corresponds to the
Upper Cretaceous-Paleocene section of the calcareous formation; its uppermost portion, corresponding to the Middle Paleocene-Eocene, has proved to
be impermeable.

A total of nine DST have been run in the well; three in open hole and
six by perforation through the 7" liner. These tests have located 4 zones
having different hydraulic characteristics:

- an upper zone, within the Scaglia Fm., down to a depth of 3448 m RT,
  which is practically impermeable;

- a zone extending from 3448 down to 3570 m RT, which includes the oil bea
  ring section and part of the capillary fringe, which has some permeabili
  ty;

- a zone with good permeability, extending from 3570 down to 3630 m RT, corresponding to the lower part of the capillary fringe and the aquifer;

- a compact zone at the base of the aquifer.

Within Scaglia Fm. the fluids are overpressured with a gradient of 0.012 MPa/m.

3. ACID JOB AND MHF IN THE OIL-BEARING LEVEL OF PIROPO 2 WELL

A series of tests have been carried out in the oil-bearing level, and more precisely between 3477 and 3497 m, in order to evaluate the maximum oil rate which could be obtained from this formation through stimulation jobs. The first DST performed in this zone without stimulation has not been followed by any production of fluids.

Following preliminary flushing with gas-oil (1 m^3 per meter of pay), an acid job was then carried out using 1.5 m^3 of 28% HCl per m of pay. This test produced oil with PI = 0.2 m^3/d x MPa.

The acid job was followed by a massive hydraulic fracturing (MHF) carried out with the following sequence:

- 76 m^3 of sea water followed by 265 m^3 of thickened water,

- four stages, each consisting of 38 m^3 of 28% HCl solution and each followed by 19 m^3 of sea water gel,

- a flushing cushion of 133 m^3. The total amount of fluids adds up to 684 m^3.

A pumping rate of 40 bbl/min was used and the wellhead pressure during pumping ran between 52 and 58 MPa.

MHF programming was based on laboratory tests on cores and formation fluids and on a numerical simulation program of fracturing which led to predicting a 4.7-fold increase in productivity, with a fracture length of 290 m.

After running the MHF, the well was put into production. The resulting PI had a value of 0.9 m^3/d x MPa, that is 4.5 higher than it was prior to MHF.

Though MHF proved to be a technical success, the oil rate obtained (12 m^3/day) was still too low to justify the development of the field.

4. STUDIES FOR DECREASING RESERVOIR OIL VISCOSITY

At reservoir conditions (43.25 MPa and 82°C) the oil is highly undersaturated (bubble point pressure = 4.51 MPa), with a viscosity of 16.7 mPa.s. This condition points to the possibility of increasing the PI of Piropo wells by reducing the oil viscosity. In fact, by saturating reservoir oil with gas it is theoretically possible to lower its viscosity.

Carbon dioxide is a gas showing high solubility into heavy oils (1); for this reason a phase diagram for the Piropo reservoir oil/CO_2 system

was determined at reservoir temperature (82°C) by laboratory tests using
a PVT cell.

From the resulting diagram (Fig. 4) the critical point corresponds to
a 73% CO_2 concentration and to a pressure close to 36 MPa; this is lower
than the reservoir pressure value (43.25 MPa).

The reservoir pressure isobar is therefore supercritical and, with in
creasing CO_2 amounts, it goes through the single-phase oil region, a two-
phase liquid area (light-oil + a semisolid phase, mostly asphaltenes), a
triple-phase area (gas + the two liquids mentioned above), and lastly a ga
seous single-phase region.

The semisolid phase brought about by CO_2 extraction of the light and
middle oil fractions shows a potential plugging danger for the rock. This
has been investigated by simulating the reservoir process in a laboratory
column.

In order to evaluate the effect on oil characteristics of injections
of gas into the reservoir, a series of PVT tests has been performed on the
oil as such and on the oil saturated at reservoir conditions with nitrogen,
natural gas ,and carbon dioxide. The results are presented in Fig. 5 which
shows that addition of natural gas (99.7% C1) and of carbon dioxide brings
about a substantial lowering of the oil viscosity; this makes them potential agents for increasing the productivity of Piropo wells through huff-
and-puff methods (gas injection alternated with oil production).

Some oil displacement tests with CO_2 under reservoir conditions were undertaken in a column; this was done to check the miscibility of CO_2
into the reservoir oil and the possible damage due to the formation of a
semisolid asphaltene phase within the rock. The column, three meters long
and with a diameter of 0.015 m, was packed with sand having 35% porosity
and permeability of 6 μm^2.

All of the tests were performed in a horizontal column slowly rotating to minimize the effects of gravity. Equipment details were given in
a previous paper (2).

In all three tests displacement always resulted in multiple-contact
miscibility with breakthrough oil recovery of about 80% and final recovery higher than 90%. For one of the three tests the results are shown in
Fig. 6.

Examination of the sand pack at the end of each run has revealed the
presence of semisolid asphaltenes within the initial portion of the column; in the first 15% section of its length the permeability was down to
half its original value. This may cause considerable damage because of a
possible skin-effect around the CO_2 injection well.

5. CONCLUSIONS

The most important results obtained during these studies may be summarized as follows:

1. Processing the seismic surveys in terms of acoustic impedance has proven a valuable means for pin-pointing the areas of highest fracturing

within the Piropo oil deposit;

2. Piropo oil is mostly present within the secondary porosity of the rock (micro- and macro-fractures, vugs and stylolites). It is also present within the primary porosity of the turbidite interbeddings when good matrix characteristics are present. The Sonic Log and Circumferential Microsonic combination has proven well suited for identifying the most fractured sections of the reservoir-rock;

3. MHF, run after an acid job, led to oil production from a previously tight interval. The increase in output by MHF has matched the value calculated in the engineering stage of the job. It is unfortunate that the final PI value (0.9 m^3/day x MPa) is still too low for commercial exploitation;

4. by dissolving CO_2 into the reservoir oil, which is highly undersaturated, viscosity is considerably lowered and, in theory, this turns out to be an efficient method for increasing productivity;

5. it is however unfortunate that CO_2 solution into the oil results in the formation of semisolid substances (asphaltenes) which could cause heavy skin effects. It is only by means of field testing that one can verify the extent by which the favorable effects due to the lowering of viscosity will be offset by this skin effect.

A CO_2 huff-and-puff test meant to give an answer to such a problem has been planned and will be carried out in 1985 in an onshore oil field having characteristics similar to those of the Piropo oil deposit.

REFERENCES

1. SIMON, R., and GRAUE, D.J.: "Generalized Correlations for Predicting Solubility, Swelling and Viscosity Behaviour of CO_2 - Crude Oil Systems", J. Pet. Tech. (January 1965) 102-106.

2. BELLA, E., BILGERI, D., CAUSIN, E., CHIERICI, G.L., GILI, V., MIRABELLI, G., and SOZZI, I.: "Piropo Heavy-Oil Accumulation, Adriatic Sea, Italy - Study of an Exploitation Process", 2nd European Symposium on EOR (Paris, 8-10 November 1982), 477-485.

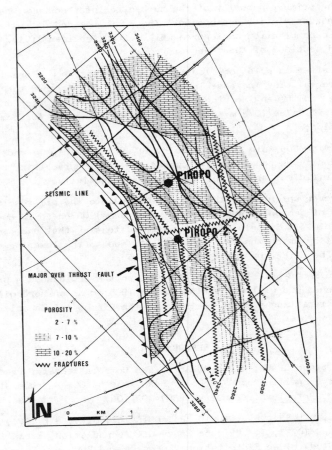

Fig. 1 – Structural map and areal distribution
of porosity in Piropo heavy-oil deposit.

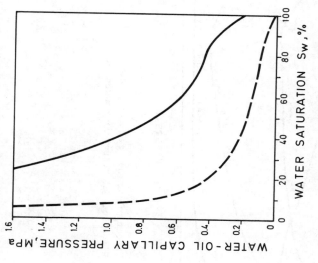

PELAGIC LIMESTONES $\emptyset = 5.8\%; k = 2\,mD$

TURBIDITES $\emptyset = 5.2\%; k = 20\,mD$

Fig. 3 – Capillary pressure curves at reservoir conditions for the two most important lithofacies.

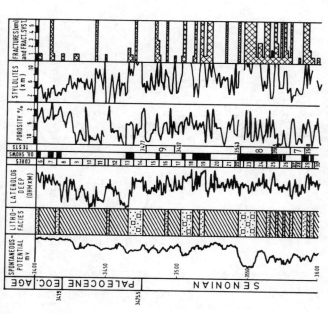

LITHOFACIES

TURBIDITES (PACKSTONES)

TURBIDITES (BRECCIAS)

PELAGIC LIMESTONES (MUDSTONE - WACKSTONE)

Fig. 2 – Piropo 2 well – Reservoir rock characteristics.

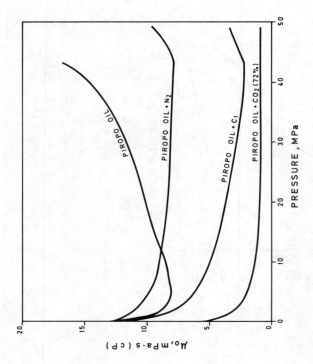

Fig. 5 – Effect of various gases on viscosity of Piropo oil at reservoir conditions.

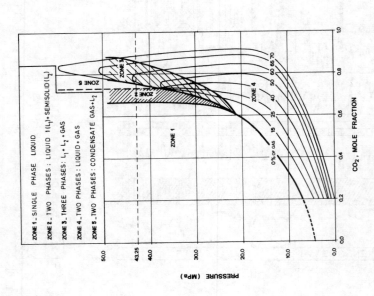

Fig. 4 – Phase diagram of the CO_2/reservoir oil system.

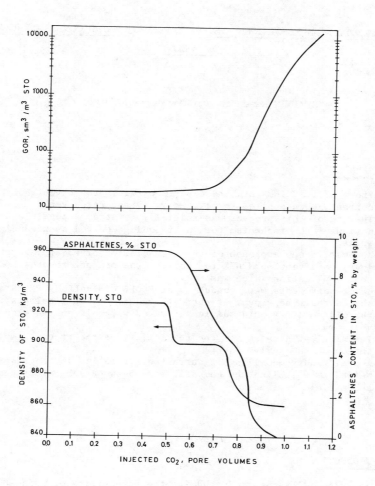

Fig. 6 – Results of a packed-column test of reservoir
oil displacement by carbon dioxide.

(05.12/78)

DEVELOPMENT OF AN ELECTRICAL SYSTEM TO PREVENT WATER CONING ON PRODUCING OIL WELLS

A. BAUDRY
SYMINEX Research and Development Director

SUMMARY :

From 1976, SYMINEX started experimenting on the effects of DC current, when applied to oil and water saturated rocks.
The aim of this project was initially to study a possible application of the mechanism for water coning control on producing wells.
However, a new approach of the mechanism observed at laboratory level, motivated SYMINEX to reorient the project towards a different application which was EOR.
Laboratory tests were conducted at ECOLE CENTRALE de PARIS and showed that the amount of current needed to electrically stimulate the oil recovery on a sample saturated with oil and water could be very small.
Technical studies showed the feasibility of the process implementation on the well.
A preliminary modelling of current distribution in the reservoir helped to establish the compatibility between laboratory results and a field application.

1. INTRODUCTION :

In 1976 SYMINEX started studying the feasibility of using DC current in water coning control for oil well application (based on previous work realised by TIKHOMOLOVA . University of Leningrad – CHILLINGAR : University of Los Angeles – HEADLEY and PIERCE : US Bureau of Mines...).
The first laboratory results showed that the action of DC current applied to an oil reservoir may, if optimized, help oil recovery.
On these findings, the initial "water coning" project was reoriented towards studying the feasibility of using DC current as a new EOR process.

- Laboratory :

Understanding of the physical mechanism when DC current is applied to an oil and water saturated sample, optimisation of results, preliminary modelling on reservoir.

These laboratory experiments, were conducted at the ECOLE CENTRALE DE PARIS, under SYMINEX's supervision.

- Technology :

Preliminary approach to the technical feasibility of field implementation of the process...

These studies showed that technological solutions exist for sending an electrical intensity in the range of 1 000 A down to a reservoir without major difficulties.

2. LABORATORY RESULTS :

Experiments have been carried out with equipment currently used for this type of tests by oil companies : HASSLER cells and imbibition cells, modified to enable the application of electrical current.

2.1. Results of the experiments :

The first laboratory tests carried out simultaneously with a water drive showed the action of electrical current on rocks of various lithologies saturated with oil and water : limestone, sandstone with different percentages of clay, and pure sand. Very rapidly, it was shown that the process was more efficient with samples exhibiting preferential axes of circulation : fractures of drains related to the heterogeneity of the porosity (natural or artificial).

Also, the tests made with a water drive against the electro-osmosis pressure showed the existence of a phenomenon different from electro-osmosis at the scale of the sample.

These results showed the need to experiment under imbibition conditions. There again, after the oil production stabilizes under spontaneous imbibition, the application of DC current triggers off a complementary recovery.

This process shows a recovery of 3 to 7 % of the oil present in many samples.

However, this recovery can be higher and can reach values up to 20 % A survey was done by the ECOLE CENTRALE DE PARIS to verify that the process is not altered by secondary effects which might interfere at laboratory level (electrode polarization, action of an alkalins ion).

2.2. Understanding of the process mechanism :

To date, there is no evidence to prove which mechanism is taking place. However, many tests have tried to demonstrate the influence of electrical current upon the interfacial tension and upon the wettability, but the results obtained do not yet enable any valid explanation of the phenomenon.

3. CALCULATION RESULTS :

3.1. Electrical field in the reservoir :

The laboratory tests helped to define the efficiency for the electrical field and energy levels.

The electrical field distribution in a reservoir was correlated with the values obtained at laboratory level. We found that the volume of reservoir submitted to effective electrical treatment, for 1 000 A injected is in the same range as the volume produced around the well.

3.2. The energetic balance :

- Energetic balance obtained in the laboratory :

With a reservoir rock of the following characteristics :
$\emptyset$ = 15 %, So = 40 %, d = .8, the overall energetic balance at laboratory level is in the range of 10 000.
- Field energetic efficiency :

The overall energetic efficiency at field level is in the range of 10 for an application in the following conditions :
- Matrix porosity : 15 %
- Oil saturation : 40 %
- Height of reservoir : 20 m
- Efficency radius of electrostimulation process = 65 m
- Extra recovery : 5 %

4. TECHNOLOGY RESULTS :

4.1. The different configurations of the set up :

At sample level, we notice that the extra oil recovery is obtained simultaneously with the current injection.
However, in the case of a fractured reservoir, we can very easily imagine that these two phases can be dissociated.

The application of the current on the reservoir matrix will enhance the migration of oil towards the fractures, then the pumping action will move this oil towards the well.

For a field application of the process, a choice has to be made :
- current application and production of the well to be consecutive or simultaneous (the latter necessitating a more sophisticated system).
Using tubing as a current conductor and as a production conductor at the same time can facilitate the operations, at the expense of a rather sophisticated well equipment.

In this case the tubing would have the following functions :
. current conductor,
. production conductor.

If the two phases are dissociated, the current injection can take place using a proper electrical completion, and subsequently this electrical completion is pulled out and well produced in a normal manner.

4.2. The electrical system :

4.2.1. Current transport in the well :

To prevent the deterioration of the conductor, and its insulation the intensity has to be controlled so as to maintain the temperature rise within a predetermined range.
A compromise is to be found between cross section of conductor and inside diameter of the casing.
For a current transport in the well, we can either use a cable conductor, or a tubing.
If a tubing is used as conductor, the electrical insulation could be achieved by proper means.

4.2.2. The electrode (anode) :

The main function of the electrode is to enable the passage of current to the reservoir.
The length of the electrode will determine to a great extent the overall electrical resistance of the circuit.
A study of heat transfer in the reservoir, near the electrode, is essential to optimize the intensity of the current.

4.2.3. The generator set :

The loop : intensity --> heating up --> resistivity change --> intensity change, makes the characteristics of the circuit rather variable.
Any drift possibility should be prevented by a control of the current intenstity.

4.2.4. Safety/equipment protection :

An efficient safety system is required to prevent accidents and equipment deterioration, with an adequate detection of working levels, coupled to an efficient protection sytem. A tight safety control procedure will have to be enforced when power is on.
This system will have to comply with standard regulations.

5. CHOICE OF SITE FOR A PILOT TEST :

The main characteristics of a reservoir suited to the electrical process would be :
- Fractured reservoir
- Shallow depth for a first application
- Sufficient thickness
- At the end of primary recovery phase
- Salty interstitial water

We have examined various European fields for a possible site for a pilot test.
At the end of this work the application of the process on the Eschau field has been considered and a feasibility study started.

(05.41/82)

ELECTRO-DISPERSION: FEASIBILITY

A.J. KERMABON, SYMINEX

Summary

Injection of dc current into oil and water-saturated reservoirs can increase the final oil recovery.
Though it appears compatible with most oil reservoirs, this method would seem particularly advantageous for cracked reservoirs.
Modelization of propagation of electric current, together with technological studies, have shown that up to 1000 Amperes and more can be injected into a reservoir. The purpose of this project was to study possible application of the method to Eschau reservoir (Alsace).

1 - INTRODUCTION

As early as 1976, SYMINEX had studied the action of electric current when applied to a rock saturated with oil and water. The previous work (TIKHOMOLOVA: Leningrad University - CHILLINGAR: Los Angeles University) had in fact shown that the mobility of the fluid in a porous medium can be changed by applying electric current.
There were three objectives to this project:
- laboratory tests to study the decisive factors in the action of the current on samples of oil and water-saturated rock,
- general technological study (surface and well),
- feasibility study of a pilot test on Eschau reservoir (Alsace).

2 - LABORATORY WORK

The aim of the laboratory test was to study the effect of electro-stimulation in the case of Eschau reservoir.
The following were the main objectives of these laboratory tests:
- optimization of the method with a rock of known characteristics,
- application of the method to a sample from Eschau reservoir,
- tests with a view to understanding the phenomena involved.

2.1 - Tests on Fontainebleau sandstone

These optimization tests carried out on Fontainebleau sandstone were conducted under imbibition conditions.

After stabilization of the oil recovery for four weeks by imbibition, a weak current is applied to the sample via two electrodes.

The following are the experimental conditions:
- type of rock: Fontainebleau sandstone (99.5% silica),
- oil: neutral oil,
- water: 35 g NaCl/litre + 2 g $CaCl_2$/litre.

The influence of the following parameters was studied:

- Oil saturation

The process would appear to be more effective when electrostimulation is applied with a high initial oil saturation.

- Water salinity

Better electrical recovery is obtained when the water salinity is high.

- Duration of electrostimulation

. the electrostimulation no longer has any effect after a few days,
. the oil recovery is mostly gained during the first five days,
. tests were conducted of variable duration, leading to the following observations:
- oil can still be produced after electrostimulation is stopped,
- electrostimulation lasting a few hours can trigger oil recovery lasting several days,
- there appears to be no direct correlation between the duration of application of the electrostimulation and the oil recovery,
- resuming electrostimulation after a "rest" period can sometimes trigger further oil recovery, though less than the previous one.

- Current level in the sample

Several series of tests were performed with current levels of 0.1 mA, 1 mA, 10 mA and 100 mA (in the circuit used, only 1% of the current flows through the sample). The test results are obtained with 1 mA, as is shown from the following table:

Milliamperes:	0.1 mA	1 mA	10 mA
Y electrical recovery	not significant	11.4%	10.3%
Z electrical recovery	not significant	47%	50%

Y : Oil recovery in terms of oil initially in place before start of imbibition,

Z : Oil recovery in terms of oil initially in place after stabilization of imbibition and before electrostimulation.

(The results obtained with 100 mA are not significant because the chlorine produced at the anode quickly deteriorates the oil, rendering it highly viscous).

The tests on FONTAINEBLEAU sandstone enabled the various parameters involved to be determined more precisely. Nevertheless, the complexity of the phenomena means that the results obtained are predominantly qualitative and that additional experiments will have to be performed.

2.2 - Tests on samples from Eschau
These tests were carried out under conditions of imbibition, with the following characteristics:
- type of rock: limestone of the Greater Oolith,
- water: 150 g NaCl/litre,
- oil: Eschau.

The current was applied for 7 weeks on two series of samples ; the first with a current of 1 mA and the second with a current of 0.1 mA.

With 0.1 mA, the mean oil recovery in terms of the oil initially in place is 4.3%.

With 1 mA, the mean recovery is 11.7%.

These tests revealed significant action of the current on the Eschau samples with relatively low electrical fields. However, the high scatter of the results means that additional tests must be carried out before envisaging applying the method to Eschau reservoir.

2.3 - Tests to understand the phenomena involved
A series of tests was undertaken by IFP on capillary tubes comprising a square-cross section constriction trapping a drop of oil.

The action of release from the trap by electricity is general and independent of the polarity of the system and the position of the oil drop trapped with relation to the most direct lines of current between the anode and the cathode.

The time needed for untrapping varies inversely with the voltage applied.

The untrapping always occurs for pH values of 11.4 near the cathode and 1, near the anode.

3 - TECHNOLOGICAL STUDIES

The purpose of the technological studies was to propose practical solutions for implementing a pilot project on Eschau reservoir.

The main conclusions of the technological studies are as follows:

3.1 - Anode
The anode is located below the production zone. Calculations have shown that for a short gap between the anode and the shoe, the possibility of a short-circuit between them can easily be eliminated.

Calculations of the temperature-rise allowing for the diffusion and convection enable the dimensions of the anode to be calculated to suit the current and the resistivity of the terrain around the anode.

3.2 - Cathode
Two types of cathode can be envisaged:
- a cathode consisting of the casing of the anode well or another neighbouring well (by means of an offtake at the wellhead),
- return via the surface.

These two solutions can be combined judiciously to balance the return currents.

3.3 - Transmission of current inside the well
Various solutions were studied:
- transmission of the current via the tubing,
- transmission of the current via cable.

3.4 - Generator
This provides the current at a voltage defined by the overall resistance of the circuit.

The overall resistance of the circuit will range from about a fraction of an ohm to several ohms. This resistance essentially depends on the resistivity of the terrain around the anode.

The objective sought is to inject maximum current into the reservoir ; however, the temperature-rises very strictly limit this current, essentially at the anode.

The voltage limitation defined by the behaviour of the insulating materials in the system is less restrictive.

A generating set of the diesel engine - synchronous alternator - type offers the flexibility necessary for the application envisaged.

In addition to reliability and robustness, this generator has the advantage that its current can be regulated in simple manner, by adjusting the field current.

3.5 - Protection of the equipment
The protection of the equipment has been studied from the point of view of galvanic corrosion. It concerns the well-anode system itself, together with the neighbouring wells and the pipelines that may be present in the immediate environment around the site. This study has brought solutions ensuring good protection of the bottom and surface systems.

3.6 - Safety of personnel
The two essential safety expects are:
- grounding: in general, with variable potential gradients in the soil, it will be preferable to ground each item of equipment locally,
- the potential gradients on the surface were studied by numerical models: the risks of encountering high gradients are to be found only in the immediate vicinity of the surface return currents.

4 - MODELIZATION RESULTS

4.1 - Electrical modelization
SYMINEX carried out a number of electrical modelizations for Eschau reservoir so as to check the feasibility of the method at this site.

These modelizations were made by means of a finite difference computing programme. This programme is fast and easy to handle, giving a fair idea of the distribution of the electrical variables in a variety of configurations.

Several electrode positions were tested, varying the distance to the shoe and the resistivities of the terrains.

Several computer runs were made in order to achieve finer modelization around the well (vertical meshing) and in the oil zone (horizontal meshing).

In addition, different insulations between the anode and the shoe were studied in order to prevent risks of short-circuits via the electrolyte in the event of poor insulation inside the well between the anode and the shoe.

4.2 - Hydraulic simulation

The purpose of this study was to select a part of the reservoir in which an electro-stimulation test would be meaningful and amenable to interpretation.

This hydraulic simulation of Eschau reservoir was performed by SNEA(P) by means of a conventional model (finite elements, solution of diffusion and mass conservation equations, flow of each phase governed by the relative permeability and cracking allowed for by the microscopic permeability).

The model was suitably calibrated on the basis of the data of the initial production period (1956 to 1968) and the data from the production tests carried out in 1982.

The following is a summary of the results of this simulation:
- in view of the inclined geometry of the structure of Eschau reservoir, a new well will have to be drilled,
- the anode could be located underground in a zone with a mean resistivity of 60 ohm-m,
- the resistance of the reservoir (2 to 3 ohms, depending on the configurations) would enable about 500 Amperes to be injected at the bottom,
- for 500 Amperes, the radius of action would be about 60 metres. The potentials on the surface remain very low and should set no problem of safety in the immediate vicinity of the well.

5 - CONCLUSIONS

The results obtained within the framework of this project are encouraging.

The work has shown the technical feasibility of the method on the reservoir and enabled adequate solutions to be defined regarding the methodology and the equipment to be used.

A specific application to Eschau reservoir was considered as difficult to optimize, essentially owing to the geometry of the structure and the small diameter of the existing wells (4 1/2"), which would make it delicate to re-use the wells.

More exhaustive studies will be needed before consideration can be given to implementing a pilot project.

(05.16/80)

DOWNHOLE STEAM GENERATOR FOR ENHANCED OIL RECOVERY

J.C. BODEN, P.J. FEARNLEY, M. MCMAHON AND F.A. RIDDIFORD
BP Research Centre, Sunbury-on-Thames, England

Summary

A Downhole Steam Generator has been developed to the stage of
detailed engineering design of a 5 MW prototype unit to inject
1 000 barrels per day of steam (cold water equivalent) at a
pressure of 70 bar. The generator incorporates a unique pulsed
burner which offers lower combustor wall temperatures than
continuous burners, and should thus overcome one of the chief
difficulties in the design of a reliable system. The major
part of the programme has been devoted to the development and
engineering of the combustor and its ignition system; the
required downhole ancillary equipment and surface facilities
have also been defined. It is suggested that the economic
success of downhole steam generators will depend chiefly on
field confirmation of laboratory predictions that the
co-injection of exhaust gases can substantially enhance oil
production rates. Other potential advantages, which could
lead to specialised applications even in the absence of an
exhaust gas effect, are also described.

1. INTRODUCTION

The development of the BP Downhole Steam Generator (DSG) was
carried out during the period November 1979 to October 1983 under
EEC Contract No TH/05.16/80.

The concept of generating steam at the foot of an injection
well instead of at the surface was originated as a method of
eliminating the wellbore heat losses which threatened to make
injection of steam into deep formations, below about 1 000 m,
impractical. It has been estimated that at this depth ordinary
injection tubing, supplied with 80 per cent quality steam, could
deliver only hot water at the sandface (1).

In addition, a DSG can offer other advantages, which depend
on which of two radically different approaches to its design is
adopted. The so-called 'indirect' or 'low pressure' DSG is in
essence a surface generator miniaturized and placed at the foot of
the well. The steam is raised in a heat exchanger and the exhaust
products are vented to the surface. However, such a generator is
relatively bulky, requiring an increased wellbore diameter, and
those units which have been tested have suffered from exhaust line
corrosion problems (2).

The BP generator is of the 'direct' or 'high pressure' type, in which steam is raised by adding water to the burner combustion products and the mixture is injected into the formation. The generator can be compact, there are no flue heat losses, and air pollution from the flue gases is reduced or eliminated. Since combustion takes place at the injection pressure, the air has to be compressed, but there is evidence that the co-injection of the exhaust gas, carbon dioxide and nitrogen, with the steam can increase oil production rates (3-5).

BP's interest in downhole burners began in the late 1960's, when a novel pulsed combustor was devised. This enabled a very high heat release rate to be achieved in a small diameter burner tube without an excessive heat flux to the walls. A 0.5 MW burner of this type was tested in a 75 m well in the United Kingdom in 1969 (6). High burner wall temperatures and consequent materials failures were one of the major obstacles to the development of a reliable DSG (7). The BP burner offered a unique method of solving this problem.

2. PULSED BURNER

Figure 1 shows the principle of operation of the pulsed burner. Fuel and air enter the combustion tube via a mixing head at its upper end (Figure 1a). Once a certain fraction of the tube is filled, the mixture is ignited near the head (Figure 1b) and a flame accelerates down the length of the tube. The pressure rise associated with the combustion temporarily halts the inflow of fuel and air, preventing continuous combustion without the need for valving (Figure 1c). The tube then refills with fresh mixture (Figure 1d), which is ignited to start the cycle again. Because the flame is in contact with any point on the combustor wall for only a very short period during each cycle, the mean heat flux from the combustion zone itself, which gives rise to materials problems in continuous burners, is greatly reduced. The thermal output is altered by varying the feed rates of fuel and air and the frequency of ignition; there is no requirement to stabilise a flame at different throughputs and operating pressures, and thus the burner head design is simplified and the turndown ratio can be high.

Flames under these conditions propagate as a more or less closely coupled combination of a pressure wave and a combustion zone, and since there is little literature data on systems of this type at the pressures encountered downhole, a special 250 kW high pressure test rig was built at Sunbury. This, and a series of lower pressure rigs, were required to optimise the design for reliable pulsed operation and to provide data for the development of a prototype DSG. Combustion efficiencies of up to 97 per cent have been measured. Because the most probable fuel for a DSG is natural gas, tests have concentrated on methane. LPG, fed as a liquid via an atomising mixing head, has also been used since it would be easier to store for a full scale trial remote from a natural gas supply. Some low pressure testing has also employed kerosine. Experimentation on the high pressure rig is continuing, in order to complete the full characterisation of the combustion process.

Since the pulsed burner requires ignition on every cycle, typically at between 1 and 5 Hz, the development of a reliable ignition system was crucial to the success of the system and received much effort. With ordinary air gap spark plugs the breakdown voltage and electrode wear increase with pressure, and no existing plug has been found which is suitable for this type of application. Specially developed plugs have now successfully achieved more than 50 million events at a pressure of 70 bar, typically equivalent to over four months' operation downhole.

3. PROTOTYPE DESIGN

A prototype generator has been designed to the specification given in Table I.

Table I - BP Prototype Downhole Steam Generator Specification

Maximum thermal output	5 MW
Steam rate	1 000 bbl/day (cold water equivalent)
Overall diameter	137 mm
Burner length	2.4 m
Overall length	8 m*
Design injection pressure	70 bar

* Including an instrumentation/telemetry package not required in a production version.

The layout of the generator is shown schematically in Figure 2. Water for steam raising is added to the exhaust gases at the lower end of the combustion tube; a steam quality of 80 per cent is assumed for design purposes. The tube itself is designed to withstand the operational conditions and has been stressed to meet the requirements of the ASME III and VIII codes.

The ignition system, instrumentation/telemetry package, supply pipes, filters and non-return valves are mounted within a sealed tubular structure which extends upwards from the burner tube.

The ignition and instrumentation packages have been built and mounted in their own pressure-tight enclosures. The instrumentation/telemetry package has been developed by a company specialising in well-logging equipment and will transmit data to the surface from downhole thermocouples, pressure transducers, an accelerometer and a spark current detector, and will also activate certain alarms and control the sparking frequency from the surface. The remainder of the generator design has been prepared in the form of detailed engienering drawings. Figure 3 is a photograph of a half-scale model of the DSG, together with the full-scale ignition and instrumentation packages.

4. DOWNHOLE INSTALLATION AND FIELD OPERATION

The generator is to be mounted near the foot of the injection well above a thermal packer, through which the steam-gas mixture

will pass before entering the formation via perforations in the
well casing (Figure 4). A DSG is typically mounted about 30 m
above the perforations, and over this distance the pressure pulses
generated within the combustor will be attenuated to less than a
ten per cent overpressure at the injection point. Thus the conditions
at the perforations will be virtually the same as for a continuous
combustor.

The packer has been redesigned from a standard unit in conjunction
with the manufacturer. Above the packer will be a jay-latch within
a polished bore receptacle, which has also been redesigned, and
this will be screwed to the foot of the generator via a criss-cross
safety joint. The generator will be lowered on the air supply
string, which is standard non-upset $2\frac{3}{8}$" tubing, to which the other
supply lines will be fixed by specially designed clamps. The
wellhead, through which the strings pass, is modified from a standard
unit. The current designs are based on $6\frac{5}{8}$", 20 lb/ft well casing.

The on-surface layout, process flow and electrical diagrams
and equipment specifications have also been prepared with a view to
testing or operation.

5. ECONOMICS AND POTENTIAL APPLICATIONS

Because of the extra cost of air compression or oxygen supply,
direct downhole steam generation is not expected to show substantial
cost advantages over surface steam generation purely as a method of
injecting a given quantity of heat or steam, if it is assumed that
insulated tubing can be used with the latter (8). This prediction
has been reiterated by recent BP studies. While insulated tubing
is not yet in widespread use, and tests have shown deficiencies in
its performance (9), it is nevertheless a serious competitor with the
DSG for the deep injection of steam.

The potential economic advantages of downhole steam generation
arise chiefly from the possible effects of the co-injection of
exhaust gases; laboratory and numerical studies have suggested that
during the early life of a field, this may increase the rate of oil
production by a large factor (3,4). Although this is confirmed by
one field test (5), it cannot yet be regarded as proven and is
likely, in any event, to be reservoir-specific. Thus its application
will probably have to await extended field tests in a reservoir of
economic importance. Tests in Canada of steam-carbon dioxide
co-injection are understood to be in preparation and further
information is likely to become available over the next few years.

Cost projections suggest that, if co-injection is established
to be advantageous, then for steam drive with gaseous fuel a
reliable, developed DSG is likely to be the cheapest method of
injecting the associated quantity of exhaust gases. Alternative
methods considered were (a) compression and injection of the flue
gases from surface generators, (b) separation and injection of the
carbon dioxide and (c) use of an alternative, surface-mounted, high
pressure generator. For relatively shallow reservoirs, it may be

more economic to operate the generators at the surface, particularly for cyclic stimulation (if workover costs are significant) but the 'DSG' type of design could still be favoured. Since none of these systems has undergone long-term proving, these conclusions are at present tentative.

Other advantages of downhole steam generation may become important in specific applications:-

(a) Atmospheric pollution should be reduced or eliminated.
(b) Sea water or other highly saline water might be used. The rationale behind this suggestion is that water temperatures upstream of the point of mixing with the exhaust gases should be low enough for antiscaling additives to be effective, thus eliminating the need for softening and that since the exhaust gases rather than the tube walls are the source of heat for steam raising, hot spots and deposition of salts are unlikely. Possible effects further downstream have also to be considered; these will probably vary with the application.
(c) There should be greater flexibility in the siting of bulky surface equipment since air compressors can be further from the well head than a steam generator before losses become serious. This could be advantageous offshore.
(d) Injection of steam into reservoirs underlying permafrost could be simplified.

6. CONCLUSIONS

A downhole steam generator employing a pulsed burner has been developed to the stage of detailed design of a 5 MW prototype. The pulsed burner should overcome one of the chief causes of failure of downhole steam generators, high wall temperatures in the combustion zone.

It is believed that the economic success of downhole steam generators will depend chiefly on the confirmation of laboratory predictions that substantial increases in oil production rates can result from the co-injection of exhaust gases, although other factors could favour the use of DSG's in special applications, such as offshore.

ACKNOWLEDGEMENT

This paper has been released by the British Petroleum Company p.l.c.

REFERENCES

1. HART, C.M. (1982). A comparative evaluation of surface and downhole steam generation techniques. SPE/DOE Third Joint Symposium on Enhanced Oil Recovery, Tulsa, April 4-7, 1982, 417-432 (SPE/DOE Paper 10704).

2. ANON (1982). Corrosion again plagues Sullair steam test.
    Enhanced Recovery Week, November 22, 1982, 2.
3. LEUNG, L.C. (1982). Numerical evaluation of the effect of
    simultaneous steam and CO_2 injection on the recovery of heavy
    oil. 1982 California Regional Meeting of the SPE, San Francisco,
    March 24–26, 613–622 (SPE Paper 10776).
4. HARDING, T.G., FAROUQ ALI, S.M. AND FLOCK, D.L. (1983).
    Steamflood performance in the presence of carbon dioxide and
    nitrogen. J. Canad. Pet. Tech., 22 (Sept–Oct), 30–37.
5. SCHIRMER, R.M. AND ESON, R.L. (1982). A direct-fired downhole
    steam generator – from design to field test. SPE/DOE Third
    Joint Symposium on Enhanced Oil Recovery, Tulsa, April 4–7 1982,
    433–440 (SPE/DOE paper 10705).
6. CHESTERS, D.A., CLARK C.J. AND RIDDIFORD, F.A. (1981). Downhole
    steam generation using a pulsed burner. Proc. 1981. European
    Symposium on Enhanced Oil Recovery, Elsevier Sequoia, Lausanne,
    563–572.
7. MUIR, J.F., WEIRICK, L.J. AND PETTIT, F.S. (1983). Field test
    of two high pressure, direct contact downhole steam generators,
    vol III: metallurgical analyses. Sandia National Laboratories
    Report SAND 83-0145/3, October 1983.
8. HART, C.M. AND MUIR, J.F. (1984). Update of comparative analysis
    of steam delivery costs for surface and downhole steam drive
    technologies. Sandia National Laboratories Report SAND 84-0050,
    April 1984.
9. MARSHALL, B.W. et al (1982). Project Deep Steam Quarterly
    Report, October 1, 1981 – March 31, 1982. Sandia National
    Laboratories Report SAND 82-1336, September 1982.

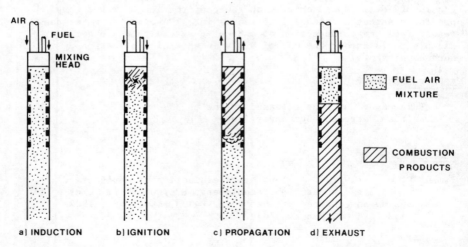

FIG 1 SCHEMATIC DIAGRAM OF THE INDIVIDUAL PHASES IN THE FIRING
CYCLE OF A PULSED BURNER

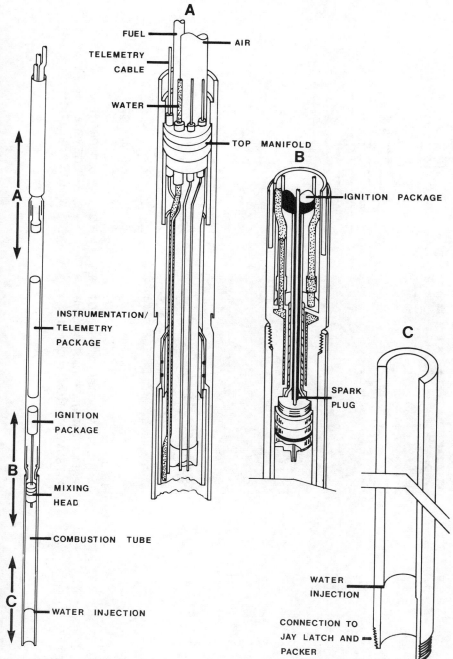

FIG 2 BP PROTOTYPE DOWNHOLE STEAM GENERATOR

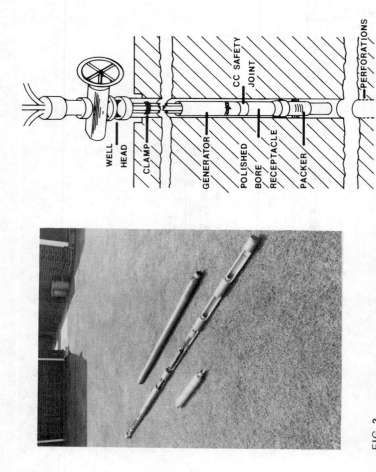

FIG 3
HALF-SCALE MODEL OF PROTOTYPE
GENERATOR WITH IGNITION AND
INSTRUMENTATION PACKAGES

WELL
HEAD

CLAMP

GENERATOR

POLISHED
BORE
RECEPTACLE

PACKER

CC SAFETY
JOINT

PERFORATIONS

FIG 4 BP DOWNHOLE
STEAM GENERATOR
WELL COMPLETION

(05.33/81 and 05.42/82)

STEAM INJECTION PILOT PROJECT INTO EMERAUDE OFFSHORE RESERVOIR

B. SAHUQUET,   )
D. MONFRIN,   )   Société Nationale ELF AQUITAINE (Production)

Summary

This Franco-Italian project to which AGIP is associated concerns a pilot project to inject steam into the Emeraude heavy and viscous oil reservoir in Congo offshore waters in a depth of water of 65 metres. The considerable accumulation (575 million tons) and the low recovery ratio (3%) obtained by primary production explain the efforts made to execute a pilot project the essential purpose of which is to prove the possibility of industrial exploitation by steam injection, despite the innate difficulties of this reservoir (shallow depths of production layers, type of reservoir, quality of oil, etc.).
The engineering work was launched at the end of 1980 and enabled the platforms and equipment needed to implement the pilot project to be defined. The equipment was installed from December 1981 to June 1983. Drilling work started without Community support in July 1983 and continued through 1984. The first water injectivity tests started in October 1983. Steam injectivity tests will be started at the end of 1984. The work will end in 1987 with evaluation of the overall results of this project.

1 - INTRODUCTION

EMERAUDE reservoir lies in an average depth of water of 65 metres about 20 metres from the coast (see figure 1). The field is separated into seven reservoirs, R1 to R6 and G-I-K with different characteristics, each reservoir consisting of several beds, lying at depths of from 190 to 540 metres. Most of the accumulations are to be found in reservoirs R1, G-I-K and R2.

At present, the field is being developed by about a hundred pumped wells, since 1973 for the Northern part and 1976 for the Southern part.

Owing to the low internal energy of the reservoir and the presence of hyper-conducting drains preferentially carrying the water and gas phases, a cumulative recovery of about 5% is all that can be envisaged by primary production, whence the interest in other methods of exploitation.

Injection of water (pilot project performed in 1974) together with studies performed on sodium water injection and in-situ combustion did not justify adopting these methods.

Laboratory research on steam injection into silty beds indicated a high recovery level (about 40% at 180°C) through the reduced viscosity, the expansion and evaporation of the effluent. This does not allow for possible sweeping of the silts by the steam. However, these results, together with experience gathered on other reservoirs can not be directly transposed to the full scale of the Emeraude field. The heterogeneity of the reservoirs (alternating hyper-conductive limestone drains and silty beds of low permeability containing the oil), the risks involved in the drilling operations and the offshore situation are all major elements in the decision to implement a pilot project, the information from which will provide a better understanding of the industrial advantage of the method.

2 - DRILLING

The upper Emeraude reservoirs are both cracked and highly depressed, and there is a considerable risk of blindly and irremediably clogging the perimeters of the wells with mud and cement. In addition, the cement will have to withstand the thermal stresses, and provide complete and tight cementations in order to isolate the various levels correctly. Tests of the long-term resistance of various categories of cement resulted in the choice of a cement lightened with glass balls protected thermally by adding silica flour. Laboratory tests enabled formulae for plugging slugs to be developed capable of being destroyed 95% by hydrochloric acid.

In addition, the shallow depth of the reservoir does not provide sufficient distance between wells from the same platform using normal deviated drilling techniques. A "tilt rig" enabling wells inclined right from the surface at an angle of up to 30% to be drilled was designed in association with FORAMER. Fifteen wells are to be drilled, of which 12 deviated, in the following configurations (figure 2):
- 1 five-spot in R1,
- 1 five-spot in R2.

These two objectives are priority for a steam drive owing to their structure and reserves.
- 2 observation wells,
- 2 huff-and-puff wells in GIK (intermediate R1 - R2),
- 1 huff-and-puff well in R3 not amenable to continuous drive owing to its small thickness and low productivity.

Levels R4, R5 (fractured and dolomitic voids) and R6 (water table close by) are not suitable for this recovery method and have hence been excluded from the pilot project.

3 - WELL EQUIPMENT

By analyzing the experience gained in primary exploitation of Emeraude, the use of steam on other reservoirs and inquiries with the suppliers of equipment and companies operating reservoirs by steam injection, the following were determined:
- definition of the completion programmes of the various injection, production, huff-and-puff and observation wells,
- the design of the compact "high temperature" inclined wellheads and a sliding joint with which the injection well tubings are equipped. An application has been filed for a patent covering this joint,
- the design of a long stroke unit for inclined pumping at 30°, with Société MAPE. The tests of these pumps carried out at the Fourc testing centre proved completely satisfactory.

4 - EMVF DRILLING PLATFORM

This is a "double pile" platform comprising a false pile in the middle of each face from level 26 m, the purpose of which is to stiffen the structure. The 20 well positions are laid out in a square. The 12 conductor tubes, inclined at an angle of 30°, are set out along the sides of the square and pointed inwards. Their overall dimensions do not exceed those of the structure. The platform has three deck levels (see figure 2).
Level + 20 m accommodates the drilling rig and a crane.
Level + 11 m mainly accommodates the long stroke pumping units and the landing of the connecting gangway to the utilities platform.
Level + 8 m carries the wellheads and their access facilities.
The detail engineering was entrusted to CG DORIS. The jacket was launched in July 1982 and the deck laid in August 1982. These two elements are built by BOUYGUES OFFSHORE.
The production module is overhung on the two levels on the NNE face of the drilling platform. It carries all the production installations, the fuel gas scrubber for supplying the boilers, the steam distribution manifold and the overhung flare.
PONTICELLI built the production module, which was installed in November 1983.
The very fast corrosion that occurred with ordinary grade steels as a result of the temperature of the crude of Emeraude necessitated numerous studies and consultations with the suppliers.
In order to test the various solutions, the installation consists of:
- 1 main production line consisting of two manifolds (1 for the tubings, 1 for the annuli), a tubular heat-exchanger in Nicrofer (effluent/fresh water) and a 3-phase separator,
- 4 test lines, each comprising 1 manifold (2 or 3 wells), a tubular heat-exchanger (effluent/fresh water or effluent/sea water and a 3-phase separator. Three of the heat-exchangers have Nicrofer tubes and the calender in ordinary steel covered with epoxy, whilst the 4th is in titanium,
- 1 heat-exchanger with sea water/fresh water plates in titanium,
- 1 flare gas scrubber, to which the gas outlets from the five separators go,
- transfer and discharge pumps,
- meters on all the separator outlets and the oil discharge line.
After processsing the production, it is sent to platform CC through a 4" hose and injected into the general network of the field.

5 - EMVU UTILITIES PLATFORM

This platform is to carry the equipment required for the following functions:
- lifting the sea water,
- desalinating the sea water
- steam generation,
- other utilities: compressed air, electricity, safety equipment, control room and helicopter deck.
The jacket is of conventional design with 4 piles and the deck has 2 levels.
The detail engineering was performed by Société TECHNIP. The jacket was launched in January 1983 and the deck laid in July 1983. BOUYGUES offshore built the jacket and deck.

Sea water desalination

This function is fulfilled by two units employing two different distillation processes:
- 1 ejection-compression system (SIDEM) operates according to the vacuum distillation principle (50°C, 0.1 bar). The energy is provided by the steam. It has a distilled water capacity of 250 m³/day.
- 1 mechanical compression unit (THERMOMECANICA). This also operates at a partial vacuum, though the energy is provided by a compressor. It has a capacity of 300 m³/day. This mixed solution was adopted because of its operating flexibility, particularly when starting, thanks to the mechanical compression unit, which is completely independent of the steam generators. Furthermore, it enables two types of equipment to be tested with a view to future development. The sea water supply is provided by five submerged pumps, each capable of an output of 100 m³/hour. There are also five filters and one electro-chloration unit. The distilled water is stored in a 30 m³ capacity buffer tank.

Steam generation and distribution

Steam production is provided by means of two STRUTHERS steam generators, each capable of 25 MM BtU/h and a maximum pressure of 70 bars. Their maximum unit capacity is 315 tons per day of 95% quality steam. The fuel gas, fed by hose from the BB, CC and HH conventional development platforms passes through a fuel-gas scrubber before use.

Down-line from the boilers there are two high pressure separators enabling dry steam to be obtained which is sent to the wells, together with a small proportion sent to the ejection-compression desalination unit. The steam distribution manifold comprises six regulating systems each consisting of: one pressure regulating valve, one flowmeter and a manual by-pass choke valve for starting injection.

The utilities comprise distribution of the electric power, since the pilot project is supplied from platform PCP, instrument and service air provided by a backup compressor, the safety equipment (fire-fighting diesel set and fire-fighting, rescue and evacuation facilities), the processing product tanks and their pumps, the storage rooms and shelter and the control room.

All these installations (EMVU and EMVF) were started up and tested from mid-1983 to mid-1984. After a few modifications, the full system is now operational.

6 - RESERVOIR STUDIES

Analysis of cores sampled in the first three wells, logs and pressure measurements have enabled the geological image of the reservoir to be refined within the pilot project mesh and the flow patterns within each reservoir to be determined. The completions described in paragraph 7 are the result of this work.

Petrophysical measurements (porosity and permeability under stress, and size distributions) and geochemical measurements (organic and mineral) are now being made to quantize a number of parameters needed for following and interpreting the injection of the steam on a numerical thermal model.

In addition, laboratory experiments on triphasic flows (oil-water-steam) and biphasic flows (oil-water at various temperatures) in the Emeraude silts, on clogging of the drains by foam and steam tracing are continuing.

7 - __WELL TESTS__

The tests started on 20th October 1983 at the end of the drilling operations and the completion work on the first three wells EMV01, EMV02, EMV03.

Level R1 - EMV01
This vertical well is the central injection well of the R1 five-spot. The pressure measurements during drilling revealed considerable vertical pressure variations in R1 and vertical communication between the base of R1 and level G.
The first tests showed the low injectivity of the silts and since the cementation towards the top had been damaged during the tests, and, since the cementation towards the top had been deteriorated during the tests, also showed the considerable continuity of the upper drains, which form a short-circuit to the wells of the neighbouring platforms. A vertical drive pattern from bottom to top was hence adopted with injection at EMV01 into the drains of bed G and production from the corner wells through the top drains (see figure 3).
Adoption of this configuration and communication between G and R1 resulted in abandoning the GIK huff-and-puff wells, since this level no longer justified separate treatment.
The initial perforations into the R1 silts were plugged and the cementation restored. The well was reperforated in bed G and the injectivity of this level tested: one cold water test and one hot water test, in order to confirm the tightness of the cementation in the upward direction.

Level R2 - EMV02 and EMV03
EMV02 is the South production well of five-spot R2. It was drilled and cored deviated (30°). EMV03 was to be the isolated huff-and-puff well of R3 ; it was drilled practically in the centre of the pattern and measurements were made (logs, cores, pressures) on all levels R1 and R2.
The pressures measured at R2 on the two wells showed that its base (10% of the accumulations) was isolated and that a vertical flow in M with reduced drain density and containing most of the accumulations was supplying the upper bed L, which has a high drain density, ensuring drainage of the oil towards the production wells. To ensure maximum sweep through the silts by the steam, completion of the injection and production wells was decided at the base of bed M (see figure 4).
In order to test the configuration adopted with respect to one of the two main objectives of the pilot project, vertical well EMV03 was completed with a central injection well of five-spot R2 following acquisition of the geological and pressure data at R3, in order to provide interferences with production well EMV02.

Steam injection test results
The purpose of this test, performed on EMV03, was to determine the steam injectivity and to test the behaviour of the cementations and silts under temperature. The steam was produced by the final steam generator of EMVU from industrial water softened by a temporary desalination unit. It was injected for 15 days at a rate of 4 tons per hour and a steam quality of 70% at the wellhead (the surface lines were partly heat-insulated during the test). The reactions at the other wells are compatible with the observations made during water injection tests and no deterioration of the cementation was detected.

Two months after going back onto production, no abnormal sand inflows are observed and the oil flow would appear to have become stable at a value above its value before steam injection.

CONCLUSION

The drilling work is to continue to the end of 1984. The same also applies to the water and steam injectivity tests.

At the end of 1984 steam injection proper will start and continue to the end of 1987, at which time the overall results of this steam injection pilot project will be determined.

Should these overall results prove positive, many sectors of the Emeraude reservoir could be placed on production with steam injection. However, it should be pointed out that the shallow depth of the reservoir limits the part from which it is possible to produce from each platform. For this purpose, an economic study should be carried out so as to determine the most appropriate of the industrial solutions derived from the Emeraude steam pilot project.

The considerable reserves in place would justify a large number of platforms, which may make Emeraude oilfield one of the largest offshore development projects in the world.

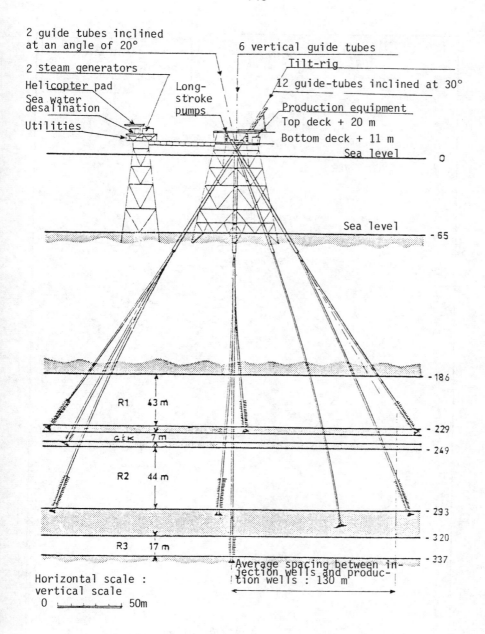

FIGURE 1 - EMERAUDE STEAM PILOT PROJECT - Overall diagram

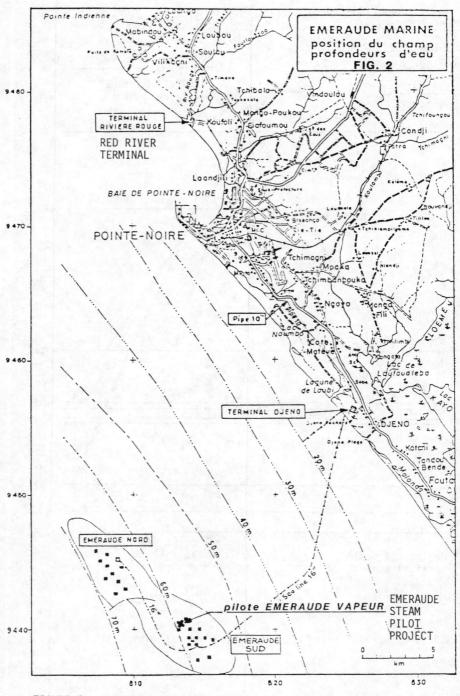

FIGURE 2 - EMERAUDE OFFSHORE FIELD - Position of field.
- Depths of water

R 1

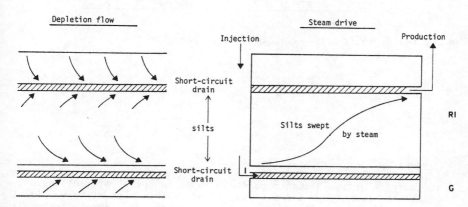

Depletion flow

Steam drive

The silts supply the two drain zones, but depletion
takes place essentially at the bottom

FIGURE 3

R 2

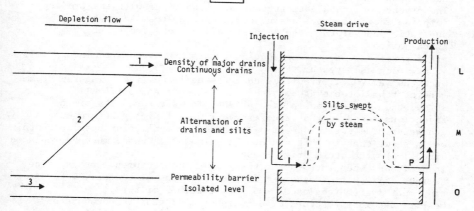

Depletion flow

Steam drive

1 _ Flow towards well

2 _ Vertical flow in bed M which feeds L as it depletes

3 _ Level O depletes independently

FIGURE 4

(05.19/80)

INJECTION OF STEAM INTO THE MARIENBRONN RESERVOIR

J. MAURY AND J. SOLIER, Compagnie Française des Pétroles

Summary

The purpose of the Marienbronn steam injection pilot project is to enable the technology and methodology of exploitation of heavy oils to be mastered. The field of application is the Marienbronn reservoir, which contains an oil with a density of 0.99 and an in-situ viscosity of 227 poises.
The reservoir consists of sands and conglomerates and lies at a depth of about 250 metres. Its structural form is of the monocline type.
19 wells have been drilled in a "FIVE SPOT" configuration on the Marienbronn pilot project.
The production installations that are associated with the project comprise feed water processing equipment, a steam generator, separation and storage tanks and an oily waters treatment installation.
After purification, the production waters are discharged into a well in a geological stratum.
Once the various difficulties encountered in commissioning the steam generator had been overcome, the initial huff and puff well stimulation cycles took place in 1984. The first results of the injections revealed that the reservoir was of highly heterogeneous character. The establishment of paths of communication between certain wells shows that in this sector the stimulation phase must be replaced directly by the steam drive phase.
In addition, when introducing the production equipment into service, serious difficulties occurred in the operation of the oily waters treatment system (coalescer and active carbon filters).
Despite the modifications made - higher operating temperature and regulation of the flows - the operation of this installation is unsatisfactory.
This being so, it is doubtful that the "oily waters" can be discharged into the geological stratum, and clogging of the discharge well can be foreseen in the medium term.
At any event, the difficulties encountered, whether at the technological level or implementation level, justify that such pilot projects be carried out before any industrial development is attempted.

INTRODUCTION

The Marienbronn heavy oil reservoir lies about 40 kilometres North of Strasburg, near Pechelbronn, the birth place of oil research in France (see figure 1).

This small size reservoir mostly lies beneath the communal forest of Lampertsloch, where many drillings had exploited the layers of Pechelbronn (tertiary formation), containing medium oils that can be extracted technically without requiring any special facilities. These wells had enabled the presence of a heavy oil deposit to be identified.

1 - DESCRIPTION OF THE PILOT PROJECT

The purpose of the Marienbronn steam injection pilot project is to exploit an Oligocene level containing heavy oils that can not be moved under the conditions inside the reservoir.

1.1 - Characteristics of the reservoir

The reservoir consists of a monocline structure (see figure 2) probably backing against the Rhine fault, the limits of which are not known with any exactitude.

The Marienbronn reservoir lies at the roof of the lower Pechelbronn layers (Sannoisian). The level exploited consists of an alternation of conglomerates and sandstone sands, more or less consolidated.

The following are the characteristics of the reservoir layer:
- depth   :   250-300 m
- total thickness   :   10 to 30 m
- useful thickness   :   10 to 20 m

	conglomerates	sands
- porosity	20%	30%
- oil saturation	80%	80%

It contains a heavy oil with the following main characteristics:
- density   :   0.99 (10.5° API)
- in-situ viscosity :   227 poises (22.7 Pa.s) at 20°C
- pour point   :   + 9°C

1.2 - Description of the installations

Only part of the reservoir will be exploited by the steam injection pilot project.

19 production wells have been set out in a "five spot" configuration over an area of 12 hectares (about 31 acres). The wells are separated from one another by a gap of about 100 metres. The well completions have been designed so that the pumping equipment can remain in place during the steam injection phases.

The production installations associated to the wells comprise:
- a steam generator feed water conditioning and treatment installation,
- a steam generator with an output of 14,000 thermies per hour capable of generating 29 tons of wet steam per hour at 80%, 70 bars and 280°C,
- an effluents treatment installation consisting of a primary decantation unit and a secondary heated treater separation unit,
- an oily waters treatment installation consisting of a coalescer and active carbon filters.

After purification, the production waters are discharged through a well into a deep geological layer.

1.3 - Exploitation strategy

The following is the exploitation method adopted:
- an initial huff and puff injection phase, during which each well of the pilot project is subjected to steam injection for a period of 4 to 5 weeks. After being closed for about one week, the well is placed on pumped production until it is no longer possible to pump the effluent,
- a second drive phase in which the steam is injected continuously into the central wells of each of the six "five spots", with production via the peripheral wells.

2 - PROJECT STATUS

2.1 - Description of commissioning of the pilot project

The wells were drilled and the production installations built during 1982. Provisional acceptance-testing of the installations took place at the beginning of 1983.

Immediately the steam generator was started up, thermal instabilities were observed on this equipment requiring two months shutdown in order to carry out certain modifications.

The injection tests were resumed in April 1983, but unfortunately again had to be interrupted in June 1983, when a tube in the steam generator economizer tube nest burst.

Following temporary repair, production and injection of steam were again resumed in July 1983. They continued until 17th September 1983 when the piercing of an economizer tube called for a further shutdown.

Examination by experts of the economizer tube revealed considerable corrosion of the tubes caused by a skin temperature on the tubes that lay below the due point of the smoke. This excessively low temperature of the fluid flowing through the economizer, as a result of an engineering error, required a major modification to the steam generator. It was decided to increase the temperature of the fluid and to build a new economizer. These modifications meant that the installation had to be stopped for 4 months.

In addition, examination of the economizer tubes revealed the abnormal presence of deposits of carbonates and silicates. Analysis showed that these deposits result from a shortcoming in control of the quality of the steam generator feed waters. The installation of a permanent control device should prevent the recurrence of such deposits, which are prejudicial to the satisfactory operation and integrity of the steam generator.

Since the beginning of 1984, steam injection has taken place without interruption.

2.2 - Operating results up to mid-1984

Eighteen months after starting up the installations, what information has been acquired concerning the exploitation of the Marienbronn steam injection pilot project ?

- Steam generation

The SULZER steam generator is an assisted circulation generator, the evaporator tube nest of which consists of a double turn.

Owing to its design, immediately it was started up, the operation of this type of generator turned out to be a delicate matter:
  . susceptibility to variations in the load incompatible with use on an oilfield, where one may have to decide very rapidly to stop injection into a well ; it should be noted that the steam

consumption of the wells and the utilities varies with time,
. impossibility to start up again following an unintentional
interruption lasting more than a few minutes.
The radiation of heat caused by the heat-resistant bricking results
in overheating of the tubes that makes it impossible to resupply
the generator with feed water before several hours have elapsed,
. the need for quality control of the feed water treatment system
is in fact just as greater constraint as for a drum type generator,
which is in conflict with the accepted concept that an assisted
circulation generator can accommodate to less rigourous treatment
of the feed waters (one less demineralization stage).

- Treatment of production
    Treatment of the oil produced has been defined according to be basic
concept that the production effluent was a diphasic liquid effluent (oil
+ water).

    In reality, the production effluent is triphasic and its composition
varies very considerably (gas slugs, oil slugs, water slugs, emulsions of
oil + water...), and the effluent treatment installation had to be
modified. A degassing drum was added at the primary decantation unit
inlet to prevent the turbulence created by the steam bubbles in the
decantation liquid.

    The separation process must be improved by adding a de-emulsifying
agent.

- Treatment of the oily waters
    At present, this has proved to be the stumbling block of the pilot
project.

    The equipment does not operate satisfactorily. The coalescer operates
more or less as a filter, despite a service temperature that has gradually
been raised from 40° to 60° and then 75°, finally approaching 90°C.

    It is probable that coalescence can not be applied as such to heavy
and viscous oils. In view of the difficulties encountered, it was decided
to add a solvent at the coalescer intake.

    Treatment of the oily waters is still delicate, and wells with too
high a percentage of water in the production have been shut down.

    Research has been undertaken on the problem of treatment of the oily
waters in order to ensure that the installations run satisfactory in 1985.

    Discharge of the purified waters into the geological stratum
    Discharge into the Buntsandstein sandstone layers occurs with
difficulty, owing to the mediocre characteristics of the host reservoir.

    In view of current regulations concerning protection of the environ-
ment, the feasibility of the discharge operations (into the natural medium
or into a geological stratum) must be ensured before starting any
exploitation to avoid encountering production interruptions owing to lack
of the means to eliminate the waste waters.

- Behaviour of the reservoir
    Because of the discontinuous operation of the pilot project, not much
information was obtained concerning the behaviour of the reservoir.

    However, the behaviour of certain wells can be partly understood from
the highly heterogeneous nature of the reservoir itself.

Indeed, during injection into seven wells, the stimulation phase did not occur, since the well in question had established a path of communication with the neighbouring wells as a result of the first cycle.

To be able to present overall results of any meaning, exploitation should be continued at least throughout 1985.

Evolution of production

Oil production was about 13 m³ per day during May and 20 m³ per day during June.

At the end of the first six months of 1984, the steam was generated by using the crude protection, with a slight excess going to the refinery. During the second half of 1984, an output of 20 to 25 m³ per day is hoped for, which, allowing for the consumption of the steam generator, justifies the expectation that 10 m³ can be sent to the refinery each day. These results are encouraging.

5 - CONCLUSION

To obtain the maximum benefit from the pilot project, the following are needed:
- first, a solution to the problem of treating the oily waters and discharging them into the geological stratum,
- second, continuation of exploitation for at least a year so as to be able to give meaningful overall results concerning the behaviour of the reservoir.

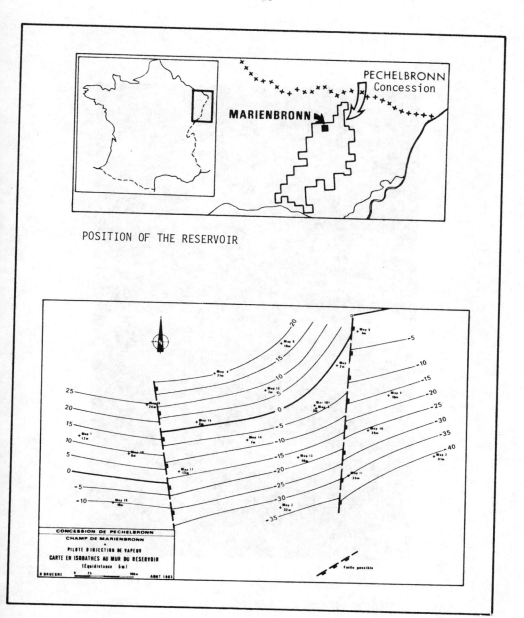

POSITION OF THE RESERVOIR

THE MARIENBRONN STEAM INJECTION PILOT PROJECT

AERIAL VIEW OF THE MARIENBRONN INSTALLATIONS

(03.67/78)

EXPLOITATION OF HEAVY OIL : THE FBH PROCESS

P.A. BORRILL and R.J. EASTERBY
British Gas Corporation, Midlands Research Station,
Wharf Lane, Solihull, West Midlands. B91 2JW
United Kingdom

Summary

The British Gas Corporation has successfully completed an R & D programme on its Fluidised Bed Hydrogenation (FBH) process. Pilot plant tests have shown that conversion of up to 62.5% to methane and ethane can be achieved with vacuum residue feedstocks (SG 1.03). Associated tests on a much larger fluidised bed model have established design parameters for individual commercial scale gasifiers. The FBH process can be included in fully integrated process schemes for Substitute Natural Gas (SNG) manufacture from heavy oils. Overall energy conversion efficiencies for such schemes are high (80%) and capital costs are modest in comparison with the best available coal-based schemes.

1. INTRODUCTION

The security of the Community's supplies of hydrocarbons depends on the enlargement or enhancement of indigenous resources and on the ability to use external feedstock supplies from the widest variety of sources with the least restriction on quality. In addition, having acquired supplies of increasingly scarce and expensive oil, it is incumbent on the Community to use these in the most efficient manner.

At present residual oil, some 35% on average of each barrel of oil, is converted into other lighter oils or directly into energy at thermal efficiencies as low as 35% in power stations.

British Gas has developed highly efficient, commercially viable processes for the gasification of distillate oils, e.g. the CRG process. In recognition of the fact that lighter oils will become less readily available, part of the long-term strategy has been to explore the gasification of heavy oils. The Fluidised Bed Hydrogenation (FBH) process converts heavy oils, including those with high metals and sulphur content, into a range of products. For atmospheric residue oils, 65% of the carbonaceous matter is converted directly to methane and ethane and 30% is recovered as aromatic liquids, of which 40% is benzene. Only about 5% of the oil appears as carbon whilst the sulphur compounds are easily recoverable in elemental form. Whole heavy crude oils have been tested in smaller assay rigs and no special difficulties are anticipated on a larger scale. Suitable further processing of the FBH products allows Substitute Natural Gas (SNG) to be the sole fuel product at an overall thermal efficiency of about 80%. Alternatively, if required, the individual products could be utilised directly.

Development of the FBH process has been carried out by British Gas with sponsorship by the Community under a three year contract. The work has also been the subject of a Joint Development Programme between British Gas and the Osaka Gas Company of Japan. This paper summarises the work carried out during this period and presents the major project advances and current status.

2. THE FLUIDISED BED HYDROGENATOR

The Fluidised Bed Hydrogenator, shown in Fig. I, can operate over a range of pressures and temperatures, although most of the work has been at 52 bar and 760°C. The bed itself consists of finely crushed coke particles which are fluidised by the hydrogen containing gas. The gas distributor is designed to encourage the fluidised coke particles, accompanied by gas bubbles, to flow up the riser tube and the more densely packed coke particles to flow down the annular region. Heavy oils are atomised into the base of the riser tube where, upon contact with hydrogen, significant amounts of methane and ethane are formed. Some constituents of the feed produce small amounts of carbon which, together with mineral matter from the oil, coat the coke particles. The resulting growth of the coke particles must be controlled. Recirculation of coke is necessary to dissipate the heat of reaction and to provide surface on which carbon deposition can occur without particle agglomeration. The product gas disengages from the bed at the top of the reactor and is quenched to condense ungasified aromatics.

3. PROGRAMME OF WORK

Although pilot plant work started some 20 years ago, progress has been interrupted by the appearance of more easily gasified feedstocks such as naphtha and by natural gas. Up to 1979 work was concentrated on whole light crude oils and included a period of joint operation of a semi-commercial scale plant with Osaka Gas in Japan between 1969 and 1973. In 1979 interest turned to heavy residual oils and this was the basis of the EEC contract.

A programme of work was established for the FBH pilot plant aimed at the gasification of atmospheric and ultimately vacuum residues. The programme included the optimisation of reactor operating conditions and the establishment of a technique for controlling the reactor particle size. Supporting work on heat recovery from the product gases and evaluation of suitable materials of construction was to be carried out. Attention was also to be given to the distribution of major and minor sulphur and nitrogen compounds in the product gas and the effect they may have on methods of gas purification.

Little published information was available on the performance of larger diameter fluidised beds operating at high temperature and pressure. It was important to identify any limits which might be imposed on the scale up of this type of recirculating bed reactor. As a second major part of the work therefore, a large diameter Fluidised Bed Model (FBM) facility capable of operating at elevated temperature and pressure was constructed to consider the problems of scale-up. A programme of tests was planned to assess the effects of various operating parameters on the quality of fluidisation and solids recirculation for a number of internal configurations.

4. DESCRIPTION OF PLANT

4.1 FBH Pilot Plant

The pilot plant is located at the Midlands Research Station in Solihull. It has a throughput of up to 12000 Nm³/d of gas with a 0.25m diameter, 10 m high reactor capable of operating at up to 70 bar and 900°C. Hydrogenating gas is manufactured and compressed on-site and fed to the reactor through gas-fired preheaters. The feedstock oil is pumped from storage and, after preheating, is atomised with some of the hydrogen into the base of the fluidised bed. The product gas is directly quenched before being cooled, metered and flared. In order to carry out the heat recovery studies, in later runs the quenched products were fed directly to the tube side of a high pressure steam boiler. Control of the increasing bed inventory caused by coke particle growth during heavy oil gasification is achieved by batch removal of bed material.

4.2 FBM Facility

The FBM facility was constructed at the Coleshill works site of British Gas in 1980. It consists of a recirculating loop of nitrogen gas. The nitrogen is preheated and supplied to the FBM vessel, 1.5m diameter, 13.7 m high, which can operate at 20 bar and 400°C. The gas leaving the vessel is dedusted and cooled before returning to the recycle compressor. The vessel, which can accommodate a variety of internal arrangements is well provided with a range of measuring devices. These measurement techniques, which were developed at the outset of the project, have been used to measure the flow characteristics of the coke bed and the gas.

5. RESULTS OF THE WORK

5.1 FBH Pilot Plant

The pilot plant has successfully gasified over 160 tonnes of residual oil feedstock in 18 runs, each typically of 70 to 100 hours duration. Typical yield data for two feedstocks, including vacuum residue with a specific gravity of 1.03, are presented in Table I.

The performance of the recirculating bed reactor has been excellent throughout tests with wide variations in oil and gas throughput. A technique for controlling the reactor coke particle size around the preferred size for this unit of 350 microns by the periodic addition of smaller sized seed coke has been demonstrated.

The heat recovery system installed on the pilot plant has operated according to design over a range of process conditions and, despite the tarry nature of the materials, has not suffered from fouling. The heat recovered by cooling from 750 to 350°C represented 5% of the original potential heat in the feedstock.

The majority of the sulphur in the feedstock (around 80%) is converted to hydrogen sulphide (H_2S). This, together with the remaining low concentration gaseous sulphur and nitrogen compounds, can be separated from the product gas by conventional acid gas removal techniques.

As a result of exposure to the corrosive reactor environment it has become clear that conventional austenitic stainless steels and nickel based alloys will not be suitable for use. A number of commercially available high alloy stainless steels have been identified which will have an operating life of at least one year despite the prevailing H_2S levels of 1%. More exotic materials have been tested which may give prolonged operation, but their use may not be economically justifiable.

5.2 FBM Facility

Experience suggested that difficulties might be expected in scaling up the FBH recirculating bed system to a commercially viable scale. In all 12 runs have been successfully carried out in which several reactor internal arrangements were tested. Measurement techniques were developed which allowed the behaviour of the oil, gas and coke particles to be simulated at intermediate conditions using only coke particles and nitrogen gas in this unit. The results showed that the original design of the FBH reactor using a single coaxial riser tube can be scaled up with confidence to a 1.5m diameter unit of the same height (10m). Most significantly the work showed that critical design features such as hydrogen/oil mixing efficiency and coke recirculation rates are fully maintained on larger scale units even at considerably higher throughputs than had been possible on the smaller pilot plant. The work has established design parameters for individual commercial reactors producing up to 0.6×10^6 Nm^3/d of SNG and has provided guidelines for the design of even larger reactors if these are required.

6. PROCESS FEASIBILITY

The experimental work carried out during and since the period of sponsorship by the Community has provided sufficient technical confidence to enable FBH reactors of a commercially acceptable size to be designed. The experimental data have been used to show that the FBH reactor can be combined with conventional hydrogen production and gas purification units to give a completely integrated SNG plant. A basic flowsheet is given in Fig. II. Hydrogen production can be achieved by partial oxidation of the reactor by-products and excess hydrogen is recovered from the product gas by cryogenic separation before recycle to the reactor. Apart from the FBH reactor, most of the other process units have already been proven at the required scale. Studies have shown a high process thermal efficiency of around 80% for an FBH-based SNG scheme. Despite the modest capital costs, the future economics of the process will depend on the price and future availability of heavy fractions and these are uncertain. A development plant is needed as part of the preparation for commercialisation and a conceptual design has been carried out.

7. CONCLUSIONS

A successful programme of research work has been carried out on the FBH process. The range of feedstocks which can be handled has been extended to atmospheric and vacuum residues. Sufficient technical data have been provided to enable FBH reactors of a commercially acceptable size to be designed. No further experimental work is considered necessary to establish large scale reactor design criteria. The decision to advance to development or commercial sized plants will depend on the projected timing of the need for SNG and the cost at that time of SNG from the FBH process in comparison with other supplies.

ACKNOWLEDGEMENT

The British Gas Corporation is grateful to the Commission of the European Communities for the grant provided for this work at a critical stage of development.

This paper is published with the permission of the British Gas Corporation.

TABLE I : <u>Typical Operating Conditions and Results from FBH pilot plant</u>

FEEDSTOCK PROPERTIES

Feedstock	Atmospheric Residue	Vacuum Residue
Density (15°C/15°C)	0.96	1.03
Carbon/Hydrogen ratio (w/w)	7.5	8.1
Sulphur content (wt %)	3.1	4.6
Asphaltenes (wt %)	4.0	8.4

OPERATING CONDITIONS

Reactor temperature (°C)	760	760
Reactor pressure (bar)	52	52
Oil/Hydrogen ratio (kg/kg)	4.9	5.2

YIELD DATA (kg/100 kg oil)

Methane	48.8	51.0
Ethane	24.9	19.0
Benzene	10.2	8.6
Heavy aromatics	17.1	15.1
Coke	4.4	9.8

RESULTS

Carbon Gasification (wt %)	65.0	62.5
Sulphur conversion to H_2S (wt %)	78.0	85.0
Hydrogen consumption (kg/100 kg oil)	8.9	8.3

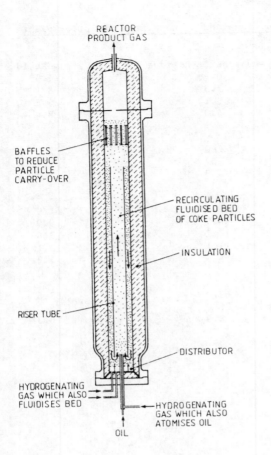

REACTOR
PRODUCT GAS

BAFFLES
TO REDUCE
PARTICLE
CARRY-OVER

RECIRCULATING
FLUIDISED BED
OF COKE PARTICLES

INSULATION

RISER TUBE

DISTRIBUTOR

HYDROGENATING
GAS WHICH ALSO
FLUIDISES BED

HYDROGENATING
GAS WHICH ALSO
ATOMISES OIL

OIL

FIGURE 1 - DIAGRAMMATIC
LAYOUT OF THE FLUIDISED
BED HYDROGENATOR

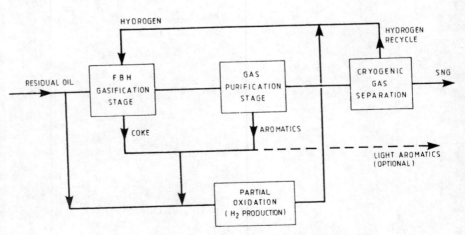

FIGURE 2 - FBH ROUTE TO SNG FROM RESIDUAL OIL

(05.20/80)

HEAVY OIL PILOT PROJECT NORDHORN

G. PROYER
Senior Res. Engineer
Wintershall AG, Kassel

Summary

In 1981 it was decided by the "WE" consortium members (BEB, Deilmann AG, Preussag AG and Wintershall AG as operator) to evaluate the feasibility of recovering Nordhorn Tar Oil by steam injection through hydraulically – induced fractures. On the basis of topographical, logistical and reservoir engineering aspects block D with the well NOH 1002 in a structurally high position of the Nordhorn Field was chosen as project area.

   The pilot project consists of 5 acre inverted 5-spot pattern with 5 wells (thereof 4 producers and 1 injector). The crude- and water treating system has a capacity of 360 tons per day of wet-oil production. With steam injection (15 t/hour) in the injector of the 5-spot pattern about 34 000 m³ of tar oil can be produced during a period of about 2 years – this represents 23 % of the OOIP. Salable products can be obtained from the heavy oil of Nordhorn by means of conversion processes. In consequence of the low proceeds for the tar oil, high investments and operating costs, the Nordhorn Heavy Oil Project is economically not attractive to date. The paper describes the planning work, including laboratory tests, simulation study, conceptual planning and the project profitability.

1. Introduction

   The Nordhorn heavy oil field is located in the Emsland area with considerable thermal recovery activities near the Dutch border. This is the second phase of a project supported financially by the Energy Commission of the European Communities. The first phase of the project consisted of drilling the five new wells on the Nordhorn structure in 1975 in order to obtain basic information on the reservoir properties and reappraise the reserves. In order to gain some insight into reservoir behavior under thermal conditions a four day steam soak in November 1977 and a thirty day steam soak in August 1978 were performed on the well NOH 1005. This phase of the project was reported previously (1).

   In 1982 injection tests with radioactive tracer in the wells NOH 1002 and NOH 1005 were performed. With production control measurements, GR-profiles and temperature-profiles and other available informations the reservoir performance during the thirty days steam soak on well NOH 1005 in August 1978 was numerically matched. Results of this history match were reported by de Grisogono (2).

2. Pilot Project Area

Concerning the heavy oil reservoir Nordhorn a pilot area appropriate in view of reservoir engineering and topographical aspects was choosen. Out of the available five wells NOH 1001, 1002, 1003, 1004 and 1005, the later three wells were topographically unfavourably located, so they were out of question as potential pilot areas. Because of immediate neighbourhood of the oil/water contact, the well NOH 1001 does as well not appear to be favourable for reservoir engineering reasons. So well NOH 1002 in a structurally high position (no free water and reprentative for about 75 % of the reservoir) proved to be an appropriate object for one of the 5 wells for the project area (Fig.1).

3. Simulation Study

The study (3) was intended to evaluate the feasibility concerning the recovery of heavy oil by means of a horizontal frac in the Nordhorn field. Within the scope of these operations calculations were made for the 5-spot pattern in line with the geological conditions for block D in the Nordhorn heavy oil field.

On the basis of former studies it was fixed that heat influx above a horizontal or bed-parallel frac heated by steam takes place through heat conduct from an uprising source of heat at steam temperature. Heat is conducted into the porous medium and as soon as steam temperature is reached, the steam rises upwards pursuant to the forces of equilibrium between steam and movable heavy oil. When the ingressing steam reaches zones with lower temperature than steam temperature, the steam condensate flows together with the hot oil in the countercurrent towards the uprising steam.

At temperatures above 135 °C the Nordhorn heavy oil has a higher density than steam condensate. As shown by numerical simulation, the hot heavy oil is thus continuously moved at high speed through the frac to the producer.

During the frac heating phase nearly all of the produced oil comes from the reservoir section overlying the frac.

An essential point of this feasibility study was the examination of the size of the pattern or the bore hole spacing respectively.

Project recovery in absolute terms is as follows:

- Original tar-oil-in-place within a 5-acre area (layer thickness 39 meters) at the pilot location (well NOH 1002) is 146 440 m^3.
- With steam injection (15 tons/hour cold water equivalent) into bed-parallel fractures induced in the Middle Bentheim Sandstone at the injector of a 5-acre (20 230 m^2) 5-spot pattern about 34 185 m^3 of tar-oil can be produced during a period of about 2 years (Fig.2).
- The tar-oil volume of 34 185 m^3 - related to total thickness (251 - 290 m depth) - represents 23 % of the original tar-oil-in-place and recovery to an assumed economic limit of 31.8 m^3/d. This corresponds to a cumulative steam/oil ratio of 6.7 or a cumulative oil/ratio of 0.15 (Fig.3).
- With the 2 1/2 acre pattern less tar-oil only 12 230 m^3 can be recovered. Related to total thickness (39 m), the recovery is only 16.8 % of the original tar-oil-in-place and the cum. oil/steam-ratio is 0.125 (steam/oil-ratio is 8.0).
- A rapid decline in tar-oil rate was noted when high-rate steam injection was decreased during matrix flow conditions.

Based on insights into the process which have come from the completed feasibility work, a 5-spot pattern size of 5 acres is recommended.

4. Technical Planning

Within the frame of the total technical concept planning work was carried out for the partial aspects of oil processing, steam generation units, storage, and transportation with the assistance of consultant firms by the Technical Planning Department of Wintershall AG, Erdölwerke.

4.1 Steam Generation

Pursuant to the results of the simulation study, modifications for the steam generator are as follows:
steam injection rate abt. 15 t/h
steam injection pressure abt. 65 bar.

4.2 Oil-Dehydration and Water-Treating System

In the preliminary study (4), three alternatives were considered and compared with each other. The maximum production rate in each alternative was assumed to be 360 tons of wet oil per day.

Alternative II, although the most costly initially, is believed to be the most practical and most able to successfully treat the heavy, viscous Nordhorn crude oil and accompanying water production.

In Alternative I the wet oil in the general-production stream would be progressively dehydrated in a free-water knock out, an electrostatic dehydrator, an electrostatic desalter, and a surge tank.

In Alternative II a third electrostatic treating vessel is proposed for the genral-production train. This vessel would be a second desalter. All three electrostatic vessels would be the dual polarity type.

The use of the third electrostatic treater in the general-production stream was based on laboratory tests which indicated that water-content reduction below 2 percent may be difficult, depending on the diluent used.

The third electrostatic treater also facilitates washing of the oil to reduce the organic-ash content.

5. Laboratory Tests

Laboratory Tests required for the simulation study were completed during the first quarter of 1982. The following measurements were carried out (Table 1).
- Rock compressibilities on 6 core samples of the well NOH 1002.
- Thermal conductivities, thermal expansion coefficients and specific heat capacities on 7 to 9 core samples resp. and one oil sample of well NOH 1002. Dependent on the type of rock, these properties were measured in different temperature ranges.
- Determination of residual saturations and end points of the relative permeabilities on two core samples from the well NOH 1002 at temperatures of 172 and 250 °C.

The lab tests comprised displacement tests with hot water, steam and heavy oil.

Ash Removal Tests were performed in the Lingen Refinery:
Nordhorn Heavy Oil contains appr. 1 % of organically bound ash, primarily calcium. As a first approach, processes for the treatment of the heavy oil were developed to meet the following assumed crude oil specifications for the refinery:

a) Water content < 0.5 %
b) Suspended inorganic matter < 0.01 %
c) Organically bound ash-forming
   compentents < 0.1 %

By treating heavy oil with solvents it was possible to reduce the water
content to values below 0.5 % by volume and to remove nearly completely
the suspended inorganic matter.

It is basically possible to remove ash from dead oil down to a
residual ash content of 0.1 % by means of formic acid and acetic acid
as well as by complex forming components. Also strong inorganic acids
might be used for ash removal. Sulphuric acid proved to be suitable.
Diluted heavy oil is brought into intensive contact with sulphuric acid
of the ph-range 2 - 4 through stirring following which calcium reacts
with the acid. A reduction of the ash of the heavy oil below 0.1 % was
obtained in the autoclave with stirrer at 130 °C.

6. Economical Analysis and Calculation of Profitability

The project will be carried out within a period of about three
years
   year 0   Phase of investment, fracs, stimulations
   year 1   Heating period, production
   year 2   Steam injection, production
Proceeds from the heavy oil including the added diesel oil
                    10.2 million DM
Investments and operating costs
                    41.0 million DM
Negative cash flow
                    30.8 million DM.
It can be concluded that the Nordhorn-Pilot-Project is uneconomic to
date. In view of the above the envisaged pilot test will not be implemented.

7. Conclusions

1. With 15 t/hour steam injection into bed-parallel fractures in a
   5-acre (20 230 m²) 5-spot pattern about 34 000 m³ tar-oil can be
   produced. This represents 23 % of the OOIP.
2. Marketing studies for the utilization of the Nordhorn Heavy Oil
   revealed the following:
   - Efforts to sell untreated oil on the market failed
   - Treated oil could as well not be placed on the market.
   - A sufficient risk-free utilization in the refinery sector
     could not be found.
   - A market price for this heavy oil could not be determined.
3. Marketable products can be obtained from the Nordhorn Heavy Oil
   by means of conversion processes. LC-Fining and Delayed Coking
   were taken into account.
4. The Nordhorn-Pilot-Project is uneconomic to date and therefore
   the envisaged pilot test will not be implemented.

REFERENCES

1. HOFMANN, H.J.: Development of Methods for Production of Tar Oil in
    the FRG.
    EC Symposium on New Technologies for Exploration and Exploitation
    of Oil and Gas Resources, Luxembourg, April 18-20, 1979 (Publ.
    Graham & Trotman, London 1979).
2. DE GRISOGONO, I.: Evaluation of the Steam Injection Test in the
    Nordhorn Oil Sand Reservoir.
3. TODD, DIETRICH and CHASE Inc.: Feasibility of Recovering Nordhorn
    Tar-Oil using Steam Injection through Hydraulically Induced Frac-
    tures.
    Prepared for Nordhorn Consortium, July 1983.
4. CREST ENGINEERING Inc.: Study of Oil-Dehydration and Water-
    Treating Systems for Nordhorn Pilot Steam-Injection Project.
    Prepared for Nordhorn Consortium, March 1983.

TABLE 1
RESERVOIR PROPERTIES
Well NOH-1002

		Porosity, % Bulk Volume	
Structural Dip	17 degrees		
Depth Top Bentheimer	250 meters	Upper Bentheim	0.205 to 0.227
Depth Base Bentheimer	289 meters	Middle Bentheim	0.205 to 0.278
Initial Temperature	75 degrees F	Lower Bentheim	0.186 to 0.227
Initial Pressure	267 to 356 psi		

Absolute Liquid Permeability, Millidarcies

	Horizontal	Vertical
Upper Bentheim	915 - 1840	368 - 610
Middle Bentheim	915 - 9690	610 - 6460
Lower Bentheim	485 - 1840	97 - 368

Temperature-Dependent Endpoint Saturations, % Pore Volume

	70 °F	200 °F	482 °F
Connate Water Saturation	5.3	11	23
Residual Oil Saturation	65	42	27

Temperature-Dependent Rock Thermal Conductivity, BTU/day/ft/deg F

75 °F	100 °F	140 °F	200 °F	300 °F	400 °F	500 °F
67.6	65.5	58.3	53.5	47.5	43.2	38.8

Heat Capacity, BTU/FT^3/deg F
Reservoir Rock   33.5
Cap and Baserock   31.7

Pore Volume Compressibility, psi^{-1} 75×10^{-6}

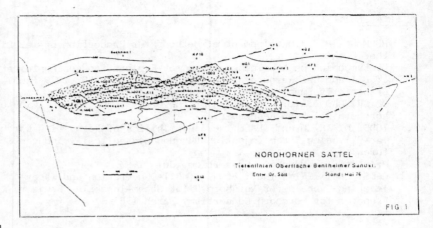

FIGURE 1

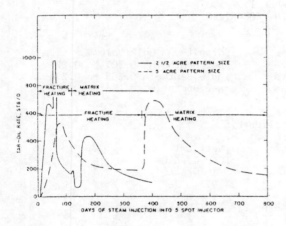

FIGURE 2 - TAR-OIL PRODUC-
TION VS. 5-SPOT PATTERN
SIZE (2000 B/D STEAM
INJECTION RATE)

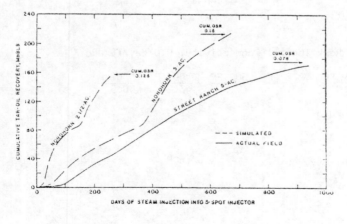

FIGURE 3 - TAR-OIL
RECOVERY FROM 5 ACRES
OF DEVELOPED ACREAGE
5-SPOT PATTERN

(05.14/79)

PILOT PLANT FOR ENHANCED OIL RECOVERY IN
A HEAVY OIL — PONTE DIRILLO FIELD — ITALY

A. CASTAGNONE, G. FIAMMENGO, E. PARTELI, M. PEREGO, A. VITALIANI

Summary

Since 1959 Ponte Dirillo field, Italy, has produced a heavy and viscous oil from a reservoir located at an average depth of 3,000 m. The undersaturated oil is within a fractured triassic dolomite reservoir affected by a fault system and featuring heterogeneous permeability. Production mechanism is by water drive. Production is at such a rate as to produce oil without water.

The purpose of this enhanced recovery pilot plant is the small scale investigation of the effects on the oil recovery of high pressure CO_2-rich gas injection for later application to the nearby larger Gela field having similar oil and reservoir rock characteristics.

A preliminary feasibility study of high-pressure gas injection in the Gela field (partially financed by the EEC — contract 16/75) had been made in 1979 (1) when, because of possible problems due to the large size of the reservoir, direct field application was not undertaken.

Laboratories tests and simulations by a three-dimensional black oil numerical model were carried out for evaluating any possible oil recovery increase by various gas injection methods.

1. INTRODUCTION

Ponte Dirillo field, close to the Gela field in south-western Sicily, was discovered in 1959; up to now it has produced about 1.32×10^6 m^3 ST of a very heavy and viscous oil from a variously fractured dolomitic formation at a depth of 3,000 m.

This small-size reservoir has been chosen as a pilot plant of an EOR process by high-pressure CO_2-rich gas injection.

The purpose is to study the results of gas injection, into a heavy-oil bearing fractured reservoir, for possible later application to the larger nearby Gela field which belongs to the same sedimentary basin and has similar reservoir-rock and fluid characteristics.

Both EOR processes have been partially financed by the EEC with contract 16/75 for the Gela field feasibility studies and with contract 05.14/79 for the Ponte Dirillo pilot plant.

2. OIL FIELD DESCRIPTION

2.1 Reservoir configuration

The Ponte Dirillo structure is an anticline trending NNW-SSE, delimited by faults on its NE and NW flanks (Fig. 1). All of these faults are normal ones, peculiar to the regional tectonic style. Two heavy-oil bearing reservoirs are present:

- The main reservoir (Taormina Fm.) consists of a fractured and vuggy dolomitic complex of triassic age. Beginning at a depth of 2,900 m, the reservoir was drilled for 200 m without reaching the bottom.
- The secondary reservoir (Streppenosa Fm.) overlies the previous one and it consists of six layers of limestone and dolomitic limestone with black shale interbeddings for a total thickness of 250 m. From top to bottom, Streppenosa Fm. shows increasing thickness of the carbonate layers together with increased dolomitization. Therefore the best oil bearing layers are the lower ones. The sixth level is vertically interconnected with the underlaying Taormina Fm.

2.2 Reservoir rock characteristics

Taking into account only the dolomitic reservoir, that is the one to be gas injected, a subdivision of the dolomites into layers by reprocessing the traditional electric logs with up-to-date calculations was made.

The results yield porosity values ranging between 6.4 and 5 % with an average water saturation value between 30 - 40 %. The available cores make up 1.6 % of the entire dolomitic section drilled and for this reason have not been taken into consideration. To define the vertical distribution of the water saturation, capillary pressure curves were used as obtained from the nearby Gela field. From all data available, the original oil-water contact has been located at 3,046 m s.s.l..

2.3 Reservoir fluid characteristics

At reservoir conditions the oil is undersaturated with the consequent absence of free gas within the reservoir. The thermodynamic characteristics of the oil indicate severe gravity segregation with °API values ranging from 15.7 to 10.4 and CO_2 content into the solution gas ranging from 8 to 61 %.

The average reservoir oil viscosity is 25 mPa.s reaching 5,000 mPa.s at surface conditions (20°C).

The reservoir water salinity has a value of 100,000 ppm for the Taormina Fm. and a value of 20,000 ppm for the Streppenosa Fm.

3. PRODUCTION HISTORY

After the initial development stage of the reservoir and construction of the "Oil Center" (1959-1967) oil production reached 100,000 m^3/yr from five producing wells and has kept at this rate for 7 years until 1974 (Fig. 2). Later on production declined to 45,000 m^3/yr in 1978, in order to avoid water cut, as it happened in the nearby Gela field where too high oil production caused an irregular intrusion of the water in some areas of the reservoir.

From 1978 to date, production has maintained an average constant rate of 45,000 m^3/yr. Initial reservoir pressure slightly declined from 30.71 MPa in 1960 to 30.14 MPa in 1980. Recently (December 1983) it has increased to 30.40 MPa (Fig. 2).

This pressure behaviour is the result of a wise production rate for the reservoir and of the subsequent active bottom water drive.

The cumulative oil production up to December 1983 was 1.32×10^6 m^3 ST, 96 % of which is from Taormina Fm.

Present oil recovery is 14.7 % of the original oil in place.

4. THE ENHANCED OIL RECOVERY PROJECT

4.1 Laboratory tests

The physico—chemical characteristics of the reservoir oil have been determined by saturation at different pressures with injected gas (75 % methane) and with the same gas enriched with CO_2 (10 - 15 %) (Fig. 3). The results show the following effects due to gas solution into the oil:
- reduction of the oil viscosity
- reduction of the oil specific gravity
- increase in the oil reservoir volume factor.

The analytical results have been utilized for designing the enhanced recovery pilot plant by high—pressure gas injection.

4.2 Numerical model

A three—phase, three—dimensional numerical model has been set up for reproducing past behaviour and for reservoir production forecast, both for natural depletion and gas injection. Its main features are:
- use of a rectangular coordinate grid system
- areal discretization of the reservoir into 14 x 13 cells and vertical discretization into 5 layers, one for the 6th layer of Streppenosa Fm. and the remaining 4 for Taormina Fm.. The resulting number of active cells is 685
- gravity segregation within each cell.

4.2.1 History match

A very good match of past history was obtained for pressure vs. time trend and for water cut evolution in Ponte Dirillo 9 well. This was possible on the basis of the following assumptions:
- layers below 3,000 m are contributing very little to oil production
- water influx inside each cell occurs only at the time when the critical value of water saturation gets to 75 %.

4.2.2 Production forecast

Future possible behaviour, up to December 1999, has been investigated on the basis of the following assumptions:

Case 1: "Natural depletion" by natural water drive (Q_o= 120 m^3/day)

Case 2: Peripheral gas injection beginning December 1983 in Ponte Dirillo 9 well at a rate of 50,000 Nm^3/day.

Case 3: Crestal gas injection at Ponte Dirillo 5 well at a rate of 50,000 Nm^3/day.

The most important results are shown in Fig. 4 and briefly presented as:

- peripheral gas injection promises increased production in the long term only, whereas, for the short term, recovery increase will only be due to lack of water production
- crestal gas injection seems less promising only due to the fact that the well, to be converted into injecting well, is actually producing 35 % of the total daily oil production.

5. PILOT PLANT

Following the laboratory tests, numerical model simulations and preceding studies concerning Gela field, a pilot plant has been set up for enhanced oil recovery by high-pressure gas injection (Fig. 5).

All the 5 wells of the project were recompleted and all pipelines connecting the well head to the Oil Center have been replaced. Ponte Dirillo wells No. 5 and No. 9 have been set up for crestal and peripheral gas injection respectively.

Gas from Gagliano field, 80 km north of Gela, is mixed at the Gela Oil Center with carbon dioxide produced at the same Center and is sent to Ponte Dirillo Oil Center. After a double compression stage it is injected at a bottom well injection pressure higher than the actual reservoir static pressure (30.40 MPa).

Gas injection began in December 1983 and, after a 3 months period of plant test, it is now definitely operating.

By June 1984 a total of 6×10^6 Nm^3 of gas has been injected. At the same time the reservoir oil production has slowly increased, beginning March 1984, from the initial 120 m^3/day of dry oil to 180 m^3/day by the end of June 1984.

During the gas injection and oil production phase, production tests have been performed on the producing wells together with lab tests on the injected gas, dynamic temperature and pressure profiles, GOR measurements at each well separator, etc.

The technical results relative to the high-pressure gas injection operations have been successful, while further advantages on production enhancement, toward improved final recovery, are to be considered valid only because of lack of water production, during pilot testing at least.

Beginning on July 1984 testing on a huff-and-puff natural gas process will be undertaken at Ponte Dirillo 5 well.

This is done to check any improvement of well productivity mainly because of viscosity reduction due to the gas dissolution into the oil.

6. CONCLUSIONS

The pilot plant for high-pressure gas injection at Ponte Dirillo reservoir is now operating. During the next years it will be possible to conclude whether the process being investigated is actually effective in improving field productivity, as indicated by the studies which have been presented throughout these notes.

REFERENCES

1. Chierici G.L., Dalla Casa G. and Terzi L.,: "Enhanced Oil Recovery by Gas Injection in a Heavy-Oil, Fractured Reservoir - Gela Field, Italy", EEC Symposium (Luxembourg, April 18-20, 1979) Vol. 1, 501-518.

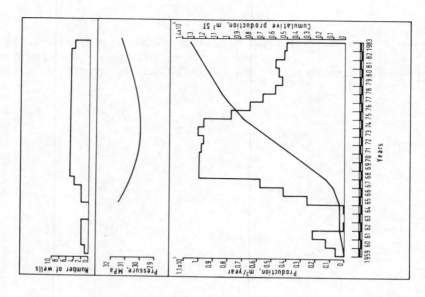

Fig. 2 – Production hystory

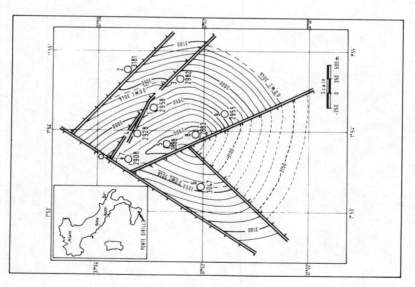

Fig. 1 – Top of "TAORMINA" Fm.

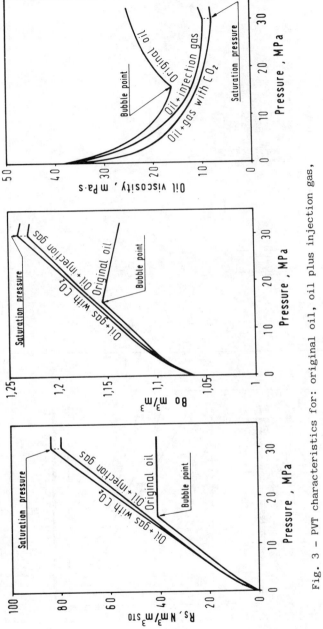

Fig. 3 – PVT characteristics for: original oil, oil plus injection gas, oil plus rich injection gas with CO_2 (10 – 15%).

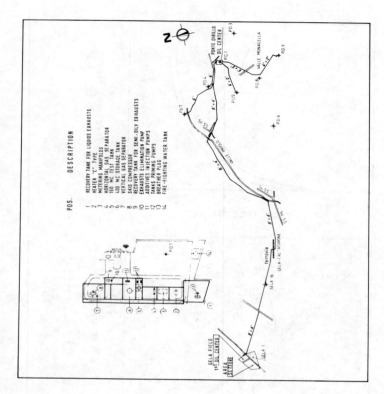

Fig. 5 - Field pilot plant

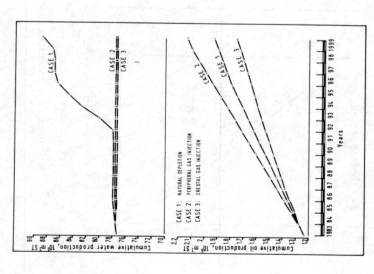

Fig. 4 - Production forecast

(05.22/81)

UPGRADING OF HEAVY OIL AND BITUMEN

R. Holighaus, K. Niemann, K. Kretschmar
VEBA OEL Entwicklungs-Gesellschaft mbH
Postfach 45, D-4650 Gelsenkirchen 2

Summary

VEBA's technology for the upgrading of heavy oils, bitumen and
refinery residues - namely the VEBA-Combi-Cracking (VCC) - and
the VEBA-LQ-Cracking (VLC) process - is derived from the old
german high pressure hydrogenation technology for the liquefaction
of coal. After World War II more than 1.3 Mio. t/years of top and
vacuum residues were converted by these processes to light, marketable products. Since 1978 more than 20 different feedstocks were
tested in bench scale units and an overall reaction model could
be developed on a statistical basis. Correlations between temperature, specific throughput and conversion level were calculated.
The prediction of product slates and qualities is possible by this
model for different feedstocks. The bench scale experiments show
that conversion levels up to 98 % can be achieved for all types
of feedstocks at temperatures between 440 and 480 °C and pressures
between 150 and 300 bar. The required pressure level varies with
the feedstock quality. All the results gathered in bench scale
equipments are checked now in a larger scale. For this reason VEBA
OEL together with LURGI GmbH, Frankfurt has built a 200 bbl/d VCC
demonstration plant at VEBA's Scholven refinery. Since start up in
May 1983 the plant has been on stream for more than 5000 h, the
longest test run lasting 1500 h. The average conversion level
during this long term test was above 80 % with a maximum conversion
rate > 95 %. Feedstocks processed up to now are top and vacuum residues from heavy venezuelan crudes (e. g. Bachaquero 17) and arab
heavy crude. The test runs demonstrated that residue conversion up
to 95 % can be obtained in a once through operation mode under
excellent stable condition in a cascade of liquid phase hydrogenation
reactors. The specific throughputs are comparable with VEBA's bench
scale experience. Product yields and properties fit the expected
values. Of special importance is the fact that pressure reduction
from 280 to 180 bar could be achieved thus indicating the large
development potential of the VCC process. Future test runs will
explore further pressure reduction, throughput increase and the
enlargement of the feedstock base to support the basic design of
a commercial VCC plant.

1. BACKGROUND

VEBA's technology for the upgrading of heavy oils, bitumen and
refinery residues - namely the VEBA-Combi-Cracking (VCC) and the VEBA-
LQ-Cracking (VLC) process - is derived from the high pressure hydrogenation technology for the liquefaction of coal (1, 2, 3, 4). In 1972

the first commercial plant for the liquid phase hydrogenation of lignite
went on stream in Leuna with a production capacity of 100.000 t/year of
gasoline. In 1944 the fuel production of 12 liquefaction plants for
lignite and hard coal reached 4 million t/years. In 1952 VEBA OEL AG
reactivated the high-pressure plants at their refineries at Scholven and
Horst and put them on stream again with petroleum-derived top- and vacuum
residues for the commercial conversion of more than 1 million t/year of
refinery residues. The concept of direct combination of the Liquid-Phase-
Hydrogenation (LPH) step with the Gas-Phase-Hydrogenation (GPH) step was
realized in a technical scale in 1954 in the Scholven refinery. The total
amount of residues converted by VEBA OEL in multi train plants exceeded
1.3 Million t/year until in 1964 production was shut down for economic
reasons. A barrel of crude oil at that time cost less than $ 2.00.
    A period of sharp increase in crude oil prices and a predicted
shortage of conventional crude oils for the next century motivated VEBA
OEL to reactivate the proven technology. Since 1978 more than 20 different
feedstocks were tested in bench scale units with capacities of 1 - 3
bbl/d. Several hundred test runs on VCC and VLQ operation mode with a
comprehensive process parameter variation led to the development of an
overall reaction model on a statistical basis (4, 5).
    The next step in VCC process development was the decision to build
a 200 bbl/d demonstration plant at VEBA's Scholven refinery. Design,
engineering and construction was commonly performed by LURGI GmbH,
Frankfurt and VEBA OEL AG during 1981 to 1983. Total erection costs
amounted to 45 million DM. Erection and plant operation are financially
supported by the Federal Republic of Germany (Ministery of Research and
Technology) and the European Community. The VCC demonstration plant went
on stream in May 1983.

2. DESCRIPTION OF THE 200 BBL/D VCC DEMONSTRATION PLANT

    A simplified flow scheme of the 200 bbl/d VCC plant (exclusive tank
farm) is given in Figure 1. The heavy oil coming from the feed storage
tank is optionally mixed with a small quantity of powdered additive in
the slurrying unit. After pressurizing the slurry with the high pressure
feed pump preheated recycle gas is added and the mixtures is heated in
the electrically powered preheating train to temperatures between 420
and 450 °C. The mixture than enters the LPH reactor system, which consists
of three tubular reactors without any internals. There the residue is
converted at temperatures between 440 °C and 480 °C and pressures between
150 and 300 bar. The reactors are not externally heated. Temperature control of the highly exothermic conversion reactions is done by cold gas
quench, as indicated in the flow scheme. The basic lay out data for this
part of the plant are given in Table 1.
    The effluent of the third LPH reactor enters a hot separator system,
which is operated under the same pressure and with a temperature somewhat
below the LPH temperature. The separation between converted and uncon-
verted material is almost completed in the first hot separator. The
second hot separator serves mainly as a gard vessel for the gas phase
catalyst in case of plant off sets.

The hot separator bottom product is depressurized into a low pressure flash drum. The flash distillates are condensed and send to the tank farm. The flash residue becomes vacuum flashed for the recovery of the heavy VGO fraction and afterwards pumped to the residue storage tank. In case of high conversion rates with additive addition this recovery is limited by the viscosity and the solid content of the flash residue.

The hot separator top product is send to the catalytic fixed-bed reactors without depressurization or intercooling in between. Like the liquid phase reactors the gas phase reactors are also well insulated and temperature control is done by cold gas quench. At temperatures between 350 and 430 °C the product is deeply raffinated and cracked to a certain extend. The effluent from the GPH is condensed, cooled, separated from the recycle gas in the cold separator and pumped to the syncrude storage tank. Because of the relatively small size of the plant no heat recovery system is installed (5). An overview picture of the plant is given in Figure 2.

3. RESULTS

Basis for the lay out of the 200 bbl/d VCC demonstration unit and the primary test parameter sets have been data which were evaluated in small pilot plants. During the last years, more than 20 different heavy oils and residues were converted to light products in the bench scale units of VEBA OEL. Fig. 3 gives a summarization of these feedstocks with some characteristic features. Top- and vacuum residues with residue contents from 50 to 100 % (560 °C$^+$) were processed. Asphaltenes from 5 to 40 %-wt are also typical as metal contents up to 1700 ppm, sulfur from 2 to 6 %, and nitrogen up to 1 %. At densities mainly higher than 1,0 ton/m³ (< 10 °API) the molar H/C ratio is around 1,4. The tests in small pilot plants showed that the VEBA Cracking Processes are suitable for all these different feedstock and that no restrictions exsist concerning high sulphur, nitrogen, asphaltene or metal contents.

Of most importance is the fact, that for all heavy oils (except those thermally pretreated materials like tar, catcracker or visbreaker residues) the conversion rates are very simular at a given throughput and temperature. This is demonstrated in Fig. 4, which shows the correlation of the conversion rate with the specific throughput at standard reaction conditions. This correlation indicates, that there are only small differences in cracking reactivity for residues from very different heavy oils. Of importance however, is a hydrogen partial pressure high enough to avoid coking reactions in the liquid phase reactors. This minimum pressure level strongly depends upon the crude oil. To process an Arabian residue for example needs a higher pressure than Athabasca or Cold Lake bitumen.

At a given conversion rate, the product yields too are similar for very different heavy oils. In Figure 5 results of the small pilot plant tests are summarized. Hydrocarbon gas yields, the naphtha and the middle distillate yields are plotted versus conversion. All tests have been performed at the same standard reaction conditions.

The full lines in Figure 5 indicate the range of VEBA's correlation system which is derived from more than 300 test runs in small pilot plants with various feedstocks. This correlation system is capable to predict conversion rates, product yields and qualities for LQ- or Combi Cracking of virgin residues. As input data boiling curve and ultimate analysis are needed. The computer model calculates closed mass and elemental balances for the desired conversion level together with product properties and calculated boiling curves. The effectivness of this tool was demonstrated elsewhere (6).

The main goals for the test series in the 200 bbl/d demonstration unit are now the verification of the experience gained in small pilot plants, the determination of scale up factors and the study of the behaviour of a not externally heated cascade of liquid phase hydrogenation reactors in a one through operation mode.

The demonstration unit went on stream in May 1983. During the first year the plant was under operation for more than 5000 h thus giving an availability of 61 %. The longest test run lasted 1500 h until the plant had a regular shut down for some security checks.

Top- and vacuum residues from Venezuelan Bachaquero 17 crude and a vacuum bottom from Arab heavy crude served as feedstocks up to now. For these feed materials a secure and stable process performance could be established achieving residue conversion levels of more than 95 % for longer periods. These high conversion levels were verified at specific throughputs similar to the VCC Simulation Model. The LPH reactor temperatures which had to be applied are very close to the expected values. It could be demonstrated that the cold gas quench system of the plant is flexible enough to gain a nearly isothermal temperature distribution over the reactors thus improving the efficiency of the LPH system.

Due to the fact that the investment costs for high pressure hydrogenation technology are higher than those for comparable low pressure processes which produce less valuable products, however, it is one important aim of VEBA's process development to reduce the pressure of the VCC process under maintenance of the very good product properties. The test series in the 200 bbl/d VCC demonstration plant could prove that a pressure reduction from 280 bar (~ 400 psig) down to 175 bar (~ 2500 psig) is possible for both tested feed oils. This pressure reduction does not effect the process performance. Conversion levels of 95 % are obtainable under these conditions, too. A further reduction of the pressure - and all tests look promising for this approach - seems to be of a more academic interest as this reduction is not accompanied by a further cost decrease.

It is of further importance that for all test runs the VLC/VCC product yield distributions were as expected with VCC naphthas having reformer feed specification and VCC middle distillate qualities better than the required standards for Diesel fuel or extra light heating oil. The properties of a typical VCC syncrude in total are given in Table 2. Depending on the GPH conditions and for 90 % residue conversion in the LPH naphtha yields between 15 and 30 %, middle distillate yields between 30 and 40 % and vacuum gas oil yields between 15 and 35 % were obtained. The gas phase reactors which serve for the hydrotreating of the primary LPH syncrude have been loaded with a commercial catalyst containing Nickel/Molybenum on a neutral carrier. This hydrotreating catalyst is able to shift the product yield distribution for the minimization of heavy gas oils if desired. In this case higher temperatures have to be applied. This "mild hydrocracking" is possible as the hydrogen partial pressure in a Combi Cracker is a considerably higher than in normal hydrotreaters.

4. CONCLUDING REMARKS

Up to now the 200 bbl/d VCC demonstration unit has demonstrated that the VLC/VCC processes are most effective tools for the high conversion of heavy residues. During the first year of operation enough data were collected to provide broad information for the basic engineering of a commercial VCC plant with a design capacity of 1,5 Mio. t/year which is in progress now.

Furthermore, the test runs indicate a large development potential of the processes. The important pressure reduction could be proven within the first year, the increase of LPH reactor efficiency is the next goal together with an expansion fo the feedstock basis especially to residues of other upgrading processes (e. g. visbreaker or FCC residues). Of further interest will be a screening of GPH catalysts and conditions and the study of long term effects.

REFERENCES

1. URBAN W., Erdöl and Kohle 8, 780 (1955)
2. GRAESER U., NIEMANN K., OiT Gas Journal 80 Nr. 12, 121 (1982)
3. GRAESER U., KRETSCHMAR K., NIEMANN K., Erdöl und Kohle 36 Nr. 8, 362 (1983)
4. GRAESER U., NIEMANN K., Preprints, Div. Petro. Chem., Am. Chem. Soc. 28 Nr. 3, 675 (1983)
5. NIEMANN K., KRETSCHMAR K., AOSTRA Symposium, Advances in Petroleum Recovery and Upgrading Technology, June 14 - 15, 1984, Calgary, Alberta
6. NIEMANN K., GRAESER U., LISCHER H., KRETSCHMAR K., Symposium on Resid Upgrading, AIChE Spring National Meeting, May 20 - 23, 1984, Anaheim, California

TABLE 1

Capacity :   0.3 – 1.5 t/h   (45 – 225 bbl/d)
               Top – and Vacuum Residues

               $\leq$ 1000 Nm³/h   (35000 scf/h)
               Fresh Hydrogen

               $\leq$ 5000 Nm³/h   (176000 scf/h)
               Recycle Gas

Pressure :   100 – 300 bar   (1450 – 4350 psig)

Temperature :   < 500 °C   (< 932 °F)

Design Data

TABLE 2

VCC Syncrude Qualities
(Feedstock : Bachaquero 17 Vacuum Residue)

Density	15 °C	g/cm³	0,83 – 0,81
Carbon		wt.–%	86,4 – 85,7
Hydrogen		wt.–%	13,5 – 14,2
Sulfur		ppm	30 – 4
Nitrogen		ppm	20 – 1
Oxygen		wt.–%	0,1
Metals		ppm	1
Asphaltenes		wt.–%	0,05
Boiling Range			C_5 – 485 °C

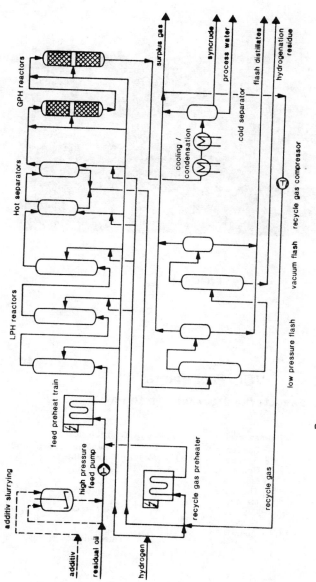

Process scheme of VCC 200 bbl/d demonstration-plant

FIGURE 1

FIGURE 2

VCC Pilot Plants

Feedstocks processed up to now

Canada
- Lloydminster
- Cold Lake
- Athabasca

Middle East
- Arabian Light
- Arabian Heavy

Others
- Tar from Lignite and Hardcoal
- Catcracker Residue
- Visbreaker Residue

Venezuela
- Boscan
- Bachaquero 13/17
- Ceuta
- Morichal
- Tia Juana
- Merey
- Orinoco
- Guahibo Lache
- Zuata

Characterization
Residue >560°C : 50-100 wt.-%
Asphaltenes :   5- 40 wt.-%
Metals :   120-1700 ppm

Total Number of Runs : 300

FIGURE 3

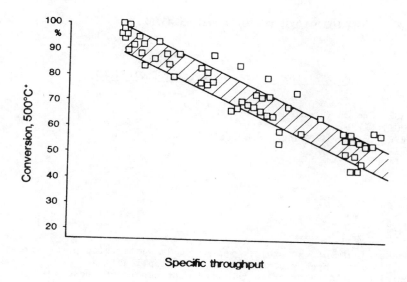

FIGURE 4

Small Pilot Plants : Summary of results

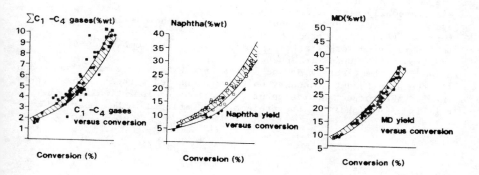

FIGURE 5

(03.94/80, 05.30/81 and 05.43/82)

PRETREATMENT OF HEAVY OILS AT THE OILFIELD

J.F. LE PAGE, French Petroleum Institute
J.C. CHARLOT, ELF France
J.L. COUDERC, Compagnie Française des Pétroles

Summary

The general aim of the project was to study and develop pretreatment
processes for non-conventional heavy oils applicable at the
production field in order to facilitate transport of these oils by
pipeline by lowering their viscosity.
Thanks to an initial feasibility phase, an experimental platform was
defined capable of pretreating many types of heavy oil and in
particular the following: BOSCAN, LAGUNA ONCE, TIA JUANA HEAVY,
ATHABASCA, COLD LAKE and NORTH BATTLEFORD.
It was decided to set up this platform at the SOLAIZE site, near
Lyons, and the work carried out with support from the Community
covered the construction engineering by Société TECHNIP and the
acceptance-testing of the installations by the operator: ASVAHL
(Association pour la valorisation des huiles lourdes - Association
for the valorization of heavy oils).
The project concerned only the thermophysical process involving the
following units: desalination, atmospheric distillation, vacuum
distillation, de-asphalting, visbreaking and hydrovisbreaking.
The work concerning catalytic hydroprocessing was dealt with
simultaneously without EEC support, since this relatively expensive
hydroprocessing operation would more probably be carried out in the
refinery than at the production field.
The engineering work started in March 1981, with completion of the
erection of the units in June 1983.
Aside from the visbreaking and hydrovisbreaking, the various units and
all the appurtenant secondary units were charged with oil. These tests
enabled the correct operation of the various equipment units to be
verified. This phase of the project was completed in November 1984.

1 - INTRODUCTION

The aim of the "PRETREATMENT OF HEAVY OILS AT THE OILFIELD" project
was to design, build and operate an experimental centre in order to
finalize, develop and test the technologies and processes for pretreatment
of heavy oils on the production fields.

This major project forms part of a policy common to the three partners – CFR, ELF France and IFP – grouped within the ASVAHL (Association pour la valorisation des huiles lourdes – Association for valorization of heavy oils) Economic Interest Group.

The overall project in fact consists of several parts:
- the first covers a research and development programme on heavy oil treatment methods capable of providing a basis for designing the experimental platform,
- the second essentially concerns the construction of this platform,
- the third concerns its startup,
- the last covers the experimentation proper.

In order to tackle the various stages of the project, we considered it advisable first to define the heavy oils concerned, and the problems they set.

Figure 1 shows their situation with relation to conventional crudes on a curve indicating the viscosity at 20°C in terms of the fraction of product that distils below 350°C ; in this classification, the term "heavy oil" designates oils with a viscosity at 20°C of equal to or greater than 1000 cSt, the fraction distilled at 350°C of which is less than 0.25 ; as a corollary, these heavy oils generally have a specific weight of over 950 kg/m³ and can not be conveyed by pipeline on the basis of the specifications applicable today in Europe.

These heavy oils hence set production problems owing to their viscosity, and also problems of transport and valorization, all of which must be solved. For instance, in the case of Venezuela, several solutions, illustrated in figure 2, have been envisaged for conveying the heavy oil of "Faja Petrolifera" as far as the coast: dilution by a low viscosity product, formation of an emulsion of oil in water, heating the pipe, succinct refining to lower the viscosity, simply rendering the oil transportable, or further refining intended to produce a very high quality synthetic crude ; in the latter case, one variant consists in implementing the first part of the operation at the production field in order to render the oil transportable and to end the operation near the loading port or point of use.

2 - FEASIBILITY AND BASIC ENGINEERING

2.1 - General studies

The following is a list of the heavy oils generally studied: BOSCAN, LAGUNA ONCE, TIA JUNA HEAVY, ATHABASCA, COLD LAKE, NORTH BATTLEFORD. The objectives were:
- to provide the necessary data to complete the specifications of the platform units as comprehensively as possible,
- to determine the conditions for making best use of the various units of the platform,
- to study various process flowcharts, on experimental bases,
- to verify the soundness of certain innovations and technological improvements.

In view of the multiplicity of the charges treated, and the number and diversity of the tests performed, not all the results will be described in detail, but only the most typical examples.

Distillation - Desalination - Dehydration

In distillation, particular mention must be made of the various distillability studies for determining the maximum distillation temperature that can be considered, allowing for the following requirements:

- the viscosity of the residue in a vacuum must not exceed 10^6 centistokes at 100°C,
- the metals content (Ni + V) in the distillate must be less than 1 ppm (parts per million),
- the thermal cracking must remain negligible.

The results of these tests have given the distillate yields that can be expected within these distillability limits.

In dehydration-desalination, the pilot tests performed either statically or dynamically have established the influence of the fluxing rate for each heavy oil within the limit of precipitation of the asphaltenes, the influence of the process transit time and of the temperature, together with the effect of various types of additives.

Parallel to these specific studies, one must also mention the research campaigns concerning desalination and in particular distillation in order to prepare charges of vacuum residues, atmospheric residues or head-cut oils, for subsequent studies of processes and process flowcharts.

De-asphalting

Numerous test campaigns have been formed on the various heavy oils mentioned above, with two different objectives in view:
- preparation of charges for studies of processes further downstream (hydrovisbreaking, hydroprocessing),
- study of the influence of the operating conditions, namely: the nature of the solvent, propane, butane, pentane and light gasolene were used. The tests using as solvent the light gasolene extracted from distillation of a heavy oil for de-asphalting the same head-cut heavy oil deserve particular mention,
- the operating parameters for a given solvent: solvent ratio, pressure, process transit time and temperature,
- the geometry of the unit (rate of rise and number of stages).

Thanks to these tests, not only where the oil and asphalt efficiencies and the quality of the products define, but also all the minimum conditions enabling a de-asphalted oil to be obtained capable of being processed without problems during the subsequent catalytic processing stages (asphaltenes, metals and Conradson carbon contents).

Visbreaking – Hydrovisbreaking

The object was to obtain a transportable oil, i.e. one meeting the following kinematic viscosity standards: Venezuela, 400 cSt at 37.8° ; Europe, 200 cSt at 10°C and Canada, 120 cSt at 20°C.

The studies carried out on the above-mentioned heavy oils have shown that mere visbreaking of the desalinated and head-cut heavy oils was insufficient to attain the objective sought. For all these heavy oils, visbreaking will have to be combined with distillation, as is illustrated diagrammatically in figure 3 for the BOSCAN and ATHABASCA crudes.

On the other hand, under optimized conditions, hydrovisbreaking enables interesting performances to be obtained, in other words, the hydrovisbroken and stabilized heavy oils (C5$^+$) all have a viscosity below 400 cSt at 37.8°C. Accordingly, they could be transported on the basis of Venezuelian standards and even for some of them (Lyodminster) on the basis of Canadian standards. For the other oils, one has to conceive a process identical to that shown in figure 3, but with an efficiency that is 10 to 15 points better compared to the process involving visbreaking.

When the charges treated are de-asphalted oils, the products obtained can readily be transported, even in the Canadian regions.

Catalytic hydroprocessing

All the hydroprocessing tests were conducted under fixed bed conditions, i.e. in accordance with a known technique and on catalysts developed outside the scope of the present contract. In actual fact, these tests did not aim so much as to study hydroprocessing proper, but rather to determine the interest of flow processes involving the hydroprocessing either of atmospheric or vacuum residues, or of de-asphalted oils.

2.2 - Process sequencing

We propose to confine ourselves here to mentioning the process schemes subjected to experimentation and to state general conclusions concerning the interest of the process scheme considered.

Integration, desalination, de-asphalting and distillation

The desalination solvent would be the same as the de-asphalting solvent, namely the light gasolene present in the heavy oil, recovered following distillation and recycled partly at the desalination unit and partly at the de-asphalting unit.

The results show that such an operation would be difficult to carry out on industrial scale, for the following reasons:
- the risk of precipitation of the asphalts in the desalination unit,
- the variation in the quality of the solvent and hence the variation in the quality of the DAO.

Desalination-visbreaking-distillation sequence

This process sequence applied at the field gives transportable synthetic oil efficiencies of from 70 to 85%, for the various heavy oils studied.

It offers a certain interest, provided that the following two conditions are satisfied:
- the synthetic oil has adequate stability,
- the production of the oil calls for production of steam in harmony with the combustion heat of the residual pitch.

Desalination-hydrovisbreaking (distillation) sequence

The tests showed that this simple sequence was sufficient to make the various heavy oils transportable, at least under Venezuelian conditions. Compared to the previous sequence involving visbreaking, the synthetic crude efficiencies are invariably better and the synthetic crude more stable.

Desalination-distillation-de-asphalting-hydrovisbreaking-hydroprocessing sequence

The results obtained on TIA JUANA HEAVY and BOSCAN are shown in figure 4 where the term H T C groups together the operations of hydrovisbreaking and hydroprocessing. As this figure shows, both with TIA JUANA and BOSCAN crude, the sequence considered enables a desulphurized synthetic crude to be obtained of much better quality than the conventional crudes.

Desalination-distillation-hydrovisbreaking-de-asphalting-hydroprocessing sequence

Comparatively, this sequence gives better synthetic crude efficiencies that the preceding one. On the other hand, certain difficulties are encountered in discharging asphalts with a very high softening point to the de-asphalting stage.

Desalination-direct hydroprocessing sequence
This sequence enables a transportable crude to be obtained of very high quality, but with the following three disadvantages:
- the cost per ton is very high owing to the considerable consumption of hydrogen,
- the operation takes place at very high pressure,
- the synthetic crude still contains asphalts.

Desalination-distillation-hydoprocessing of the atmospheric residue-de-asphalting sequence and recycling of the hydroprocessing asphalts
This sequence, which is the most expensive per ton of heavy oil processed, is the one leading to the best yield figures for very high quality synthetic oils.

Lastly, it should be mentioned that some of these flow processes have been subjected to comparative cost studies, in particular the last four, which were compared amongst one another and against a coking process.

Amont the possible methods of processing these oils, we have eliminated coking, which appeared both less well adapted to this problem and well mastered by the American competition. On the other hand, we have chosen to be able to study and compare on the platform the other direct and indirect methods of valorization of heavy oils. The result is the general diagram shown in figure 5.

On the basis of existing know-how before the beginning of the project, filled out by the results acquired in the preceding experimental studies, seven process files have hence been compiled:
- study bases,
- desalination - atmospheric distillation,
- vacuum distillation,
- de-asphalting,
- hydroprocessing - aminos,
- visbreaking - distillation of the products,
- utilities - storage facilities.

These files have been compiled on a basis of a mean capacity of 20,000 tons per year, whilst ensuring considerable operating flexibility in order:
- to process a wide variety of charges,
- to compare several technical variants for certain processes (desalination, de-asphalting, hydroprocessing),
- to apply a wide range of operating conditions to each process,
- to experiment and compare several process layouts.

3 - ENGINEERING AND CONSTRUCTION

3.1 - Engineering
This part of the project was entrusted to the TECHNIP engineering company, and on the basis of the process files, resulted in drafting of rules and general design and construction specifications, putting out requests for tender for the main equipment and cost estimation of the installation.

The first stage of the project led to experimental determination of its feasibility, the development of certain innovations and technological improvements, the selection of the best-suited methods of treatment and the corresponding processes, the selection of the characteristics of the unit, the compilation of the process files and layout files, the execution of the basic engineering and the estimation of the investment costs. Once these results had been obtained and files compiled, it logically followed that the second stage of the project was confined exclusively to building the heavy oil pretreatment platform in accordance with the general diagram shown in figure 4 and on the basis of a capacity of 20,000 tons per year.

The project did not include the cost of procuring the equipment and the cost of construction.

Furthermore, it concerned only the "thermophysical" method, i.e. involving all the main and appurtenant units (storage facilities, utilities, links with the refinery) shown in figure 5, except for the catalytic hydroprocessing.

Initially, in fact, construction of the platform was planned to take place in two stages.

In the first stage, construction of the thermophysical process units intended for pretreatment at the production field, confining the problem to making the heavy oils concerned capable of being transported.

In the second stage, construction of a hydroprocessing unit in order to study the deep conversion of the heavy oils either at the production field, or at the refinery. This hydroprocessing was designed to operate in accordance with several technologies: fixed bed, counterflow moving bed, boiling bed and "carried" bed with initially soluble catalyst ; it is coupled with an amino wash and distillation of the products.

Consequently, even though the construction of the hydroprocessing unit was not covered by this contract, this unit will not be dissociated in the present exposé from the other units of the platform.

In continuity with the basic engineering carried out under the first contract, the detail engineering was performed by Société TECHNIP within the framework of a contract taking effect from 1st January 1981.

This work took place from March 1981 to August 1982, essentially involving the preparation of the worksite, the design of the units, and the distribution and followup of the orders. From the practical standpoint, the result was the preparation of 34 engineering files, 9 operating manuals, 200 maker's files corresponding to the 200 main items of equipment of the platform, 10 cost and expenditure control reports and the construction of a 1/25th working model for each main unit, used to establish the layout and pipework drawings and for installation of the equipment and pipework on the worksite.

Reality proved that this engineering work was more complex than in the case of an industrial operation, for the following reasons:
- the small size of the installations,
- the flexibility sought, in order to be capable of trying out and comparing a variety of types of equipment and flow processes,
- the need to heat most of the liquid circuits,
- the resolve to use existing European equipment, or even European prototypes,
- the practice of going through a models stage.

To illustrate the work accomplished during this phase of the project, figures 6 and 7 give photographs of the models of the de-asphalting and hydroprocessing units.

3.2 - Construction

This second part of the work of TECHNIP started in October 1981 with the gradual introduction of a worksite supervision team which consisted of up to 25 people, under the responsibility of a site manager. The mandate of this team, under the responsibility of the site manager, was to check the work performed by the different contractors on the worksite, to synchronize this work, to see that the work programme, drawings and construction rules were complied with until the work was mechanically complete.

Figure 8 gives a view of the platform on completion of construction.

Construction terminated at the end of June 1983 with the acceptance-testing of the installations by ASVAHL.

4 - STARTUP OF THE UNITS AND VARIOUS TECHNOLOGICAL TESTS

All the appurtenant units, together with the following main units were started: desalination, atmospheric distillation, de-asphalting, hydroprocessing and distillation of the products. Visbreaking, hydro-visbreaking and vacuum distillation still remain to be started. For each main unit, we shall confine ourselves to giving just some of the typical results obtained.

De-asphalting

For the startup tests, Safanya vacuum residues were used as charge. About 1000 tons of asphalts were produced either in the form of a mix with solvents to make up a heavy fuel, or pure in the form of extruded asphalts. Figure 9 gives a number of typical results obtained concerning the de-asphalted oil efficiency (DAO) and the quality of the asphalt.

The following general conclusions can be drawn from an analysis of the results:
- the operation of the flocculation mix and the decantation stage proved satisfactory,
- the asphalts washing stage at the foot of the decantation unit must be modified to make it more effective and selective,
- the falling film evaporator turned out to be easier to use and capable of high performance: the results obtained are in conformance with what was expected as a result of the feasibility stage.

Desalination and atmospheric distillation

3000 tons of BOSCAN Venezuelan crude and Italian ROSPO MARE crude were processed. During desalination, the influence of the following parameters was studied: fluxing agent type and ratios, washing water ratios, additives nature and ratios, desalination temperature and resistance time, and nature of the mixing system (water-hydrocarbons). Under optimum conditions, the performances hoped for were achieved, namely the following with BOSCAN crude:
- desalination efficiency: 90% by weight, i.e. less than 5 ppm of sodium remain in the residue,
- output water: 0.6% by weight,
- hydrocarbons in water: 160 ppm.

As for atmospheric distillation, this set no major problem during startup. Head-cut crudes were successfully prepared, together with atmospheric residues, whilst the distillate limits of the charges and the operability of the column were studied. The residues obtained will undergo de-asphalting and a variety of conversions during the second 6-month period.

Catalytic hydroprocessing

After allowing for some of the technical problems related to the high pressure used, the small capacity of the installation of the considerable flexibility required in terms of operating conditions, the unit was filled with its oil charge at the end of April. Tests were performed in succession on Safanya de-asphalted oil, Kir-kouk atmospheric residue and since the 20th of May on Safanya vacuum residue. With the catalyst system used, the performances observed are excellent. Following this startup test, the unit will be used for processing head-cut and de-asphalted heavy oils which have already undergone desalination and atmospheric distillation operations.

Adaptation of the hydroprocessing unit to processing of heavy oils with high metal contents

For heavy oils with high nickel and vanadium contents (Ni + V 250ppm), the cycle time of the demetallization catalyst becomes very short ; boiling bed, mobile bed or carried bed processes become advantageous compared to the conventional fixed bed inasmuch as they make it possible to add fresh catalyst and to withdraw the spent, metal-poisoned catalyst. The reactors installed on the platform were initially designed to operate in these different modes, though with the addition of the necessary equipment to enable the catalyst to be added and withdrawn during operation. As regards the method of flow of the fluids, it was decided to operate in an initial period with a counterflow mobile bed in which the catalyst circulates from top to bottom of the reactor, whilst the fluids (charge and hydrogen gas) circulate from bottom to top. The engineering of this hydroprocessing unit is planned to be completed by the end of 1984.

5 - CONCLUSION

After a technical and economic feasibility study concerning the processes and the sequencing of the processes covering the direct and indirect methods of processing heavy oils, the Solaize platform has now been built. It groups together seven main processing units and features a capacity of 20,000 to 30,000 tons per year and considerable operating flexibility enabling several technical variants to be compared for each process and several flow processes to be compared for each type of heavy oil. Most of these units have been successfully started up and the technological tests forming part of the programme carried out. The programme of tests planned to start on 1st January 1985 now has to be carried out.

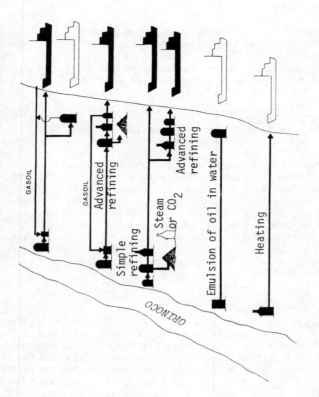

FIGURE 2

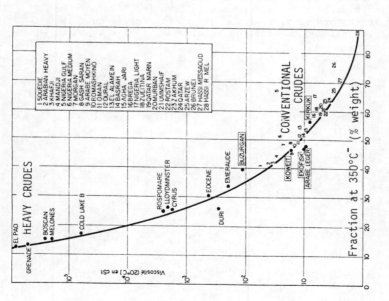

FIGURE 1 - VISCOSITY OF CRUDES AT 20°C IN TERMS
OF FRACTION DISTILLED AT 350°C

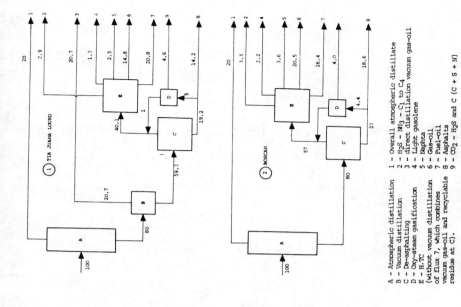

A – Atmospheric distillation
B – Vacuum distillation
C – De-asphalting
D – Oxy-steam gasification
E – H.TC
(without vacuum distillation
of flux 7, which combines
vacuum gas-oil and recyclable
residue at C).

1 – Overall atmospheric distillate
2 – H₂S – NH₃ – C₁ to C₄
3 – direct distillation vacuum gas-oil
4 – Light gasolene
5 – Naphta
6 – Gas-oil
7 – Fuel-oil
8 – Asphalts
9 – CO₂ – H₂S and C (C + S + N)

FIGURE 4 – PROCESSING OF TWO HEAVY OILS

FIGURE 3 – VESBREAKING OF DESALINATED HEAVY OILS

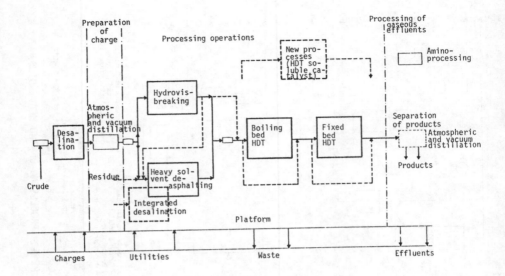

FIGURE 5 - GENERAL FLOW PROCESS DIAGRAM OF HEAVY OIL PROCESSING
EXPERIMENTAL PLATFORM

FIGURE 6 - DE-ASPHALTING UNIT

FIGURE 7 - HYDROVISBREAKING AND HYDROPROCESSING

FIGURE 8 - HEAVY OILS PLATFORM
Beneral view

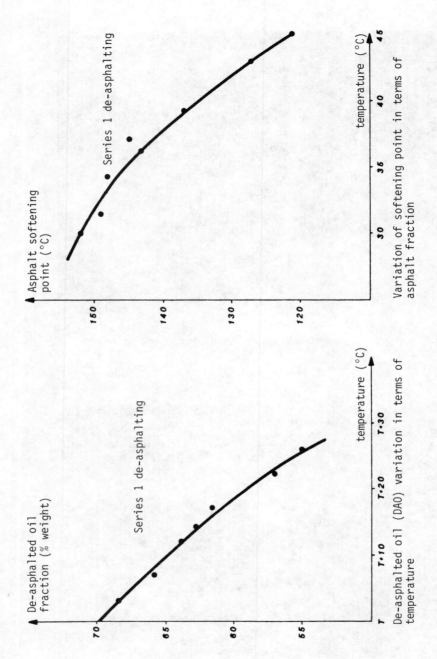

FIGURE 9 – DE-ASPHALTING UNIT

(05.02/76, 05.21/80 and 05.28/81)

METHOD PILOT PROJECT AND INDUSTRIAL
PILOT PROJECT FOR INJECTION OF MICROEMULSION
AND POLYMER INTO CHATEAURENARD FIELD

D. CHAPOTIN, A. PUTZ, Société Nationale Elf Aquitaine (Production)

Summary

The first attempt to achieve tertiary recovery by chemical means in
Europe was made by Elf Aquitaine, with the assistance of the French
Petroleum Institute, on Chateaurenard field, from 1978 to 1980. A
plug of microemulsion displaced by a concentrated solution of polymers
and then by water was injected in the centre of an inverse five-spot
with sides measuring 100 metres.
A very clear response was observable at the production wells 2 months
after starting injection. In all, the cumulative production of oil
amounted to 4520 m³, of which 2550 m³ represent a gain compared to the
primary production: the ultimate recovery level of the part swept is
estimated as 73%, whereas predictions of recovery by natural depletion
were 43%. Study of the efficiency of the microemulsion has shown the
predominant role played by ion exchanges in the oil mobilization
mechanism ; ion exchange is responsible both for the continued
efficiency of the products and the considerable trapping of the
surfactants. In economic terms, the overall result is less favourable,
and the cost of the tertiary oil is 3 to 4 times that of the market.
A second project was started in 1980 with a view to attaining economic
validity of the method ; this operation, involving the same techniques,
was carried out on a larger scale and under strictly industrial
conditions, representing the intermediate stage before going to
generalized application.
The configuration adopted is a pattern of 4 adjacent inverse five
spots ; amongst other things, it enables the area swept between the
4 injection wells to be determined and the overall volume of the
central production well to be accurately specified.
Optimization of the choice of the products has resulted in the
adoption of a microemulsion with a higher water content and a
surfactant plug size over twice that chosen for the first test.
Operation started in June 1983 and so far 29% of the porous volume
has been injected. The increase in the oil production so far accounts
for about 60% of the total production.
The total cumulative production expected (forecast by numerical
simulations) is 47000 tons, of which 37000 tons of tertiary oil.

1 - CHARACTERISTICS OF THE CHATEAURENARD FIELD

This field consists of a combination of 3 structures lying at the same
stratigraphic level (the Neocomian).

The sandy reservoir rests on the base of a limestone plateau, of the topographical form of channel ravines filled by transgression with sand from the Valanginian age.

The sandy body is hence subject to rapid variations in thickness (average thickness 2.5 metres).

The sand are in general clean:
- low clay content (0 to 12%),
- porosity: 25 to 30%,
- absolute permeability: 1 to 2 Darcy.

The oil produced contains no dissolved gas.
- density: 0.89 - 27°API,
- in-situ viscosity: 40 cp at 30°C,
- water salinity very low: 0.5 g/litre.

The reservoir has been on production for over 20 years. 28% of the accumulation in place has been produced. At present, the wells are flowing under edge water drive, which is responsible for a high WOR (90%), owing to the mobility ratio that is very favourable to water (M = 12) and the low angle of dip of the strata.

Under the present conditions of exploitation, the production should tail out over 30 years, reaching a final recovery ratio of 43%.

During this exploitation closing phase, the opportunity of testing new tertiary recovery technologies is justified by:
- acces to new oil (enhanced recovery),
- an appreciable speeding up of the rate of production.

2 - MICROEMULSION METHOD PILOT PROJECT (Project 05.02/76)

From 1978 to 1980, Elf Aquitaine applied an experimental pilot project on the methodology on Chateaurenard field so as to evaluate the method of recovery by injection of microemulsion and polymers. The project was already the subject of a paper read at the Luxemburg Symposium in 1979. Here, we propose essentially to develop the findings and the overall results of the project.

2.1 - Characteristics and description of the project

a)   The configuration adopted (Figure 1) is an inverse five-spot panel with sides measuring 100 metres consisting of:
- one injection well CHU 16 I,
- one observation well CHU 164,
- four production wells.

The porous volume concerned is 10300 m³, with an estimated accumulation of 5700 m³ at the beginning of operation. An interference test proved the continuity of the reservoir. Owing to the exploitation of the field surrounding the mesh, there is a high pressure gradient running from East to West (3 bars per 100 metres).

b)   The following formula was adopted for the microemulsion:
- 30% reservoir water,
- 40% oil,
- 30% commercial sulfonates + alcohols.

Laboratory tests on a porous medium have shown that a 5% plug of the porous medium would recover 95% of the oil in place. This result can be explained by the drop in the oil-water interfacial tensions, freeing the oil traps in the porous medium by capillarity. This action will mobilize a bed of oil in front of the microemulsion plug, with its progression degraded:
- by the chromatographic scatter of the constituents,
- by retention of the surfactants in the porous medium (1 to 1.5 mg/rock).

c) Operation started in February 1978 with the injection of 974 m³ of microemulsion: the solution was made up in batches from a predetermined volume of the three main constituents. The single phase solution was then filtered through a 5 micron retention mesh and then injected.

After the microemulsion, the following injection sequence was applied:
- 8500 m³ of concentrated polymer solution (polyacrylamides in powder form),
- 4800 m³ of polymers of decreasing concentration,
- 6700 m³ of flushing water.

The role of the plug of polymers is to ensure control of mobility and to push ahead the microemulsion and bed of oil by piston effect.

2.2 - Results

The production wells displayed a distinct response: the percentage of oil in the output of 2 wells CHU 161 and CHU 162 rose from 10% to 65% (figure 2).

The influence of the static pressure gradient was decisive in the response from the other wells, since the wells higher up the gradient CHU 163 and CHU 16 displayed a less marked response.

In all, 4520 m³ of oil were produced, of which 2550 m³ represent the oil gain compared to the extrapolation of primary production.

To estimate the efficiency of the method, we have based ourselves on numerical simulations which combine all the static data and provide the production profiles of each well.

The main results that emerge from this computation basis show that:
- owing to the pre-existing pressure gradient, only 2/3 of the fluids injected contributed to the production of the pilot project,
- the useful volume V_u effectively swept was 10300 m³, corresponding to an accumulation of 5700 m³,
- this useful volume contributed to 3760 m³ of the oil produced by the pilot project,
- the volumes injected finally amounted to:
  . 0.06 Vpu of microemulsion (i.e. 610 m³),
  . 0.50 Vpu of concentrated polymers,
  . 0.30 Vpu of diluted polymers,
  . 0.40 Vpu of water,
- the average residual oil saturation at the end of the sweep operation is $S_{orf} = 0.19$.

In terms of overall material consumption, these results show that the weight of recovery of the zone swept amounts to 73% of the initial oil in place. Compared to the 43% to be expected by water flooding action, and for the 20 year horizon, the gain is 30%. This corresponds to an increase in production of 70% compared to conventional exploitation methods.

This interpretation involves uncertainties related to the estimation of the porous volume, the "open" configuration of the panel and the imprecision in assessing the average thickness of the reservoir.

It should be emphasized that the pilot project benefitted from the following favourable factors:
- the close spacing (100 metres),
- the short time period during which the products remained in place.

These good technical results are hence not directly transposable with a view to industrial generalization of the method.

From the costs standpoint, the overall results are less favourable. The basic research and additional investment (surface installations and drilling the wells) result in a cost for this tertiary oil that is 3 to 4 times greater than that of the market.

3 - MICROEMULSION AND POLYMER INDUSTRIAL PILOT PROJECT
(CONTRACTS 05.21/80 and 05.28/81)

The encouraging results of the methodology pilot project open prospects for additional reserves in a tight world energy context.

In October 1979, Elf Aquitaine started the industrial pilot project, representing an indispensable stage before envisaging general application to the field as a whole.

3.1 - Purpose

To exceed the point of economic return, application of the chemical recovery method has to be envisaged with:
- a wider spacing between wells,
- products of lower costs and involving smaller quantities.

It is to assess the effect of the change in dimension in technical and cost terms that the industrial pilot project was undertaken.

3.2 - Configuration (Figure 3)

Generalized application of the method must reside on the existing network of production wells. Accordingly, the following configuration was adopted: four inverse five-spot injection meshes (spacing between production wells: 400 metres). Four wells: CHU 181 - 182 - 183 - 184 were drilled at the barycentre of each mesh.

This configuration was changed as the project was carried into effect, reducing the dimensions of each mesh to 270 metres owing to:
- the application of infill drilling on the entire Neocomian field in order to reduce the spacing so as to enhance recovery and speed up the rate of production from the reservoir,
- the difficulties we encountered in arriving at a microemulsion formula that is cheaper than that proposed with the methodology pilot project.

New injection wells: CHU 519 - 505 - 506 - 525 were drilled in the centre of the redefined meshes.

The pilot panel was situated in a region with extensive water flooding: the output expected by natural depletion was low and it will be easier to estimate the gains in recovery by the microemulsion.

The static pressure gradient across the pilot zone is low (1 bar/100 metres).

The continuity of the layer was confirmed by interference tests.

The following are the static characteristics of the panel obtained by drilling and interference tests:
- area of panel: 288,000 m²,
- average useful height: hu = 2.6 m,
- porous volume included within perimeter: Vp = 224,420 m³,
- accumulation at beginning of operation: 102,340 m³ (Soi = 0.456),
- porous volume actually swept (based on numerical simulations) Vpu = 290,000 m³.

3.3 - Microemulsion formula

The costs of the microemulsion account for 30% of the total budget, whence the interest in optimizing the choice of the products and the volume injected.

In the search for less expensive microemulsions, the new idea was to use non-ionic surfactants that are cheap, added to sulfonates. The laboratories would appear to have made appreciable progress in this area by proposing a microemulsion with a high water content and a much lower cost.

The initial conception of the pilot project complied with the specifications of this initial formula.

Displacements in a porous medium representative of the reservoir revealed a high retention of the non-ionic surfactants by the clay in the medium, resulting in a considerable loss of sweep efficiency.

Accordingly, the non-ionic based formula was discarded and we we reverted to a microemulsion similar to that used for the methodology pilot project. The work essentially involved:

1) Optimization of the concentrations

Formula	Methodology pilot	Industrial pilot
Water	30%	55%
Oil	40%	15%
Commercial surfactants + alcohols	30%	30%

2) The optimization of the volumes to be injected: a plug of microemulsion equal to 2.4% of the effective porous volume (compared to 6% in the methodology pilot project) was injected into the meshes of the industrial pilot project.

The following were next injected:

0.4 Vp of concentrated polymer solution,
0.4 Vp of solution of polymers with decreasing concentration,
0.4 Vp of purified water.

3.4 - Surface installations

The installations must be the forerunners of the industrial stage where the method is put to generalized use. They are fully automated. In particular, we have designed a line production method for the micro-emulsion.

The infrastructure of the pilot project station comprises:

1) An injection station with a displacement pump for each injection well ; it will be used for each phase of the method (microemulsion, polymers and water).

2) A microemulsion production unit featuring programmable automated control. The solution passes through static mixers. Quality control is ensured by a double flow measuring system and a continuous viscositymeter.

The microemulsion passes through a 35 m³ retention tank. The single phase mixture is then filtered and sent to the injection station.

3) A polymer solution production unit: 1700 ppm liquid poly-acrylamides (viscosity 75 cp at 10 sec^{-1}). This dissolution takes place in two stages:

- production of a concentrated solution in a branched-off circuit,
- dilution in the main circuit, by means of an adjustable pressure loss system.

Each injection well is thus equipped with its own high pressure dissolution unit up-line from the injection displacement pump.

The general contracting firm Ponticelli built the station.

The following are the main suppliers:
- displacement pumps: Pmh - Dosapro,
- centrifugal pumps: Sihi - Grosclaude - Guinard,
- fire pump: Camiva,
- tanks: Carlier Plastique - Plastover,
- purification unit: Wemco,
- filters: Sofrance,
- cocks and valves: Mapegaz - Trouvay - Cauvin - Phocéenne - Fischer,

- control unit and regulator: Foxborro,
- instruments: Air service – Otic Fischer – Sereg – Metra,
- measuring equipment: Metra – Mecilec – Anacom,
- automation system: S.M.C.,
- electric generators: R.V.I.,
- buildings: Sotraco and Cobap.

4 – OPERATION

4.1 – Microemulsion phase

From June to July 1983, 7860 m³ of microemulsion were injected.

The representative point of the microemulsion in the phase diagram closely approaches a gel zone, so a tolerance of ± 3% in the concentration of each constituent had to be set. The main quality factor was that of the viscosity (operating range 51 cp ± 5 cp) with automatic cutoff if ever this tolerance was exceeded with possible correction of the concentrations at the retention tank. The operation was conducted under difficult conditions ; only the wellhead filtering encountered problems owing to the presence of particles in the oil from the Neocomian supplied by the field.

The injectivity index of all the wells improved considerably during the injection period (a rise of about 50 %), thanks to the detergent capacity of the microemulsion, reducing the skin effect near the injection wells.

Extraction at each production well was determined so as to obtain overall balance between injection and production on each mesh.

4.2 – Concentrated polymer solution injection phase

This started in August 1983 and ended in May 1984. 56 650 m³ of polymer solution was injected (i.e. 0.29 Vp).

Operation is linked to the production conditions of the central well CHU 18.

In September 1983, sulfonate and alcohol shows were observed in the production water of CHU 18. This resulted in deconsolidation of the sands and clogging of the wells by creep of the clays.

This breakthrough revealed the existence of a high permeability drain between injection well CHY 525 and central production well CHU 18. These results were confirmed by interference tests. Subsequently, a close relation was noted between the sulfonates content in well CHU 18 and the injection rates at well CHU 525.

To avoid problems of sand inflows, a gravel pack was installed on production well CHU 18.

To reduce the permeability contrast around CHU 525 and to improve control of mobility, the viscosity of the solution injected into this well was increased from 75 to 150 cp. The sulfonates content at CHU 18 decreased at the same time, becoming negligible from January 1984 onwards.

4.3 – Initial results

The percentage of oil in the production started to increase in August 1983 at CHU 183, after 0.12 Vp had been injected.

This particularity of CHU 183 may be explained by the hydrostatic pressure gradient through the pilot zone, causing viscous plugs to drift towards the South-West. This assumption was confirmed by the rise in the percentage of oil at the two production wells in the South and South-East of the panel. To prevent this drift, we increased the output from the wells further upstream, CHU 19bis, 181 and 25 by an amount equivalent to the flow caused by this pressure gradient.

The rise in oil production then occurred at CHU 18 - 182 - 181 - 6 - 5 - 184 - 25 in the order indicated. The following table shows how the percentage of water (BSW) and its production varied at each production well (up to the end of April 1984).

Conclusion

The results obtained are encouraging, but one must wait until completion of the water injection phase that is to follow the polymer injection phase to come to any conclusion as to the economic validity of the Chateaurenard industrial pilot project.

WELL	Initial BSW %	BSW at end of April 84 %	Oil production (tons) at end of April 1984 Total oil	Tertiary oil
CHU 19bis	100	100	0	0
CHU 181	82.6	67.6	2454 tons	719 tons
CHU 182	95.6	81.7	1037 "	583 "
CHU 25	100	96.7	33 "	33 "
CHU 18	89.2	64.5	2454 "	1219 "
CHU 5	94.0	79.8	465 "	144 "
CHU 184	97.8	94.9	277 "	61 "
CHU 183	72.8	58.2	2903 "	1024 "
CHU 6	94	79.4	331 "	163 "
Total of 30th April 1984			9954 tons	3946 tons

The tertiary oil has been calculated as the oil increment above the initial oil percentage threshold (1 - BSWI).

i.e. tertiary H = total H $\dfrac{(BSWI - BSW)}{1 - BSW}$

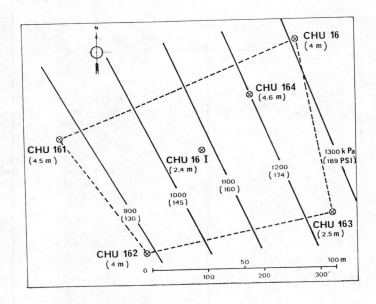

FIGURE 1 - CONFIGURATION OF METHODOLOGY PILOT PROJECT
Initial pressure gradient - The total height of the reservoir
is indicated at each well

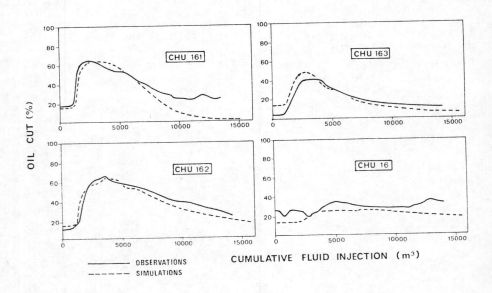

FIGURE 2 - COMPARISON OF OBSERVED AND SIMULATED OIL CUTS

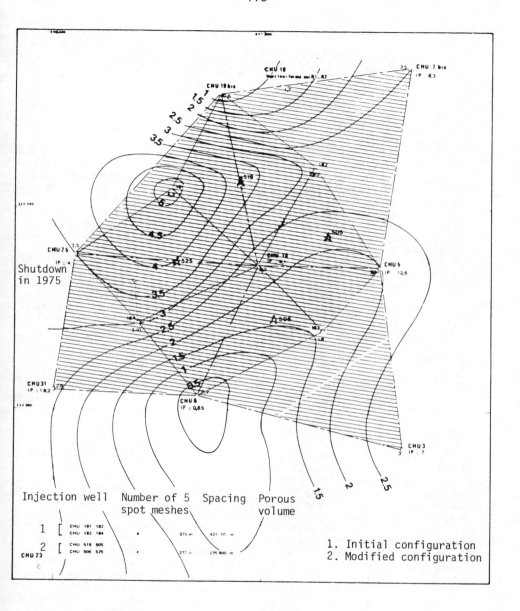

FIGURE 3 - GENERAL DIAGRAM OF INDUSTRIAL MICROEMULSION PILOT PROJECT
NORTH CHUELLES PANEL - (Isopaques of reservoir R3)

(05.06/77)

DEVELOPMENT OF BRINE SOLUBLE POLYMERS AND ASSOCIATED CHEMICALS FOR THE ENHANCED RECOVERY OF PETROLEUM

I.G. MELDRUM
BP Research Centre

Summary

The enhancement of oil recovery in North Sea reservoirs by a polymer solution flooding process requires the polymer to viscosify sea water at low concentrations and high temperatures for years. Unfortunately the acrylamide polymers which are most commonly used as mobility control agents were found to lose viscosity and precipitate within tens of days. Biopolymers showed more promise and the performance of xanthan polymers came close to meeting the viscosity/stability requirements when they had been purified. However the best performance was obtained with a scleroglucan polymer which gave the highest viscosity and maintained this viscosity without loss for a year at 90°C. The filtration and injectivity characteristics of the scleroglucan polymer were also very satisfactory. Model flow studies indicated that polymer solution flooding could significantly increase the rate of displacement of viscous oil during the early part of a flooding process but that a field trial would require detailed reservoir characterisation and thorough laboratory evaluation. No suitable opportunity for a field trial has yet arisen.

1. INTRODUCTION

The development of new ways of improving oil recovery from North Sea reservoirs is made particularly challenging by the combined features of an offshore location, high temperature, high salinity and reservoir heterogeneity (1). Operating from an offshore platform restricts the amount of any chemical treatment to quite low levels. This favours the use of polymers because their long, threadlike molecules can exert their influence over a large volume in solution. Polymer additives are believed to be capable of improving oil recovery by reducing the mobility of the floodwater through viscous and permeability effects. This results in less by-passing of oil on both macroscopic and microscopic scales. Depending on the reservoir situation, the polymer treatment can be applied uniformly or restricted to certain locations such as high permeability channels.

Laboratory studies and field tests on polymer flooding had previously been carried out in Europe and in the USA (2, 3) but at lower salinities and temperatures than in the North Sea. It was therefore important to investigate the effect of the more severe conditions on polymer performance. There was also considered to be a need for further studies of the flow behaviour of polymer solutions in porous media containing crude oil so that reservoir performance could be predicted with more confidence. The last stage of the development was to make the polymer available in a suitable form for offshore application with respect to transport, handling, storage, dissolution and injection.

2. EXPERIMENTAL

A large number of polymer samples were screened for their viscosifying properties and stability at high temperatures. The main polymer types were: acrylamide, vinyl pyrrolidone, hydroxyethyl cellulose, cellulose sulphate, guar gum, xanthan and scleroglucan. Solutions were made using filtered sea water containing hypochlorite biocide (< 0.2 ppm) which was added on collection. The polymer solutions were deoxygenated and the pH was adjusted to 6 - 6.5 with carbon dioxide to prevent scale formation. The oxygen level was shown to be less than 50 ppb. Formaldehyde (> 200 ppm) was added to biopolymer solutions. Long term tests were made in sealed glass capillary viscometers as illustrated in Figure 1. The viscometers were immersed in a thermostatted bath and tilted to fill the capillary. The rate of flow of liquid was measured electrically and was converted with the aid of a computer programme into viscosity/shear rate data in the range 20 - 2000 s^{-1}.

3. RESULTS

3.1 Viscosity Retention at High Temperatures

Most of the viscosity measurements were made on polymer solutions in sea water at a temperature of 90°C which was considered to be typical of North Sea reservoirs. Acrylamide was the first polymer to be examined because it had been used more than any other in the laboratory and field as a mobility control agent. In preliminary experiments in the presence of air, the viscosity of poly(acrylamide) solutions decreased by 50% after 40 days at 80°C. This highlighted the possibility that the deterioration in performance might be caused by oxygen which is present only at very low levels (< 50 ppb) in the sea water injected into reservoirs (4). Exclusion of oxygen to this extent would have been impractical with conventional viscometers for large numbers of samples and long periods of testing and so special sealed glass capillary viscometers were constructed for this purpose.

Even after modifying the test procedure in these ways, all the sea water solutions of various types of poly(acrylamide) lost viscosity rapidly at 90°C as shown in Figure 2 and formed a gelatinous white precipitate within 60 days.

On changing the base fluid for the poly(acrylamide) solutions from sea water to distilled water or 3% sodium chloride, no precipitation occurred, even after 500 days. However polymer solutions containing magnesium chloride (0.5%) or calcium chloride (0.1%) quickly formed precipitates (5 days). A possibility that the glass surface was participating in the precipitation process was investigated by using other materials such as PTFE but no effect was observed. The dependence of precipitation on polymer composition was also investigated. It was found that the time taken for precipitation to occur became shorter as the anionic level of the poly(acrylamide) was increased. Polymer solutions with more than 43% anionic monomer units precipitated immediately when calcium or magnesium chlorides were added. A poly(acrylamide) which was almost completely nonionic was heated at 90°C in distilled water at a concentration of 1% for 130 days. Ammonia was detected when the sealed tube was opened and titration showed that the level of hydrolysis had increased from 1 to 42%. This was consistent with the calcium content of the precipitate which was formed on adding calcium chloride to the polymer solution. Another poly(acrylamide) with a high anionic level (31%) was heated in a calcium chloride solution at 90°C until it precipitated. Analysis of the precipitate showed that the sodium in the original polymer had been largely replaced by calcium and that the proportion of metal acrylate groups had increased from 31 to 36%.

As shown in Figure 2, poly(vinyl pyrrolidone) was found to be much more stable. After a small initial decrease, the viscosity was almost unchanged after 500 days at 90°C. However a high concentration of polymer was required: its viscosity number was about 1/5th of the most effective poly(acrylamide).

Guar gum, hydroxyethyl cellulose and cellulose sulphate lost their viscosifying properties rapidly like poly(acrylamide) but they did not precipitate.

Initially, biopolymers of the xanthan type gave cloudy solutions which formed deposits on heating. Filtration through glass fibre media and cellulose acetate membranes greatly improved the clarity of solutions and allowed measurements to be made in sealed glass capillary viscometers. As a precaution against biological activity formaldehyde (1000 ppm) was added to the solutions but similar results were obtained in its absence. Some of the polymers showed an initial increase in viscosity but after about 20 days all the solutions slowly declined in viscosity as in Figure 2. The most stable xanthan polymer lost 30% of its specific viscosity after one year at 90°C. Its viscosifying performance, as measured by its initial viscosity number at 90°C, was more than twice that for the best poly(acrylamide). Two purified scleroglucan polymers were still more stable with little viscosity loss after a year. The viscosity number for one (CS11) was six times as high as for poly(acrylamide).

3.2 Flow Behaviour in Viscometers

Whereas biopolymers were known to have very pronounced shear thinning properties (5), poly(acrylamides) can exhibit shear thickening effects on flowing through model porous media such as layers of fine wire screens (6). A comparison of their flow behaviour in capillary and screen viscometers was undertaken to assess the implications for mobility control. The viscosity number measured in the capillary viscometer for an anionic (30%) poly(acrylamide) in sea water (1.7) was much lower than in distilled water (13.5). In the screen viscometer the viscosity number was more than twice as high for the distilled water solution (29.6) as in the capillary viscometer and it was reduced to a lesser extent on going to the sea water solution (8.3). The viscosity numbers for hydroxyethyl cellulose and xanthan polymers showed much less variation. The xanthan and scleroglucan polymers were found to shear thin much less at 500 ppm than at higher concentrations and increasing the temperature to 90°C made the xanthan polymer solutions almost newtonian but the scleroglucan polymer was almost unaffected (power law exponent = 0.6 - 0.7 at 90°C).

The stability of the polymers against mechanical stresses was investigated by forcing the solutions at high pressure through a capillary and measuring the change in viscosity after a number of cycles. As shown in Table I, the loss in viscosity for poly(acrylamide) was much greater (80%) than for xanthan and scleroglucan polymers (7 - 12%).

The injectivity of the polymer solutions was assessed by measuring the time taken for 100 cm^3 of solution to pass through a 5)m filter. The shortest flow time (7 s) was obtained for solutions prepared from the solid form of the scleroglucan polymer CS11. Some of the xanthan polymers completely blocked the filter. In general, long filtration times corresponded to large losses in viscosity after filtration. Hydroxyethyl cellulose gave short filtration times and small viscosity losses. The rate of filtration for poly(acrylamide) was very slow.

3.3 Flow Behaviour in Porous Media

Prior to core testing, polymer solutions were subjected to a screening test in which solution was pumped through a column packed with ballotini glass beads (100 - 150 μm) and the pressure differential was measured. Biopolymer solutions gave high resistance factors (1.5 - 20) at low concentrations (100 - 1000 ppm) and low residual resistance factors (< 1.5 at 1000 ppm). The resistance factors were converted to porous medium viscosities which are compared with low shear bulk viscosities in Figure 3. The frontal advance rate was quite high (0.3 mm s^{-1}). At low polymer concentrations the viscosities were proportional to each other but with increasing concentration the porous medium viscosity rose much less rapidly. For high molecular weight poly(vinyl pyrrolidone) the viscosities were proportional to each other over the whole range and the residual resistance factor was lower than for the biopolymers but the polymer concentration was much higher (0.5 - 3%). The scleroglucan polymers gave resistance factors in the order liquid CS11 > L21 > solid CS11. Residual resistance factors did not seem to be related to the resistance factors themselves.

A preliminary comparison of the oil displacing performance of sea water and polymer solutions was made using thin rectangular beds of fine glass beads (100 - 300 μm) with linear fluid displacement. In a typical run the permeability to sea water at residual oil saturation was 20 D and the oil permeability at irreducible water saturation was 6 D. Decreasing the value of oil viscosity/aqueous phase viscosity from 7.7 for sea water as the aqueous phase to 0.8 for a CS11 scleroglucan polymer solution increased the yield of oil by 20% after the first pore volume. Thereafter the increased yield gradually diminished to a final value of about 10%. When alternative pathways differing in permeability were created in the bed, the polymer solution did not compensate significantly for the difference. Changing to a triangular bed enabled radial displacement in one quarter of a five spot flooding pattern to be modelled. Experience with this system suggested that the most useful comparison between floods was the fraction of oil originally present which was recovered at the point of water breakthrough. As in the case of linear displacement, the main benefit from the polymer was obtained within one pore volume. In one series of runs a much more viscous oil was used giving a value for oil viscosity/sea water viscosity of 132 instead of 7.7 for the linear displacement. This resulted in additional oil recovery from polymer flooding compared with sea water being increased from 20% at one pore volume for the less viscous oil to 50%.

Preliminary model studies of the flow of floodwater through heterogeneous reservoirs were carried out using two layers of ballotini differing in bead size. Pumping CS11 scleroglucan polymer solution into both layers from a common feed resulted in more flow through the high permeability bed than for sea water. The adverse effect of the polymer was even greater when the polymer solution was followed by sea water flooding. Only when polymer solution was pumped into the high permeability layer at the same time as sea water was pumped into the other layer did the polymer correct for the difference in permeability.

Initial attempts to investigate the flow of polymer solutions in sandstone cores met with face-plugging problems. This was overcome by more thorough filtration of the solutions. As shown in Table II the resistance factors for cores without oil at 90°C were in the range 2.2 - 2.3 for bipolymers and 1.4 for poly(acrylamide) at a polymer concentration of 500 ppm. As was observed for columns packed with ballotini, the viscosities of the biopolymer solutions determined from resistance factors (0.76 - 1.00) in the core were considerably lower than the bulk, low shear viscosities (1.2 - 4.2). Acrylamide differed from the biopolymers in that its residual resistance factor was as high as its resistance factor. Similar results were obtained on flooding a core with residual oil. Changing from the high permeability Clashack core to a sub-millidarcy core resulted in large irreversible increases in pressure drop across for scleroglucan and vinyl pyrrolidone polymer solutions.

3.4 Polymer Dissolution and Logistics

The candidate polymers examined in this work were supplied as dry powders which were relatively slow to dissolve in water or as dilute solutions which were bulky. The possibility of producing them in a dispersion form which combined ease of handling with good storage stability, fast dissolution and high solids content ($>30\%$) was investigated. The most promising route was to grind dry polymer in a low toxicity diluent in the presence of a surfactant. Moderately stable dispersions were obtained but further development would be required.

4. DISCUSSION

From an early stage of the development the suitability of conventional acrylamide mobility control polymers for high temperature and salinity seemed doubtful. The absence of any significant improvement in polymer stability after greatly reducing the oxygen level suggested that oxygen was not involved in the main degradation process. Instead, subsequent studies of the mechanism of degradation were entirely consistent with the amide groups hydrolysing to carboxylate anions which paired with divalent metal cations in the sea water and reduced the solubility of the polymer. Since the polymer fell so far short of the stability requirement there was little hope that it could be modified within the time scale of the development. Poly(vinyl pyrrolidone) was extremely stable but far too much polymer would have been required to viscosify the flood water.

Initially the stability of xanthan polymers also seemed to be unsatisfactory but in this case performance could be improved relatively easily by removing impurities such as cell walls and proteins. Although the level of stability achieved in this way was very much better than poly(acrylamide), the rate of viscosity loss was high enough to limit the duration of a polymer flooding process at 90°C to about a year. Still greater stability was achieved by scleroglucan polymers which showed little if any loss in viscosity after a year at 90°C. Unlike their synthetic equivalents, the xanthan and scleroglucan biopolymers were highly resistant to mechanical degradation. The capability of the biopolymer solutions to permeate rapidly through porous media would be of considerable benefit at the filtration and injection stages in polymer flooding.

It is likely that shear thinning was at least partly responsible for the poor performance of scleroglucan polymer in the model studies of flow through heterogeneous reservoirs. To offset the difference in permeability the flooding solution should ideally increase in viscosity with increasing flow rate rather than decrease. Otherwise, injection of polymer solution would have to be controlled so that the solution only entered the high permeability zone.

In uniform, idealised porous beds polymer solution flooding was demonstrated to be of considerable potential benefit in displacing a viscous oil. The enhancement of oil displacement was greatest at the early stages of flooding, particularly before water breakthrough. This would be particularly desirable for an offshore location because of high operating costs. In high permeability core tests the polymer solutions behaved in much the same way but on reducing the permeability to a sub-millidarcy level, the polymer solutions were found to be causing blocking problems despite careful pre-filtration. It would appear that polymer flooding might not be feasible for such a low permeability.

Although scleroglucan polymer satisfied the primary viscosifying requirements for a North Sea reservoir application, before it could be used in practice, further development would be required to make it available in a compact, handleable, readily dissolvable form. Preparing the polymer as a dispersion in oil showed considerable promise but greater stability against sedimentation would be necessary.

5. CONCLUSIONS

Acrylamide mobility control polymers hydrolyse far too rapidly to be recommended for long term flooding processes where the floodwater contains divalent metal ions and the reservoir temperature is in the region of 90°C. In addition to losing viscosity, the polymers form a gelatinous precipitate which would be likely to cause blocking problems in the reservoir. Scleroglucan biopolymer gives excellent viscosifying and stability performance and is superior in most respects to other polymers. It would appear to be capable of retaining its flow properties at 90°C for much longer than a year. Its shear thinning characteristics would greatly assist the filtration and injection steps in a flooding process but might adversely affect sweep efficiency, particularly in a heterogenous reservoir. Because of the complex flow behaviour of polymer solutions in the reservoir, a polymer flooding process should be based on detailed reservoir characterisation and thorough laboratory evaluation. Otherwise the process is likely to behave very differently from what is expected. For offshore applications further development would be required to produce the scleroglucan polymer in an acceptable liquid form. Although there was no evidence of biological degradation in the laboratory evaluation, this is a potential threat to biopolymers and should be checked. Provided no major obstacles like this are encountered, scleroglucan polymer could be used to increase the rate of production of oil from viscous North Sea reservoirs during the early part of the flooding process. So far, no suitable opportunity for a polymer flooding field trial has arisen.

REFERENCES

1. DAVISON, P. and MENTZER, E. (1980). SPE 9300.

2. JEWETT, R.L. and SCHURTZ, G.F. (June 1970). J Petrol Technol, p 675.

3. UZAIGWC, A.C., et al. (1974). J Soc of Petrol Engr, 14, 33.

4. MITCHELL, R.W.   OE-77.1.SPE 2.6677.6.

5. JEANES, A.   J Polymer Sci, Symposium No 45, 1974, 209 - 227.

6. JENNINGS, R.R., et al (1971).   J Petrol Technol, pp 391 - 401.

ACKNOWLEDGEMENTS

   This paper describes work carried out by Dr E. Mentzer and Mr P. Davison on EEC contract TN/05.06/77.

TABLE I

STABILITY OF SEA WATER POLYMER SOLUTIONS IN HIGH SHEAR TEST

Polymer	Concentration ppm	Relative Viscosity		Viscosity Loss %
		Initial	Final	
CS11 (liquid)	1750	844	789	7
Xanthan	2000	439	417	5
Scleroglucan L21	2000	772	683	12
Experimental B5	2000	411	333	19
Acrylamide	5000	68	14	80

TABLE II

CLASHACK SANDSTONE CORE TESTS AT 90°C AND 170 bar
FRONTAL ADVANCE RATE 24 mm min^{-1}.   POLYMER SLUG SIZE 4 PV (500 ppm)

Polymer	Sea Water Permeability mD	Resistance Factor	Residual Resistance Factor	Viscosity (mNs m^{-2})	
				Core	Brookfield
Experimental B5	245	2.4	1.3	0.76	1.5
Scleroglucan L21	212	3.3	1.4	1.00	4.7
Scleroglucan CS11	252	2.6	1.4	0.80	1.35
Xanthan	396	2.9	1.2	0.91	6.0
Acrylamide	368	1.4	1.4	0.43	0.4

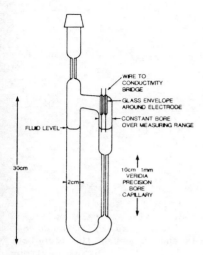

FIGURE 1 SEALED GLASS CAPILLARY VISCOMETER

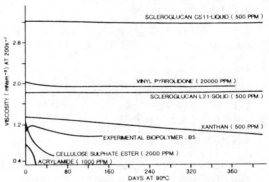

FIGURE 2 STABILITY OF POLYMER SOLUTIONS IN SEA WATER AT 90°C

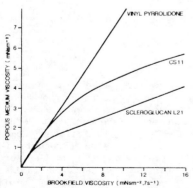

FIGURE 3 COMPARISON OF BULK AND POROUS MEDIUM VISCOSITIES
OF POLYMER SOLUTIONS AT 55°C

(05.15/80)

ENHANCED OIL RECOVERY FROM THE EGMANTON OILFIELD BY CARBON DIOXIDE MISCIBLE FLOODING

C. BARDON
Institut Français du Pétrole
I.A. WOODHEAD
D.M. GRIST
British Petroleum Company Ltd.

Summary

The project was initiated by BP Petroleum Development Ltd. and undertaken in conjunction with GERTH members. The overall objective was to evaluate carbon dioxide miscible flooding as an enhanced oil recovery process for the Egmanton oilfield and to assess the technical viability of carbon dioxide flooding in the Western Europe environment. A phased approach was adopted so that decisions to proceed could be made on the basis of the data and understanding gained at particular stages.

Activity included the review of available data, laboratory and theoretical investigations, plus an assessment of the logistics of large scale CO_2 supply. Much was learned from these preliminary studies. A subsequent phase of detailed work consisted of numerical reservoir modelling and associated laboratory support. As part of the integrated project the CO_2 injection well was drilled which provided further data and material essential to the studies. After a phase of engineering design work, it was decided to construct and commission the project site facilities in order to conduct production well tests and CO_2 and water injection trials. The low CO_2 and water injectivities measured, combined with the results of simulating the oil recovery from the project area, led to the decision to terminate the project on the grounds of extended duration and escalated cost. Much valuable technology was obtained up to this point including the predictive model developed and the engineering and well design concepts that were successfully tested.

1. INTRODUCTION

The Egmanton CO_2 flood project was initiated to assess the engineering aspects of oil recovery by CO_2 injection and to develop a predictive numerical model. The project was a joint venture between BP Petroleum Development (BPPD) and Groupement European de Recherches Technologiques sur les Hydrocarbures (GERTH) which includes Compagnie Francaise des Petroles, Societe Nationale ELF-Aquitaine (Production) and Institut Francais du Petrole (IFP). Financial support from the European Economic Commission Directorate General for Energy was available from December 1980. The Project has been reported in previously published papers (1,2).

2. RESERVOIR AND PILOT AREA DESCRIPTION

The UK East Midlands oil reservoirs are thin estuarine sands of low porosity and low permeability (1-3 MD), which are normally pressured and produced by pumping. The reservoirs occur at the base of the Westphalian (Carboniferous) sequence. The Egmanton field is a north-east to south-west trending anticline (Fig. 1). It was discovered in 1955 and is now largely watered out.

The CO_2 flood pilot area is approximately a rectangle defined by production wells EG 1, EG 20, EG 35 and EG 46. As part of the preparation for the pilot, and to provide material for studies, a new well, EG 68 was drilled as the central injector of the inverted 5 spot configuration. The reservoir section was fully cored and a full suite of logs run.

The main characteristics of the pilot area within the Egmanton No 1 sand group reservoir (mainly the Crawshaw Sand) are given in Table 1. The original oil-in-place estimation within the pilot area was 8.5×10^4 m^3 but the remaining oil in place after waterflooding is estimated as 7.6×10^4 m^3 using a remaining oil saturation of 0.40. A sedimentological study was conducted by BP using data obtained from EG 68 as well as older core and gamma ray log data from all wells in the field. This geological study confirmed that the Central Crest area and the 5-spot well configuration chosen for the pilot test coincides with the thickest and best area of reservoir development in the Egmanton field.

3. LABORATORY STUDIES

Egmanton reservoir oil was reconstituted from wellhead oil and gas samples, and PVT studies conducted to establish the phase behaviour of the CO_2/oil system. The presence of water was found to have no significant influence on the phase behaviour. Close observation revealed that two separate liquid phases existed above 85 barg (1230 psig) the point at which a separate gas phase vanished. Some paraffinic deposit also occurred. This is similar to the findings of other workers [1], [3]. Studies were completed utilising slim tube displacement experiments [4]. These indicated a relative increase in the recovery of oil at pressures above that at which the gas phase disappeared and two liquid phases remained. This pressure limit is commonly called the minimum miscibility pressure and is 85 barg (1230 psig) for the Egmanton oil/CO_2 system. The reservoir pressure is approximately 138 barg (2000 psig) so that this apparent or dynamic miscibility condition can exist when CO_2 is injected. Over a large pressure range CO_2 swelled the oil and reduced its viscosity. These latter two effects will assist recovery even where miscible conditions are not achieved.

Seven experiments were conducted with 2 m long vertical sand packs in order to assess the efficiency of oil displacement with CO_2 under tertiary conditions and to compare the efficiency of continuous CO_2 injection versus slug injection (Table 2). Using core samples from the EG 68 core, experiments were then undertaken to assess the efficiency of CO_2 as a tertiary recovery agent under reservoir conditions at a post waterflood oil saturation. These core tests, results from three of which are shown in Table 2, confirmed the conclusions from the sand pack tests and provided the data base for the simulation of the displacement process in the reservoir. It can be seen that "continuous" injection of two pore volumes of CO_2 reduces the post waterflood oil saturation by approx. 80% in both sand packs and cores.

4. PREDICTION OF PERFORMANCE

Based on the laboratory results, numerical simulation of field performance was completed using a combination of 1-D and 2-D models (2). More sophisticated reservoir engineering models were found at the time (1981) to have shortcomings (2), (5). The 1-D displacement model $TUBCO_2$ is a simple 3 phase, 4 component simulator accounting for the main CO_2 displacement mechanisms of oil swelling, oil viscosity reduction, oil stripping, miscible drive and CO_2-water solubility. Oil is represented by two pseudo-components, one only allowed to vaporize. Miscible and immiscible displacements can both simultaneously be present, depending on linear position.

The 2-D stream-line model TUBVOL can perform areal flow calculations, especially in the case of pressure maintenance and various slug injections involved in enhanced oil recovery processes. It accurately computes the drainage limits and stream-lines in any heterogeneous reservoir and well pattern case. It can integrate various linear, variable section displacement models, with a more detailed space discretisation than the rectangular mesh models. This gives reduced numerical dispersion and computing time.

The 1-D model was first used to match the laboratory displacement experiments performed on the EG 68 core samples. Fluid densities and viscosities were taken directly from laboratory measurements. Preliminary waterflood results were easily matched. CO_2 flood simulations then closely reproduced the laboratory results (Fig. 2).

The prediction of the pilot test performance comprised the following steps:-
- changing the initial laboratory core oil saturation in $TUBCO_2$ to the present pilot area saturation.
- scaling the linear $TUBCO_2$ model to the well spacing and vertical dimension.
- deriving the 2-D areal performance of the pilot flood from the linear performance by introducing the linear model into the stream tube model after selection of proper flow rates and pressure gradients.

The base case selected for the predictive pilot simulations was based on homogeneous sand properties with average thickness 16 m (52.5 ft), a uniform oil saturation of 0.40 and a uniform average mobility of 1.6 md/cp.

Well volumetric rates (reservoir conditions) were:-

	m^3/d	bl/d	
EG 1	7.1	45	production
EG 20	6.0	38	production
EG 35	7.1	45	production
EG 46	3.6	22	production
EG 68	23.0	150	injection

The amount of tertiary oil over and above that produced by waterflood was predicted (see Fig. 3). A 7,500 tonne slug of CO_2 followed by one pilot area pore volume of water would yield an incremental 4,200 barrels of oil. This equates to 0.9% of remaining oil in place. Modelling predicted that a CO_2 slug of at least 12,000 tonnes followed by water would be required to yield an observable increase in oil cut of 5% to 10% at the production wells. At a 1983 UK price of £80 per tonne this amount of bulk liquid CO_2 would have cost £960,000.

5. INFORMATION DERIVED FROM FIELD TESTS ON THE EGMANTON OILFIELD

Prior to undertaking the EEC funded joint CO_2 Project BP independently conducted a single well CO_2 injection test on nearby well EG 51. This test evaluated CO_2 injectivity and the general performance of surface equipment as well as determining well behaviour on back production. The test indicated that adequate CO_2 injectivity into the Egmanton reservoir was achievable and that carbon dioxide mobilised oil in the reservoir. Some of this oil was recovered when the well was placed on back production (2). The surface equipment used performed well and operational problems with ice and CO_2 hydrates were overcome. The corrosion inhibition methods employed (coatings & inhibitors) were successful.

Within the joint project and in preparation for the later 5-spot pilot test, a series of pressure interference tests were conducted in the pilot area. It was concluded from the initial tests that full communication existed between EG 68 and production wells EG 35 and EG 46. There was uncertainty about connection between EG 68 and wells EG 1 and EG 20. A subsequent reversed pulse test showed that there is reservoir connection between EG 68 and EG 20. It is suspected that some permeability barrier may exist between EG 20 and EG 35.

6. SURFACE FACILITIES

The process facilites (Fig. 4) were designed on the basis of safety, functionality and low cost.

The production handling facilites consisted of a produced fluids heater, a stainless steel separator (Robert Jenkins Oil & Gas Ltd.) and storage tanks, the water tank being epoxy lined. The separated oil and water flowstreams were metered by high turn down ratio positive displacement meters. These meters made it possible to use the separator for individual well tests as well as pilot area production.

Produced water was pumped from the storage tank, passed through a tilted plate separator to remove residual dispersed oil, and filtered prior to re-injection at EG 68 via a high pressure, oscillating positive displacement, 4 cylinder variable stroke, LEWA Herbert Ott Gmbh. pump. This train appeared to produce adequate water quality and no water injectivity loss was observed. The equipment was fabricated in 316L alloy stainless steel to avoid carbonic acid corrosion in the presence of chlorides. Water could also be injected from the Egmanton field high pressure water main.

The CO_2 injection facilites were designed and installed by the Distillers Company (CO_2) Ltd. They included a 50 tonne refrigerated CO_2 storage tank in which bulk liquid CO_2 was stored at 20.5 bar and $-17°C$. Liquid carbon dioxide from this tank was pumped to the wellhead by a 4 cylinder high pressure LEWA pump (as described above) via a heat exchanger which raised the temperature of the high pressure CO_2 stream to $+10°C$ to prevent freezing of the injection well annulus and the formation. During operation of the CO_2 injection facilities, difficulties were encountered with the CO_2 turbine flowmeter due to the fluctuating fluid temperature and hence density. It would have been better to monitor the cold liquid flow prior to heat exchange. The LEWA pump was prone to vapour locking but this problem was rectified by providing additional insulation on the pump casing.

Hydrogen sulphide monitoring equipment manufactured by Maihak A.G. (lead acetate ribbon photo cell type) was installed to provide a reliable long term continuous measure of the hydrogen sulphide levels in the produced gas streams. Iron sponge pellet beds were identified as the contingency method for vent gas scrubbing.

The four production wellheads were fitted with stainless steel trim McEvoy annulus valves and corrosion resistant polish rods and insert pumps (two were 2-stage pumps). Injection well EG 68 was equipped with Cameron stainless steel trim wellhead valves and the well was completed with AMF TK2 epoxy resin coated (2 3/8 inch, 60.3 mm) tubing to resist corrosion. The annulus was isolated with a Baker HRP-4-SP hydraulic set packer with HYCAR seals. The annulus fluid was diesel oil.

7. INJECTION TRIALS

The injection trials (autumn 1982) were carried out to establish the injectivity of CO_2 and water and to provide reservoir engineering data from a variety of tests. These trials were essential in order to check the results of the earlier single well test and to establish via the model whether the entire 5-spot field trial could be implemented on a reasonable time scale. The injection trials were also conducted to assess the performance of the installed facilities (see above) and the EG 68 well completion. There were no problems with the latter.

The CO_2 injectivity attained during the trials was 0.23 res. m^3/ day/bar (0.1 reservoir barrels/day/psi). This is significantly lower than the 0.7 res. m^3/day/bar (0.3 reservoir bl/day/psi) achieved during the previous single well test at EG 51. Maximum injection rate at just below reservoir fracture pressure was 12.4 m^3 pd (80 bwpd) for water and 14 res. m^3/day (89 res. bpd) for CO_2 - equivalent to 11.5 tonnes per day of CO_2. Bottomhole fracture pressure is (184 barg) 2672 psig. with CO_2 as working fluid.

A total of five significant pressure transient tests were conducted during the injection period. In chronological order these were a pressure fall off (PFO) test during preliminary water injection, two PFO tests and a two rate test during CO_2 injection and a PFO test carried out during a follow up water injection period. These tests were analysed (6) by classical type curves, modified Horner plots (2 rate test) and analysis of the dual slopes on the Miller Dyes Hutchinson plots of the PFO tests during CO_2 injection to assess the Darcy flow parameters for both the CO_2 and the water radial flow regimes in the reservoir around the well. Results are shown as Table 3. Little change in skin factor is observed showing that the near well reservoir rock condition (slight fracturing - length 4 m) was not altered significantly. The effective permeability to CO_2 at reservoir conditions is one tenth that to water but appears to increase during the injection period. The final pseudo steady state permeability to CO_2 - calculated independently - was 0:22 mD. The early two rate test gives an anomalously high permeability to CO_2. Comparison of the water injection PFO tests shows the expected increase in near well permeability to water after oil has been displaced from this region by the carbon dioxide injection (well stimulation effect). CO_2 and water injectivities calculated from these tests are referenced above.

8. DISCUSSION
8.1 Project Timeframe

The relatively low well injectivity found by testing showed that the project timeframe would extend beyond that anticipated. It was originally estimated that a water injection rate of 64 m^3/day (400 bpd) could be attained in the pilot area. Subsequent testing has shown that such an injection rate is rarely achieved without fracturing in the East Midlands reservoirs that remain on production. At the maximum CO_2 injection rate attained (11.5 tonnes/day) it would take 2.5-3 years to inject the required minimum of 12,000 tonnes of CO_2 at below fracture pressure. It was likely to be at least six years before any part of the oil bank thus generated appeared at a production well and over 20 years before the full effect of the CO_2 flood could be estimated.

8.2 Options to reduce the Timeframe

One approach considered was to drill a conventional observation well close to the bottom hole location of EG 68 and observe and sample the fluid banks as they passed by. This was not acceptable due to the additional cost of £0.5 million (in 1983), the uncertainty in identifying a location in full connection with the injection well, difficulties in log interpretation due to mixed salinity in the waterflooded reservoir and the probable onset of interlayer crossflow if the observation well was perforated for sampling.

The second approach considered was to run a reduced volume test by drilling three or four slimhole (60 mm) injection wells close to on site production well EG 20 and conducting a normal 4 or 5-spot flood. Due to the paraffinic character of the Egmanton oil, these slimhole wells could only be used as injectors so that much expensive CO_2 would be lost outside the treated area. Directional control of slimhole wells is difficult, making the creation of a small area symmetric pattern uncertain. This may then have given rise to serious flow distortion within the reservoir. Log information would be restricted due to the limited suite available for slimhole wells. BP and GERTH also estimated that, if the slimhole option was pursued, the total project cost would still considerably exceed the original project budget.

8.3 Termination

In May 1983 the Project Management, having considered the technical, time and cost ramifications of the options, concluded that the Project should be terminated.

8.4 CO_2 Sources

A study into the cost and availability of CO_2 in Western Europe was conducted by IFP (7). It concluded that the major sources are industrial waste streams (Urea plants, ammonia plants, lime and cement kilns) and electric power stations (see (3) also). The latter offer the most potential although the process of extraction from flue gas is energy intensive. The cost of CO_2 from power station sources is high - about £35 pt in 1982. It was generally concluded that CO_2 flooding will not be an economic enhanced oil recovery process unless a large subterranean source of the gas is available close to the oilfield.

9. ACKNOWLEDGEMENTS

The authors wish to acknowledge the contribution of the late Dr. N.J. Webber to this project. This paper is released by The British Petroleum Company and GERTH.

10. REFERENCES

1. The East Midlands Oil Project Two Years On. D.M. Grist, F. Musgrave and R.W. Mitchell. European Petroleum Conference, London. 25th-28th October, 1982.

2. Egmanton CO_2 Pilot Injection Project. J.E. Bradley, P. Clyne, J. Combe and C. Bardon. 2nd European Symposium - Enhanced Oil Recovery, Paris. 8th-10th November, 1982.

3. Miscible Displacement. Soc. Pet. Engrs. Monograph 8, 1983. F.I. Stalkup.

4. The Use of Slim Tube Displacement Experiments in the Assessment of Miscible Gas Projects. B.J. Skillerne de Bristowe. 1981 European Symposium on EOR, Bournemouth. September 1981.

5. Numerical Study of a CO_2 Flooding Pilot Test in the UK, East Midlands. U. Sellier and J. Combe. Budapest CO_2 Symposium. March 1983.

6. Advances in Well Test Analysis. Soc. Pet. Engrs. Monograph 5, 1977. R.C. Earlougher.

7. Availability and Cost of Carbon Dioxide from Industrial Sources in Western Europe. A. Chauvel and S. Franckowiak. Budapest CO_2 Symposium. March 1983.

TABLE I

EGMANTON PILOT AREA PARAMETERS

Datum Depth	945 m ss
Average Reservoir Pressure	141 barg.
Average Reservoir Temperature	43°C
Original Oil Saturation	0.55
1982 Average Oil Saturation	0.40
Porosity	0.15
Permeability	1-10 mD
Oil Gravity	39° API
Paraffin Wax Content	0.18

TABLE II

POROUS MEDIA DISPLACEMENTS

2 m Long Sand Packs

Run Number	Pressure Bar	Initial Condition Sor	Waterflooding		CO$_2$ Flooding			Post CO$_2$ Waterflooding	
			PV Injected	Sor	Rate ml/hr	PV Injected	Sor	PV Injected	Sor
1A	103	81.2	1.51	25.1	30	2.00	5.1	–	–
1B	103	79.9	1.51	24.0	30	0.10	23.9	2.04	15.5
2A	151	81.3	1.52	25.1	30	2.00	4.2	–	–
2B	151	81.1	1.50	24.5	30	0.10	24.4	2.00	14.0
2C	151	80.2	1.50	23.6	30	0.20	23.5	1.10	9.2
3A	207	81.7	1.49	27.4	30	1.99	2.6	–	–
3B	207	81.7	1.50	19.7	30	0.10	19.5	2.01	12.1

EG 68 Core Tests

Run Number	Pressure Bar	Initial Condition Sor	Waterflooding		CO$_2$ Flooding			Post CO$_2$ Waterflooding	
4A	151	51.1	1.50	23.0	1	0.20	20.1	1.70	12.4
4B	151	50.1	1.50	23.1	5	0.10	22.8	1.20	18.8
4C	151	53.5	1.50	25.6	1	3.17	6.0	–	–

TABLE III

INJECTION TESTS AT EG 68

Test and Analysis Method	Test Date 1982	Well Skin Factor	Inner Fluid Bank		Outer Fluid Bank (Water)	
			Calculated Mobility mD/cP	Effective Permeability mD	Calculated Mobility mD/cP	Effective Permeability (mD) to water
PFO – Water injection pre CO$_2$ (MDH method)	28-10	-4.0	0.86	0.60	–	–
2-Rate Test CO$_2$ injection (modified Horner method)	22-11	-2.0	–	0.56	–	–
PFO – CO$_2$ injection (Ramey Type Curve)	25-11	–	2.2	0.14	–	–
ibid (MDH method)	25-11	-3.4	1.3	0.09	1.7	1.2
PFO – final CO$_2$ injection (Ramey Type Curve)	4-12	–	2.4	0.17	–	–
ibid (MDH method)	4-12	-3.2	1.7	0.12	1.8	1.3
PFO – Water injection post CO$_2$ (MDH method)	29-12	-2.2	2.2	1.50	–	–

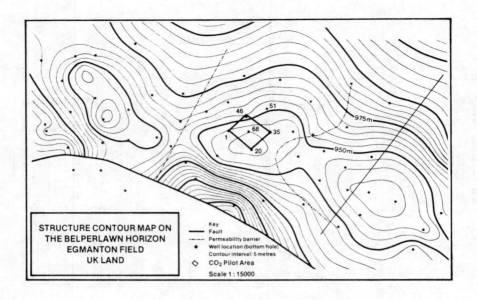

FIGURE 1

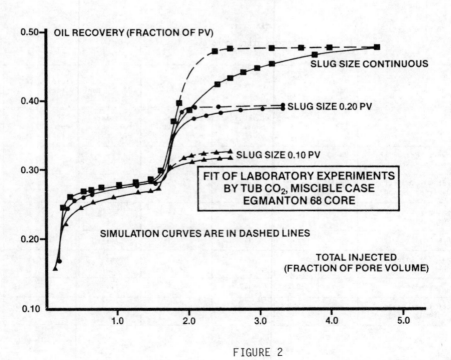

FIGURE 2

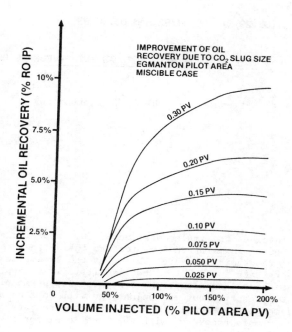

FIGURE 3

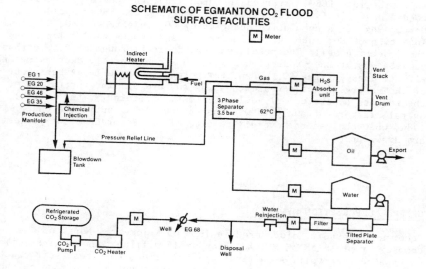

FIGURE 4

(05.29/81, 05.44/82 and 05.48/83)

PILOT PROJECT FOR INJECTION OF MISCIBLE GAS INTO PECORADE RESERVOIR

J.L. MINEBOIS, Société Nationale ELF AQUITAINE (Production)

Summary

This project involves the implementation of CO_2 injection at Pecorade, the objective being to maintain the pressure inside the reservoir above critical reservoir pressure, whilst at the same time improving the oil recovery rate in a sector that is difficult to exploit by conventional methods (compact porous zone, but low permeability).
A reservoir study carried out in conjunction with work on a petrophysical model has revealed a possible recovery rate of 80% using CO_2 for primary recovery, with the same level for a water drive, followed by a CO_2 sweep.
Thanks to in-situ CO_2 injection tests, in 1982, the choice was directed towards a "inverse five-spot" pilot project with a spacing of 200 metres between the wells. This project comprises one injection well PCE 13 and three production wells PCE 04, PCE 22 and PCE 23. Injection into PCE 13 started in March 1983, but the exploitation conditions quickly deteriorated and injection was interrupted in March 1984 for a cumulative quantity of CO_2 injected of 10,700 tons. Injection of CO_2 into PCE 13 in fact caused cracks to develop around the injection well. The result was rapid breakthrough of free carbon dioxide gas at the production wells, thus considerably decreasing the effectiveness of the CO_2 sweep. Furthermore, double-layer behaviour of the reservoir was revealed, making any overall interpretation of the results a particularly delicate task. Accordingly, the initial overall factors of analysis will not be available until the end of 1984, when the general overall results of the Pecorade pilot project are drawn up.

1 - GENERAL DESCRIPTION OF PROJECT

Pecorade reservoir, discovered in 1974 65 kilometres North of the Lacq complex, represents an accumulation of 15 million tons of light oil in a carbonated reservoir. As a first approximation, this reservoir can be divided into a fissured summit zone with a compact unfissured zone on the flank of the structure displaying poor petrophysical characteristics.
The oil is saturated with gas, allowing only a small drop in the reservoir pressure if excessive inflows of free gas are to be prevented and

degassing inside the layer. Accordingly, peripheral injection of water was very quickly combined with production.

This initial production configuration, whilst enabling the fissurized summit zone to be exploited under normal conditions, affected the flank zone only little, owing to the very low water injectivity of the wells in this sector. The quantity of hydrocarbons that may thus be trapped has been estimated at 5 million tons, so it is to improve the drive in this zone that a new enhanced recovery method was sought.

In view of the depth of the reservoir (2700 metres) and the problems of injectivity, it was decided to employ methods of injecting low viscosity gaseous products, the composition of which would make it possible, on contact with the oil, to generate a miscibility process conducive to enhanced recovery.

2 - CHOICE OF THE CO_2

Laboratory work has revealed that carbon dioxide gas can considerably improve drainage of the carbonate formations near the injection wells, thus helping to improve water injectivity in this type of reservoir. Furthermore, experiments on test-samples have proved carbon dioxide gas to be a major factor in mobilization of the oil.

Lastly, the availability near the injection site of considerable production (at the Lacq plant) first of hydrocarbon products of the LPG type and second carbon dioxide gas (in the form of a mixture of CO_2 and H_2S) provided the motivation for comparison of the performances of these two methods. The initial results revealed the advantages of CO_2 compared with the hydrocarbon mixtures, since CO_2 is used both in secondary recovery and tertiary recovery, and is also available at lower cost.

This work as a whole has hence resulted in choosing CO_2 for the Pecorade enhanced recovery pilot project.

3 - RESERVOIR STUDIES

3.1 - Petrophysical model

In order to study the CO_2 performances in a real-life medium, a petrophysical model was built from samples of full-diameter cores from the Baremian reservoir of Pecorade. Three experimental sweeps were carried out in the SNEA(P) laboratories, each one following restoration of the initial oil and water saturation levels in the reservoir (petrophysical characteristics $\emptyset$ = 7 to 8% and 5 < K < 50 md).

The final oil recovery after sweeping to 1.5 times the pore volume, is 80% for a CO_2 primary drive, of which 1/5th originate from condensates evaporated by the CO_2.

In the case of an initial water drive followed by CO_2, the recovery resulting from the water drive is 49%, and from the CO_2 31%.

Lastly, the final recovery is identical, regardless of whether one is operating on secondary or tertiary drive.

These results, which confirm the interest of CO_2 as displacement agent should however be qualified, since the recovery values depend on the microscopic sweep without allowing for the natural horizontal and vertical heterogeneities within the reservoir.

3.2 - Sweep efficiency

The CO_2 available at the Lacq plant contains 3% of impurities, of which 1% of H_2S. Tests were therefore carried out on an artificial porous medium in order to determine the sweep efficiency under the reservoir conditions for mixtures of CO_2 and H_2S. These tests also enabled the miscibility pressure to be determined.

The work performed showed that the H_2S content in the CO_2 is not prejudicial to the recovery, but quite on the contrary, appreciably improves the results obtained with pure CO_2 alone. However, the maximum acceptable H_2S content in the CO_2 is kept at 1000 ppm in order to remain in conformance with legislation concerning transport by road tankers.

3.3 - Numerical model

To account for the phenomena observed locally within the pilot-tested zone, and possibly to extrapolate these results to the full scale of the field, several numerical models were tested for this particular application.

The COMPAKIT 4-component compositional model transpired to be the best suited and following de-bugging, the model will be adjusted on the basis of the experimental results during exploitation.

3.4 - Other studies

PVT work on oil-CO_2 mixtures was carried out in order to determine the main characteristics of the phases present. This study provides a physical reference for setting a thermodynamic model based on the PENG.ROBINSON equation. The same model will subsequently be used to generate the equilibrium constants used in the compositional model (see paragraph 3.3).

One of the main parameters that should be continuously followed throughout the exploitation phase concerns the detection of CO_2 breakthrough and following the CO_2 content in the effluent produced.

Two non-radioactive tracers, sulphur hexafluoride SF6 and a natural isotope of the carbon of the CO_2 (C-13), were finally adopted and will be tested in turn by adding them into limited slugs of CO_2 (of about 300 tons).

4 - INJECTIVITY TESTS

In order to determine more precisely the CO_2 injectivity in-situ, two limited injection operations, one of 20 tons, the other of 200 tons, were made in injection well PCE 20, near the zone set aside for the pilot project.

The injection conditions complied with during these two tests are well above those necessary for exploitation of the pilot project and in the present state-of-the-art, do not appear to set any major exploitation problems.

First test: flow = 150 to 200 l/mn. Wellhead pressure 400 to 475 bars.

Second test: flow = 270 to 330 l/mn. Wellhead pressure 480 to 535 bars.

5 - PREPARATION OF THE PILOT PROJECT

5.1 - Definition of the pilot project zone

As an initial approximation, the Pecorade reservoir can be divided into a cracked summit zone and a peripheral zone with mediocre petrophysical characteristics, representing an accumulation of 6 million tons of oil. Accordingly, the CO_2 pilot project is set up on the latter sector, where the water injectivity is practically zero.

The pilot zone was initially defined as a "inverse five-spot", based in the East on an already existing well (PCE 4). The spacing between the wells was set to a maximum of 300 metres in order to ensure good representativity of the reservoir as a whole, whilst at the same time limiting the response time between the injection wells and the production wells.

5.2 - Pilot project wells

- PCE 13 injection well was drilled from the Pecorade processing centre in order to group together all the high pressure injection installations whilst complying with safety and production requirements. This surface layout called for a 1750 metre offset of the well for a depth of 2600 metres, i.e. an average deviation angle of 45°. The difficulty in maintaining the profile together with instrument-control "side track" workover made it particularly difficult to drill this well, which was finally completed in December 1981.
- PCE 22 production well run into excessively high pressures in a Mano dolomite layer (namely 730 bars at a depth of 2715 metres), resulting in abandoning the profile, and side track workover. It was completed in June 1982, after 6 months of drilling.
- PCE 23 production well was drilled without encountering any major incidents, from July to October 1982. However, a production test made after acidification of the reservoir did not yield encouraging results, consequently leading to cancellation of the fracturation operation that had initially been planned. A well workover operation was required to recover a defective pressure probe in a side pocket at the bottom of the string of tubings.

It should be noted that the two production wells PCE 22 and PCE 23 are located on the surface on a production cluster situated 2 kilometres from the processing centre (cluster III).

Following the difficulties encountered in drilling wells PCE 13 and 23, the final well of the five-spot (PCE 24) was not drilled, in order to keep to the budget and planning requirements.

5.3 - Surface installations

Since it would have been too expensive to lay a gas line between Lacq and Pecorade, trucking of the CO_2 by road tanker was adopted as the best solution for this project.

The installations are dimensioned to ensure production, transport and injection of 80 tons of CO_2 per day.

The CO_2 is extracted from Lacq gas and then liquefied at a temperature of $-20°C$. It is then trucked by road tanker to the injection site 65 kilometres away, where it is again stored at $-20°C$. The injection is carried out by means of two high pressure pumping units.

Lastly, a separation installation specific to the pilot project was installed at cluster III for production wells PCE 22 and 23, enabling the pilot project wells to be gauged continuously, without disturbing the day-to-day exploitation of Pecorade reservoir.

6 - DESCRIPTION OF THE PROCESS

(Construction and erection of the installations is not covered by the EEC contract).

6.1 - Installations at the Lacq plant

The raw gas processing at Lacq provides a mixture of 60% H_2S and 40% CO_2, from which are extracted in the 22000 unit enriched acidic gas for the thiochemical units and the associated CO_2 gas containing 12 to 15% of H_2S. A modification of the process from MDEA sub-pressure washing has made it possible to produce CO_2 in accordance with the standards required for road tankers, namely 2000 Sm^3 of CO_2 containing less than 1000 ppm of H_2S. These characteristics were obtained in December 1982.

Next, the purified CO_2 is compressed to a pressure of 24 bars, dried and then liquefied at $-20°C$ by an ammonia circuit before being stored in two 175 m^3 tanks (note that all the storage and loading operations are controlled directly by an automated system). This unit, which was started in February 1983, was fully operational by mid-March at rated conditions, namely 100 tons of CO_2 per day.

6.2 - Installations at the Pecorade centre

The CO_2 pilot project storage and injection installations are grouped together at the processing centre (Cluster I) in order to satisfy the operating and safety requirements so as to be capable of permanently following the injection parameters.

The storage units (2 x 175 m^3) are identical to those of the Lacq plant and are supplied by a tanker truck shuttle service. The tank emptying and storage operations are also controlled by an automatic system.

The two pumping sets, one of which is kept on standby, are dimensioned so as to ensure an injection rate of 5 tons per hour, for a service pressure of 500 bars and an intake temperature of $-20°C$. At the outlet from the pump unit, the CO_2 is heated to $40°C$ and then injected into the well. This makes it possible to avoid excessively high thermal stresses on the equipment of the injection well.

Lastly, a recycling circuit with a return line to the storage facility has been installed in order to ensure a certain degree of flexibility with respect to the injection pressure and flow.

6.3 - Operation

These installations apply conventional industrial methods and neither production nor maintenance set any major problems.

However, following a gradual deterioration of the conditions of injection into well PCE 13 (mid-March, wellhead injection pressure: 420 bars for 80 tons of CO_2 per day ; mid-August, wellhead injection pressure: 480 bars for 60 tons per day), by the end of September 1983, a crack was observed in the hydraulic blocks of the CO_2 injection pumps requiring the process to be completely shut down on 2nd October.

The parts were examined in November and December, the conclusion being that the mechanical characteristics of the material were insufficient and that fatigue failure had occurred without any trace of corrosion. Pending replacement blocks, a repair was made in January 1984, enabling CO_2 injection to be continued from the end of January, until final shutdown on 7th March 1984.

The two new substitute blocks were available at the end of June for the final part of the Pecorade CO_2 pilot project programme: water drive in the reservoir.

7 - STARTING THE PRODUCTION WELLS AND FOLLOWUP OF THEIR OUTPUT

(Drilling these wells does not form part of the EEC contract).

7.1 - Starting production wells PCE 22 and 23
Following perforations and acidification, well PCE 22, which was completed at the end of May 1982, showed little eruptive pressure, with a high WOR. Clearing operations were carried out by lightening the fluid column with "coiled tubing", enabling production tests to be carried out in January and February 1983: 13 to 23 bars at the wellhead for a production of 24 m³ per day, of which 15% water. This well was closed during the second quarter of 1983 in order to follow the development of the head pressure and had to be resumed by pumping when it was opened again in June.

Drilling of well PCE 23, completed in October 1982, was followed by perforation and acidification of the reservoir in November. These operations had to be interrupted in December in order to retrieve a defective bottom pressure sensor located in the side pocket at the bottom of the production column. The failure of this instrumentation then required that the tubing string be fully extracted. This well was cleared and production tests carried out in March 1983 resulting in 16 to 20 m³ of anhydrous oil per day being produced for a very uneven wellhead pressure varying from 10 to 60 bars.

7.2 - Finalization of the separation installations
The low eruptivity of wells PCE 22 and 23 makes it impossible to send the effluents from Cluster II (where these wells are located) to the processing centre 2 kilometres away (30 bars required at the wellhead). Accordingly, these low production conditions required the separation process to be modified in order to bring the oil output as far as the centre.

It was hence decided to install a new low pressure separator (3 bars) at Cluster III, combined with recovery of the liquid effluents in two storage tanks (1 bar) for degassing and repumping of the effluents to send them to the processing centre.

7.3 - Followup of the wells - Conclusion
The work performed in production wells PCE 22 and 23 revealed the presence of a twin layer in the reservoir, also to be found at well PCE 4:
- a top limestone drain (thickness 4m ; K = 10 to 50 md),
- an intermediate non-permeable zone 20 metres thick (K = 0.1 md),
- a porous zone (thickness 20 m : K = 1 md).

The problem of placing this twin layer on production with single completion in the production wells was moreover a major handicap in any quantitative interpretation of the well results during operation of the pilot project, and this handicap was only partly overcome with the help of the logging carried out later.

It should also be noted that well PCE 22, the output of which is still low and irregular, had to be assisted by "coiled tubing" lifting and pumping. Despite these operations, the well could not be kept on production beyond 18th October 1983, and has been closed since this date.

8 - CO_2 INJECTION

This injection was carried out from March 1983 to March 1984. The total amount injected was 10700 tons, representing 10% of the porous volume subjected to CO_2 drive under the bottom conditions. It should be noted that the 1984 injection period is now undergoing interpretation and that the results will be recompiled during the final report at the end of the year.

8.1 - Production logs

A systematic series of measurements at the bottom of the wells was programmed in order to followup the month-to-month development of the main production parameters at the horizons perforated.

Measurements of the bottom pressure revealed the influence of water injection in peripheral well PCE 18, one kilometre South-East of the pilot project, with a pressure response lag to two months. On the other hand, on interrupting this injection of water, the production wells appear to remain affected only little by the CO_2 injection, despite its proximity, and the bottom pressure declined at a gradual rate.

The main information provided by the temperature measurements concerns the detection in the two production wells followed regularly (PCE 4 and 23), of a cold point at the lower perforations. This part of the reservoir corresponds to the porous zone of the Barremian and this temperature anomaly is interpreted as the effect of arrival of free CO_2 at the well.

However, it should be pointed out that whilst a general trend is to be seen for these measurements as a whole, they often revealed wide scatter and considerable divergence from well to well. Furthermore, during long interruptions of injection (the first two weeks of August 1983 and then from October 1983 to February 1984), considerable changes in the CO_2 drive were detected and made any correlation test for the injection period as a whole a delicate matter.

8.2 - Gaugings and samples

The objective of this work is by correlating the samples taken and the bottom measurements, to follow the evolution of the composition of the fluids produced. In this way, the samples taken at regular intervals on the separator outlets of the effluents have made it possible to check the CO_2 contents, the arrival of the tracers at the production wells and the pattern of change in these parameters with time.

These analyses confirmed that CO_2 drive does not propagate uniformly, but develops along a network of cracks in well defined directions.

For instance, in PCE 04, a steady increase in the CO_2 contents was observed up to the end of June 1983 (31% CO_2), after which this parameter gradually declined throughout the entire 6-month period, levelling out at the end of the year to a CO_2 content in the unprocessed oil of 6.5%.

It would seem that resumption of injection of CO_2 in mid-August (following two weeks interruption) was of no significant effect on the production of this well. At the same time, well PCE 23 kept a 30% CO_2 mole content until October, then steadily declining, apace with interruption of injection, in PCE 13. Nonetheless, the CO_2 content of the separator gas was still 18% in terms of moles at the end of December.

At PCE 22, no appreciable change in the CO_2 contents was observed. However, the low production level, the damage to the layer during the drilling operation and the absence of production logs mean that one can not be affirmative in stating that the CO_2 has not spread further down the dip.

8.3 - CO_2 breakthrough

One of the main parameters followed continuously throughout the exploitation period concerns the detection of CO_2 breakthrough at the production wells. With this in view, right from the first days of injection, a 300 ton CO_2 slug containing SF6 as tracer was injected in March 1983. As for the production wells, they were kept closed for 3 months, whilst following the head pressures regularly, in order to check the pressure response resulting from CO_2 drive in the matrix conditions.

As soon as the well was opened in June 1983, apart from the major difficulties in placing wells PCE 22 and 23 on production, considerable CO_2 contents traces with SF6 were detected in the effluents produced. Accordingly, it transpires that in less than 3 months, the CO_2 had already exceeded the limits of the pilot project, making any truly accurate estimate of breakthrough impossible. This very rapid CO_2 breakthrough at the production wells hence calls into question the previously accepted theory of a matrix sweep from the injection well.

When the pilot project was resumed in February 1984 (after 4 months interruption), CO_2 breakthrough occurred and was detected 48 hours after starting injection at well PCE 23 and after 10 days at well PCE 4.

9 - CONCLUSIONS

Injection of CO_2 into well PCE 13 (bottom pressure about 700 bars) resulted in fracturation of the porous zone of the reservoir.

Allowing for the breakthrough times, this fracturation appears to have extended as far as the vicinity of production wells PCE 23 and PCE 4.

These observations would point to the fact that CO_2 drainage of this part of the reservoir is liable to be of little effect, if it be confirmed that it has already reached the confines of the pilot project in gas form.

The production logs clearly revealed a twin layer behaviour of the production wells, but did not enable figures as to the distribution of the flows to be determined between the drain and the porous zones. The presence of CO_2 in the upper drain was nonetheless distinctly revealed when the lower perforations of the porous zones were plugged by laying a cement plug (operations performed in April 1984).

All the facts observed so far (fractured conditions, presence of CO_2 in the drain, arrival of free CO_2 at the production wells...) make all overall interpretation of the results a particularly delicate matter. Accordingly, the initial factors of synthesis will not be available until the end of 1984, when the general consolidated results for the Pecorade CO_2 pilot project are established.

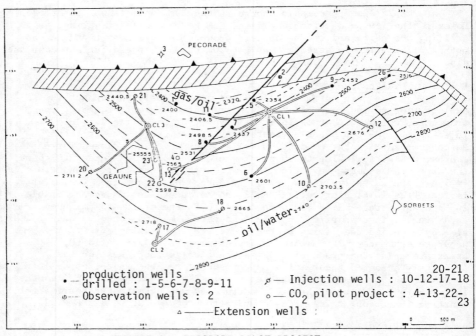

FIGURE 1 - PECORADE PILOT PROJECT
Isobaths at roof of B2 dolomite Barremian

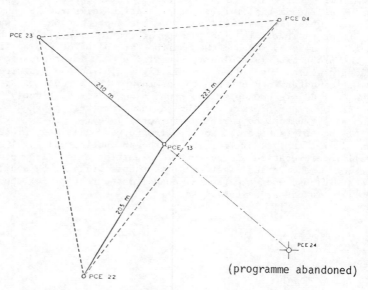

FIGURE 2 - PECORADE PILOT PROJECT
Layout of wells

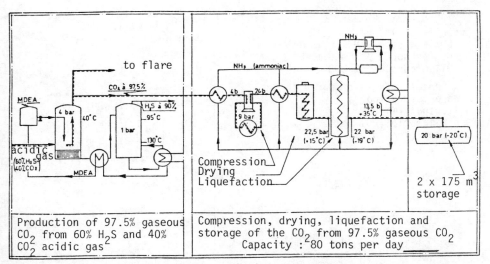

Production of 97.5% gaseous CO_2 from 60% H_2S and 40% CO_2 acidic gas	Compression, drying, liquefaction and storage of the CO_2 from 97.5% gaseous CO_2 Capacity : 80 tons per day

FIGURE 3 - PECORADE PILOT PROJECT - Liquid CO_2 production unit at Lacq

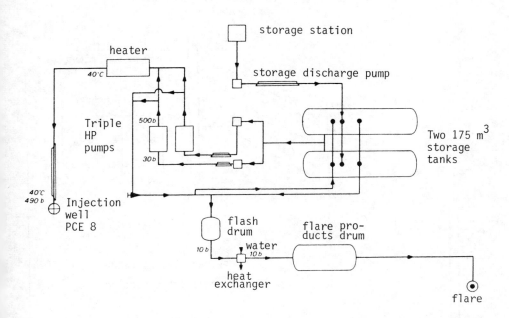

FIGURE 4 - PECORADE PILOT PROJECT - Flow process diagram of the Pecorade installations

PECORADE pilot project - CO_2

Overall results on 6.12.1983
    CO_2 injected   9 150 t
    CO_2 produced   450 t
    Oil produced   7 000 t

CO_2 content of gas produced (moles)
Head pressure (bars)
Oil output (m^3/day)
Gas-oil ratio (vol/vol)
CO_2 flow (tons/day)

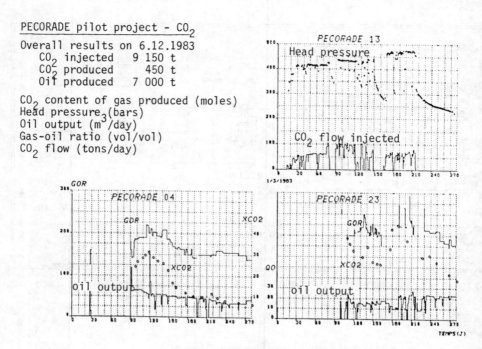

FIGURE 5 - PECORADE PILOT PROJECT

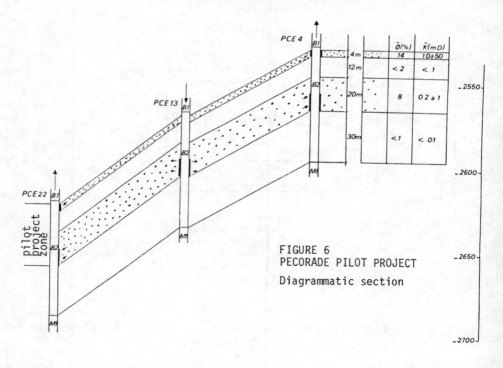

FIGURE 6
PECORADE PILOT PROJECT

Diagrammatic section

(05.34/82)

NITROGEN INJECTION IN NORTH SEA RESERVOIRS

R.S. HEARN and M.G. BAYAT
Britoil plc

SUMMARY

Studies are in hand to examine the technical and economic feasibility of using nitrogen as an EOR medium in North Sea reservoirs : this paper outlines the status reached in the studies and discusses a proposed plan to continue the work.
The initial work consisted of preliminary studies carried out in three phases (each based on real field data), viz:-
- PVT experimental work
- reservoir simulations
- process design, technical evaluation and cost estimating.
The PVT experimental work has been completed and has provided physical data for use in the reservoir simulations.
An initial reservoir simulation indicated that further work was required to overcome the limitations of reservoir modelling. A preliminary probabilistic reservoir analysis was therefore carried out using the results of a previous simulation and of the PVT experimental work. The results of this indicated that additional, detailed, reservoir simulation studies were justified.
The process studies have so far been limited to assessing the technical feasibility and making provisional cost estimates of suitable facilities. These provisional estimates might suggest that in economic terms nitrogen injection is only a borderline proposition for the field studied. However, as the estimates were based on preliminary data they must be considered as incomplete. From discussions with other companies there are indications that economic considerations would favour a cooperative scheme in which nitrogen is supplied from a central source to a number of different injection projects simultaneously.
It was generally concluded from the various preliminary studies that the results were sufficiently encouraging to justify continuing the work. Further studies have therefore been initiated, aimed at reducing the various uncertainties by means of a full-field 3-D model and, hopefully, leading through a field trial to an optimised and viable nitrogen injection scheme. The steps in the proposed programme are listed and discussed.

1. INTRODUCTION

Background
     Of the oil originally in place in the reservoirs of the UK
Continental Shelf, only about 40% can be recovered by conventional
production methods. To recover any part of the remainder would
necessitate using more-sophisticated techniques, such as Enhanced
Oil Recovery. There are a number of such methods for Enhanced Oil
Recovery (EOR) already in operation in other areas, or under study,
including the injection into the reservoir of a suitable gas, the
most common used being natural gas and CO_2.
     Nitrogen has also been used as an alternative injection gas in
several applications in the USA. It has many advantages over CO_2,
being relatively inert, non-corrosive and recoverable from the
atmosphere, and, unlike natural gas, has no alternative fuel value.
     This paper describes the work carried out to-date on a study to
investigate the feasibility of injecting nitrogen in North Sea
reservoirs as a means of increasing ultimate oil recovery. It also
outlines a proposed programme of future work that is considered
necessary to bring the studies to a hopefully successful conclusion.
Scope of the Studies
     It was originally envisaged that the study would be carried out
in three separate, but simultaneous, phases, viz:-
I.   PVT experiments and measurements (at Heriot-Watt University);
II. reservoir identification and simulation;
III. process engineering, cost and economic evaluation.
The main steps of the work undertaken as part of the original study,
and referred to herein as the "Screening Study", have been
completed, and the outcome of each of the various steps is described
in the paper.
     It became apparent as the study progressed that it would have
to be extended beyond what was originally envisaged. The results of
the Screening Study are therefore discussed and used to develop a
proposed programme of future work on Nitrogen Injection, which is
also described. Reference is also made in the paper to the views
and activities of other companies and organisations where these are
considered relevant.

2. THE SCREENING STUDY

A. PVT EXPERIMENTAL WORK AT HERIOT-WATT UNIVERSITY

     Phase behaviour studies of nitrogen and field crude oil were
carried out in the PVT Laboratory at Heriot-Watt University
Petroleum Engineering Department. These were to provide data on
fluid properties and on changes resulting from nitrogen contact.
     Nitrogen dissolving in the reservoir fluid will firstly reduce
oil density and viscosity and lead to some oil swelling. Further
ingress of nitrogen will eventually result in the formation of a
gaseous phase, once saturation conditions have been reached. This
gaseous phase will contain some hydrocarbons, through vaporisation
from the oil, again resulting in changes in density and viscosity of
both phases. Flow of the enriched gas through the reservoir system
will cause additional hydrocarbon vaporisation. Close to the
injection wells gas stripping will result in increased residual oil
viscosity and density.

The experimental programme first involved examining nitrogen solubility at different pressures at reservoir temperature and measuring the resulting changes in density and viscosity. In order to study reservoir fluid behaviour, experiments were performed at a variety of simulated conditions. Multiple contacts of gas with fresh batches of oil allowed the changes in gas properties on moving through the reservoir system to be measured. Multiple liquid contacts were also performed to examine the changes in liquid properties after repeatedly stripping the oil with fresh nitrogen gas.

In general, results were beneficial, with reductions in density and viscosity of the oil, increases in gas viscosity and build-up of an enriched gas front. Although nitrogen could not achieve miscibility with the oil under reservoir conditions, it moves in the direction of miscibility in what amounts to a reduction in the differences in properties between the two phases. Repeated stripping of the oil with nitrogen led to increases in liquid density and viscosity. However, these effects were not extreme and, importantly, no asphaltene precipitation occurred. It is not considered that these results would mitigate against a gas injection scheme.

The fluid properties and their changes were represented by a comprehensive phase behaviour package based on equations of state. These were adjusted to reproduce the experimental results with acceptable accuracies.

B. RESERVOIR MODELLING AND SIMULATION

Attempts were made to simulate the nitrogen displacement process by means of a 2-D, cross-sectional, numerical model using a modified black oil simulator. In the model, N_2 was assumed to be injected downdip into the reservoir when the water injection scheme had reached a mature stage, indicated by the producing watercut.

Although some indications could be obtained regarding the displacement and recovery process, e.g., N_2 was clearly seen to be sweeping the top layer rapidly, it was not easy to quantify the process for the following reasons:-

(a) There was difficulty in understanding and modelling accurately three-phase flow because of lack of suitable three-phase relative permeability data. Simplistic assumptions had to be made which could be grossly wrong. Another difficulty encountered with three-phase relative permeability was in modelling the reduction in residual oil saturation.

(b) The PVT study had indicated that 224 standard cubic feet of N_2 would dissolve in a barrel of oil, on the assumption that there was complete mixing between the two phases. However, not all the N_2 injected would contact the oil in the reservoir to allow complete solution to take place. Furthermore, the difficulty in quantifying the extent of contact between N_2 and the oil makes it correspondingly difficult to interpret the breaking through of N_2 to the producer.

(c) PVT phase behaviour suggests that as a result of mixing N_2 with field crude, the PVT properties of the resultant crude would be altered. This change in properties could only be modelled by using a compositional simulator and not with a black oil simulator.

Consequently, it was decided that the estimate of additional recovery and of N_2 required should be conducted by means of analytical methods. A probabilistic approach would enable sensitivities to be examined and a most likely case to be identified.

Additional reserves resulting from N_2 injection would be made up of three components:

i)   unswept oil in the attic reservoir;
ii) by-passed oil in unswept relatively low permeability sands;
iii) waterflood residual oil that would be mobilised.

Some preliminary estimates could be made by incorporating waterflood results from previous simulation studies.

The additional recovery obtainable by nitrogen injection, as estimated by the probabilistic method, was considered encouraging enough to justify further, detailed, 3-dimensional, 3-phase reservoir studies.

C. TECHNICAL FEASIBILITY AND COSTS

Bases Used

During the early stages of the screening study a preliminary evaluation was made of (i) the technical feasibility of recovering and compressing nitrogen and injecting it into an offshore reservoir and of (ii) the costs and economic attractiveness of doing so. As, at the time, there were no firm data available from the reservoir simulation studies on which to base these, they had to be started from assumed conditions and the cost estimates made for a wide range of plant sizes and other configurations.

For the purpose of assessing the technical feasibility of nitrogen injection four methods of making nitrogen available on an offshore platform were selected and examined:

a) Installation of a cryogenic plant on an existing offshore platform.

b) Onshore nitrogen separation with pipeline distribution to an offshore platform or platforms.

c) Extraction from turbine exhaust gases or by inert gas generators. Production of high purity nitrogen from turbine exhaust gases is technically complicated and more expensive than conventional cryogenic separation of air to produce nitrogen gas. For these reasons the method was not considered further.

d) Transport of cryogenic liquid nitrogen by tanker to an offshore location.
The quantity of liquid nitrogen which would have to be available and transported offshore to a North Sea field was estimated to be in the region of 1600 tons/day and this quantity is more than the present total UK production. Furthermore, the production of liquid nitrogen for a gaseous application is inherently expensive and was considered unlikely to be viable. It too was not studied further. (But the use of this route for limited field trials cannot be ruled out.)

For the purpose of providing the basis for assessing the economics of nitrogen injection, two of the foregoing four production methods (a and b) were costed in some detail. Equipment weights and sizes were obtained informally from UK suppliers, and the costs developed on the basis of supplying nitrogen with an oxygen content of not more than 10 ppm.

For case (a) – nitrogen generation on a platform – it was assumed that the facilities would be installed on an existing platform.

For case (b) – nitrogen generation onshore – the costs were based on the assumption that the nitrogen production facilities would be located at Sullom Voe and that high pressure nitrogen would be transported 130 miles by 12-inch pipeline to a typical steel platform in the Viking area, on which nitrogen receiving and injection facilities had been installed.

Technical Conclusions

The conclusions from the feasibility assessment were that :-
- the combination of nitrogen generation, whether onshore or offshore, and nitrogen injection from an offshore platform is technically feasible,
- the preferred generation process is likely to be cryogenic distillation of air,
- in the case of offshore generation, some aspects of this would constitute an extension of proven technology to an untried application and operating environment.

The implication of the last of these conclusions is that there could be special technical problems associated with offshore nitrogen generation and injection that have not yet been fully evaluated, but which would require detailed investigation before a project were implemented.

The following are among the technical parameters that need defining more-closely and the possible problem areas that need investigating that we have identified :
- the acceptable level of nitrogen purity,
- the ingress of chlorides and hydrocarbon gases into an offshore cryogenic plant,
- the zoning of equipment for safety considerations,
- the disposal of oxygen-enriched air,
- train type and configuration for nitrogen compression,
- structural implications of the installation, wind loading, contact with cryogenic liquids, etc...

Cost Estimates

On the bases of the capital costs for the schemes outlined in the foregoing, it was estimated that the cost of nitrogen injected into the reservoir ranged from about £1.30 to £2.60 per thousand cubic feet. The corresponding costs of additional oil would depend, of course, on the applicable oil/nitrogen recovery factor, which at the time was – and is still – unresolved. Preliminary indications based on general assumptions then valid were that these costs could range from £10 to over £30 per barrel.

General Conclusions

From the foregoing preliminary indications of the possible cost of additional oil, a general conclusion could be drawn that for the cases and conditions examined nitrogen injection might be economically viable, but is unlikely to be very much better than borderline.

However, it was also recognised that the cost estimates were based on data of a very provisional nature and, as such, any conclusions drawn from them must themselves be considered provisional.

It was decided that no further work of this nature would be undertaken until reservoir engineering studies indicate that a reasonable amount of additional oil can be recovered. Such data as were available were sufficiently encouraging to warrant a continuation of the reservoir engineering studies.

Joint-Industry Injection Schemes

By the time that Britoil's Screening Study was being evaluated, information was becoming available on the views and activities of other Companies and of Government Organisations. These tended to support the view that small, individual nitrogen injection schemes are less likely to be as economically attractive as multiple schemes, in which several simultaneous injection projects are supplied with nitrogen from a central recovery plant.

This aspect has been discussed industry-wide through the aegis of the British Department of Energy, who has undertaken to study this aspect further.

One of the main problems that might confront a central generation scheme could be if the participating projects have widely different starting-up dates.

3. CURRENT AND FUTURE WORK
BACKGROUND TO THE PROPOSED PLAN

It has been established in earlier sections of this paper that the results of our Screening Study, though inconclusive, were sufficiently encouraging to justify further work on the evaluation of nitrogen injection.

It was therefore agreed within Britoil that the studies should be continued. We are very pleased to say that these views were shared by the Commission, who agreed to continue to help fund this work. But in deciding on the future programme, due cognizance had to be taken of the experience gained in the Screening Study and from information gained from outside sources. In particular, we had to pay particular attention to :-

i)   the indications that economic considerations would probably favour a scheme in which nitrogen from a single recovery plant is supplied simultaneously to a number of different injection projects, possibly even to different operators;

ii) the lack of any precise understanding of the mechanism of the fluid flow through the reservoir for a nitrogen/oil system; and because of this

iii) the difficulty of simulating the dynamics (and thermodynamics) of the flow mechanism in a computer model;

iv) the similar difficulty of checking any assumptions made in the absence of a field pilot test.

As regards the first of these aspects, we recognised that this introduces features that can only be resolved in consort with others. It was decided, therefore, to keep in contact with those other Companies who have a potential interest in nitrogen injection and, in particular, with the Department of Energy, and await the outcome of any proposals for cooperative schemes sponsored by the latter.

But as regards the other aspects listed above, since these are reservoir-oriented and predominantly field-specific we felt that they had to be resolved for each specific field by each respective Company concerned. For Britoil, this implied continuing this work through a series of in-house reservoir studies, possibly leading up to a field pilot test. At each key stage of the work an economic check would need to be made to ensure that we were justified in carrying on with the studies.

The work would be based initially and primarily on the Etive and Tarbert sands that had already been identified as potentially suitable. Later work could be aimed at trying to identify other suitable sands.

It has been mentioned that a logical major step in assessing the viability of nitrogen injection would be to carry out a field pilot test. But to do so offshore, with nitrogen, would require installing very expensive facilities that might have no further use.

A possible alternative would be to use natural gas, which would be available at the time in the area, as the injection medium, instead of nitrogen. But it would be necessary, of course, to ascertain by means of the reservoir study that the effects of nitrogen injection could be assessed with the reservoir model from the results of a pilot test using natural gas.

Part of our proposed study, therefore, is to establish in the model, and rigorously check, parameters that would enable us to interpret the effects of nitrogen injection from results obtained in field trials with natural gas.

BRITOIL'S PROPOSED PLAN

Field observations and earlier reservoir studies have indicated that oil is being under-run by water in the Etive and possibly the Tarbert sands. The ultimate objective of the studies now being undertaken is to ascertain whether it is possible to displace this attic oil by injecting natural gas or nitrogen and thereby increase oil recovery. A detailed reservoir simulation study is a necessary prerequisite of such an investigation.

We propose to tackle this investigation in a number of stages, as listed below, with economic checks carried out at each stage :-

1. Develop a Model of a Fault Block where Gas Has been Injected

: To history-match the gas injection that had already been in operation at one time. Data from this review will be used in stages 2 & 4. This stage of the work has been started.

2. Modelling of the Pilot Area : To select a suitable pilot area for testing the injection process in the Etive sand and, using the simulation model, predicting its performance against reservoir parameters derived in Stage 1.

3. Evaluation of Stage 2 : To assess the results of Stage 2 to ensure that these justify continuing the study.

4. Full-Field Modelling : To predict the performance of the Etive sand over the entire field under injection.

5. Economic Evaluation : To carry out an economic assessment of the results so far, so as to ascertain whether nitrogen injection appears viable and that a pilot test is warranted.

6. Design and Cost Pilot Test : If the pilot test is shown to be useful, to design and cost the facilities and procedures for it.

7. Engineer and Install Pilot : Mainly pipework.
   Test Facilities

8. Carry out Pilot Test : Using natural gas as substitute.

9. Evaluate Pilot Test :

10. Modify Full-Field Model : As may be necessary.

11. Carry out Detailed : To decide whether nitrogen
    Economic Evaluation : injection is viable.

The time phasing of the proposed work programme allows for a decision as to whether to go ahead with a pilot test being taken in about May/June 1985 and, if justified, implementation of the test by late 1985. Evaluation of the field trial could commence in early 1986 and this would complete the work carried out under the EEC contract.

We believe that the programme presented represents the most effective procedure that we can devise while trying to keep costs within reasonable bounds. We also believe that this work could make a very valuable contribution towards assessing whether nitrogen would be a suitable medium for increasing the ultimate recovery of North Sea hydrocarbons economically.

(05.09/78)

SEARCH FOR DRAINAGE METHODS FOR HEAVY AND VISCOUS
OIL RESERVOIRS INVOLVING THE RISK OF WATER
INFLOWS - APPLICATION TO ROSPO MARE

R.H. COTTIN   ) ELF AQUITAINE Research and Applications Division
A.G. BOURGEOIS) and ELF ITALIANA Operations Division

Summary

The objective of the studies was to select the best way of draining
the accumulation of heavy oil at ROSPO MARE, situated in a karstic
reservoir with no matrix porosity and an active underlying aquifer.
After geological synthesis and elaboration of a structural image of
the karst, the studies on a high performance mathematical model
resulted in encouraging simulations for exploitation through a
horizontal well, compared with conventional exploitation methods. It
was decided to add a third horizontal well to the two planned on the
experimental production platform, where tests were started in mid-1982.
The present results of these tests corroborate well the advantage of
the horizontal well, which, by increasing the chances of intersecting
the oil fractures in the karst without approaching the water level,
permits anhydrous oil extraction at a high critical flow rate. The
additional studies originally planned were relegated to a lower level
of importance owing to the very unattractive "cost/efficiency ratio"
(injection of carbon dioxide gas, steam, ...).

INTRODUCTION

Rospo Mare is a special case of a reservoir owing to the conjunction
of the following three main factors:
- situation in relative deep offshore waters (60 to 100 metres - 20 kilometres from the coast),
- a karstic reservoir without a permeable porous matrix and with an active
underlying aquifer,
- its heavy and viscous oil.
The risks of extremely fast breakthrough of water with flooding of the
production wells were evident and would suggest that conventional
development would not be economically profitable. An ambitious research
programme was hence decided to select the most appropriate drainage method
and was started in 1978 when only very little data was available on one
well (RSM1). Accordingly, the various phases of this research programme
were dealt with in widely differing ways ; the first two "elaboration of
a specific model of the reservoir" and "studies of horizontal drillings"

absorbed most of the effort, whereas the following phases were hardly touched upon, owing to lack of information and/or interest. The studies never took on a general nature, since they dealt with the specific case of ROSPO MARE, but the main results can be transposed to any reservoir of similar characteristics (heavy and viscous oil with a risk of motor breakthrough).

1 - ELABORATION OF A SPECIFIC MODEL OF THE RESERVOIR

The initial idea was to generate from a highly thorough geological study synthesis a specific mathematical model of the karstic reservoir capable of processing dynamic problems ; this idea was abandoned and we consider as a first approach, simulation of the microscopic flows can be undertaken validly with conventional models, since there is no fundamental difference between flow in a conventional porous medium consisting of solid grains and pores and flow in a karstic medium consisting of compact boulders and voids. In addition, the general equations of thermodynamics and fluids mechanics (Darcy, diffusivity) on which the conventional models are based, appear to be legitimately applicable to the case of karsts, with the following provisos:
- one remains in laminar conditions;
- the meshes of the model have dimensions that are distinctly greater than those of the karstic blocks;
- the parameters be adapted to the karsts (low porosity, high permeability, zero capillary pressures, sudden relative permeabilities resulting in a "piston" front, initial water saturation and residual oil saturation low or zero;
- it be checked that this adaptation is compatible with the internal operation of the model.

The geological synthesis work was maintained, since it made a contribution towards the decision taken to operate in stages, to prove or to disprove the possibilities of economic development of this reservoir : Assessment well - Production pilot - Long duration tests.

1.1 - Geology

The ROSPO MARE reservoir is situated in limestone massifs of the lower cretaceous, more than 350 metres thick, and in eocene breaches resulting from the break-up of these limestones, of variable thickness (0 to 150 metres). This combination, together with the underlying jurassic, forms a limestone mass of over 1000 metres. These carbonates have a non-impregnated impermeable matrix, but are karsticized at the summit for about 100 metres, the karstic voids and the fractures alone being impregnated with heavy and viscous oil. The genesis of this karstic massif is complex.

The plateau is isolated by the rise of the miocene waters (transgression). At the top, the karst is limited by an irregular erosion surface, the reservoir may even disappear, being replaced by fossil valleys, filled with miocene sediments that do not generally form a reservoir. The roof of the karst drops at a gentle gradient (about 2%) towards the north. On the edges, the karstic plateau is limited in the same way by the paleo-topography at the edge of the miocene sea. The area of 150 to 200 km² of the karstic plateau was needed to collect the waters producing the karsts, though this does not guarantee that these 150 km² are karsticized, and the useful area may be less. We hope to be able to delineate the spread of this reservoir by means of three-dimensional fine seismics and a number of assessment drillings: an anhydritic term of the Messinian of the Miocene series indeed considerably detracts from the effectiveness of two-dimensional seismics.

A – Characterization of the karst

By comparison with other karsts, the ROSPO karst is a:
- gravific karst, the essential driving elements of the flows being gravity (the breaches and fractured limestones being easily penetrable),
- hot climate karst, the predominant corrosion of which matches the oligocene karsticization that is well developed in the perimediterranean area,
- tabular karst of the karstic layer type, in the FORD classification,
- evolved karst that has reached this maturity by the extent of the purely karstic porosity and the hierarchization of the system from upstream to downstream.

B – Organization of the karst

The karstic reservoir comprises the following from top to bottom:
- the upper infiltration zone or epikarst (15 to 20 metres) offering the maximum karstic voids, but also the maximum number of clogged voids,
- the lower infiltration zone, or vertical transit zone (35 to 80 metres) in which there are little and few dissolutions,
- the upper flooded zone (20 to 30 metres) formed by more or less clogged galleries winding along the lines of distension of the massif (bayonet sub-horizontal path) moving towards their exit at the edge of the karstic plateau,
- the lower flooded zone, or deep karst, which may be up to 200 metres thick, in which water flows are slow, resulting in a very low rate of dissolution.

C – Drainage polarity

Thorough sedimentological study of the various wells drilled leads to the conclusion that there is a gradual reduction in the marine influences from North to South, the veritable mark of the transgression of the Miocene on the drainage paleo-network of the karst. The correlation of the vertical zonings of the karst reveals this privileged direction of drainage of the functional karst from South to North. However, there is another direction of fracture perpendicular to this, namely directed from East to West.

1.2 – Modelization

In view of what has just been stated, various simulations were made on conventional models, though all came up against the same stumbling block: the calculations became unstable* immediately on breakthrough of the water at the production wells and it became impossible to continue the simulations for hydrated production. Despite several attempts to reduce the time steps and the size of the meshes in the production zones, two high resolutions have to be used:
- the TRITRI in its IMPIMS version (implicit pressure – implicit saturation),
- the very high performance MEPHISTO model with fully implicit resolution.

The first model (in the RZ 2C biphasic cylindrical version) was used to interpret the few days of production tests made when drilling the assessment well RSM2 (second half of 1978), where the karsticization is unusual in that it is interrupted over a certain height of the reservoir

---

* The conventional models lent themselves to the extreme conditions of the ROSPO MARE karst only with difficulty (porosity = 1%, permeability: several Darcy's or even several tens of Darcy's, mobility ratio over 500 and relatively abrupt permeabilities).

with the existence of an impermeable bed (clay): the simulation indicated breakthrough after 80 days of extraction at a rate of 250 m³ per day, i.e. 20,000 m³ of production, immediate flooding of the well and extremely low reabsorption of the water cone owing to the low density contrast between the oil and the water phases. These scarcely encouraging results guided the studies towards the solution of a horizontal well in which one could a priori hope to delay the breakthrough of the water thanks to a lower aquifer pressure than in the case of the vertical well.

2 - STUDY OF SUBHORIZONTAL DRILLINGS

A major preliminary study on a numerical model (1979 and 1980) the purpose of which was to compare the efficiency of horizontal wells against a conventional solution yielded sufficiently encouraging results to justify the decision to drill an experimental platform to comprise two partial penetration wells (one vertical, one deviated) and a third horizontal well. The main results of the comparative study are listed in the tables below, and can be summarized as follows:

VERTICAL WELLS

PARAMETER STUDIED	N° OF CASES	NUMBER OF WELLS N	DATA FLOW PER WELL Q, m³/day	THICKNESS OF TEAR E, m	BREAKTHROUGH TIME Tp, day	RESULTS BREAKTHROUGH RECOVERY Np, m³	RECOVERY RATE %
NUMBER OF WELLS N	7	1	250	1000	47	11 700	0.71
	8	4			46	46 700	2.83
FLOW	11	1	50	1000	226	11 300	0.68

HORIZONTAL WELLS

PARAMETER STUDIED	N° OF CASES	DATA OVERALL FLOW Q, m³/day	LENGTH OF WELL L, m	THICKNESS OF TEAR E, m	BREAKTHROUGH TIME Tp, day	RESULTS BREAKTHROUGH RECOVERY Np, m³	RECOVERY RATE %
	2	1000			63	63000	3.82
FLOW Q	3	250	500	1000	253	63200	3.83
	1	100			670	67000	4.06
	9	50			1635	81700	4.95
LENGTH OF WELL L	2	1000	500	1000	63	63000	3.82
	4	1000	1000	1000	117	117000	7.09
WELL SPACING E	2			1000	63	63000	3.82
	5	1000	500	500	56	56000	6.79
	6	1000	500	250	56	56000	13.58
	10			100	43	43000	26.06

2.1 - Study on TRITRI model with semi-explicit resolution (1979)
- The critical flow from a horizontal well 500 metres in length is twice that of a vertical well ; under hypercritical conditions, the recovery on breakthrough in this well is 5 tons greater than that of the vertical well.
- For both horizontal wells and vertical wells, the recoveries increase when flows close to the critical value are adopted.
- As the well spacing decreases, the interference effect causes the absolute recovery of each well to drop, but increases the overall recovery rate in the reservoir zone concerned.

2.2 - Study on TRITRI model with implicit resolution (IMPIMS version - 1980)

This study is intended to simulate without numerical instability what occurs following breakthrough in the water and its results are collated into the following table.

It is assumed that the vertical well and the horizontal well each produce 250 m³/day.

WOR	CUMULATED OIL PRODUCTION (m³)		HORIZONTAL/VERTICAL
	HORIZONTAL WELL	VERTICAL WELL	RATIO
BREAK-THROUGH	53 500	10 500	5,1
1	56 500	13 600	4,2
2	58 800	15 400	3,8
3	60 500	16 800	3,6

It would appear that the advantage of the horizontal well over the vertical well tends to diminish following breakthrough of water, where production is almost similar in either case, whereas before breakthrough, a ratio of 5 to 1 in favour of the horizontal well existed.

2.3 - Tests on experimental platform and interpretation

After installing an experimental production platform, three wells, one vertical (RSM4), the other deviated (RSM5) and the third horizontal (RSM6), were drilled, with the vertical penetration beneath the roof of the reservoir not exceeding 30 metres (see layout of wells). Pumping was started on RSM4 on 21st August 1982, RSM5 on 23rd September and RSM6 on 26th October. The initial interference tests of the production on wells 4 and 5 (*), the pressure surges and the critical flows yielded a series of interpretations, the main results of which are shown in the following table and lead to better knowledge of the petrophysical characteristics of the reservoir.

A - The average total porosity of the karsts would be about 1.5% and their mean dynamic porosity $\emptyset$ (1-SWI-SOR) would be from 0.7 to 1.2%.
B - The permeability of the karsts would vary widely:
. about 8 Darcy in the vicinity of RSM1,
. about 2 Darcy in the vicinity of RSM2,
. about 30 to 50 Darcy in the vicinity of RSM4,
. about 5 Darcy in the vicinity of RSM6.
C - The conductivity of the aquifer in the vicinity of RSM4 would attain 320 Darcy x m.

(*) The tests of well 6 were studied under contract 05.36/82.

WELL	PRODUCTIVITY INDEX m³/day/bar	MEAN POROSITY Ø %	Ø(1-Sw-Sor) %	MEAN PERMEABILITY Darcy	TEST	TYPE OF INTERVENTION
RSM 1	2.3			8	build-up test N° 2 1,395 - 1,405 m	Diffusivity Eq
	8			9.5	build-up test N° 3 1,435 - 1,446 m	Diffusivity Eq
RSM2	5.2			2	extended test	TRITRI-RZ2C model
		1.5				Electrical logs
RSM4	7			100	initial build-up	Diffusivity Eq
			1.2	31	extented tests	Water-coning laws
			0.7	50	extended tests	MEPHISTO RZ2C model
RSM5	1.5			5	initial build-up	Diffusivity Eq
			1.0	5	extended tests	Water-coning laws

D - An initial output of 300 to 600 m³/day was obtained by pumping on experimental well RSM6 and resulted in breakthrough after a cumulated production of 75000 m³. For RSM4, which was pumped at 230 m³/day, break-through took place after producing 8000 m³ of oil. These results are in quite remarkable agreement with those obtained by simulation (see table in paragraph 2.2). After breakthrough, the simulations turned out to be pessimistic, since the extended tests on the experimental platform wells revealed the existence of critical flows below which it was still possible to produced anhydrous oil. The critical values at the end of 1983 were 110 m³/day for RSM4, 20 m³/day for RSM5 and 440 m³/day for RSM6. In addition, it appeared that the water inflows are remarkably reversible. As soon as one drops below the critical value, one can immediately revert to anhydrous production. Unfortunately, the lack of dehydration and desalination facilities at the site did not enable the extended evolution of the water percentage after breakthrough to be explored.

E - In the theoretical studies of F. GIGEL (French Petroleum Institute Review - May/June 1983), a 600 metre horizontal well should have a productivity index (PI) 6 to 7 times greater than that of a vertical well perforated through 30 metres, assuming that the reservoir is homogeneous and isotropic. Now, the PI of RSM6 is around 300 m³/day/bar, i.e. 40 times that of RSM4. This hence proves that RSM6 has crossed through intense karsticization zones, the permeability of which is above the mean value around RSM4. This result is logical, since the probability of intersection of the drainage network of the karst is distinctly higher for the horizontal drain (a 500 metre drain) than for the vertical well (a 30 metre drain).

F - At the end of 1983, RSM6 had produced 4 times more oil than RSM4, which is a remarkable performance, confirming the major advantages of horizontal wells in the case of ROSPO MARE.

3 - OTHER STUDIES

A - Despite a good fluidization and inflation effect of the CO_2 on the crude of the reservoir, injection of carbon dioxide gas at ROSPO MARE is scarcely attractive:
- an enquiry revealed the existence of numerous sources in the region of the Abbruzzy, but none in the immediate vicinity of ROSPO MARE,
- overall, the efficiency of a CO_2 sweep should be less than that of water, owing to the yet less favourable mobility ratio and the "umbrella" phenomena at the roof of the layer.

B - Study of steam injection was hardly touched upon, owing to the depth of the reservoir, its situation offshore and its very low porosity. The presence of an aquifer that would "gobble up" calories would lower the thermal efficiency of the operation yet more.

C - The "study of variants" and "study of aquifer slowdown processes" were left aside, since the necessary information to orient the work was lacking.

4 - CONCLUSION

As a whole, the studies carried out under this contract enabled a method of drainage, the horizontal well, remarkably well adapted to optimum extraction of a heavy and viscous oil situated in a relatively thin karst with an active underlying aquifer, which is the case at ROSPO MARE. The long duration tests still going on on an experimental platform are a striking demonstration of this.

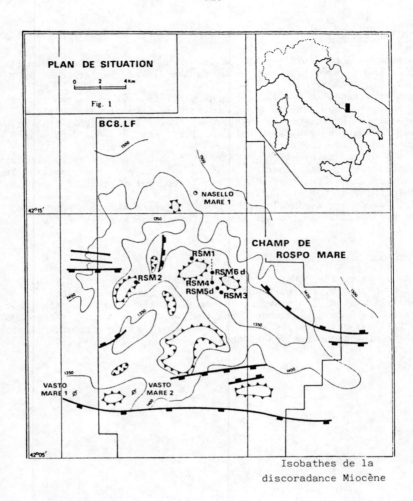

PLAN DE SITUATION

0   2   4 Km

Fig. 1

BC8.LF

NASELLO MARE 1

CHAMP DE ROSPO MARE

RSM1
RSM6 d
RSM2
RSM4
RSM5d   RSM3

VASTO MARE 1

VASTO MARE 2

42°15'

42°05'

Isobathes de la
discoradance Miocène

FIGURE 1 - SITUATION MAP

(05.24/81)

PRODUCTION TEST ON ROSPO MARE
HEAVY OIL RESERVOIR (ITALY)

M.L. LEBIHAN   )
A.G. BOURGEOIS)   ELF ITALIANA Operations Division

Summary

The lack of dynamic data concerning ROSPO MARE reservoir involves a risk of early water breakthrough into production and has made a production test at the site practically inevitable, from an experimental platform with two wells. A test programme was elaborated on the basis of the production of two wells, one vertical, the other deviated, both partially penetrating into the reservoir and operating on open-hole production to prevent any risk of artificial inflow of water from the aquifer, for instance as a result of any defects in the cementation. The project took place in two phases. The first consisted in the engineering design stage for the installations, in view of the particularly heavy and viscous oil present at ROSPO MARE. The second, representing the test proper, was mainly aimed at observing the water breakthrough times into the two wells, pumped at different flows, and observation of how the hydrated production evolved following breakthrough. Observation of the pressures enabled a slight interference between the wells to be detected in the beginning, and measurements to be made showing that the drop in static pressure with the extraction is low, thus confirming, where it necessary, the activity of the underlying aquifer. ROSPO MARE reservoir is operated by ELF ITALIANA on behalf of a consortium consisting of AGIP and SAROM.

The main danger at ROSPO MARE is that of early water inflows owing to a combination of several unfavourable factors: a relatively small total height of the oil reservoir, a low net oil height ($H\emptyset S_p$ = 1.5 m) owing to the nature of the karst with no matrix porosity, a high contrast of the mobilities (375) with a very small density difference between the oil and the water (0.05 gr/cm³) and the fear of a powerful aquifer.

Two essential questions arose: Do water inflows render primary production impossible, and how can one combat such water inflows ?

Since virtually none of the dynamic characteristics of the reservoir were known, there was practically no way to attempt to solve this problem other than by a long duration production test, knowing that at the same time a research programme has been launched to select the best means of drive for this type of reservoir.

It was hence decided to carry out a long duration test from an experimental platform where two wells would be drilled in order to discover how the pressures and flows on anhydrous production developed, the water breakthrough time, and the evolution of the pressures, flows and water percentages in the production following breakthrough: in other words, the behaviour of the aquifer.

The two essential phases of this contract are the engineering studies and the test proper. Addition of a third horizontal well during the test and the general interpretation of the results of the test are not dealt with here.

PHASE 1 - ENGINEERING STUDIES

1 - Production support

The production support consists of a jacket with four piles and six slots surmounted by a two level deck. This platform must act as guide support for the six wells possible, as support for the drilling rig and its utilities, and as preprocessing and product discharge support.

Design of the jacket. The following is the design flowchart (analysis in three phases).

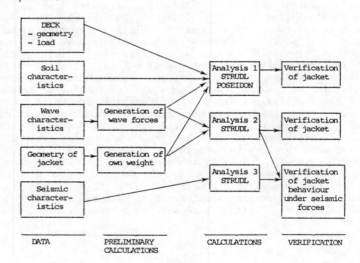

ROSPO MARE
Jacket design flowchart

The programme used for the statistical and dynamic calculations is the STRUDL programme processed on an IBM 370/138 for providing the input data for the second programme for the elasto-plastic computations for the jacket on its piles (POSEIDON). The secondary structures (boat landing, skidding, 6 conductor tubes, 3 risers, 2 sump-casings, 2 J-tubes) were subjected to separate modelization.

Design of the deck. The deck is designed independently from the jacket, assimilated to a lattice of beams.

. During the drilling phase, the top deck consists of a helicopter pad with living quarters and all the drilling equipment, and the bottom deck comprises the various production equipment. Four loading combinations were considered, each corresponding to a position of the drilling rig on a wellhead with action of a wind at 90° to the major face (35 m/s).

. During the exploitation phase, the top deck comprises extraction pumps, whilst the same equipment remains on the bottom deck as during the drilling phase ; only one loading case was examined.

. During a phase intermediate between production and drilling, another loading case was studied with both the derrick and the production pumps installed.

2 - Surface installations

Processing of the effluent is kept to the minimum. The production extracted by the CERIAB long-stroke pumps is separated from its gas content, and then repumped to discharge it to the floating storage support. The gas is burnt at the flare and the water is not separated from the oil, but metered from wellhead samples (measurement of the BSW). Accordingly, the production system was at first sight simple, but the high pure oil viscosity and the yet higher viscosity measured on the oil-water emulsion produced in the laboratory had to be considered. It is intended to heat the wellhead production (60°/EC 101 and 102), enabling the viscosity to be reduced and hence improving the gas/liquid separation and facilitating discharge by a screw pump towards the storage facility. Either one of the wells can be tested in alternation on a three-phase test separator DS101. The flow from the other well and the liquid effluents are collected in a degassing unit DS102 operating at the lowest possible pressure to prevent any risk of degassing in the compartments of the tanker, and forming part of the load circuit of the discharge pumps GA 103 A and B. These pumps are driven by electric motor and are of the screw type, specially adapted for viscous oils. The discharge flow is set by adjusting the rotary speed of the pumps by means of a speed variator controlled by the level of separator DS102.

During the production test phase, it was intended for the utilities to use the installations of the drilling rig (a POOL compact rig). In actual fact, this did not work out, since it was necessary to design and supply these utilities, namely living quarters for a crew of 22, a helicopter pad capable of receiving 2 "PUMA" class twin-turbine helicopters, a telescopic crane, 2 electric generating sets, 2 compressors, a fresh water storage capacity, 2 flooded pumps for the firefighting and service water networks, and 1 high pressure pump for working in the wells. The design of these elements was generally modular or even of the one-piece type, when compatible with the light hoisting equipment, and hook up at sea was generally kept to the minimum. It should be pointed out that the fully-enclosed electric generating sets and switchboard installed in a container acting as cabin were fully accommodated beneath the helicopter pad and hence protected and on the opposite side of the platform to the living quarters.

3 - Evacuation terminal

The system adopted is a floating storage facilities anchored by hawsers to a buoy connected to the platform by a 900 metre sea-line. The floating storage facility consists of a 25,000 ton tanker (hold capacity 19,000 tons) with its tanks heated by steam coils. The CALM type buoy is anchored by 4 anchor chains. The sea-line is an electrically heated 6" hose. The oil will be discharged to refineries in the Adriatic by means of a 6,000 ton shuttle tanker. A number of other details are worthy of indication. The buoy is a fully-welded flat cylinder surmounted by a turntable with trunnion and held by means of four 2 1/2" diameter anchor lines 500 metres in length. As regards the oil product circuit, the underwater hose arrives at the buoy through a "trumpet", crosses the trunnion and is locked by two half-shells and connected to the 12" swivel joint. The pipework of the turntable sends the product through a cross and vertical offtake and at the gangway level, the circuit divides into two parts with a 12" diameter that is symmetrical with relation to the mooring line, and situated perpendicular to it. The floating loading hose to the tanker has a diameter of 10" and a length of 200 metres and acts as link between either parts of the gangway of the buoy and the manifold amidships on the tanker.

PHASE 2 – PRODUCTION TEST

During the second half of 1981, the RSMA platform was laid and the drilling ring installed. Drilling of the wells started in December 1981 and continued during the first quarter of 1982 whilst the terminal was being installed. The chartered tanker arrived only on the day the tests were to start in August 1982. Let us remind the reader of the general context of the test. We shall only concern ourselves with vertical well RSM4 and deviated well RSM5D, which penetrated into the reservoir for only about 30 metres at impact points 700 metres away. The purpose of this partial penetration was to avoid any artificial inflow of water (for instance, behind cementation not correctly sealed), knowing that in order to discover the architecture of the karst down to the water level, following the test, the wells were to be sunk to a greater depth. The two wells produced with an 8 1/2" open hole (9 5/8" casing at the roof of the reservoir) and are equipped with two tubings, one 4 1/2" production tubing with a long-stroke pump and one 2 3/8" measurement tubing with circulation sleeve (above the double packer) for heated fluxing gas-oil, if necessary. These wells were also equipped with permanent pressure and temperature measuring gauges (DPTT - Flopetrol stress gauges) positioned below the pump about 80 metres from the reservoir.

1 – Production sequence
- Production by pumping at a rate of 230 m³/day from well RSM4 at the end of August 1982, whilst keeping a watch for any eventual interference, on RSM5D.
- Production by pumping at 50 m³/day from well RSM5D at the end of September.

On the basis of these flows, observation of the pattern of bottom pressures and the percentage of water in the output was the major objective of the test. Owing to the high viscosity of the crude which slows down degassing during separation, the flow can not reliably be measured by a meter ; one had to base oneself on the pump operating parameters (length of stroke, number of strokes per minute, efficiency) and monitor the full production by gauging on the tanker everyday. The percentage of water in the output was determined by AQUATEST (measurement of release of hydrogen with a specific reagent) for the low values and by centrifuging above 4%.

2 – Production history (see diagrams)
RSM4. Initially, production is anhydrous, though during the drilling 8 months earlier, the reservoir had absorbed more than 2,000 m³ of sea-water. Water breakthrough took place early in October after a cumulative anhydrous production of 8700 m³ and at the end of October, the BSW was 5.5%.

Next, during the winter, depending on the sea and weather conditions and the precarious mooring system, an alternation of periods of production at over 200 m³/day and halts occurred, leading to the observation that following a closed-down period, when restarting, the production was slightly hydrated, so the water cone was reabsorbed rapidly, though the BSW falls off quickly down to 10% within 15 days at these flow levels.

From March to May 1983, various flows were tested in order to situate the critical flow: at 135 m³/day and beyond, the BSW drops, at 114 m³/day, the BSW does not drop and the production remains anhydrous. From May to mid-July, the well produced 115 m³/day of anhydrous product. From mid-July to the end of September 1983, tests were carried out in order to determine the behaviour of the well with respect to inflows of water. Fine determination of the critical flow yielded a value of from 112 to 117 m³/day. During the second half of September, for a flow of 150 m³/day, the BSW rose to 25% without showing any tendency to level out. Unfortunately, the experience on anhydrated production could not be extended beyond this owing to the lack of dehydration installations at the site and for reasons of frequent production interruptions mainly caused by the unfavourable weather conditions from October to December 1983. On 31st December 1983, the cumulative production from RSM4 was 58,700 m³.

RSM5. Observation of a slight interference with RSM4 when this well was producing alone, at the beginning of the test. The production from RSM5, pumped at a rate of 50 m³/day from the end of September 1982, was initially hydrated (BSW reaching 10 to 15%), although this well had absorbed no seawater whilst the reservoir was being drilled. During a month, the BSW diminished for no apparent reason, levelling out at about 1.3%. From November 1982 to February 1983, owing to the recent production halts, it was observed that the BSW oscillated from 0.6 to 1.6%. Water breakthrough is difficult to define, since it occurred fairly gradually: it would appear to have taken place in the beginning of March 1983, after producing 6500 m³. From March to the end of June 1983, a fairly anachronic behaviour of the well was observed, with BSW values of from 0 to 3%, not always following the flow which oscillated from 20 to 50 m³/day. After being closed for 2 months, water breakthrough took place a week after opening in September, at a rate of 25 m³/day. From September to December 1983, production was not uniform, a flow of from 20 to 80 m³/day, interspersed with production halts and BSW values of from 1.5 to 3.5%. The cumulative production was 12,000 m³ at the end of 1983.

3 - Pressure history

During the interference tests, which are dealt with in another ROSPO MARE contract, the pressures were read on the surface by means of the multi-channel computer system (MCS Flopetrol), providing a resolution of 1/1000 psi. Next, the readings were made by means of a surface reading panel (SRP Flopetrol), the limited number of reading diodes of which meant that the resolution was only 0.15 psi, but which was adequate for the pressures under flow, but insufficient for the pressure surges. The accuracy of the instruments was seriously affected by the drift phenomenon that is inherent in the strain gauge system. For the pressures under flow, it was necessary to cope with extremely widely varying minimum and maximum mean pressures, owing to the effect of the pump strokes. Plotting of this mean pressure in terms of the flow results in a scatter of points, though linear extrapolation is possible, giving values of the productivity index that are compatible with other measurements (amongst others, the pressure surges).

RSM4   IP = 7.6 m³/day/kg/cm²   RSM5D   IP = 1.3 m³/day/kg/cm²

For the static pressure (see diagram), the drop due to the extraction is less than the positive drift of the sensor, since the pressure appears to rise. Making an assumption as to the value of this drift, one arrives at a pressure decline that would be in the range of 30 psi for a cumulative production of 155 000 m³ (August 1983). This figure is certainly erroneous, but it reflects a certain degree of reality, namely that the ROSPO MARE reservoir is subjected to active water flooding maintaining an almost constant pressure.

As was stipulated in the contract, the information drawn from this test was used within the framework of contract TH/05/09/78 entitled "SEARCH FOR DRAINAGE METHODS FOR A HEAVY OIL RESERVOIR INVOLVING THE RISK OF INFLOW OF WATER - APPLICATION TO ROSPO MARE".

It can however be concluded that this test, conducted in accordance with the programme set, with the sole restriction with respect to a highly hydrated production and the depths of the wells that had reached an irrealistic level, had reached its objective, particularly by making it possible to determine the water breakthrough times on two wells worked in different ways, the critical flow beyond which the production is anhydrous, and demonstrating the reversibility of the coning phenomenon resulting in water inflows.

(05.36/82)

HORIZONTAL DRILLING IN ROSPO MARE
OIL RESERVOIR

J. VENTRE, Reservoirs Department of SNEA(P)
A.M. DORMIGNY, Reservoirs Department of ELF ITALIANA

Summary

The ROSPO MARE karstic limestone reservoir contains heavy and viscous oil together with a highly active underlying aquifer. The preliminary studies and the tests on an assessment well give rise to fears of early breakthrough of the water, followed by very fast flooding, but also predicted that recovery could be enhanced by means of a horizontal drain.
A pilot scale production platform with three wells: 1 vertical, 1 deviated and 1 horizontal, revealed a performance ratio between the horizontal well and the vertical well that was yet better than expected, owing in particular to the heterogeneity of the reservoir and to the good adaptation of the horizontal drain to this type of reservoir.

1 - ROSPO MARE RESERVOIR

ROSPO MARE lies in the Adriatic sea, about twenty kilometres from the coast, in a depth of 60 to 100 metres of water, and about 1300 metres below the sea floor. It is operated by ELF ITALIANA in association with AGIP and SAROM. The accumulation could be very considerable, amounting to up to 200 Mt. The reservoir consists of massive fractured and karsticized limestone massifs of the cretaceous, and a breach of variable thickness (0 to 150 metres). This karstic network has been covered and to a great extent filled with sediments from the miocene to pliocene ages.

1.1 - The reservoir

Thoroughgoing studies have been carried out to attempt to delineate the spread, organization and petrophysical characteristics of the karst. The extension of the karsticized zone bears no relation to the interest of the horizontal drilling operations, and it should simply be noted that it may be very wide and extend to an area of 150 to 200 km².

a) Organization

During the emergent phase or phases affecting the combination of limestone massif and breach, a "gravific" karst infiltrated into the series.

Its development was controlled by hydrological, sedimentological and tectonic factors.

The main fault lines are organized around a hectometric mesh and, all other things being equal, from zones in which the fractures are more frequent, privileging the karstification. A vertical zoning exists, reflecting in particular the "hydrological" organization of the karst, although disturbed locally by sedimentological factors. The water flows generating the karst determined:
- at the top, an infiltration zone in which the water flows into the karsticized cracks, forming a system of vertically predominant conduits. Throughout the first metres of this zone (the epikarst), almost all the cracks are karsticized. As one descends, the system is hierarchized and only a few fractures lend themselves to the infiltration, the others not being karsticized ;
- at the base, a deep karst which during the action of the karst was saturated with water (flooded zone) ;
- at the limit between the two zones, the flows become sub-horizontal and create large dimension voids, though few in number.

It should be noted that in the case of ROSPO MARE, there is no clay barrier of any extent, and the discharge point of the infiltration waters was hence sea level.

When this base level changed, the old conduits were gradually abandoned and new ones created. It would appear that a beating motion zone of about 25 metres exists at ROSPO MARE.

In order to exploit the wells with a sufficient margin above the water table, the maximum penetration that can be accepted is about 30 metres. The drain will therefore take place exclusively through the top of the infiltration zone, with conduits that are essentially almost vertical.

The karstic zone thus formed was plugged by:
- recrystallization,
- the clay residues of dissolution,
- inflows of continental sediments,
- filling by marine sediments during transgressions.

During and after formation of this ensemble, the faults continued to adjust, possibly shifting the position of the different zones with relation to one another.

These have resulted in the creation of small grabens filled with miocene sediments not generally acting as reservoirs as encountered in horizontal well RSM6D.

b) Petrophysical characteristics

The effective porosity is difficult to estimate. The dynamic porosity $\emptyset$ (1 - Swi - Sor) deduced from interference tests and the evolution of the critical flows observed would be about 0.7 to 1%. The porosity of the small voids (vacuoles) is about 1.8% at the summit of the reservoir. The total porosity including the large voids (conduits) could attain 4%. The permeability is very considerable, varying from 2 to 50 D.

1.2 - Characteristics of the oil

The main difficulty in exploitation stems from the characteristics of the oil. There is a very low (ρ_0 = 0.93) water-oil density contrast and a highly unfavourable mobility ratio (μ = 300 cp).

Coupled with the presence of a highly active underlying aquifer, this gives rise to fears that the well will very quickly be flooded.

2 - ADVANTAGES OF THE HORIZONTAL WELL

Simulations on meshed models made in 1979-1980 are reported on elsewhere in the paper "The search for drainage methods for heavy and viscous oil reservoirs involving the risk of water inflows: application to ROSPO MARE". In order of magnitude, the expected recovery before break-through rose from 10-12 kt for a vertical well to 50-60 kt for a horizontal well. Recovery beyond breakthrough of the water was difficult to simulate (owing to numerical instabilities) and in either case is very low. The critical flows expected were also very low: 10 to 20 m³/day, but this was first and foremost due to an underestimation of the permeability. The simulations and calculations were made with homogeneous media "equivalent" to karst. One of the additional arguments raised in favour of the horizontal well was however the greater probability of encountering large dimension voids, and it was decided to add the RSM6D horizontal well to the pilot project planned with two other wells: RSM4 vertical well and RSM5D deviated well.

3 - DRILLING THE HORIZONTAL WELL

The horizontal well was drilled in 88 days without encountering any particular technical difficulties. Most of the objectives set were attained, namely:
- piercing the reservoir through a distance of 604 metres, most of which horizontal,
- excellent control of the deviation, complying with the constraint of not descending more than 30 metres below the roof of the reservoir,
- extraction of 8 cores with a total length of 46 metres and a recovery of 37.4 metres (81%),
- recording of a complete set of logs:
  . by the SINFOR method   GR/BGT
                                      GR/ISF/BHC
                                      GR/DLL
                                      GR/CBL of the 9 5/8" casing
  . by the SCHLUMBERGER method   GR/D/N

A production logging programme is to take place in November 1984.

On the other hand, the characteristics of the reservoir meant that an uncemented strainer liner had to be run in. The problem of selective completion has hence not yet been solved.

4 - PRODUCTION HISTORY

4.1 - Interference tests
RSM4 vertical well was placed on pumping alone on 21st August 1982, the receiving wells being deviated well RSM5D 700 metres to the South-East and horizontal well RSM6D 700 to 1200 metres to the North.

After a month of observation, RSM5D was in turn set on production, RSM6D remaining alone on observation until 22nd October, when it placed on production at a rate of 280 m³/day. No interference was observed at RSM6D, either because it was beyond the resolving capability of the measuring instruments (1/100 psi), or because a permeability barrier exists between RSM4 and RSM6D.

On the other hand, between RSM4 and RSM5D, the interference is weak, but almost immediate.

4.2 - Hypercritical flow
From the 20th of November onwards, the flow was increased to 540 m³/day. Until the end of January 1983, production was interrupted by numerous halts of from 8 to 10 days, mostly due to bad weather conditions.

From February 1983, production can be considered as almost continuous, gradually increasing from about 485 m³/day to 610 m³/day (mean values).

Breakthrough of the water appears to have taken place on 6th June 1983. Before this date, the BSW values were below 0.1%. From 6th to 12th June, the BSW rose to 0.6%.

The cumulative production was then 75,000 m³, that is almost 9 times more than that of vertical well RSM4 on breakthrough: 8700 m³. These values agree fairly well with the predictions on meshed models.

In contrast, the critical flow was distinctly better than that planned. The production was then gradually reduced in order to remain in the neighbourhood and slightly below this critical flow.

4.3 - Paracritical flow
From 13th June 1983, after closing down for 5 days, the well was reopened at flows of from 400 to 525 m³/day. The BSW dropped to its initial values of below 0.1%.

In July and August 1983, production was regular at an average flow of about 465 m³/day. In mid-August, a slight change occurred in the percentage of water for a flow of 485 m³/day, whereas at 475 m³/day, production was anhydrous, thus enabling the "critical flow" to be situated between these two values.

From 15th to 25th September 1983, a hydrated production test led to producing at a mean flow of 483 m³/day. The percentage of water rose to 2.4%, without any apparent onset of stabilization.

After closing down for 4 days at the end of September, the well was reopened at a flow of 4 to 5 m³/day, and its BSW was again below 0.3%.

Up to the end of March 1984, production was irregular, ranging from 400 to 450 m³/day, with peaks at 500 m³/day, with many intermittent halts, particularly owing to unfavourable weather conditions. The BSW mostly lies below 0.1%, with a few peaks at 0.8%, corresponding to peak flows.

These frequent halts probably allowed partial resorption of the water cone and masked the critical flow trend.

Production again became regular in the Spring, enabling a critical flow of about 350 m³/day to be observed in May 1984.

4.4 - Hydrated production and production logging
Hydrated production tests are scheduled for October 1984. As regards the production logging, tests were successfully performed both for the operation of the drilling rigs in a fairly viscous oil, and for running in the tools in a horizontal well at Lacq in France. The logging on RSM6D should take place in November 1984.

4.5 - Pressures
Permanent pressure sensors were installed on all the pilot project wells. The RSM6D sensor operated up to January 1983.

During production, the pressure variations caused by the pumping lie in the range of 50 psi, distinctly greater than the few psi of mean ΔP imposed on the layer. This makes any exact estimation of the IP difficult. The final measurements are about 1200-1300 m³/day/kg/cm², that should be compared against the value of 1.8 on the deviated well RSM5D and 9 on the vertical well RSM4.

The latest static measurements showed a decline of 1.4 kg/cm² for a cumulative pilot project production of 300,000 m³, confirming the activity of the aquifer.

5 - SIMULATIONS OF THE BEHAVIOUR OF THE RESERVOIRS

Two methods were used to attempt to simulate the evolution of the critical flow and establish production forecasts.

5.1 - Simplified method using the coning law of Sobocinsky

Principle: for a vertical well, the critical flow can be expressed as:

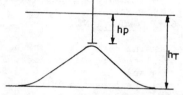

$$Qc = 1.52 \times 10^{-3} \frac{\Delta \rho \cdot K \cdot h_T (h_T - h_P)}{\mu o \cdot Bo}$$

i.e. $Qc = B \times K \times hT \times (hT - hp)$

If it be assumed that the flow of a horizontal well originates from a single or from a limited number of production points, one can expect that the error is not too great by approximating to an equivalent number of vertical wells:

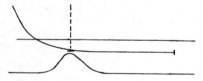

1 equivalent well

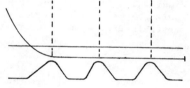

3 equivalent wells

And for n wells:

$$\sum_{i=1}^{n} Qc = B \times hT (h_T - h_P) \sum_{i=1}^{n} Ki$$

This brings us to the case of a single well, provided its geometry (hT and hp) is the same for all the wells.

It is also assumed that the wells drain a limited surface in a fixed time, which is fairly close to reality in the case of the discontinuities in the karstic network or interferences between several wells or several platforms, and which remains an acceptable approximation for a sufficiently short period of time in the case of an infinite medium.

$$Np (t) = A \emptyset (hT^{(0)} - hT^{(t)})$$

$$Qc (t) = \lambda hT^{(t)} \times (hT^{(t)} - hp)$$

$$Np (t) = \int_{0}^{t} Qc^{(t)} dt$$

The calibration on the evolution observed in the critical flow enables A∅ to be determined (to be precise, one should speak more of a A∅ (1 - Swi - Sor) and extrapolate the flows.

5.2 - Forecasts on model

The simulations on meshed models turned out to be difficult owing to the numerical instabilities created by the extreme conditions of ROSPO: low porosities, very high mobility ratios and very high permeabilities. Following unsuccessful tests on two other modules, the MEPHISTO model of SNEA(P) used in the RZ form (without its thermal option) provided satisfactory calibration setting on the initial history of vertical well RSM4. To make predictions, we then proceeded to a three-dimensional model. The initial results appear to be highly comparable to those obtained by the first method. Sensitivity tests are continuing in order:
- to test other sets of parameters,
- to simulate a longer history, allowing for all the wells.

6 - CONCLUSIONS

The performances of horizontal well RSM60, both in terms of the absolute performance and that relative to the neighbouring vertical well RSM4 and the deviated well RSM5D, are distinctly better than those that could have been predicted by calculation in the case of a homogeneous medium. The reason, in our view, is the karstic nature of the reservoir and in particular its organization into fractured zones measurable in thickness in metres and karsticized, with spacing in the hectometric range. Within such a reservoir, a horizontal drain has a considerable chance of finding itself in direct communication with several of the better zones, whereas a vertical well would in most cases be linked to the neighbouring karsticized zone only by cracks with a much less conductive capacity.

	RSM4 vertical well	RSM5 deviated well	RSM6 horizontal well
IP $m^3/day/kg/cm^2$	9	1.8	1200-1300
Initial flow	250	50	600 (maximum pump capacity)
Critical flow estimated in March 1983	120	20?	480
Cumulated production on breakthrough	8500	(6500)? (1)	75000
Cumulated production on 30/6/84	73000	15000	212000
Anhydrous flow on 30/6/84	100	–	350

We only have the comparison of 2 conventional wells (vertical well RSM4 and deviated well RSM5D) and one horizontal well (RSM6D), which could appear a narrow basis for generalizing, considering the heterogeneity of the karst.

Nonetheless, we consider that the difference in performance is such that it is indicative of the geometry of the well, if not with respect to detail points, at least in its major features.

(1) Difficult to indicate - this well had a production that was slightly hydrated as soon as it was opened.

(05.23/81)

METHOD OF ENHANCED PETROLEUM RECOVERY
BY INJECTION OF CO$_2$ INTO COULOMMES-VAUCOURTOIS FIELD

E. COUVE DE MURVILLE, PETROREP S.A.

Summary

Coulommes-Vaucourtois field lies 45 km East of Paris and is reaching depletion. In 25 years, about a quarter of the reserves in place have been extracted. Theoretical studies and laboratory measurements have shown that the characteristics of the reservoirs were compatible with injection of carbon dioxide and that the additional recovery rates could be encouraging.

Experimental injection of a small quantity of CO$_2$ into the Vaucourtois upper reservoir, which has a very low permeability, was made at the end of 1982. For this test, a suspended well was used for several years. After injection, the well remained closed for four months and was then returned to production. The average daily quantities produced are appreciably the same as those that had led to abandoning the well. This experiment is still continuing today.

In 1983, an injection well was drilled within a triangle formed by three existing wells. 2,600 tons of CO$_2$, followed by 900 tons of nitrogen, were injected in early 1984. The three wells were interconnected by pipes to a collection centre enabling the production from each to be measured separately. The first shows appeared two and a half months after starting injection in a well twice as far away as the three surrounding wells.

INTRODUCTION

PETROREP, holder of the Coulommes-Vaucourtois permit and the operating company, the French Petroleum Institute, SNEA(P) and TOTAL-CFP set up a joint venture to implement the pilot project for injecting CO$_2$ into the Vaucourtois structure (Fig. 1).

Today, the reservoir is operated by pumping and is now nearing depletion. Most of the production wells are flooded. When production reaches the economic limit, nearly three quarters of the initial reserves will still remain in the ground. Most of the oil still trapped consists of thin layers at the roof of the reservoir and in the space between the wells. Accordingly, the oil is no longer driven by the upward movement of the water from the aquifer. In the zones swept by the water, nearly half the initial reserves remain trapped, by capillary forces.

By using carbon dioxide gas (CO_2), the inflation of the oil is enhanced when the gas comes into contact with the oil, in a ratio of from 1.5 to 2, accompanied by a considerable reduction in its viscosity, thus improving its mobility within the reservoir. Furthermore, in a drive system comprising a gas injection well and one or several production wells, the gas will tend to move towards the upper part of the structure owing to its low density, and hence to drain the oil which had never been mobilized by the movements of water inside the reservoir.

Initially, the project consisted of two phases:

. A theoretical phase comprising:
  - study of the natural fracturation of the reservoirs,
  - thermodynamic measurements of the crude oil - CO_2 mixture enabling the conditions of miscibility, mobility and recovery to be verified,
  - a reservoir study enabling the past history of the production periods to be resumed and the reserves updated,
  - simulations enabling the injection parameters to be determined.

. An injection phase and production phase through the same well. The upper reservoir (R1) of Vaucourtois field containing about a quarter of the total reserves, but featuring very low permeability, was selected for this test.

It very quickly appeared that it would be highly difficult, not to say impossible, to extrapolate the results of this injection and production test through the same well to a sweep between wells for the structure as a whole. Accordingly, it was decided to reduce phase 2 to an injectivity test and to carry out the third phase involving a CO_2 sweep of the oil. For this purpose, an injection well was drilled into the lower reservoir of the Vaucourtois structure (R3), which consisted of detritic cracked limestone containing most of the reserves.

The rareness and remoteness of the CO_2 sources available at the time the injection operation was carried out contributed towards the high cost of this gas. Nitrogen, which was more readily available and cheaper in price, was injected after the CO_2, so as to push the carbon dioxide gas towards the production wells.

1 - PHASE I - STUDIES AND MEASUREMENTS

1.1 - Natural fracturation of the Coulommes-Vaucourtois reservoirs

Situated 9 kilometres South-South East of Meaux, the Coulommes-Vaucourtois reservoir was discovered in 1958 in the Dogger of the Paris basin, lying at a depth of from 1840 to 1890 metres. It consists of a very flat anticline with two combinations, one at Coulommes, the other at Vaucourtois, separated by a slight saddle. The reservoir consists of limestone formations comprising the following from top to bottom:
- A layer of oolithic limestones 6 to 12 metres thick (reservoir R1), the upper part of which is very often compact and unimpregnated, with the remainder being of very low permeability in any case ;
- A layer of compact and cracked sublithographic limestones about 10 metres thick, with no oil in the matrix, except in the South-West part of the field, where a porous lens formation that can be up to 2 metres thick is to be found (reservoir R2) ;
- The third reservoir R3 which consists of gravel and vacuole detritic limestones, sometimes cracked. This is the one containing the major reserves of oil. The total thickness of reservoir R3 is 200 metres. The summit of this structure is impregnated with oil through a thickness of about 10 metres at Coulommes and reaches over 30 metres at Vaucourtois in the wells situated in the centre of the sector. The remainder of R3 lies

below the water level. The aquifer is fed permanently, the result being that no pressure drop has been recorded since starting to exploit the reservoir.

Observation of cores shows that the layers are practically horizontal and that the fractures are generally vertical, parallel to each other and oriented in an NW-SE direction. This direction is that of the Bray fault, which crosses most of the Paris basin and borders the Coulommes structure in the South ; the fractures are not continuous and are laid out in relays, thus dividing the reservoirs into distinct elongated blocks.

Reservoir R1 is more fractured towards the North-East, displaying a better thickness and porosity than in the SOuth-West, where it is thinner, more clayey and less porous.

The compact limestone is the most fractured formation, and probably does not form a seal between the two reservoirs.

Reservoir R3, which is much more porous than permeable, is less fractured, but the fractures are visible in all the exploration boreholes, in which this formation was cored. Well BG-2, situated in the extension of a secondary fault, has the highest fracturation index of the entire field.

1.2 - Petrophysical study of reservoir R1

The purpose of this study was to attempt to follow the porous and permeable lens type formations of the oolithic limestone, through the field, and to establish a relationship between the porosity and the permeability. This aim was not achieved owing to the very structure of the limestone, which displays a double porosity system: microporosity inside the ooliths, generally filled with water, and intergranular macroporosity representing the useful porosity of the reservoir, the interconnections of which form its permeability.

The microporosity can vary from 2 to 8%. This variation is of considerable importance when compared with the total porosity (2 to 15%). Accordingly, the useful porosity is generally below 5% and can vary very quickly, as also can the permeability, with the variations in the constituents of the rock.

1.3 - Laboratory measurements - Simulations

Determination of the thermodynamic properties of the various oil-CO_2 mixtures has shown that under the conditions of pressure and temperature prevailing in the formation (186 bars, 71°C), the CO_2 should become soluble in the oil at Coulommes in a proportion of 32% (the fraction by weight of the mixture) for a gas-oil ration (GOR) of 220 m^3/m^3. The inflation factor was 1.52 ; the viscosity of the oil dropped from 3.2 to 0.8 cp.

An experimental sweep in a porous medium inside a glass tube packed with sand and saturated with oil showed that in theory a recovery of 85% of the oil in place can be obtained at the reservoir pressure after injecting a quantity of CO_2 equal to about 1.5 times the volume of the pores.

The sweep on a core was then made. The core was "reconstituted", namely restored to the actual conditions of pressure, temperature and residual saturation following a water sweep. This experiments showed that one could expect to double the oil output after injecting a quantity of CO_2 equal to twice the pore volume.

These initial and highly encouraging results were used to make simulations the intention of which was to predict the evolution of the dissolution front, the duration of the sweep, the date of breakthrough into the surrounding wells, the quantities of additional oil produced and the quantities of gas to be injected, depending on the different assumptions adopted. Several simulations were made varying the ratio of the vertical and horizontal permeabilities and the criteria determining when to stop the production at the profitability limit values. The spacing between the wells was that of the Vaucourtois well, i.e. 400 metres on the average.

In the least favourable case (kv/kh = 0.01, and breakthrough at the nearest well in 200 days), a recovery of 1 m³ of additional oil could be expected for 1.5 tons of CO_2 injected. However, to obtain this result, 60,000 tons of CO_2 would have had to have been injected, which is a quantity totally incompatible with the budget initially adopted.

To solve this problem, after carrying out the short duration injectivity test already planned in the programme, it was decided to drill an intermediate well inside the triangle formed by the three existing wells, slightly "offcentered", so that the closest well is then only 150 metres from the injection well.

By extrapolating the results of these simulations with this configuration, one could expect a reaction at the nearest well after injecting about 5500 tons of CO_2 or 2600 tons of CO_2 followed by 900 tons of nitrogen. Under the conditions at the bottom, the nitrogen occupies three times more volume than does the CO_2. In the pilot project, the nitrogen is only involved as a means of pushing the carbon gas, since it is not miscible at reservoir pressure. Furthermore, nitrogen costs less than CO_2 and is available throughout the year, whereas the carbon dioxide gas produced in France is reserved for the food consumption industry as soon as Spring arrives.

2 - PHASE II - INJECTIVITY AND PRODUCTION TEST IN A COMMON WELL

Well F-19 South of the Vaucourtois structure was selected because its production history shows that the oil comes from the upper reservoir R1, which is completely isolated from the lower reservoir. The well was shut down in 1979 at a pumped production rate of about 700 litres per day.

300 tons of CO_2 were injected in 15 days at the end of 1982. The well remained close for about four months to enable the gas to diffuse properly.

On 30th June 1984, the well had produced a total of 160 m³, i.e. an average of 350 litres per day. Production is obtained by "eruptions" of 1 to 2 m³ separated by several days, during which the well remains quiescent. Use of a pump spreads out the period of eruption and increases production. Nonetheless, this never exceeded 700 litres per day, which was the well abandon flow.

Technically, the test proved positive, since it demonstrated that:
- CO_2 could enter the formation without entirely rejecting the oil in the vicinity of the well (no free gas zones were revealed),
- the dissolved CO_2 was in fact capable of moving the oil towards the well when it is opened and assisting production by "gas-lift" through the tubing,
- the phenomenon of production by expansion of the dissolved CO_2 is lasting in effect ; critical gas saturation was not reached, probably through a balance in the redissolution and constant expansion of the CO_2.

In addition, it was shown that:
- the productivity index of the well did not rise appreciably. Probably, the advantages gained by lowering the viscosity of the oil and the expansion of the dissolved gases are offset by the reduction in the permeability to oil around the wells caused by the presence of the gas,
- the gas-oil diphasic mixture that leaves the well is difficult to pump if the gases can not escape through the annulus.

- until today, there is no indication of the formation of H_2S, nor of emulsions liable to hinder production,
- no decline was observed. The eruptive production drive source is indeed the CO_2 injected,
- despite these positive results, it seems clear that this production mode can not be envisaged on industrial scale. The test is continuing today.

3 - PHASE III - SWEEP TEST BETWEEN WELLS

3.1 - Injection well IG-1

The position of the injection well on the Vaucourtois structure close to the summit of the anticline was selected owing to its vicinity to three other wells: BG-11: 150 m, BG-22 and BG-61: 250 m.

This well, drilled in October 1983, entered the initial water level of the field and featured water saturations in the range of 60 to 85%. The productivity from the well was nonetheless greater than expected, despite its position in a zone that had been assumed to have been well drained by the surrounding wells. During four months of production, it produced four times more oil than each of its neighbours.

3.2 - Injection of carbon dioxide gas

From 12th March to 9th May 1984, 2600 tons of CO_2 were injected into the top of reservoir R3. The following facilities were used: 4 tanks each with a capacity of 30 tons, 2 booster pumps and 2 cryogenic pumps, one heater and various temperature regulation and automatic shutdown systems should the pressure of the injection gas and the temperature exceed certain maximum values (Fig. 2).

The CO_2 was supplied by 20 ton trucks. The daily flow was 43 tons, the injection pressure close to 45 bars and the temperature at the wellhead 10 to 20°C.

3.3 - Injection of nitrogen

Immediately following the CO_2, 900 tons of nitrogen were injected, using the same injection system and a storage capacity of 43,000 litres of liquid nitrogen at a temperature of - 196°C. The daily flow was 14 tons and the injection pressure 150 bars at the wellhead.

3.4 - Production installations

The three wells closest to the injection well are linked by fibreglass pipelines (to prevent any corrosion) to a central system for separating and sampling enabling measurements to be made on the gaseous and liquid effluents. Each well can be sampled independently from the two others.

3.5 - Results

On 24th May, i.e. a few days after interrupting the injection of CO_2, an initial reaction appeared at well BG-2 lying 540 metres NNW from the injection well, in the form of production of free gas and an increase in its production of oil and water.

It had not initially been expected that this well could be attained by the gas and no special equipment was available for its completion.

A "gas-lock" quickly appeared in the bottom pump, cutting off production. The well was completed as an eruptive well.

It should be noted that with relation to the injection well, BG-2 is situated in the direction of the network of cracks revealed during the geological study and it is hence normal that manifestations occur in the NNW-SSE direction. At present, no show has been detected towards the SSE.

On 13th July, the gas appeared at BG-22 situated 250 metres North of the injection well, accompanied by twice the production of water and oil.
It is too early as yet to draw conclusions concerning this break-through.

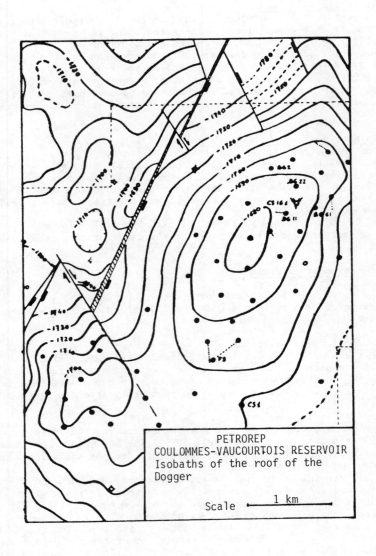

PETROREP
COULOMMES-VAUCOURTOIS RESERVOIR
Isobaths of the roof of the Dogger

Scale ⊢——— 1 km

FIGURE 1

FIGURE 2 - INSTALLATIONS FOR INJECTING CO_2 INTO THE COULOMMES-VAUCOURTOIS
WELL (Photograph IFP)

(03.112/81)

DESIGN AND DEVELOPMENT OF A HOMING-IN DEVICE FOR BLOW-OUT CONTROL

B.C. LEHR and W. BROEKHUIZEN
Koninklijke/Shell Exploratie en Produktie Laboratorium
Rijswijk, the Netherlands

Summary

Investigations have been carried out into the development of a
homing-in technique based on passive acoustic principles.
Theoretical models were worked out to determine both, the relative
distance and direction between a relief well and a blowing well.
Distance determination is based on time delay measurements in the
acoustic radiation field along the relief well axis, whereas
direction determination is based on the scattering of sound by the
relief wellbore. Two engineering-model homing-in devices, one for
distance and one for direction determination, have been
constructed according to these principles. Test results for both
tools obtained by acoustic cross borehole measurements in shallow
test holes are presented and discussed.

1. INTRODUCTION

The drilling and production of oil and gas wells is subject to
strict safety procedures, but nevertheless control over a well is
occasionally lost ('blow-out'). The resulting uncontrolled flow of
formation fluids or gas can have serious consequences for the safety of
personnel and for the environment, which may become extensively
polluted. Also, the economical consequences resulting from loss of
hydrocarbons and production drop out may be considerable. Hence, it is
of paramount importance to control a blow-out with minimum loss of time.
If access to the wellhead cannot be gained quickly, or if there is an
internal blowout, a relief well may be drilled to intersect the blowing
well near the interval where the uncontrolled flow originates, as
indicated schematically in Fig. 1. Generally, the relief well is
required to intersect the blowing well within a distance of about 10
metres or less in order to establish good hydraulic communication. In
most cases, however, the downhole trajectories of the wells, as
calculated from directional survey data, are not known with sufficient
accuracy to drill a relief well successfully (Ref. 1). Special
instruments, referred to as 'homing-in tools', must be lowered into the
relief well to determine the relative distance and direction between the
bottom part of the relief well and the blowing well.
At present three different homing-in techniques are operational
and commercially available, namely the Ultra Long Spaced Electric Log
(ULSEL) from Schlumberger, which only indicates distance (Ref.2,Ref.3),
the Magnetic Ranging instrument (MAGRANGE) from Tensor, Austin, Texas
(Ref.4), and the recently available ELREC tool from Gearhardt Industries
(Ref.5). These three techniques are based on electric, magnetic and
electromagnetic induction principles, respectively, and have in common
that they require steel to be present in the target well. They are
therefore not suited for directing a relief well into an open hole far

below the casing shoe. This inherent limitation of the available homing-in tools initiated research efforts to identify alternative methods that do not require steel as a target.

During acoustic measurements (noise logs) in various blow-out situations the noise generated by the blowing well could be detected in a relief well approaching the blowing well (Fig.1). Measured noise levels indicate that the blowing well may be audible over distances up to 100 metres in the subsurface, which considerably exceeds the range of conventional homing-in tools. Hence, acoustic methods were felt to have some potential for homing-in.

In 1981 an EEC-supported project aimed at the design and development of a homing-in device for blow-out control based on acoustic principles was started at the Koninklijke/Shell Exploratie en Produktie Laboratorium, Rijswijk, Netherlands. Passive acoustic techniques have been identified to determine both the relative distance and the direction to a blowing well (Sections 2 and 3, respectively) by measuring, in a relief well, the acoustic radiation field generated by the uncontrolled flow of liquid and/or gas in the blowing well. These principles have been incorporated into engineering-model homing-in tools which are being tested under simulated blow-out conditions.

2. DISTANCE DETERMINATION

The underlying principle of distance determination by passive acoustic means is illustrated in Fig. 2. Parts of the wellbore (e.g. where the violent influx of fluids from the reservoir into the wellbore or flow restrictions occurs) act as localised sound sources. The existence of such sources has been confirmed from noise logs taken in actual blow-out situations. The measuring device consists, in principle, of two acoustic sensors located on the axis of the relief well. Both sensors record the continuous noise emitted by the blowing well. As a result of the different sound paths, a time lag, Δt, is observed between the output signals of the two sensors. This time lag can be measured by cross-correlation techniques or related methods to estimate a time delay between random signals. The travel time difference, $\Delta t(z)$, can be desdribed by the geometrical relationship (Fig. 2):

$$(1) \qquad \Delta t(z) = \frac{1}{c} \cdot \left\{ \sqrt{D^2 + (z_o - z - S)^2} - \sqrt{D^2 + (z_o - z + S)^2} \right\}$$

where D denotes the relative distance between the relief well and the blowing well, z_o the depth position of the sound source, z the midpoint depth of the two acoustic sensors, c the average sonic velocity of the formation and 2S the spacing between the sensors. Eq. (1) also holds for deviated but (approximately) straight relief wells if z is taken as along borehole depth and D as pertaining to an orthogonal projection of the source location on the borehole axis.

Travel time differences, $\Delta t(z)$, can be measured at a large number of depth positions, z_n, in the relief well by logging the interval above and below the expected depth of the sound source, resulting in a data set $\Delta t(z_n)$. An estimate for the source depth z_o and distance D can then be obtained by varying the parameters z_o and D in eq. (1) and calculating the curve $\Delta t(z)$ that best fits the data points $\Delta t(z_n)$. The average sound velocity, c, to be entered in eq. (1) would usually be derived from sonic logs.

To verify this method experimentally, a sensor array containing four hydrophones was constructed. This tool can be run in shallow holes on standard seven-wire logging cable and contains downhole electronics for amplification and transmission of the four hydrophone signals through the logging cable. At surface the signals are processed with a waveform analyser and also recorded on magnetic tape for off-line processing on a mainframe computer.

To evaluate the distance-determining device, and also that for direction determination (see Section 3), acoustic cross-borehole measurements have been carried out. A sound source was lowered into a borehole, simulating a blowing well, while either the distance- or the direction-determining sonde (see Section 3) was run in an adjacent borehole simulating the relief well. For this purpose, a configuration of three test holes was drilled in the chalk formations of Zuid Limburg, the Netherlands. The surface locations of the holes constitute the vertices of a triangle with sides of 20, 30 and 40 metres, hence enabling cross-borehole measurements with three different relative distances and directions to be carried out. The holes are virtually vertical, are completed with 10 3/4" casing set to 134 m, and have open-hole intervals from 134 m to 265 m (T.D.). Formation properties, in particular sonic velocity and density, have been thoroughly investigated from petrophysical logs taken in one of the holes as well as by core measurements. The Danian and Maastrichtian chalk formation encountered is a sequence of highly porous chalk layers with an alternating degree of cementation and shows an acoustic impedance profile of considerable complexity with impedance contrasts up to a factor two.

An example of the results of passive acoustic distance determination obtained in the test holes is shown in Fig. 3 and Fig. 4. The sound source which emits random noise was lowered to the depth of 170 m in test hole no. 1, while in test hole no. 2, at 20 m lateral distance, the interval between 210 m and 140 m was logged with the hydrophone array. Fig. 3 shows cross-correlation functions (positive parts only) calculated from a pair of hydrophone signals at various depths. The correlation peak due to travel time difference can easily be traced against depth as indicated in Fig. 3, crossing zero around the source depth. Figure 4 shows the best fit, according to a least-squares technique, of a curve, according to eq. (1), to the set of data points $\Delta t(z_n)$ taken from the correlation peaks in Fig. 3. The resulting best estimates for source depth and distance, 175 m and 19.9 m, respectively, agree well with the actual values of 171 m and 20.3 m. As can be seen from Fig. 4, however, the measured time delays can differ significantly from those calculated according to eq. (1), in particular at depths around the source depth. This scatter is mainly caused by sound velocity differences in the various layers of the formation, whereas a constant velocity c was assumed for the curve $\Delta t(z)$ calculated from eq. (1). Our results show, however, that the use of an average sound velocity, c, as an approximation yields satisfactory estimates for the distance between a localised sound source and a borehole, even in the rather inhomogeneous chalk formation penetrated by the test holes.

3. DIRECTION DETERMINATION

The direction from a relief well towards the blowing well is the second parameter that a homing-in device must be capable of determining. We have developed a passive acoustic technique that can be applied inside a borehole to determine the direction to a remote sound source. The noise generated by a blowing well is propagated through the surrounding formation in the form of stress waves, i.e. compressional and shear waves. When such stress waves impinge on the (fluid-filled) relief wellbore, which represents a discontinuity in the (solid) formation, they are scattered and give rise to a generally complicated vibrational motion of the borehole wall. This vibrational pattern of the borehole wall contains information on the propagation direction of the incident wave and, therefore, on the direction towards the blowing well. Calculations of the scattering of stress waves obliquely incident to a fluid-filled cylindrical borehole have shown that the distribution of acoustic intensity around the borehole wall exhibits a maximum at the spot where the waves first hit the borehole wall, both for incident compressional and incident shear waves. Hence, detection of the maximum in the angular distribution of acoustic intensity, measured at the borehole wall, gives the direction to the blowing well.

This direction-determining principle has been incorporated in a second engineering-model tool shown in Fig. 5. The downhole sonde contains three acoustic sensors mounted on a revolving pad assembly. It is operated while stationary and, by extending the arms, can clamp the sensors against the borehole wall and measure the acoustic intensity. The angular distribution of the intensity is sampled by alternately clamping, measuring, retracting and rotating the sensors around the borehole wall, as indicated in Fig. 5. The orientation of the pad assembly with respect to magnetic north is determined by means of a flux-gate compass housed in the tool body. Hence the angular distribution of acoustic intensity is referenced to magnetic north. The sonde contains electronic circuitry for communication with a control panel at surface and for control of the various tool functions, i.e. mechanical operation, compass reading, conditioning and transmission of the output signals of the sensors. A microcomputer at surface processes the data and controls the plotting of the resulting intensity distribution.

An example of the results obtained from cross-borehole measurements in our test hole configuration is given in Fig. 6. It shows a polar plot of the acoustic intensity measured at the borehole wall with the direction-determining tool in test hole no. 2 while the noise source was lowered in hole no. 1. The true direction from hole no. 2 towards the sound source is indicated by the arrow in Fig. 6. Also given in Fig. 6 is the theoretically expected distribution that has been calculated according to the scattering model described above. Note that the measured and calculated curves are given in arbitrary units so that only the shape of the distributions may be compared. The measured distribution is found to be in reasonable agreement with theory, particularly in that it gives the direction towards the sound source.

Acoustic vibration measurements at the wall of a borehole as they are described here may, of course, be strongly influenced by the conditions of the borehole wall and mud cake. The reliability of the measurement depends mainly on the quality of mechanical contact between sensor and formation. Our results indicate, however, that this type of measurement in a borehole is possible by suitable tool and detector design and that it is feasible to determine the direction towards a blowing well by passive acoustic means.

4. CONCLUDING REMARKS

A new homing-in technique based on passive acoustic principles has been developed to determine both the relative distance and direction between a relief well and a blowing well. Engineering-model tools have been constructed according to these principles and evaluated in a configuration of shallow test holes. The results obtained so far are promising in that they agree with theory and prove the feasibility of the method. It is planned to improve the interpretation software for distance determination by including sound-velocity changes in the formation. It is also the intention that, with the cooperation of the European logging industry, the engineering-model tools will be further developed and lead to the construction of a prototype homing-in tool which can be run under actual field conditions.

REFERENCES

1. WOLFF, C.J.M. and WARDT, J.P. de, Borehole position uncertainty. Analysis of measuring methods and derivation of systematic error model.
SPE 9223, 55th Ann. Fall Techn. Conf. of SPE, Dallas (1980).
2. MITCHELL, F.R. et al., Using resistivity measurements to determine distance between wells.
J. Petr. Tech., June 1982, pp. 723-740.
3. RUNGE, R.J. et al., Ultra Long Spaced Electric Log (ULSEL).
Trans. SPWLA 10th Ann. Log. Symp., Houston (1969), H-1-22.
4. MORRIS, F.J. et al., A new method of determining range and direction from a relief well to a blow-out well.
SPE 6781, 52nd Ann. Fall Techn. Conf. of SPE, Denver (1977).
5. WEST, C.L., KUCKES, A.F. and RITCH, H.J., Successful ELREC logging for casing proximity in an offshore Louisiana blowout.
SPE 11996, 58th Ann. Techn. Conf. of SPE, San Francisco (1983).

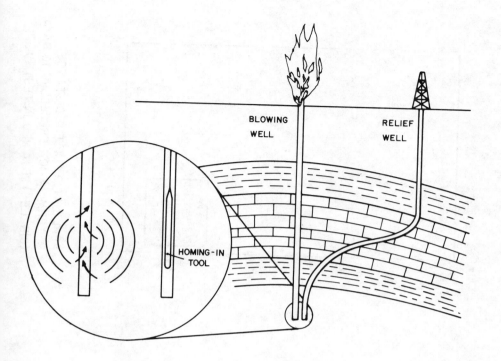

FIG. 1 - PASSIVE ACOUSTIC HOMING-IN PRINCIPLE

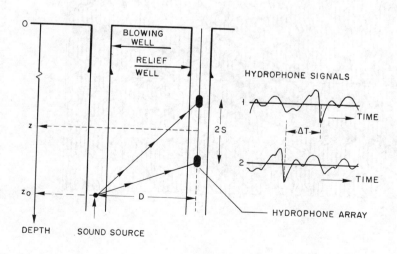

FIG. 2 - PRINCIPLE OF DISTANCE DETERMINATION

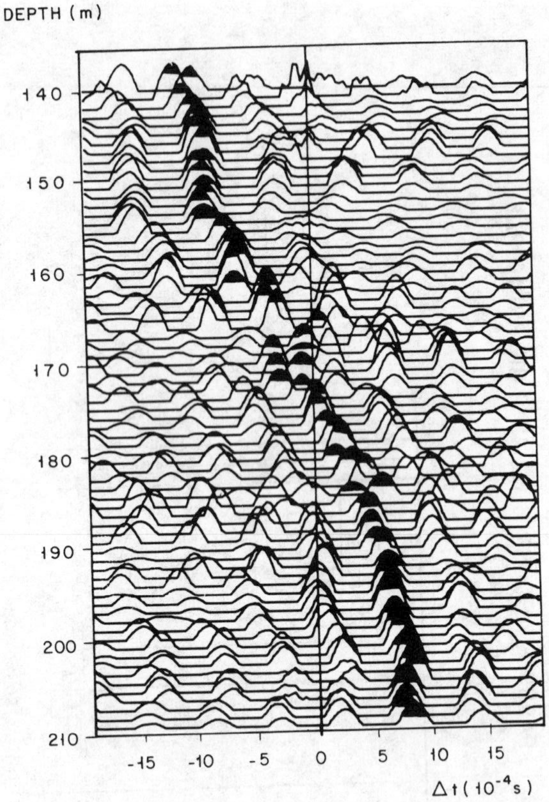

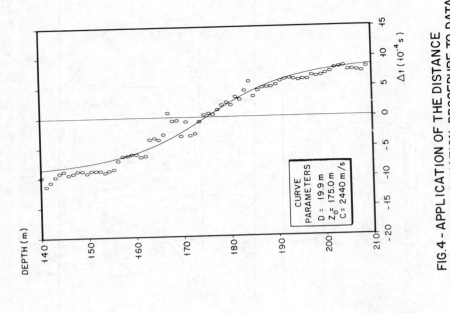

DEPTH (m)
140
150
160
170
180
190
200
210
-15
-10
-5
0
5
10
15
Δt (10^-4 s)

FIG. 3 - CROSS-CORRELATION RECORDS INDICATING ARRIVAL-TIME DIFFERENCES.

DEPTH (m)
140
150
160
170
180
190
200
210
-20
-15
-10
-5
0
5
10
15
Δt (10^-4 s)
CURVE PARAMETERS
D = 19.9 m
Z0 = 175.0 m
C = 2440 m/s

FIG. 4 - APPLICATION OF THE DISTANCE DETERMINATION PROCEDURE TO DATA (CIRCLES) OBTAINED IN TESTHOLES

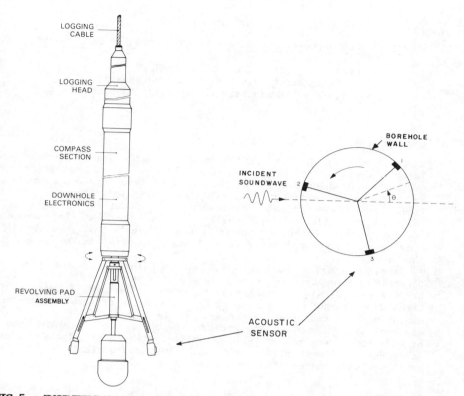

FIG. 5 – ENGINEERING–MODEL HOMING–IN DEVICE FOR DIRECTION DETERMINATION

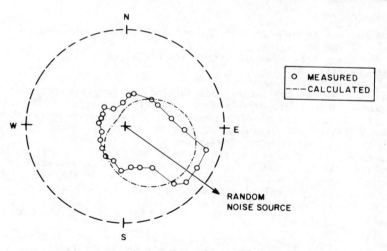

FIG. 6 – POLAR PLOT OF THE ACOUSTIC INTENSITY DISTRIBUTION AT THE BOREHOLE WALL

(05.04/76)

RESEARCH ON IMPROVED HYDROCARBON RECOVERY FROM CHALK DEPOSITS

T.L. VAN WAART AND R. HARTLEY
Shell Internationale Petroleum Maatschappij;
Koninklijke/Shell Exploratie en Produktie Laboratorium

SUMMARY

Chalk rock is characterised by high porosity, possibly containing large amounts of oil-in-place, but also by low permeability and productivity, which often make this type of reservoir non-commercial or marginally productive. Conventional stimulation methods have failed to give sustained improvement in productivity owing to the specific properties of soft chalk: ductile behaviour, low hardness and homogeneity. Potential rewards would accrue if this rock could be made commercially productive since it is widespread in Europe and in the Middle East.

Research carried out by Shell and subsidised by the European Economic Community has resulted in: the development of a new stimulation technique, consisting of an acid-fracturing treatment in which the fracture is subsequently propped: a drainhole technique employing ultra-high-angle wells: a geological model of chalk reservoirs; and methods of detecting fractures and determing their orientation in the reservoir. Combination of these may give sustained improvement in well productivity and thus substantially reduce the costs per unit volume of oil produced from chalk.

1. INTRODUCTION

Chalk rock is widespread in Europe and forms an exploration objective in parts of the North Sea.

Characteristics of chalk reservoirs are:
- high porosity (20-40%);
- low permeability (0.1-10 mD);
- thick pay zones with large accumulations of hydrocarbons.

Except where secondary permeability is developed, wells in these reservoirs have a low natural productivity and conventional stimulation treatments give poor results. Generally speaking, most of the chalk reservoirs are therefore marginally profitable.

Because of the potential rewards which would accrue if these reservoirs could be made commercially productive, a research effort was sponsored by the EEC between mid-1976 and end 1980 at the Shell Laboratory in Rijswijk, the Netherlands, to develop suitable methods to increase the individual well productivity in tight-chalk reservoirs.

The chalk project comprised three basically independent lines of investigation:
1. development of improved techniques for well stimulation, aimed at obtaining sustained improvements in well productivity;
2. development of techniques for drilling and completion of more effective drainholes, aimed at increasing the inflow from the reservoir into the wellbore;

3. development of logging techniques for improved evaluation of productivity, both to avoid continuation of non-commercial ventures and to provide information for optimum drainage of productive reservoirs.

The results of the research and of initial field testing of the stimulation technique are the subject of this paper.

2. RESEARCH ON STIMULATION

Conventional techniques of stimulation of wells in soft chalk reservoirs, where no secondary permeability is present, have given disappointing results: typically, the wells have low post-treatment productivity and a rapid production decline.

The failure to obtain sustained productivity improvements in chalk reservoirs can be attributed to the physical and mechanical properties of soft chalk:

1. the ductility of the chalk will cause 'pore collapse' leading to a pronounced reduction in permeability beyond a critical stress level. This will result in a strong decline of the productivity index with increasing well draw-down and depletion;
2. the homogeneity of the chalk will give evenly etched fracture walls in conventional acid-fracturing treatments, resulting in low fracture conductivity;
3. the low yield strength of the chalk will cause healing of both natural fractures and induced fractures with acid-etched grooves. This will also result in a strong productivity decline with increasing well draw-down and depletion;
4. the low hardness of the chalk will lead to excessive proppant embedment, resulting in low fracture conductivity.

2.1 Hydraulic Fracturing

Chalk is a very homogeneous, 'fine' carbonate rock. It has a low yield strength and a low Brinell hardness. Soft chalks experience a sharp increase in compressibility with increasing rock pressure at a critical stress level. This phenomenon, referred to in the literature as 'pore collapse', can lead to a pronounced reduction in permeability. The failure mechanism of soft chalk is ductile at high rock pressure. In tension, on the other hand, failure is brittle, similar to that of sandstone.

The fracturing behaviour of soft chalk has been tested in the laboratory on large cylindrical chalk samples under confining rock pressure. The samples were hydraulically fractured from an open borehole in the centre, drilled either along or at 45° with respect to the sample axis. The fracturing behaviour of chalk was found to be 'normal', resulting in tensile fractures along a plane perpendicular to the least in-situ principal stress direction. In those cases where the direction was perpendicular to the wellbore axis, two-winged tensile fractures developed. The dimensions of these fractures were in accordance with the Geertsma-de Klerk theory.

In those cases where the borehole axis was not perpendicular to the least principal stress, the fracture initiated from a short interval of the open borehole and propagated perpendicular to the least principal stress. The open borehole length intersected by the fracture was very short. It ranged between one and five borehole diameters.

The important implication of this result is that a fracture induced through a perforated interval from a deviated well may be in contact with only a few perforations, which could severely restrict oil or gas production. Some degree of shear failure and stepwise re-orientation may be associated with the propagation of these fractures, which could further restrict fracture conductivity. All this can be avoided by drilling deviated wells along planes perpendicular to the least in-situ principal stress.

2.2 Fracture Conductivity

Fracture-conductivity measurements under simulated closing pressure have been made in chalk samples with etched grooves and various types of proppant. The test results can be summarised as follows:

- a strong decline in the conductivity of unpropped fractures with etched grooves was observed in soft chalk with increasing closing pressure. At a closing pressure of 240 bar the conductivity was practically zero. Visual inspection of the fractures after the experiments showed that the etched grooves had collapsed,
- in more competent chalk with a Brinell hardness of more than 100 MPa, the grooves did not collapse and the decline in fracture conductivity was not so severe,
- non-etched fractures in soft chalk, propped with a non-crushing proppant such as sintered bauxite also showed a strong decline in conductivity with increasing closing pressure,
- proppant embedment in all the samples proved to be limited to approximately one-half grain diameter per fracture wall. It was found to be necessary to place at least three and preferably more layers of proppant in the fracture to maintain adequate fracture conductivity.

2.3 Reaction Kinetics

As a further subject of research on stimulation, the reaction kinetics of various acids on different types of chalk have been measured under dynamic flow conditions as a function of shear rate in a rotating-disc cell. Special attention was paid to the influence of the saturating pore fluid in the chalk on the reaction rate. From these tests it can be concluded that the rate of reaction of HCl with brine-saturated chalk is adequate, while that of the same acid with oil-saturated chalk can be too low for acid-frac treatments. A surfactant may therefore need to be included in the acid formulation to help flush away the oil.

2.4 Conclusions

The conclusions derived from the stimulation research can be summarised as follows:

- Competent Chalk Reservoirs

These can be stimulated by an acid fracturing treatment using a viscous pre-flush followed by acid with a water wetting agent.

- Soft Chalk Reservoirs

Here a propped fracture is required to maintain fracture conductivity. As about one-half grain diameter is embedded in each fracture wall sufficient width should be created to allow three or more proppant layers to be placed.

In case fracture growth has to be limited to avoid extension of the frac into nearby fluid contacts, a combination of propped and acid fracturing is applied. Such a propped acid frac is wider (width is a combination of hydraulic fracture width and acid-etched width) than a conventional hydraulic fracture and can accept multi-layers of proppant.

3. STIMULATION FIELD TESTS

Initial field testing of the propped/acid stimulation technique took place in the Danian chalk formation of the Dan and Gorm fields, offshore Denmark, and of the Albuskjell field, offshore Norway. Mechanical problems and sand-outs precluded complete execution of the stimulation programme in some wells.

Vertical growth of the fracture towards the gascap has resulted in early GOR build-up in various wells.

Some 50 chalk completions have now been stimulated using the viscous pre-flush/acid technique, either with or without proppant. The results have generally been good. The productivity improvements obtained were close to prediction and the production levels showed no strong decline. Such results are in strong contrast to those obtained with various other treatments in chalk completions, which typically showed poor productivity improvement and a rapid production decline.

4. RESEARCH ON DRAINHOLE DRILLING

In view of the expectation that more effective drainholes in a tight reservoir might substantially contribute to the improvement of its productivity, an effort was made to determine the productivity of a number of alternative drainhole configurations in comparison with that of a single vertical hole.

The production performance of various drainhole configurations was investigated with both an electrolytical analogue model and an analytical approach. Multiple drainholes, highly deviated wells and the effect of fractures were included in this investigation.

The most important conclusion of this study is that a drainhole with a single branch extending almost horizontally over a large distance through the reservoir will yield a considerable improvement in productivity as compared with that of a non-stimulated vertical well. Improvement in productivity due to deviation begins to manifest itself only after a deviation of some 40° has been reached. Increases in productivity of a factor 5 of or greater are observed at deviation angles in excess of 80°. Stimulation will approximately double the productivity improvement already obtained from deviation alone.

An analytical method was developed for calculating the productivity of any arbitrary drainhole configuration. Using this method it was calculated that one selectively stimulated high-angle well with five small vertical fractures can replace both the total productivity and the total drainage area of 4-5 vertical fractured wells. The orientation of the fractures in the horizontal plane does not affect the calculated productivity of the drainhole. It is stressed, however, that to ensure optimum inflow from the fractures the well should be drilled, as has been mentioned, perpendicular to the least in-situ principal stress.

Reservoir stratification will not affect the production rate of such a deviated well, provided the fractures extend from top to bottom of the pay zone.

An experimental and computational study was made of borehole stability during drilling of high-angle wells in a soft-chalk reservoir. That study revealed that deviated boreholes in a typical soft-chalk reservoir in the North Sea will neither collapse nor fracture, regardless of deviation angle, provided the proper mud weight is used. However, reservoir depletion leads to an increase in effective stresses. Hence, a high-angle open borehole might not remain stable, which hampers application of this technique in partially developed fields.

A review of high-angle hole drilling practices in the USA shows that drilling and completion of such wells (up to 85°) is technically feasible. All the equipment required was considered conventional.

It is estimated that the total drilling, completion and stimulation costs of a highly deviated drainhole offshore will be at least twice those of a conventional well. Since a highly deviated well can replace 4-5 vertical wells, it is possible that there is a potential to reduce the development drilling costs. On offshore platforms, where space is at a premium, reduction of the number of well slots could also yield considerable savings.

The main conclusions of the drainhole drilling research can be summarised as follows:

- one selectively stimulated high-angle well drilled perpendicular to the least in-situ principal stress can replace four vertical, fractured wells in tight chalk;
- borehole stability during drilling of high-angle wells in undepleted reservoirs can be avoided by using a proper mud weight;
- savings of some part of offshore development drilling costs can be expected.

No field tests have been carried out as no suitable candidate wells have been identified.

5. RESEARCH ON PRODUCTIVITY EVALUATION

A study has been made of natural fracture systems in a chalk reservoir over a domal uplift with a central graben.

The natural fracture system of strata uplifted over a roughly circular salt dome is expected to consist of a system of horizontal fractures and a system of vertical and radial fractures. The orientation of hydraulically induced fractures is expected to be vertical and radial. If the structure has a central graben, the vertical fracture system is expected to be approximately perpendicular to the major graben faults in the central region of the structure, but to tend to become radial (with respect to the centre of the domal uplift) towards the end of these faults. The least compressive stress trajectory will be roughly elliptical, with the major faulting direction as the long axis of the ellipse.

Several petrophysical tools for the detection of natural and induced open fractures in uncased boreholes were investigated. For fracture detection in highly porous chalk and chalky limestone formations, a sonic tool, such as the Circumferential Microsonic Log (CMS) and the Circumferential Acoustic Device (CAD), is expected to give the best results. Both tools are based on the same principle. Acoustic signals are generated and propagated circumferentially around the borehole. The attenuation of the shear wave is measured in four quadrants between two transmitters and two receivers. If an open fracture is present in any one of the quadrants, the shear wave attenuation in this quadrant will be more severe.

The CAD, designed by Shell Development Company in Houston, has been adapted to detect fractures in highly porous chalks. In these chalks the shear wave arrives too soon after the direct fluid-wave to be detectable. The attenuation of the latter, and much stronger, pseudo-Raleigh wave is detected instead. A compass has been incorporated in the CAD to determine the orientation of hydraulically induced fractures intersecting the borehole wall.

During various field tests a number of technical problems were identified and solved. An encouraging response to formation characteristics and repeatability was obtained in the last field trial. Owing to the few fractures encountered this test was not conclusive as to the fracture finding capability of the tool.

In the meantime other developments to detect fractures, such as the borehole televiewer, were being developed. These were considered more promising, and consequently further testing and development of the CAD tool has been stopped.

6. CONCLUSIONS

The final conclusions of the various studies were:
- competent chalk reservoirs can be stimulated by an acid fracturing treatment utilising a viscous pre-flush followed by acid with a water wetting surfactant;
- in soft chalk reservoirs the induced fracture must be supported with proppant, as the formation is too weak to support the stresses and keep the fracture open;
- at least three layers of proppant are required to allow for embedment of grains in the fracture wall. Sufficient fracture width is created either by high pumprates or, in case the lateral growth of the fracture is to be limited, by widening the fracture using acid;
- high-angle wells with selectively stimulated intervals may lead to a significantly higher production rate and drainage area per well, but this has not been field tested;
- the CAD tool showed an encouraging performance in detecting fractures, but development was not progressed because more promising methods became available.

(05.26/81)

EFFICIENT EXPLOITATION AND UTILIZATION OF SCHANDELAH OIL SHALE

P. Wenning and Dr. Kruk
VEBA OEL Development Company
Coal Mining Company of Braunschweig

Summary

VEBA OEL AG developed a process for oil shale pyrolysis - called cyclone pyrolysis - with high efficiency even for oil shales yielding only small oil gains of only five per cent. VEBA OEL AG and Coal Mining Company of Braunschweig initialized a development program in 1975 aiming the exploration of the largest oil shale reservoir in Western Germany, pilot plant tests of the cyclone pyrolysis, and the construction of a commercial plant according to this process. The reservoir exploration was partly completed. For the beginning of 1985 the put - into - operation of the pilot plant is scheduled with a capacity of 300 kg oil shale per hour. The pilot plant is at Gelsen-kirchen Scholven location of VEBA OEL AG.

1. STATUS AND RESULTS

In addition to the limited oil reserves of Western Europe oil shale - besides coal - plays a major part as alternative energy source.

The largest oil shale reservoir in Western Germany - the reservoir "Schandelah" - is located between the towns Helmstedt, Wolfsburg, and Braunschweig, see Fig. 1. Claims for this reservoir belong to the Coal Mining Company of Braunschweig.

By thermally treating the oil shale, pyrolysis oil - shale oil - is obtained which favourably can be processed to receive motor gasoline and fuel oil.

In order to exploit the "Schandelah" oil shale with economic benefits, VEBA OEL AG and Coal Mining Company of Braunschweig initialized in 1975 the following development programm:

Phase I   : Exploration of the reservoir
Phase II   : Process comparison of pyrolysis processes,
             laboratory pyrolysis experiments, shale oil treatment
             experiments
Phase III : Oil shale exploitation study,
             development of a pyrolysis process with beneficial em-
             ployment of the energy of the oil shale and
             preparatory tests for own development
Phase IV   : Further reservoir explorations and
             process development for pilot plant
Phase V   : Construction of pilot plant
Phase VI   : Construction of commercial plant

Phases I through III have jointly been investigated with the Coal Mining Company of Braunschweig. They were supported by the Ministry of Technology of the Federal Republic of Western Germany. These investigations have been completed . The development and plant test of the oil shale pyrolysis process is carried out by VEBA OEL AG with financial support of the Ministry of Technology and the Commission of the European Community. The oil shale to be processed is donated by the Coal Mining Company of Braunschweig.

From the reservoir exploration the following results were derived: The reservoir is divided into a mean - and a side - reservoir, see Fig. 2. The capacity of the rservoir is about two thousand million tons of oil shale. The oil schale has a thickness of 32 - 42 m, the sedimentation of the shale is through-shaped and exploitation can be done by surface mining, see Fig. 1.

Bench scale pyrolysis runs with samples from drilling showed that about one hundred million tons of shale oil (i. e. 5.2 per cent of the shale) as well as three per cent of pyrolysis gas (C_1 - C_4) can be obtained.

Furthermore, the residue still contains about five per cent carbon. Provided beneficial utilization of all three pyrolysis products - gas, oil and carbon - about 13 per cent of the oil shale can be utilized for energetic purposes. This is the requirement for economic utilization of oil shale, from which only small product yields can be obtained as in the instance of "Schandelah" oil shale.

The process comparison of pyrolysis processes for oil shale showed, that no process satisfies technical as well as economic requirements. Because of this, the development of an own process was initialized. This process enables to use the complete energy of the oil shale, which has been proved by preparatory tests.

In the phase IV which is now in progress this process shall be tested and improved by long-term pilot plant runs with a throughput of 300 kg/h.

The mechanical completion of the plant is scheduled for the end of 1984. Running the plant can start in the beginning of 1985.

2. DESCRIPTION OF PYROLYSIS PROCESS

In the process investigated by us, heating up the oil shale to pyrolysis temperature is carried out by direct heat exchange with hot, recirculating pyrolysis gas in a two-step cyclone system; the residual carbon of the pyrolysis residue is burnt in a fluidized bed.

Fig. 3 shows a simplified process scheme of the pilot plant. The oil shale is delivered as ground material. When passing the first cyclone step the oil shale already predried in the mill is heated by flue gas to a temperature of about 250 °C. The heating in the first step is carried out that way, that pyrolysis does not start yet.

In the second step (pyrolysis cyclone) the oil shale is heated to a temperature of 450 - 500 °C to start pyrolysis. Instead of flue gas, hot pyrolysis gas is employed as heat transfer medium. By this, mixing of pyrolysis gas with alien gases can be avoided.

The oil shale pre-carbonized that way passes from the pyrolysis cyclone to the pyrolysis drum. Residence time of the oil shale in the drum is approximately half hour and can be modified by altering the inclination and the revolutions per minute of the drum. The pyrolysis drum is equiped with an indirect heat supply. By this, specific throughput of the pilot plant shall be increased.

The hot pyrolysis residue from the pyrolysis drum is fed directly to a fluidized bed oven, where the residual carbon is burnt. The combustion air is heated in a two-step cyclone system by direct heat exchange with the residue from the fluidized bed oven. The oven is furnished with a heat plate for steam generation.

Condensation of pyrolysis gas is carried out in two separate steps. First, the circulating gas/pyrolysis gas - mixture is cooled by quench oil to about 300 °C. Pyrolysis oil condensed from the pyrolysis gas serves as quench oil. For separating oil shale dust the pyrolysis gas/ circulation gas mixture passes a high efficiency dust removal cyclone before passing the gas cooler. The heavy pyrolysis oil fraction can be mixed with refinery residues.

In a second condensation step the pyrolysis gas is cooled to 35 °C. The condensate which is pyrolysis oil and water passes to a seperator. The heavy oil is processed by hydrogenation in the technical laboratory of VEBA OEL Development Company or mixed with crude from the refinery. The pyrolysis water is fed to an existing waste water treatment plant.

One part of the circulating gas/pyrolysis gas-mixture - the generated pyrolysis gas - passes to an internal gas desulfurization. The other part of the gas is heated by the flue gas from the pyrolysis drum and the fluidized bed oven in the circulation gas/flue gas heat exchanger as circulating gas to 600 °C, and employed for heating the oil shale.

3. <u>CONCLUSIONS</u>

By cyclone pyrolysis as developed by VEBA OEL AG best utilization of the oil shale energy is obtained. Further advantages of the pyrolysis process are:
- by direct heat exchange between gas and solid high throughput is possible with small equipment sizes. This results in lower investment costs.
- fast heating step of the oil shale provides high oil yields.
- during residue combustion in the fluidized bed sulfur is integrated as calcium sulphate in the combustion residue.

If the development can be completed successfully the process can also be used for exploitation of other oil shale reservoirs inside as well as outside the Common Market. Simultaneously this project serves as an extension of the common European technology.

Oil Shale Field in the Area of Shandelah

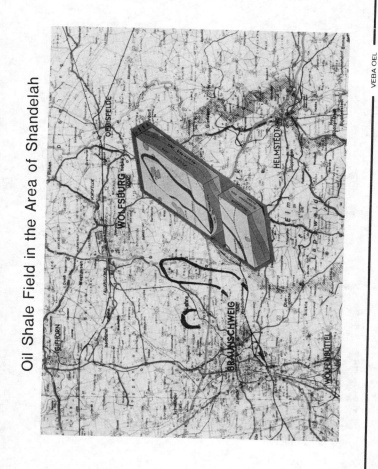

VEBA OEL

Fig. 1

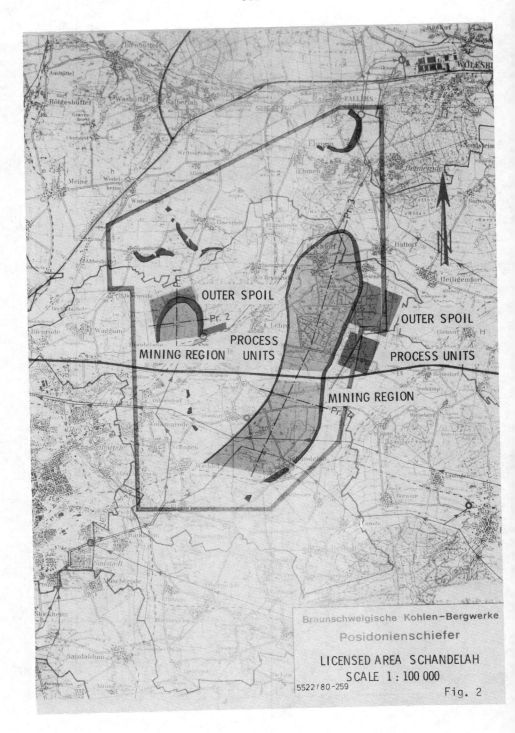

Fig. 2

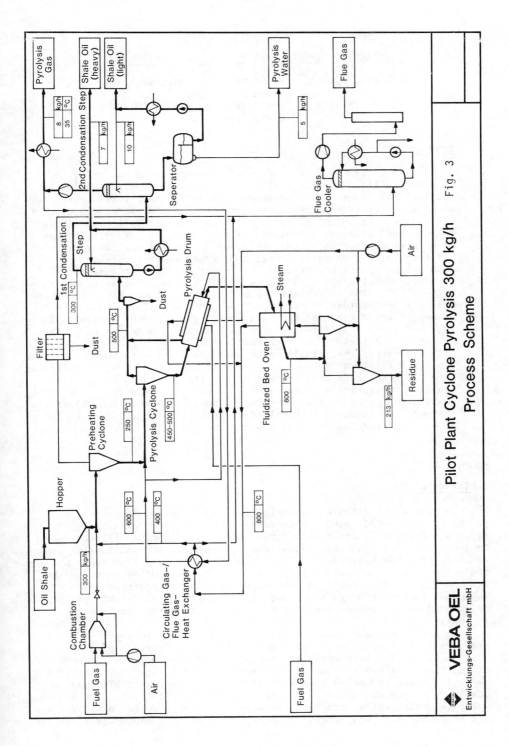

Pilot Plant Cyclone Pyrolysis 300 kg/h   Fig. 3
Process Scheme

VEBA OEL
Entwicklungs-Gesellschaft mbH

(05.01/76)

TREATMENT FLUIDS TO IMPROVE SEA WATER INJECTION

D.E. GRAHAM, J.P. HYATT AND A. STOCKWELL
BP International plc
P. DAVISON, D.C. GARDINER AND D.M. GRIST
BP Exploration plc

Summary

A micellar fluid has been developed for stimulating sea water injection wells which are limited in capacity by residual oil in the critical reservoir flow zone around the well. In initial laboratory work, much effort was devoted to establishing the basic mechanism of oil displacement by micellar fluids in porous media. In particular the physical properties controlling residual crude oil displacement from pores were established. Sand bed tests with various geometries and core tests were carried out, the latter under reservoir conditions. These demonstrate the effectiveness of the micellar fluid for raising effective permeability to sea water at temperatures up to 100°C. The injectivity of a Forties Field sea water injection well has been raised by treatment with this micellar fluid. The 50 tonnes of fluid required for field trial was blended, shipped and injected without difficulty. The initial improvement in injectivity declined to a level lower than that which had been predicted. Although rather inconclusive with only small gains in injection rate, the field trial was encouraging. The breakthrough of micellar fluids into produced crude oils is beneficial with regard to crude oil emulsion resolution with and without chemical demulsifiers. A brief description is presented of sea water soluble thermally precipitable polymers for in-depth vertical flood profile correction and their effects on produced emulsion resolution.

1. INTRODUCTION

Sea water injection to maintain reservoir pressure is currently practised in several North Sea reservoirs viz: Beryl, Forties, Montrose, Piper and Claymore. Eventually direct sea water flooding will take place in some cases. If injection wells are completed in the oil-bearing zone the rate of water injection is restricted by the presence of crude oil in the pores of the rock particularly in the region immediately around the well. This oil is immobile despite continued water injection. The injection rates can be raised, however, if the oil in this critical region is removed as the conductivity (permeability) of the formation will be increased.

Very considerable increases in injection rates were reported by previous investigators who used micellar solutions (complex surfactant solutions) to stimulate wells by residual oil removal (1,2). Reduction of the number of wells required and improved flexibility in operating the reservoir water injection programme are particularly desirable ends in costly North Sea operations. Micellar fluids for stimulating low salinity water injection wells have been available for some time (3) but these are not effective with sea water. The first part of this paper describes the development through to field testing of a sea water compatible micellar fluid for well stimulation at temperatures up to 100°C as well as monitoring the competitive effect on crude oil demulsification processes.

Micellar fluids (or microemulsions) contain dynamically structured arrays of surfactants which retain aqueous and oil phases within the arrays. Usually they are clear liquids but can be slightly opalescent and they have a high but controllable viscosity. A high proportion of surfactants is required in the formulation of such fluids and their function is controlled by interfacial activity, rheology and ability to absorb in-place brines.

The second part of this paper describes the use and effects of polymers that precipitate from solution above a certain temperature. Such systems could be used to modify water injection profiles in-depth and therefore to adjust the areal or vertical sweep of a reservoir.

2. MICELLAR FLUIDS

2.1 Mechanism of Oil Displacement

Using some initial micellar solutions, observations of residual oil displacement in model capillary systems similar in concept to those used by Jones (4) showed that the efficiency of oil displacement depended on the viscosity of the micellar solution and the interfacial tension between it and the crude oil. It was clear that the fluid must absorb the aqueous phase encountered in order to reach and displace many of the isolated oil drops. A Winsor type II+ micellar fluid was hence desirable and not a complex middle phase forming system (5). In the capillary array (6) action commenced with bulk displacement of most of the trapped residual oil with mobilised oil connecting with further oil and the whole moved forward by the viscous micellar fluid. This stage is followed by entrainment of oil from films and small adhering drops left behind. This second stage displacement is assisted by the high surface drag force exerted by the micellar fluid and the reduced interfacial tension ($\sim 10^{-2}$ mN m^{-1}). Almost complete oil displacement occurred if the silicate capillary surfaces were preferentially water wetted whereas thin oil films remained in the oil wetted case. The latter would not be expected to restrict permeability significantly – as borne out by subsequent sand pack tests.

Sand column tests showed that the use of a limited slug of micellar fluid was quite feasible. Distinct oil bank formation and good oil removal was achieved. Fig 1 records the halfway point of a vertical section sand bed test (100 mm x 420 mm x 4 mm) in which sea water flood residual crude oil is being displaced by micellar fluid injected at the centre of the bed through the inclined tube. The black outer zone is the oil bank (zone of high oil saturation) and in the grey zone behind it oil drops travel forward in the fluid – corresponding to stage two of the displacement mechanism. The sand bed tests also showed that the use of a biopolymer solution in sea water to drive the micellar slug forward more efficiently reduces expected improvement in permeability due to polymer remaining trapped in the bed. A micellar fluid viscosity of 20-30 cP at room temperature was found to be sufficient for good oil displacement (for 8 cP oil) and yet not give rise to excessive drive water penetration. Sea water dilution without immediate and rapid viscosity loss and without breakdown into two or more phases was essential. Any such breakdown leads to permeability reduction due to the re-entrapment of one of the phases and due to emulsion flow (7).

2.2 Micellar Fluid Composition

Many studies of micellar fluid compositions have been reported (8,9). In our case a rapid development programme based on the construction of partial ternary diagrams by titration methods established a small range of micellar fluids stable at temperatures up to 100°C. The fluids comprised kerosine as oil phase, sea water and approximately 20%w/w of surfactants. For reasons not fully understood increased alcohol concentrations reduced thermal stability but 2-6% alcohol (2-butanol) was essential for viscosity control. It was also essential to have the ammonium salts of anionic surfactants. With the sodium salts precipitation of insoluble derivatives from the micellar fluid occurred in the presence of sea water at temperatures above 30°C. The formulation selected for the field trial was coded FT1/3 and its ternary diagram is given as Fig 2. The manufactured composition is that of point X on the diagram. We are not concerned with the viscous compositions lying to the upper right of X as in use the fluid dilutes towards the sea water apex. Salient properties of FT1/3 are given in Table I. The composition at X is freeze-thaw cycle stable with a freezing point of -10°C. Since the transport of sea water to a blending plant is undesirable, the practical formulation to which the Fig 2 diagram relates was made using 5% rock salt brine and in use was diluted with sea water.

2.3 Evaluation Tests

In sand column (520 mm x 10 mm dia) tests the effective permeability to sea water (K_{ew}) was raised by a factor of 2.5 over that at residual crude oil conditions at 22°C, 60°C and 90°C. A slug volume of 12.5% was found to be a reasonable minimum. A radial flow test in a packed triangular sand bed at 22°C showed that a 200% increase in injectivity was possible. Equipment was set up to conduct tests with core samples under reservoir conditions (175 bar and 90°C – Forties conditions). The 25 mm dia core samples were held in a Hassler sleeve variable length core holder developed at the BP Research Centre. The back pressure control system was also a special development consisting of a transfer barrier vessel connected to the exit line from the core on one side and to a hydraulic oil circuit on the other. Standard back pressure control was applied to the rapidly flowing oil and the steady pressure was exerted on the exit fluid from the core. Corrosion of the unit was avoided by construction in 316 grade stainless steel and the absence of welded connections. Cores were obtained from a UK outcrop source (Lower Cretaceous, Hastings Beds) or were consolidated reservoir cores.

Core tests were conducted by first saturating the cores with reservoir brine, measuring absolute permeability (K_a) and then flushing with 5 pore volumes of separator crude oil at 80°C. The cores were left to equilibrate at this temperature for 20 hours. A sea water flood was then commenced at the stated temperature in Table II until a steady K_{ew} is increased due to residual oil displacement and a 100% improvement could be expected with confidence. The reservoir core tests suggest that a considerably greater improvement is possible – ca 150%. Permeability improvement is independent of flow rate as would be expected and good results were obtained at temperatures up to 91°C. The higher flow rates are comparable to those in the formation around an injector and the lower temperatures are typical of an established injector due to the cooling effect of the prolonged injection of sea water. The relative amount of micellar fluid used is large because of flow line losses which were difficult to estimate.

2.4 Injection Well Stimulation Trial

The first well available on the Forties Field for a test was injector FB26. This had a very long injection interval – 6.5 m of oil sand and 52.6 m of aquifer and an inclination of 63.5°C. The amount of fluid required to treat a reasonable 5 m radius zone (150 tonnes) exceeded the platform deck load limit. However the target oil bearing sand was a 5.4 m interval at the top of the completed zone interval and this zone accepted water very readily. It was decided to blend and inject ca 50 tonnes of FT1/3 in the expectation that this would mainly enter the higher region and displace oil. The well was an established injector with a bottom temperature of 16°C. Calculation, assuming a negligible change in tubing friction and a K_{ew} rise of 150% predicted an inflow gain of 40% into the upper oil sand (10) or an overall injectivity gain of 9%.

Downwell flowmeter (Production Combination Tool incorporating flow, pressure and temperature measurements) surveys and a pressure fall-off (PFO) test were conducted before and after the trial to establish the change in injectivity index (injection rate/injection pressure) and skin effect (if any). The flowmeter surveys were difficult to evaluate because of high angle well deviation and a restriction two-thirds of the way down the injection interval which obstructed the flowmeter. Centralising the flowmeter in the inclined well was also a problem.

Fifty tonnes of FT1/3 were blended at the BP Detergents plant at Pumpherston, Livingston, Scotland. The fluid was shipped from the BP base at Dundee to the Forties Field in acid transport tanks. On the platform the fluid was pumped into the well at 5.0 barrels per minute (0.8 m³/min) and the well was then reconnected to the sea water injection manifold. The injection rate was observed to fall over some 20 minutes as the viscous FT1/3 entered the formation but it then rose (just as in core tests) to a higher level than before treatment. Over the first two days a 25% gain was recorded (Fig 3) and 5 months later this had increased to 40%. The injectivity index rose from 11 to 13 based on downwell flowmeter and BHP measurements. In general the trial was a qualified success limited by the small extent of oil bearing sand in the long injection interval.

In any subsequent test it is hoped to reduce the amount of micellar fluid required by shifting the composition up the ternary diagram away from the brine apex. Blending of a concentrate with sea water on the platform would be desirable but requires a special mixing tank. Use of separator crude oil as oil phase is impossible because of the solid wax content of cold crude oil which would filter out on the sand face.

2.5 Resolution of Micellar-Containing Simulated Emulsions

The resolution of simulated water-in-crude oil emulsions containing various concentrations of the micellar fluid FT1/3 and the individual components have been determined in both laboratory tests (11,12) and on a pilot plant, designed and built at the BP Research Centre, which attempts to simulate production conditions (13). Emulsions were prepared with oil which had previously been displaced from sand in the displacement test loop by the micellar fluid. Interestingly, the resolution of such emulsions without addition of a proprietary demulsifier was both faster and quantitatively better than when demulsifier chemicals are added at the 30 ppm level.

The inhibitive effect of chemical demulsifiers resolving emulsions containing residual micellar fluids is likely to be a consequence of the micellar components preventing demulsifier molecules adsorbing to the oil-in-water interface. Hence, it is reasonable to postulate that the microemulsion components can (i) adsorb to displace some of the indigenous crude surfactants, or (ii) adsorb to form a mixed interfacial film with a lower saturation interfacial tension than the indigenous surfactants alone. Consequently, this interdigitation with crude surfactant molecules already adsorbed at the oil-water interface, which reduced the intermolecular cohesion of the adsorbed film and thereby reduced the interfacial rheological properties, could well account for the enhanced coalescence of water droplets and thus emulsion resolution.

The principle component in the formulations to affect simulated emulsion resolution is the commercially available alkyl benzene sulphonate surfactant (ABS). This component is used in a neutralised form (NABS) and is available in forms ammonium neutralised ABS (A/NABS) and diethylammonium neutralised ABS (D/NABS); both these forms were monitored in resolving the simulated Forties emulsions and compared with similar resolution by chemical demulsifiers.

In these experiments the demulsifier was solubilised or dispersed in the crude oil prior to being added to the simulated Forties emulsion and the ABS was dissolved in xylene which was a preferred solvent to heptane or other hydrocarbon solvents. The emulsions were usually prepared at 40°C and the demulsifier or ABS added to the prepared emulsion which was then shaken to disperse the chemical before being placed in the heating bath maintained at 40°C for emulsion resolution (11).

Table III indicates the resolution of 10 and 5 per cent Forties emulsions, respectively by Petrolite demulsifier RP 968 at 30 ppm level and A/ABS at 30 and 60 ppm levels. The main points of interest are firstly that it would appear the resolution of emulsions with A/ABS is concentration dependent with overdosing problems and secondly, whilst the A/ABS is not as fast as RP 968 in resolving the emulsions, it is almost as effective over the longer residence time of several hours.

Overall, the A/ABS chemical is not as effective as Tetrolite RP 968 in resolving the simulated emulsions. In addition, the A/ABS is not as effective as the microemulsion formulations in resolving the emulsions. The ammonium neutralised ABS would appear to be less effective than the diethyl ammonium neutralised ABS surfactant in resolving the crude oil emulsions. In addition there is also some evidence for overdosing the emulsions; resolution at the 15 ppm level is clearly greater than at either 30 or 60 ppm levels. It would seem that 15 ppm level of addition would be near an optimum concentration for both A/NABS and D/NABS when resolving Forties emulsions.

The water separated from the simulated emulsions by RP 968 showed little evidence of oil content compared with the water separated by the addition of ABS. This water was particularly oily in content containing values as high as 200 ppm oil. The higher the concentration of ABS used in emulsion resolution, the higher the oil content in the separated water. However, the lower the ABS concentration, the lower the oil content of the separated water. This factor, though, offers potentially the most serious problem to microemulsion component contamination of crude oils. It is important to understand that the concentration and also the neutralised form of the ABS will be controlled by the contamination level of the microemulsion in the crude oil and so cannot be predetermined.

3. THERMALLY PRECIPITABLE POLYMERS

3.1 Concept

In the later life of many waterfloods it is frequently desirable to nullify the effect of high permeability zones in the reservoirs. This is particularly important for post water breakthrough performance and would have the effect of improving the sweep factors and reducing the produced water to oil ratios. Conventional injection well treatments with gelled polymer (14), or polymer gelled in place, are well established (15). It is well known however that many water soluble polymers precipitate from solution above a critical temperature. By the use of an appropriate polymer it should therefore be possible to adjust the injection profile in depth around an injector without recourse to gelation. Thermally precipitable polymers would have the advantage that the plugging effect is in-depth.

The near-well zone around an established injector is cold. Water entering a high permeability zone is transmitted rapidly to a higher temperature region. Precipitation of the injected thermally-precipitated polymer solution would occur as the injected polymer reaches the part of the reservoir at the precipitation temperature of the polymer. This would reduce the flow of fluid through this section of the reservoir. The cold ring extends further for high permeability thief zones. However most of the fluid injected into the well will tend to flow down thief zones, and so most blocking by precipitation will occur here. The temperature upstream of the restriction will then rise and gradually a long plugged region will form, limited by the extent of crossflow into surrounding regions. Such crossflow may cause displacement of oil from zones which would otherwise be inefficiently swept. However a flow restriction will exist in the high permeability zone.

The limiting solution temperatures (T_o) for polymers depend on the salinity of the solution and to some extent on the molecular weight range. For practical polymers T_o is a range of temperature rather than a fixed point. Hydroxypropylcellulose (mol wt average MW = 10^6) precipitates from water solution at 38 – 40°C and from sea water above 32°C. Another potentially suitable polymer, soluble in cold sea water but precipitating below 75°C is poly (vinylalcohol).

Poly (acrylamides) may give a similar effect for different reasons. Hydrolysis to polyacrylic acid is significant above 50°C and if divalent cations are present bridging causes precipitation. This gives rise to interesting speculation on the real mode of action of poly (acrylamides). For practical operations in this mode, very high molecular weight viscosifying polymers are not necessary.

3.2 Laboratory Results

Initial sand column tests were carried out using sea water and solutions of hydroxypropylcellulose HPC-M (Klucel grade M supplied by Hercules Powder Co) in sea water. Results quoted in Table IV show that at 45°C and above the flow of sea water is reduced and that on cooling down the column the original sea water permeability is nearly restored. Polymer tracing with the I_3^- complex ion showed that the effect in run 1 (Table IV) was not due to face plugging and run 3 demonstrates this as well because the column was first flooded with polymer solution and then heated in this case. Table IVb displays results from high and low permeability cores connected in parallel to the flow stream. Only the high permeability core was heated. The apparent permeability ratio K_H/K_L demonstrates the flow diversion effect obtainable – test 4. Unfortunately a large slug is required to achieve a small final reduction in permeability ratio. The temperature of the high K core was too low in test 5 but some effect occurred in contrast to test 6 in which poly (vinylalcohol) (PVA) solution was employed. This last test probably

highlights the importance of crystal size and form since PVA is
precipitated and remains as a fine dispersion at 55°C in sea water whereas
HPC separates as larger crystals.

3.3 Resolution Polymer Stabilised Emulsions

The effects of these viscosifying polymers on the resolution of
simulated emulsions have been studied using a series of polyacrylamides,
under various conditions, and selected North Sea crude oils. The
polyacrylamides selected were Alcomer 80 to Alcomer 120 at concentrations
ranging from 0 - 300 ppm and the crude oils include Forties and Buchan in
emulsions of various water cuts. The main points arising from this study
demonstrate that the small but significant enhancement or inhibition of
emulsion resolution in the presence of several demulsifiers depends on
the oil type, the demulsifier type and degree of hydrolysis of the
polyacrylamide. Consequently, this complicated the attempt to obtain a
general rule regarding the likely effect of viscosifying chemicals on
crude oil dehydrating and desalting processes.

It is evident, that at very high concentrations, the polyacrylamide
may induce partial emulsion resolution, even in the absence of a
demulsifier. Unfortunately, the separated water is oily and the
interface, between the oil and separated water, is poor. This condition
also exists when demulsifiers are used to resolve the emulsions in the
presence of high polyacrylamide concentrations. In the special case of
Forties crude oil and a cationic demulsifier, the anionic polyacrylamides
tend to inhibit demulsifier resolution of the emulsion. However, on the
other hand, in the case of Buchan crude oil, nonionic demulsifiers are
enhanced by nonionic polyacrylamides in emulsion resolution. But, in
general, the resolution of North Sea crude oil emulsions is not seriously
impaired by polyacrylamides in the emulsified water, at levels less than
200 ppm.

3.4 Possible Application

Unfortunately calculation shows that indiscriminate injection into
an injector would not achieve flow diversion unless an extremely high
permeability (thief) zone exists. In most cases the solution will enter
the low permeability zones too and the thermal gradient is steeper in
these zones so that T_O is reached quite quickly. Also with lower
permeability such zones are more susceptible to particle plugging.
Thermally precipitable polymers may have application where changes in
areal sweep are required in a well cooled reservoir or where specific
injection into a particular zone is mechanically and economically
feasible.

4. CONCLUSIONS

A micellar fluid for sea water injection well stimulation has been
developed. This is effective in core tests at temperatures between 21
and 91°C under reservoir conditions. It functions by displacing pore
blocking residual crude oil to raise effective permeability to sea water.
A 50 tonne quantity of FT1/3 has been manufactured and used to stimulate
a Forties injection well without difficulty, with a reasonable benefit
and with no serious problems to downstream emulsion treatment processes.

A new concept in the use of polymers to control waterflooding has
been introduced. The precipitation of polymers above a critical
temperature can be utilised.

REFERENCES

1. GOGARTY, W.B., KINNEY, W.C., and KIRK, W.B. (Dec. 1970). Injection Well Stimulation with Micellar Solutions. J. Petroleum Technology p 1577.
2. DAUBEN, D.L. and FRONING, H.L. (May 1971). Development and Evaluation of Micellar Solutions to Improve Water Injectivity. J. Petroleum Technology p 614.
3. MORAN, J.P. and MAXWELL, W.N. (1970). Field Results of Injection Well Stimulation Treatments using Micellar Dispersions. SPE Spring Symposium 1970 SPE 2842.
4. DAVIS, J.A. and JONES, S.C. (Dec. 1968). Displacement Mechanisms of Micellar Solutions. J. Petroleum Technology p 1415.
5. HEALY, R.N., REED, R.L. and STENMARK, D.G. (1976). Multiphase Microemulsion Systems. Soc. Petroleum Engineers J. 16, p 147.
6. GRIST, D.M. and BUCKLEY, P.S. (1977). Petroleum Displacement in Porous Media by Saline Micellar Solutions. Enhanced Oil Recovery Symposium London, May 1977, published by BP Educational Services.
7. BOYLE, M.J. and GRIST, D.M. (1978). Emulsion Flow Through Porous Media. Proceedings from European Symposium on Enhanced Oil Recovery, Edinburgh July 1978.
8. SHINODA, K. and FRIBERG S. (1975). Microemulsions : Colloidal Aspects. Advances in Colloid and Interface Science 4, 281, 1975.
9. DREHER, K.D. et al (1976). Rheological Properties of Fluids composed of an Alkylbenzene Sulfonate, Decane, Cyclohexanol and Water. J. Colloid and Interface Science 57, 379.
10. DAVISON, P. GRIST, D.M., HYATT, J.P. and GARDINER, J.C. (1979). Treatment Fluids to Improve Sea Water Injection. Presentation at EEC Symposium "New Technologies for the Exploration and Exploitation of Oil and Gas Resources", Luxembourg, April 1979.
11. GRAHAM, D.E. and STOCKWELL, A. (1980). Selection of Demulsifiers for Produced Crude Oil Emulsions. European Offshore Petroleum Conference, London, October 1980, pp 453 - 458, EUR 191.
12. GRAHAM, D.E., STOCKWELL, A. and THOMPSON, D.G. (1983). Chemical Demulsification of Produced Crude Oil Emulsions. "Chemicals in Oil Industry" edited by P.H. Ogden, pp Royal Society of Chemistry, special publication No 45, pp 73 - 91.
13. STOCKWELL, A., GRAHAM, D.E. and CAIRNS, R.J.R. (1980). Crude Oil Emulsion Dehydration Studies. Oceanology International 80, Brighton, March 1980, Section F, pp 9 - 14.
14. MAZZOCHI, E.F. and CARTER, K.M. (1974). Pilot Application of a Blocking Agent. J. Petroleum Technology Sept 1974 p 973.
15. US Patent 4098337, 1977.

TABLE I

PHYSICAL DATA ON FT1/3[a]

Viscosity[b] (cSt)	Temp (°C)	Total Brine[d] Content (%)	Viscosity[c] (cSt)
57.0	14	50	33.8
39.6	20	55.5	37.5
33.5	22	58.5	39.8
11.3	40	63.5	27.9
5.5	60	69.5	10.0
4.0	70	75.0	4.1
3.3	90	97.0	1.1

[a]For composition point X on Fig 3; density 0.957 g/ml at 23°C
flash point 37°C

[b]Plant blended sample viscosity 28 cSt at 21°C, mean of 3 batches

[c]Effect on viscosity of dilution with sea water from point X (Fig 3) at 23°C

[d]All samples were stable to 90°C.

TABLE II

RESERVOIR CONDITION CORE TESTS

at 175 bar

Test No	Core	Crude Oil	Core Length mm	Test Temp °C	Sea Water Flow Rate mls/hr	Microemulsion Fraction PV	Permeabilities to Sea Water mD		
							K_a	K_{ew} at Residual Oil	K_{ew} Final After Treatment
1	Forties	F	89.5	23	94	0.67	10.4	2.8	7.8
2	Forties	F	89.5	23	140	0.76	23.0	4.0	12.2
3.	Forties	F	81.0	83	216	0.61	34.3	4.1	10.9
4.	Magnus	M	84.4	91	186	0.71	226.0	37.3	184.6
5.	Magnus	M	80.5	78	360	0.71	180.1	4.6	37.6
6.	Ashdown Sand	F	102.3	21	144	0.47	3966	645	2226
7.	Ashdown Sand	F	91.7	88	227	0.58	37.2	6.0	11.0
8.	Magnus	F*	88.0	90	198	0.82	120.3	7.8	14.6
9.	Ashdown Sand	M	139.0	21	180	0.4	389.0	30.3	57.9
10.	Magnus	F	84.4	91	186	0.66	226.0	37.3	184.6

*Live Forties crude ex 10 bar production separator used.

TABLE III

RESOLUTION 10% FORMATION WATER-IN-FORTIES CRUDE OIL EMULSIONS CONTAINING
(a) MICELLAR FLUID FT1/3 AND (b) INDIVIDUAL COMPONENTS

(a)

	Conc. (ppm) Micellar FT1/3	Vol H_2O separated (% vol emulsion) at given times (mins)					
		15	25	45	60	5 hrs	16 hrs
40	400	0.8	4.5	5.5	6.4	–	7.0
40	700	1.2	6.0	8.2	8.8	–	9.0
40	400 + 30 ppm*	–	3.3	3.5	–	3.8	–
40	400 + 30 ppm*	3.8	4.0	–	4.4	4.7	–
22	700	1.3	2.6	3.6	3.9	–	–
35	700	2.0	6.0	8.0	8.5	–	–
35	0	0	0	0	0	–	–

*30 ppm demulsifier RP 968 (ex Petrolite)

(b)

Component	Conc. ppm	Vol H_2O separated (% vol emulsion) at given times (mins)				
		20	40	60	80	15 hrs
A/NABS	15	0	3	4	6	9
A/NABS	30	0	2	3	5	8
A/NABS	60	0	0.5	1.0	1.8	5
D/NABS	15	0.5	2	4.5	7	8
D/NABS	30	0.3	1.5	4	6	7
D/NABS	60	0.7	1.5	2.2	2.5	3

TABLE IV

(a) SAND COLUMN TESTS

Sand particle size 106 − 250 microns; Column length 530 mm; diameter 10 mm

Test No	K Initial Darcy	Polymer Slug	K Darcy Sea Water	K Darcy Sea Water
1	11.3	1 PV of 0.1% HPC/M (at 45°C)	7.1 (45°C)	10.8 (23°C)
2	17.0 (45°C)	1 PV of 0.25% HPC/M (at 45°C)	3.8 (45°C)	15.4 (20°C)
3.	20.3 (20°C)	1 PV of 0.25% HPC/M (at 20°C)	14.3 (45°C)	18 (23°C)

*Column initially at 20°C but then heated to 45°C.

(b) PARALLEL CORE TESTS

POLYMER SLUG INJECTION − HIGH PERMEABILITY CORE HEATED

Using 2500 ppm polymer solutions in sea water 40 mm long 24 mm diameter Ashdown Sandstone cores

		Slug Size/ PV	Temp	K_H/K_L		
				Before Slug	After Slug	At Residual Polymer
4.	HPC−M	2.8	55°C	2.15	0.42	1.42
5.	HPC−M	2.8	40°C	2.66	1.96	2.68
6.	PVA	2.8	55°C	2.5	2.5	2.5

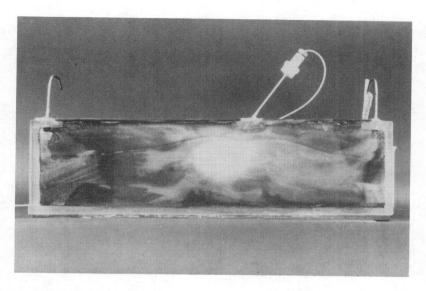

Fig. 1 - Vertical sand section test

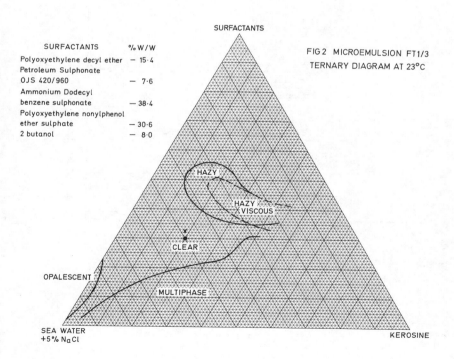

Fig. 2 - Microemulsion FT 1/3 - Ternary diagram at 23°C

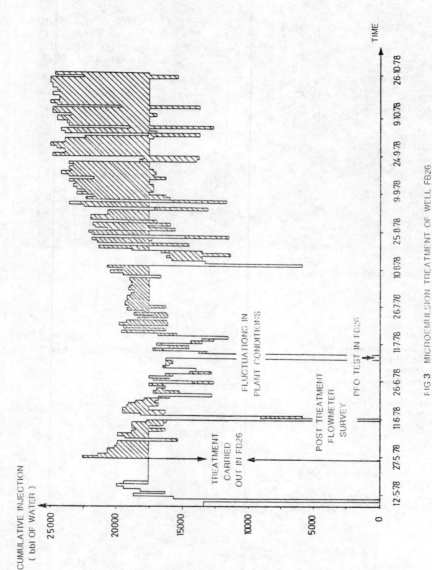

FIG 3   MICROEMULSION TREATMENT OF WELL FB26

ADDITIONAL WATER INJECTED AS A RESULT OF MICROMULSION STIMULATION OF WELL FB26

(05.05/76)

CONDITIONS OF EXPLOITATION OF TAR SHALES OF THE TOARCIAN

A. COMBAZ, Compagnie Française des Pétroles

Summary

Despite the relatively low mean oil content (40 to 50 kg/tons), the Toarcian shales in Europe represent such considerable potential reserves of gaseous and liquid hydrocarbons that Compagnie Française des Pétroles, the French Petroleum Institute and Société Nationale Elf Aquitaine (Production) have combined together to study the feasibility of exploiting these shales.
Aside from field reconnaissance by sounding, and laboratory analyses, the emphasis has been laid on study of pyrogenation technologies.
Test on the micropilot project level have enabled the optimum conditions of processing these shales to be defined, yielding an output in the same order of magnitude as the Fischer test.
Pyrolysis-cracking tests have shown that it is possible to obtain a gas rich in ethylene-based hydrocarbons, of which 70 % of ethylene itself.
In addition, industrial tests have enabled the capital and operating costs of an open-cast working to be assessed, on the basis of 100,000 tons of shales per day.
Furthermore, the combustion of shales in pellets ($\leq$ 5 mm) and blocks ($\leq$ 250 mm) has been tested by simulating the conditions of combustion by in-situ pyrolysis. Only in the second case did the results obtained prove satisfactory, since they resulted in outputs attaining 90 % of the Fischer test.
The future prospects for such exploitation schemes would appear to be essentially tied to the possibility of applying in-situ methods of exploitation.

1 - EXPLOITATION

In the conditions prevailing today, one can not afford to ignore the thousands of millions of potential tons of oil represented by tar shales in Western Europe, and especially the "card shales" of the Toarcian. In the past, these have been exploited on the very small scale: in France in the Doubs département and the region of the Causses and in Germany (Württemberg).

European card shales involve the disadvantage compared to other reservoirs in the world (USA, Brazil, USSR, etc.) of existing in only thin layers: from a few metres to a few tens of metres thick, and in particular of comprising only a low oil potential per ton of rock.

However, their outcrops are relatively widespread and justify open-cast workings at certain privileged sites.

The aim of the work carried out from April 1975 to December 1977 was to assess the exploitability of these rocks in the following two areas:
- ex-situ pyrogenation techniques,
- in-situ technologies.

2 - PYROGENATION TECHNOLOGIES (EX-SITU)

2.1 - The laboratory tests were intended to define the optimum conditions of pyrolysis of the shales at Fécocourt (South of Nancy) and to produce the oil needed for analyzing cuts obtained by fractionated distillation.

The micropilot project rendering the pyrogenation possible receives 500 gram charges of shales with a size distribution of from 1 to 5 mm.

This micropilot installation consists of:
- a vertical stainless steel reactor,
- an articulated heated shale furnace (3500 Watts), together with the necessary heating, control, measuring and recording devices.

The oil efficiencies obtained approach those of the "Fischer test". (The Fischer test is the usual reference employed, consisting in the pyrolysis of 100 g of crushed rock at 550°C in a steel "Fischer flask", with recovery of the condensable elements).

In the presence of steam, slightly more oil is obtained, but the best efficiency (107% of the "Fischer test", as compared to 98% with N_2) is obtained with hydrogen.

The basic compositions of the oils obtained are very similar, regardless of the gas used (see table 1).

Table 2 proposes a comparison between shale oil (average sample) and a Middle-East crude.

2.2 - In addition, the possibilities of valorizing the organic matter of tar shales were examined, though not in the form of oil, but that of a gas rich in ethylene compounds. The solution resorted to to achieve this was to build a pilot plant for the pyrolysis and cracking of the oil, consisting essentially of a pyrolysis reactor and a cracking reactor.

The cracking temperatures ranged from 800 to 950°C, in successive atmospheres of CO_2, H_2 and H_2O.

The quantity of oil produced falls off rapidly as the cracking temperature rises, becoming zero at about 950°C.

The ethylene-based hydrocarbons are present in considerable quantities, at least up to about 850-950°C, though their compositions vary considerably with the cracking temperature (see figure 1). At around 500-600°C, the higher ethylenes predominate, though their production diminishes as the temperature rises, whereas on the contrary, the production of ethylene increases to a maximum at about 820°C.

At this temperature, the total output of ethylene hydrocarbons represents by weight about a quarter of the oil production of the Fischer test: 70% by weight of these ethylene-based compounds consist of ethylene itself, i.e. 17 % of the oil production of the Fischer test.

By replacing the nitrogen with a variety of other sweep gases (CO_2, H_2, H_2O) during the tests made with the cracking column filled and at a temperature of 800°C (pyrolysis at 550°C), only very little change in the quantities of ethylene, methane, carbon dioxide and hydrogen would appear to occur.

The most marked effect is that of hydrogen on the non-cracked oil production, which appreciably doubles, whereas the production of gas, all in all, is modified only little.

2.3 - Industrial tests

To define the industrial conditions of processing the Toarcian tar shales of the East of the Paris Basin, the process developed by LURGI, in Essen (RFA) was used.

After working on a micropilot installation, tests were made on the "LURGI-RURHGAS" (L.R) pilot installation on a sample of 30 tons of homogenized crushed shale. The results obtained enabled an economic study to be performed that was extrapolated to an industrial modular installation involving a total quantity of 100,000 tons of rock per day.

The predicted output would be in the range of 95% of that of the Fischer test.

Principle of the process

The originality of the Lurgi-Ruhrgas pyrolysis process lies in the use of the residual shale produced by the unit as heat-transfer agent. The crushed raw shale is mixed in a pyrogenation reactor equipped with an Archimedes screw with a fraction of baked shale that has first been superheated.

The temperature of the pyrolysis reactor can be modulated between 500 and 600°C.

The outlet from the reactor communicates with a first separator, the function of which is to collect most of the solid material corresponding to the pyrolysis residue. A second separator of the cyclone type, intended to eliminate the fine materials carried over by the pyrolysis gases is followed by a condensor in which the oil + water are separated from the incondensable products.

Part of the baked shale is conveyed to the waste discharge unit, whereas the other part is led to the base of the lifting pot, where it is heated by combustion of the carbon remaining in the shale following pyrolysis (fixed carbon), or if necessary combustion of additional fuel. The residual superheated shale the purpose of which is to act as heat transfer agent is then conveyed via the lifting pot to a hopper where it is held before being mixed with the fresh shale charged into the unit.

The top part of the hopper is connected to a cyclone so as to eliminate the dust of the residual gases to be discharged to atmosphere after passing through a heat recovery system.

Investment

According to an evaluation to within an estimated accuracy of 15% in 1979, the capital investment required for the pyrogenation, condensing and dust-catching installations as a whole was 700 MDM for an output of 1.6 Mt/year of shale oil.

3 - IN-SITU TECHNOLOGIES

This involves exploiting shale in depth (in-situ), thus eliminating the stages of mining, crushing and storing the fresh shales and the problem of disposing of the processed shales. The other advantages are just as important, namely: the very limited water requirements, protection of the environment and access to depths containing very considerable reserves of shales.

In this area, therefore, the limitations are technological in nature, the main obstacle being the impermeability of tar shales.

Application of in-situ pyrolysis combustion methods hence first requires fracturation of the rock in place.

The programme was confined to combustion-pyrolysis simulation tests in the laboratory - firstly on crushed shales and secondly on shale blocks.

3.1 - Combustion on ≤ 5 mm pallets

The test covers a charge of 76 kg (of which 30% fines) introduced into a tube 20 cm in diameter and 210 cm long, at a pressure of 10 bars, in a nitrogen atmosphere ; the test was not satisfactory owing to the deformation of the combustion front and the fact that most of the oil produced burnt.

3.2 - Combustion of ≤ 250 mm blocks

The objectives set for this exploratory study to simulate "in-situ" combustion and pyrolysis were achieved using a furnace processing only 3 tons of Toarcian tar shales of the 0-250 mm grade. The results obtained correlate with those of the "Bureau of Mines", recorded in its 10 ton and 150 ton furnaces at Laramie when operating on air alone.

The first CERCHAR test, however, gave results that were remarkably better as regards the maximum temperature reached, the oil recovery and the decarburation ratio.

The following are the main results (under the aeraulic and thermal conditions of the best of the two tests):
- the average rate of advance of the combustion front is 0.1 m/h,
- the maximum local temperature can reach 1220°C in the intergranular voids ; in this way, the heat exchanges and the rate of the pyrolysis and combustion reactions are speeded up ; the result is a good average decarburation ratio,
- an oil recovery efficiency of 87% is possible under favourable conditions, first at the core of the pallets or blocks (notably a fast rise in temperature), and secondly in the charge (fairly uniform rate of progression successively of the pyrolysis front and then the combustion front),
- the width of the blocks, parallel to the strata, which can be completely burnt when the combustion front passes, is about 170 mm,
- the thickness of the blocks perpendicular to the strata, which can be completely burnt under the same conditions, is no more than about 30 mm.

Clearly, these two exploratory tests, carried out on a pilot project furnace did not make it possible to discover the optimum aeraulic and thermal conditions for processing this Toarcian tar shale.

4 - CONCLUSION

Two types of in-situ pilot project experiments exist in the USA:

- the semi in-situ process, or GARRETT process applied by the OCCIDENTAL
RESEARCH CORPORATION.
It consists in partial mining, with excavation of a gallery and
fracturation of the rock by explosives in a mining chamber.
The final tests performed are of industrial size, in strata measuring
40 m along the sides and 90 m in height, with a production of 4800 m³ of
oil (60 % of the Fischer test).
This method would be attractive primarily for shales with a high carbon
content in thicknesses of at least 50 metres, which is not the case for
European shales.
- the in-situ method that has been used by ERDA in Wyoming for several
years. The work involves the application of hydraulic fracturation
techniques, followed by explosives, with pyrolysis combustion in the
layer.
Whilst new tests undertaken since 1977 appear promising, much research
still remains to be performed before the industrial stage can be tackled.
   In addition, cost calculation shows that the methods envisaged result
in shale oil prices above those of the market today.
   In addition, the ex-situ methods also involve the need to invest large
amounts of capital and seriously affect the environment.
   In-situ methods accordingly appear the most promising, all the more
so since they employ known petroleum techniques. Further research work is
therefore being continued along this path.

Table 1

	Nitrogen pyrolysis	Hydrogen pyrolysis	Helium pyrolysis	CO_2 pyrolysis	$N_2 + H_2O$ pyrolysis	$CO_2 - H_2O$ pyrolysis
Density d_1^{20}	0.9352	0.952	0.954	0.958	0.970	0.968
Flow point °C	− 15	− 24	− 36	− 27	− 15	− 27
Kinematic viscosity Cst at 20°C	25.7	23.2	25.6	36.5	54.8	46.2
at 37.5°C	10.8	10.2	11.5	12.6	19.3	18.2
Productivity index (g/100 g oil)	44.3	44.6	45.0	46.5	46.0	45.0
Conradson carbon content % weight	2.5	2.2	2.4	2.3	3.4	2.9
Elementary analysis C% weight	83.7	82.4	82.9	83.0	83.1	83.0
H% weight	10.2	10.15	10.15	10.05	10.0	10.0
N% weight	1.1	1.0	1.0	1.1	1.1	1.0
S% weight	3.75	3.8	3.75	3.7	3.75	3.7
O% weight	-	2.65	2.2	2.15	2.05	2.3
CH	8.2	8.12	8.17	8.27	8.31	8.3

Table 2

CHARACTERISTICS	CRUDE (Middle-East)	SHALE OIL
Density at 15°C	0.869	0.958
Viscosity at 20°C (Cst)	17	23
Pour point °C	− 25	− 24
Sulphur % weight	2.5	3.5
Nitrogen % weight	0.12	1.1
Carbon/hydrogen	7	8.3
Brome index (elefins)	nil	45
Nickel ppm	7	5
Vanadium ppm	27	1
Gasolene % weight (PI-150°C)	15.5	6.0
Fuel % weight (150-400°C)	37.5	54.0

Table 3

1 - Raw materials

Tar shales	t/h	4,167
Makeup naphtha	t/h	2.8

2 - Products obtained

Purified heavy oil	t/h	76
Purified medium oil	t/h	81.36
Naphtha	t/h	18.3
Purified gas	Nm³/h	54,420
CO_2	Nm³/h	26,600
Sulphur	t/h	0.365
Pyrolysis residue	t/h	3,464
Superheated steam (75 bars, 475°C)	t/h	390

3 - Consumptions of utilities

Electricity	kWh/h	40,080
Steam at 4 bars pressure	t/h	114
Water	m³/h	5,587
Heating gas	Mk cal/h	33
Inert gas	Nm³/h	800
Instrument air	Nm³/h	1,400

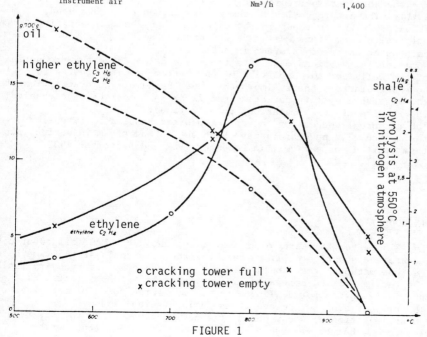

FIGURE 1

(05.07/77)

PILOT PROJECT FOR RECOVERY OF OIL BY STEAM INJECTION
(UPPER LACQ FIELD)

B. SAHUQUET, Société Nationale ELF AQUITAINE (Production)

Summary

The UPPER LACQ steam injection pilot project is original in being the
first steam injection operation in a fractured carbonate reservoir.
Preceded by experimental studies which showed that positive results
could be obtained by applying steam drive to this type of reservoir,
a pilot project was hence designed and set up. Steam injection started
in October 1977 and continued till April 1982. The enhanced recovery
amounted to 35,600 tons for a total of 251,000 tons of steam injected.
Various techniques were used to determine the progression of the hot
zone: the results obtained show that heat transfer took place
efficiently within the reservoir: no steam breakthroughs were observed
and the bottom temperatures at the production wells varied only
little. The heterogeneity of the reservoir, on the other hand,
resulted in uneven sweep in the pilot project zone. In addition,
original observations were made concerning the decomposition of the
carbonates during this steam injection.
Numerical simulations made on a thermal model enabled the production
of the wells and the evolution of the bottom temperatures to be set
on a satisfactory basis. Improvement of the numerical description of
the reservoir by introducing a "matrix drain" configuration enabled
the results of the simulations to be refined.
The technical success of this pilot project has led to extension of
the method. Two new panels have been set up and steam injection into
them started respectively in December 1982 and September 1983.

1 - INTRODUCTION

   In 1975, after 25 years of exploitation, the UPPER LACQ oil reservoir
still contained about 16 million tons, the recovery ratio at the time being
17%. Various methods that could be applied to enhance this recovery ratio
were then envisaged and continuous injection of steam was selected as the
most promising method for dealing with the fractured zones of the reservoir.
   The project adopted, however, was an innovation, inasmuch as
continuous injection of steam had so far never been applied to this type
of fractured carbonate reservoir ; a pilot project phase thus seemed
indispensable to test the efficiency of the method and to define certain

application parameters such as the optimum dimension of the panels, the injection rate, etc..

The project was carried into effect from October 1977 to April 1982 ; technically it represents a success and a number of observations have led to hitherto unpublished results concerning steam drive in a carbonate medium.

2 - DESCRIPTION OF THE RESERVOIR

Discovered in 1949, the UPPER LACQ reservoir originally contained about 20 million tons, of which only 17% have been recovered.

The trap is an anticline, the summit of which lies at a depth of about 600 metres, the impregnated thickness reaching 120 metres and the area covering 5.7 km². A large underlying and lateral aquifer has kept the pressure in the field practically constant throughout the exploitation period.

The limestones in the reservoir, of the lower Senonian, offer two types of facies with very different characteristics (see table 1): first, the "compact" limestone zones with good porosity but low permeability, and second, fractured dolomitic, and sometimes karstic, limestones, of low porosity and matrix permeability, but in which the network of cracks endows the wells with a high productivity index.

Table 1

- COMPACT LIMESTONES ZONE

Porosity	15 to 20%
Permeability	1 to 10 mD

- CRACKED DOLOMITIC ZONES

Matrix porosity	10%
Matrix permeability	0.5 mD
Cracked porosity	0.5%
Cracked permeability	5000 to 10000 mD

The compact zones are located in the North-West and South-East, with the central part of the deposit being fractured dolomitic (Fig. 1) ; the limits of the facies are still imprecise in certain portions of the reservoir, all the more so since overlays of the two types of facies occur in certain places.

The compact zone wells mostly have low productivity indices and are being worked by pumping and the product contains a high proportion of water (> 95%). The fractured zone wells have high indices (up to 1800 m³/day.bar) with natural flow and a choke at the wellhead so as to produce only the oil tha reaches the well via the network of cracks.

The UPPER LACQ oil is asphaltic, relatively dense (0.925 g/cm³, 21.5° API), its viscosity at the bottom conditions (60 bars, 60°C) is 17.5 cPo, its GOR is low (11 vol/vol) and it is sub-saturated (saturation pressure: 8 bars).

3 - CHOICE OF AN ENHANCED RECOVERY METHOD

In 1975, after 25 years of exploitation, the production level at UPPER LACQ had fallen to 50,000 m³/year. To step up production and increase the recovery ratio, it appeared necessary to apply of method of enhanced recovery, particularly for the fractured dolomitic zones where the water from the aquifer had already swept most of the network of cracks and where imbibition had become the predominant production mechanism. An efficient method well suited to the characteristics of the reservoir and the nature of the oil was therefore sought. Methods calling for drive at pore level of the fluid in place by the fluid injected appeared to be sure to fail in view of the considerable contrasts between the permeability of the network of cracks and the matrix medium. On the other hand, injection of steam seemed to be a sure means of shifting the oil trapped within the matrix blocks. This is because the steam imparted to the network of cracks by the fluid injected is transmitted by conduction to the matrix blocks, gradually raising their temperature.

The thermal expansion of the fluids saturating the porous medium then causes them to move towards the surface of the matrix blocks and be expelled into the network of cracks. The thermal expansion of the skeleton of the porous medium can also result in a reduction in the pore volume leading to additional expulsion of the fluids contained in it. The improved imbibition as the temperature rises would also be a factor enabling recovery to be increased or speeded up. The other mechanisms recognized as active in thermal methods (lower viscosity, change in the wettability and interface tensions, stripping when the temperature reaches a sufficiently high level) also play their part in the same way as in a homogeneous porous medium, though with relatively less effect than in thermal expansion.

Accordingly, injection of steam was adopted as the appropriate enhanced recovery method, and a series of laboratory experiments carried out to check the validity of this approach and to quantize the recovery ratios that could be expected where the method to be applied under the conditions prevailing in the UPPER LACQ field.

Three types of experiment were carried out:
- natural imbibition at the reservoir temperature (T = 60°C),
- hot water drive (150°C),
- injection of steam at 290°C.

The encouraging results obtained (ref. 3) were an encouragement to proceed to tests on the field, though in view of the novel character of the operation, with a preliminary pilot project before development of the process.

4 - POSITION OF THE PILOT PROJECT WELL AND SURFACE INSTALLATIONS

It was decided to locate the pilot project well in the central part of the reservoir (figure 1), where the density of the existing production wells enabled a panel with the right dimensions, necessitating only the pilot well to be drilled, and which would require specially adapted completion.

The wells of the pilot panel are situated as shown in figure 2.

The production wells apply two production methods, depending on the type of facies around them:
- in fractured zones, they are eruptive: LA 2, LA 4, LA 79, LA 85,
- in compact zones, conventional rod pumping is used.

A steam generator (Babcock) was specially installed for this pilot project, near LA 87 ; it has a capacity of 8 tons per hour and supplies the steam at 86 bars at a steam quality of approaching 100%.

5 - EXECUTION OF THE PILOT PROJECT

Injection of steam into LA 87 started on 5th October 1977 and continued until 7th April 1982. The mean daily flow was kept at 160 - 170 tons per day. The total quantity of steam injected during the project was 251,000 tons.

Throughout the pilot project, frequent measurements were made to follow its evolution and be in a position to analyze it: gaugings, pressure measurements, temperature measurements, analysis of samples, injection of tracers. It might at first sight be thought that in view of the considerable extent of the network of cracks, the influence of steam injection would first make itself felt on the quality of the reservoir waters produced or sampled (drop in the salinity owing to dilution by the condensed steam), and then by the increase in the bottom temperature, the evolution of the oil production with time being more difficult to predict.

5.1 - Evolution of the oil production

In reality, it was the evolution of the oil production that was detected first ; about three months after starting steam injection , the production from the natural drive wells (LA 2, LA 85 and LA 4, even though situated in second rank compared to the injection well) began to rise significantly. The precocity of this evolution in the output, so soon after beginning injection, confirms the recovery scheme that can be conceived for this type of cracked reservoir and that had been suggested by the laboratory tests: the steam or hot water sweeping the network of cracks gradually increases the temperature of the matrix blocks ; next, essentially owing to the expansion of the fluids, in view of the rapidity of the phenomenon, the oil is expelled from the matrix blocks into the network of cracks, where it circulates towards the production wells all the easier since its viscosity is now lower.

Figure 3 shows the pattern of evolution of the production from the wells affected by the steam injection. It shows that 3 wells situated in the distinctly fractured zone: LA 2, LA 4 and LA 85, were particularly involved and accounted for the essential part of the gains in output.

Another eruptive well, LA 30, despite its distance from the injection well, displayed a change in production in early 1978, compared with those of the above-mentioned wells of the pilot project, though this evolution was of lesser extent and shorter duration.

Well LA 37 which was placed on production only in 1979, displayed a production peak in 1980 ; however, since it had no past history, the influence of the steam injection on its output can not be assessed.

The other wells in the pilot project zone reacted only little to the injection of steam (LA 14) or even remain completely unaffected, like LA 88 and LA 79 and which, according to the interference test, was nevertheless the well best connected to the injection well LA 87.

5.2 - The additional oil and the steam-oil ratio

Calculated from the curves of decline of each well, the oil gains obtained during this pilot project amounted to 35,600 tons (figure 4). The overall steam-oil ratio is 705 tons/ton (i.e. 6.52 vol/vol).

In early 1982, the instantaneous SOR attained their economic limit values. Since elsewhere, a phase of extension of steam injection in a zone in the neighbourhood of the pilot project was planned for the Summer of 1982, this seemed the right time to avoid problems of interference between the two operations and the project was hence stopped in April.

5.3 - Progression of the hot zone due to the injection of steam

To determine the displacement and progression of the hot zone, two parameters were followed in particular:
- the path of the condensed water,
- the evolution of the bottom temperatures at the production wells.

5.3.1 - Path of the condensed water

Two types of method were used to detect the path of the water produced by condensation of the steam: first, measurements of the salinity of the reservoir water at the various production wells, and second, injections of tracer in the steam and analysis of the levels at which it reappeared at the various wells.

5.3.1.1 - Evolution of the salinity of the reservoir water

The initial salinity of the reservoir water being low at UPPER LACQ (8 g/l), the performance of this method turned out to be fairly low. Despite this, at well LA 2, a drop in the concentration of Cl^-, SO_4^{--} and Na^+ ions was observed, reflecting fairly well the dilution of the reservoir water in this Eastern portion of the pilot project panel.

5.3.1.2 - Injection of the trace element

Tritium water (^3H_2O) was chosen as trace element for the steam injected into UPPER LACQ field. A plug of tritium water (dosage 5 Curies each time) was injected twice - in February 1978 and October 1979 - together with the steam. The level of tritium was then measured in weakly samples of the water taken at the production wells. The results obtained after these two injections differed considerably.

Following the first injection, arrival of tritium water was detected about a month after the injection, essentially at wells LA 79 and LA 14, in the North part of the panel, to a lesser extent at LA 37 an very little at the other wells (LA 2, LA 4). After the second injection in October 1979, the tritium arrived very quickly (within a few days) and in considerable quantities at LA 2, whilst simultaneously dropping at the other wells, which displayed a response after the first injection. Accordingly, it would appear that a considerable change in the flow pattern of the fluids inside the reservoir had occurred between these two injections, despite the fact that the operating conditions of the field had not been changed during this period.

5.3.2 - Evolution of temperatures at the bottom

Throughout the pilot project, frequent temperature measurements were made at the production wells. During the first two years of the project, no change in the pattern of temperature was observed at any well. It was not until mid-1979, and only at well LA 2 that a gradual rise in temperature was detected. In January 1982, the bottom temperature reached only 75°C, i.e. an increase of 16°C compared to the initial value in this well.

This little change in the bottom temperatures shows:
- first, that the spacing could have been tighter, since in view of the flow injected, the hot zone is far from attaining all the area covered by the panel,

- second, that the heat transfer between the fluid injected and the porous medium was very efficient.

If this be considered in the context of the speed with which the tritium passed from LA 87 to LA 2 after the second injection of the trace isotope, one sees that there is no incompatibility between rapid circulation of the fluids and good transfer of heat between the fluids injected and the porous medium crossed, provided this good heat transfer is permitted by the high density of cracks.

The methods used to locate the hot zone generated by injection of steam all conclude as to the preferential shift towards well LA 2 situated East of the injection well.

5.4 - Action of the steam injected on the porous carbonate medium

During 1979, a distinct evolution was observed in the GOR of the crude recovered: the original GOR was about 11 vol/vol and rose to as far as 30 at this time at well LA 2. The cause of this variation was sought and analysis of the gases produced showed that this rise in the GOR was caused by CO_2 diluting the original fatal gas.

The isotopic analysis of the CO_2 produced was used to determine its origin.

Indeed, depending on whether it came from hydrocarbons or carbonates, the CO_2 presents a proportion of ^{13}C identical to that of its origin medium so this origin can be discovered through isotopic analysis.

Analysis of numerous samples taken showed that the higher the CO_2 content, the more the ^{13}C content tends towards the characteristic value of the carbonates. We were hence able to conclude that the CO_2 diluting the initial gas comes from breakdown of the carbonates (ref. 2).

Laboratory experiments using samples of reservoir rock also showed that under the conditions of injection into UPPER LACQ field, the steam sweeping this type of carbonate rock gives rise to the formation of carbon dioxide gas.

A calculation performed on all the gas associated to the oil produced during the pilot project gives a value of 275 tons of CO_2 formed from the carbonates.

This dissociation of the carbonate rock during injection of steam is hence an important phenomenon and may lie at the origin of the alterations of the network of cracks that could have caused the path of the fluids observed in 1979 to change. Equally well, the carbon dioxide gas formed was also certainly a factor assisting the recovery of the oil.

6 - NUMERICAL INTERPRETATION OF THE PILOT PROJECT

Several types of thermal models were used, both two-dimensional and three-dimensional, and various configurations were tested (radial-circular, current tubes, heterogeneous pattern: drains - matrix blocks).

Folowing adjustment of a standard configuration representative of the fractured rocks (drains - matrix blocks) (figure 5), numerical simulation of the sectors of the pilot correctly reproduced the production histories observed (figure 5).

Overall simulation of the entire pilot project operation was not fully concluded, since it quickly appeared that the considerable heterogeneity of the zone concerned, coupled with the lack of adequate fine data on the reservoirs did not provide a sufficient guarantee of the representativity of the results obtained.

CONCLUSION

Injection of steam into UPPER LACQ which involved the originality of taking place in a fractured carbonate medium has yielded much information concerning the application of this enhanced recovery method to this type of reservoir:
- provided the cracking is sufficiently intense, transfer of heat between the fluid injected and the reservoir takes place efficiently and ensures the technical success of this operation, without early breakthrough of steam or water occurring ;
- by helping the oil expelled from the matrix blocks to drain towards the production wells, the network of cracks ensures that the oil production changes very quickly ;
- laboratory and production site results have shown that thermal expansion is the predominant phenomenon involved in steam-drive in a fractured medium ;
- tritium water is a choice trace isotope for steam injection operations,
- during this test, the importance of the dissociation of the carbonate rock under the effect of the steam injection appeared, at least for the condition prevailing in this reservoir ;
- the numerical interpretation of this operation has provided good partial calibration settings. However, the pilot project as a whole could not be satisfactorily simulated owing to the extreme heterogeneity of the experimental zone ;
- the extra oil production (+ 35600 tons) the project yielded is however its most tangible positive result. It has incited SNEA (P) to continue this type of operation on this field, and an initial extension phase comprising two new panels followed the pilot project. It required drilling of 8 new wells (two injection wells and six production wells). Steam injection into the first panel started in December 1982 and into the second in September 1983. Just as with the pilot project injection, an early rise in the production was observed, only two months after each new injection. The operation is continuing and the results are promising.

REFERENCES

1. B.C. SAHUQUET, J.J. FERRIER
    "Steam drive pilot in a fractured carbonated reservoir: UPPER LACQ field"
    J. Petroleum Technol., Vol. 34, p. 873 - 880 (April 1982).

2. B.C. SAHUQUET, B. CORRE
    "Injection of steam into a cracked carbonate medium - the UPPER LACQ field pilot project"
    2nd European Colloquium of enhanced oil recovery, Editions TECHNIP, PARIS, p. 611 - 623 (November 1982).

3. B.C. SAHUQUET
    "Experimental study of the injection of steam into UPPER LACQ reservoir"
    Franco-Soviet Symposium - MOSCOW - May 1977.

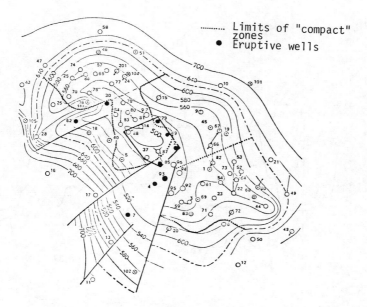

FIGURE 1 - MAP OF UPPER LACQ FIELD
Isobaths at roof of the reservoir

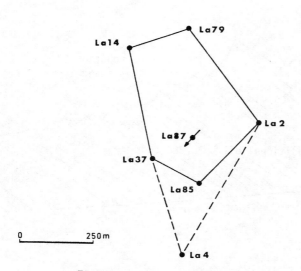

FIGURE 2 - LAYOUT OF PILOT PANEL

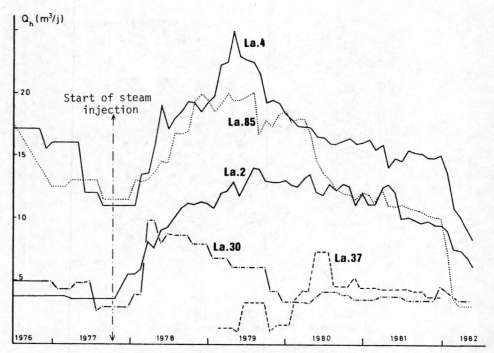

FIGURE 3 - HISTORY OF OIL PRODUCTION

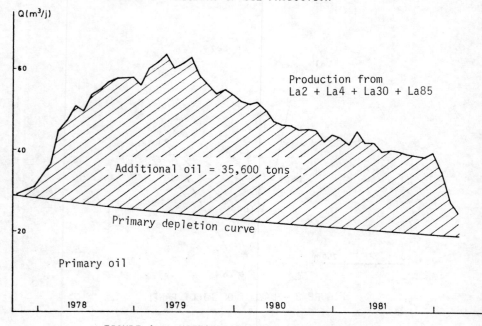

FIGURE 4 - HISTORY OF EXTRA OIL EXTRACTION

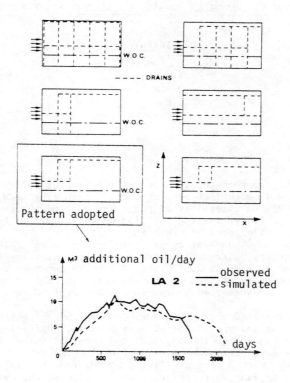

FIGURE 5 - PATTERN OF DRAINS - MATRIX BLOCKS
Application to production well LA2

(05.31/81)

TAR SHALES: THE TRANQUEVILLE IN-SITU COMBUSTION PILOT PROJECT

J.E. VIDAL, Compagnie Française des Pétroles,
Ch. BARDON, French Petroleum Institute

Summary

A research programme on the in-situ valorization of tar shales was
conducted from 1980 to 1984 by a consortium formed by BRGM, GDF, IFP,
CFP and SNEA(P). An initial study phase has enabled the general
characteristics of pyrolysis/combustion of the shale to be defined
and an initial pilot project operating strategy.
The pilot experiment was set up in Toarcian shales at TRANQUEVILLE
(Vosges department of France), where a 25 metre thick layer of
shales lie at a depth of 200 metres. Two petroleum type wells
drilled 60 metres apart from each other were placed in communication
by a horizontal unsupported hydraulic crack made from an injection
well.
Air was then injected into the crack. Combustion was ignited by
electric heating and spread for 40 days during which period the air
injection was continued. The physical and chemical parameters (flows,
pressures, temperatures, composition of effluents) were recorded
throughout the test. A core drilling 5 metres from the injection well
revealed a burnt zone 30 cm thick on either side of the crack.
Calculations show that about 75 tons of shales would be involved and
that the combustion front would spread for about 12 metres.

INTRODUCTION

Tar shales are sedimentary rocks containing over 5% of organic
matter (kerogene). When raised to a temperature of 500°, the kerogene
undergoes pyrolysis and gives coke and an oil that is comparable to
certain crudes.
Pyrogenation enables 40 to 250 litres of oil to be extracted per ton
of shale, depending on the quality of the shale. Recoverable world
reserves of shale oil are estimated at 27×10^9 tons, of which 4.10^9 tons
in Europe.
Between 1973 and 1978, the Groupe d'Etudes des Recherches Bitumineux
(GERB - Bituminous rock research group) in which participate BRGM, GDF,
CFP, IFP and SNEA(P) studied the various methods of valorization of the
shales of the Toarcian. It concluded that ex-situ exploitation is not
industrially profitable and raised major environmental problems, and that

the in-situ method could in the long term prove attractive.

The purpose of the TRANQUEVILLE pilot project was to check the feasibility of in-situ combustion of the kerogene of the shales in order to produce:
- either shale oil, or lean gas,
- or heat (subsequently recoverable by injection water).

1 - GEOGRAPHICAL AND GEOLOGICAL SITUATION

The site at TRANQUEVILLE lies about 40 kilometres South-West of Nancy and 20 kilometres North-East of Neufchateau, 1 kilometre from the village of TRANQUEVILLE.

The tar shales of the Toarcian lie at a depth of 208 and 233 metres. They are covered with Toarcian marls (121-208 metres) and Bajocian limestones (24-121 metres), with a final surface layer of limestones of the middle Bathonian. They have a very low dip (1.5° South).

The following are the main characteristics of the shale:
density: 3.35 to 2.60,
porosity: 0,
porosity after combustion: 6 to 32%,
permeability: 0,
organic carbon content: 6 to 10%,
approximate mineral carbon content: 5%,
hydrogen content: 1 to 1.5%,
sulphur content: 2 to 3%.

The oil outputs (Fischer test) are better in the bottom part of the layer. They vary from 28 to 70 kg/ton.

2 - DESCRIPTION OF OPERATIONS

2.1 - Combustion tests in the laboratory

These tests were carried out in 1981 by the IFP, the purpose being to define the pyrolysis/combustion characteristics of shale blocks (ignition temperature, penetration of fuel into the block, potential oil extraction..) depending on the operating conditions. These tests were to enable the operating conditions of the pilot project to be determined. The samples of shales used originated from Créveney quarry, near Vesoul, with a structure and kerogene content approaching that of the TRANQUEVILLE shales.

The tests were performed in a cell 10 cm in diameter and 25 cm in height, capable of operating up to a pressure of 100 bars and a temperature of 600°C. This cell is equipped with thermocouples on the wall in order to monitor the temperature settings and with thermocouples inserted into the heart of the block so as to follow the reaction temperature.

This cell forms part of apparatus consisting of:
- a gas supply system (air, nitrogene, CO_2), the composition and injection flow rate of which can be controlled and monitored,
- a trap situated at the bottom of the cell for recovering the liquid effluents (oil and water),
- a pressure regulation valve enabling the pressure of the gases to be reduced to atmospheric,
- a gas meter,
- CO, CO_2 and O_2 continuous gas analyzers,
- a chromatograph to determine the N_2, H_2 and hydrocarbon contents.

All these tests were carried out at a relative pressure of 40 bars and a temperature setting of 400°C.

Two series of tests were carried out as part of this study. In the first, the shale block was introduced into the cell alone, with a gaseous environment around every face. In the second series of tests, the gap between the block and the walls of the cell was filled either with sand, or shale debris. The main results are indicated in tables I and II.

In the first series of tests, the best oil output is obtained with pyrolysis in the presence of nitrogen. Output drops as the oxygen concentration rises in the gas injected. In air, it falls to only 25%.

In tests with combustion in the presence of air, following a pyrolysis stage, the residual carbon was burnt. For a combustion period of about 30 hours, the depth of penetration of the combustion into the block of shale was about 2 cm.

The second series of tests enabled the movement of a combustion front to be observed.

For a given oxygen content in the gas injected, better oil outputs were obtained in the tests packed with a filling, since this enables a pyrolysis zone to spread forward of the combustion zone (the oil produced is thus drained outside the cell and hence only partially oxidized by the residual oxygen).

2.2 – Environmental impact report

An environmental impact report made by BRGM in 1981 showed that the pilot operation would have only minimal effect on the environment.

2.3 – Drillings

The two wells TRA1 and TRA2 were drilled by conventional oil drilling methods. The only difficulty encountered was caused by the presence of considerable karstification of the Bajocian limestones, resulting in a complete loss of circulation.

TRA1 was equipped with a 4 1/2" casing completion at the wall of the shales, cemented up to the surface.

TRA2 was provided with 7" cemented casing at the roof of the shales and a 4 1/2" uncemented perforated liner installed from the bottom to the roof of the shales. The two wells were cored in the shale. Sedimentological, structural and geochemical analysis of the cores yielded results that complied with what was expected.

2.4 – Fracturation

Communication between the two wells was obtained by setting up an unsupported horizontal hydraulic crack from TRA1. A window was first cut into the 4 1/2" casing at elevation 225 m to enable the crack to start at the desired depth and to enable the upper part of the casing to lift freely.

Table 1
Shale block combustion tests
in the laboratory in a gaseous environment

Test	Duration h	Gas injection Mean flow 1/mn	% O_2	Oil produced as % weight of shale
SCH1 – Pyrolysis in presence of N_2 (400°C)	9	1.8	0	3.8
– Combustion	7.5	2.1	21	
SCH2 – Pyrolysis in presence of N_2 (400°C)	4.5	2.1	0	3.8
– Combustion	30	2.2	21	
SCH3 – Pyrolysis in presence of N_2 (500°C)	7.5	2	0	4.14
– Combustion	10	1.9	21	
SCH4 – Combustion (400°C)	31	1.95	1.6	2.87
SCH5 – Combustion (400°C)	29	2.2	3.8	1.13
SCH6 – Combustion (400°C)	31	1.5	8–10	1.34
SCH7 – Combustion (400°C)	32.5	1.6	13–15	1.08
SCH8 – Combustion (400°C)	29	2.0	21	0.82

Table 2
Shale block combustion tests
in the laboratory with a fill around the block

Test	Duration h	Gas injection Mean 1/mn	% O_2	Oil produced as % weight of shale
SCH10 – Combustion (400°C) (shale debris)	31	1.95	21	2.34
SCH11 – Combustion (400°C) (sand)	31	1.83	21	2.00
SCH12 – Pyrolysis in presence of N_2 (500°C)	5.5	1.95	0	3.33
– Combustion (sand)	18.75	2.0	21	0
SCH13 – Pyrolysis in presence of N_2 (440°C)	5.25	2.4	0	3.03
– Combustion	28.5	1.93	21	
SCH14 – Pyrolysis in presence of N_2 (470 to 600°C)	3	1.95	0	3.19
– Combustion (cemented block)	19	1.80	21	

To determine the geometry of the fracture, a device for measuring deformations of the surface of the soil was installed. This device consisted of inclinometers (IPG), settlement meters (COYNE et BELLIER) and a precision level (IGN).

Three fracturation tests were needed to establish the link.

During the first test, the pressure rose to 248 bars. Fracturation did not occur, since the opening in the casing was insufficient and had to be revised.

In the second test, a volume of 100 m³ was injected at a rate of 2.38 m³ per minute and a maximum pressure of 185 bars. This volume was insufficient to make a crack reaching TRA2. Analysis of the deformations on the surface reveal a circular subhorizontal fracture centered on TRA1, stopping a few metres before reaching TRA2.

Lastly, during the third test, a total volume of 1600 m³ of water was injected at 100 m³/hour and a pressure of 100 bars. The communication with TRA2 was established after injecting 400 m³ (level rise in TRA2). However, on completion of pumping, the production flow was only 1.8 m³/hour for an injection flow of 100 m³/hour. This low production can be explained by the very high pressure losses near the production well ; opening the well results locally in elastic closure of the crack, the water injected flows round the production well and continues to spread the crack.

Interpretation of the inclinometer measurements shows that on completion of the operation, the crack was circular, and centered on TRA1 with a radius of 150 metres and a maximum opening of 17 mm.

Whilst not perfect, the hydraulic link between TRA1 and TRA2 appeared sufficient to proceed with phase 2 of the project.

2.5 - Combustion tests

To be able to vary the oxygen content of the air injected, it was decided to use reconstituted air. Oxygen and nitrogen are pumped in liquid form until the injection pressure and then evaporated and mixed.

Combustion is then ignited by means of a 40 kW electrical resistor situated at the bottom of the well enabling the air injected to be heated to 400°C.

After atmospheric separation of the effluents, the gas is rejected to atmosphere through a cold vent and the liquid effluent is stored in one of the tanks used for the fracturation test.

The gaseous effluents are continuously analyzed (oxygen, carbon monoxide and carbon dioxide contents). The pressures, flows and temperatures, on both injection and production, are also recorded continuously.

There are three phases to the combustion test:

Preliminary phase (2 days): injection of 10,000 Nm³ of nitrogen at a rate of 400 Nm³/hour, followed by 3000 Nm³ of air at 150 Nm³/hour.

Ignition phase (1.5 days): the bottom temperature is kept at 400°C and 150 Nm³/hour of air is injected.

Combustion propagation phase (39 days): injection of 260,000 Nm³ of air at varying flows and composition:
- 8 days at 200 Nm³/hour,
- 7 days at a rate rising to 1000 Nm³/hour,
- 7 days at 350 Nm³/hour,
- 17 days at 110 Nm³/hour.

The results obtained can be summarized as follows:
- the injection pressures vary from 65 bars for a flow of 100 Nm³/hour, to 90 bars for a flow of 1000 Nm³/hour.

During the injection phase, a total of 12000 Nm³ of gas was produced at TRA2, i.e. 5% of the total quantity injected. The production flow depends both on the injection flow and the counterpressure maintained by means of choke of TEA2. This choking turned out to be necessary, since complete opening of the well invariably resulted in a sudden drop of production owing to local "pinching" of the crack.

The oxygen content of the gas produced varied from 8% during the initial period (injection flow 150 Nm³/h) to 17% during the high flow period (1000 Nm³/h). This corresponds to a sharp drop in the percentage of oxygen consumed. This then levelled out to 11% during the final production phase (flow 110 Nm³/h).

The carbon dioxide content remained very low throughout the test, stabilizing at 1.5%. This value is not significant, since most of the carbon dioxide produced was dissolved in the water of the crack, like that of TRA2.

No percentage of carbon monoxide was observed. Likewise the hydrocarbons content remained very low: in the order of magnitude of the sensitivity of the measuring instruments, namely 1%. The excess oxygen downstream from the combustion front is probably responsible for the almost complete combustion of the products of pyrolysis.

Coring study

Geochemical analysis of the core around the combustion zone showed that the zone visually subjected to modifications lies between elevations 225.6 and 227.0 m. The analyses made with a ROCK EVAL II apparatus equipped with an organic carbon module concerned cores between 223.3 and 228.8 m, so as to cover the zone burnt widely in order to detect the disturbances to the neighbouring layers caused by the combustion. The main results obtained are shown on the geochemical log of figure 3.

An examination of the results gives rise to the following comments:
- the main combustion zone lies in the interval between 226.0 to about 226.6 m, i.e. about 60 centimetres,
- the second combustion zone is very small (225.7 to 225.8 m), i.e. about 10 cm and situated just above the main zone,
- the combustion zones are characterized by the disappearance of all organic matter,
- on either side of these two combustion zones, there are completely pyrolyzed "lenses" about 5 to 10 cm thick,
- a very high ratio is to be observed between the residual organic carbon and the total organic carbon (figure 3, column 3), showing that the residual organic carbon consists of a "coke" that can no longer be pyrolyzed. The oil formed in these levels was probably expelled either into the cracks, where it may have burnt (the case of the upper zone), or towards the main combustion zone or the adjacent zones. The heavy products resulting from this pyrolysis certainly also participated in the formation of the coke encountered in them.

3 - CONCLUSIONS

The in-situ combustion pilot test of the TRANQUEVILLE tar shales has shown that it is possible to ignite a combustion front and then to propagate it by injecting air into a single unsupported crack.
It revealed two problems:

- the low efficiency of the link between the production well and the crack,
- the difficulty in controlling combustion to avoid burning the products of pyrolysis.

New studies will be needed to find solutions to these two problems if one is to envisage a new pilot project capable of producing shale oil.

On the other hand, the result of the test is entirely positive with respect to the generation of heat.

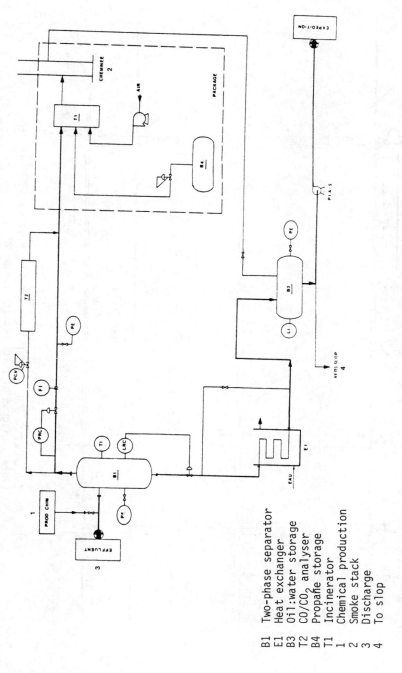

FIGURE 1 - ENGINEERING OF THE TRANQUEVILLE PILOT PROJECT
Gasification of the shales in-situ
Block diagram of processing of the effluents

B1   Two-phase separator
E1   Heat exchanger
B3   Oil:water storage
T2   CO/CO$_2$ analyser
B4   Propane storage
T1   Incinerator
1   Chemical production
2   Smoke stack
3   Discharge
4   To slop

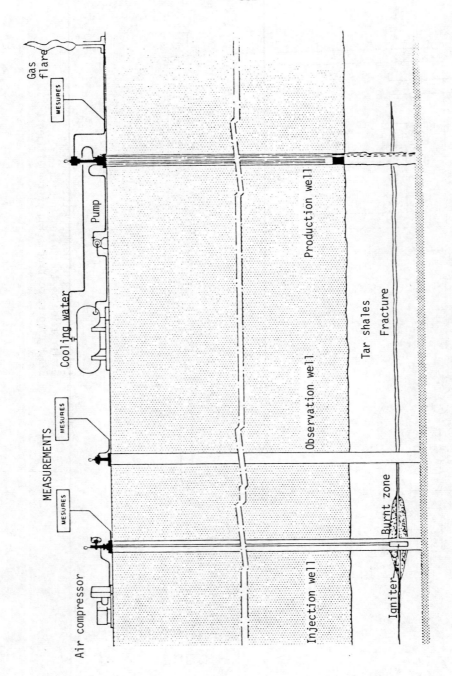

FIGURE 2 - TAR SHALES : THE TRANQUEVILLE IN-SITU COMBUSTION PILOT PROJECT
Overall diagram of pilot project (phase 2)

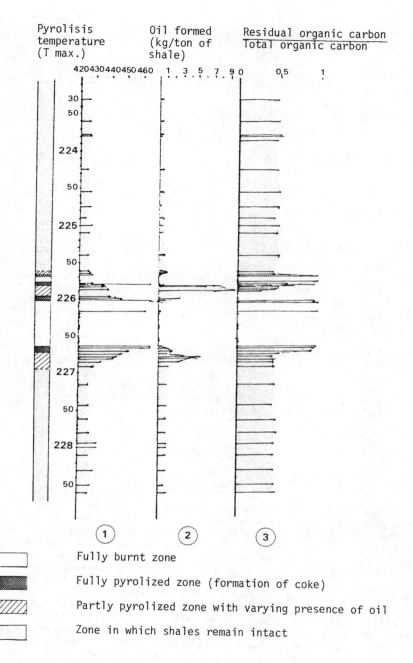

FIGURE 3
GEOCHEMICAL ANALYSIS OF THE CORE TAKEN FROM THE IN-SITU COMBUSTION ZONE

(05.11/78)

IMPROVING RECOVERY FROM VERY HEAVY OIL RESERVOIRS

J. BURGER, French Petroleum Institute

Summary

The purpose of the project was to carry out laboratory experiments to evaluate the possibilities of using a combination of injection of solvent and thermal methods to exploit very heavy oil reservoirs. Phase 1 of the project involved research outside the porous medium. A viscosity/temperature/dilution correlation was obtained for cases where the mixtures of crude and solvent result in true solutions. In addition, the compatibility of the various crudes with diluting agents was characterized by studying the precipitation of the asphaltene products of the crude.
Phase 2 consisted in testing the displacements in single dimension cells in order to determine the effect of adding solvent on the injection of steam or in-situ combustion. It was observed that injection of a plug of solvent was capable of reducing the residual saturation when injecting steam, whereas addition of solvent did not reduce the air needed for in-situ combustion, except in very few cases.

1 - COMPATIBILITY OF A CRUDE AND A SOLVENT

The addition of a hydrocarbon diluting agent to a crude can result in the precipitation of certain constituents of the crude. By convention, the compounds precipitated by n-heptane under standard conditions are termed asphaltenes. Under the conditions of the contract, the quantity of products that can be precipitated by a given diluting agent was determined by using this diluting agent instead of the n-heptane in accordance with an operating procedure identical in all other respects to that of the standard test for determining the asphaltenes content.
The effect of various parameters on the precipitation of the asphaltene products of a crude was studied: nature of diluting agent, ratio of crude to diluting agent, contact time, temperature and redissolving of the precipitate by another diluting agent. The favourable effect of the aromatics content of the diluting agent on its compatibility with an oil product was confirmed. It was also verified that the quantity precipitated by diluting agents of the n-alcanes family decreases as the length of the chain rises.

In addition to these studies made outside the porous medium, the possible loss of permeability of the porous media as a result of precipitation of asphaltene products by certain solvents was also studied.

The following results were obtained (reference 1:
- for each crude and each porous medium, a correlation is obtained between the loss of permeability and the quantity precipitated (figure 1),
- for the least permeable medium, it is found that the permeability seriously starts to deteriorate beyond 1% of precipitate (figure 1),
- the sensitivity to clogging is related to the size of the particles of the asphaltene products.

This part of the project, based on the use of various methods of analysis, had shown how important it is that the solvent envisaged is compatible with the crude.

2 — VISCOSITY OF CRUDE-SOLVENT MIXTURES

After study of the various installations that can be used to determine the viscosity of very heavy crudes either alone or mixed with various solvents, its was decided to build a prototype viscositymeter capable of operating at a temperature of up to 300°C and a pressure of 100 bars. Its principle is based on determining the time needed for a fluid to flow in a capillary over a distance between two marks on a cell fitted with a sight glass under the effect of gravity. The flow is assumed to be laminar and to obey Poiseuille's law. A set of capillaries enables a wide range of viscosities to be measured. In this way, the viscosity of many mixtures of crudes and solvents were successfully determined in a temperature range of from ambient to 260°C, and a pressure range of from atmospheric to 90 bars.

The diluting agents used were pure hydrocarbons or light petroleum cuts ; several crudes were studied. The ratio of the viscosity of the crude and the diluting agent for the various cases examined was from 10 to 10^6 (reference 2).

The purpose of the measurements was to determine whether the behaviour of the crude-solvent mixtures depended on the nature of the constituents of the mixture or whether a general law could be found. The effect of the temperature on the kinematic viscosity is represented in the ASTM diagram. One can see the linearity of the viscosity curves in this representation, both for the crudes and the solvents and for the crude-solvent mixtures, provided the dilution does not cause the asphaltene constituents of the crude to precipitate. Typical results are shown in figure 2.

The effect of dilution at a given temperature is shown in the ASTM diagram using the 200-300° temperature scale as scale of the percentage volume of the solvent in the mixture (figure 2). We see that this representation enables a linear correlation to be obtained with the diluting agent concentration. This is verified even for diluting agents acting as precipitants of the asphaltene compounds of the crude, as long as the diluting agent content remains below a certain critical value resulting in flocculation of these compounds.

It hence appears that knowing the viscosity of a crude and a diluting agent at two temperatures, one can predict with fairly good accuracy the viscosity of the constituents and their mixtures, at leas as long as the crude and diluting agent remain compatible, in other words provided none of the constituents of the crude precipitate. The effect of the dilution on the viscosity of a crude is all the greater, the less viscous the diluting agent.

On the basis of the results summarized in paragraphs 1 and 2, the low viscosity aromatic compounds, which are well compatible with the crudes, would appear to be the best suited for use as diluting agents to facilitate the exploitation of very heavy oil reservoirs. Amongst other things, it appears that one can envisage using injection of a light catalytic cracking cycle oil cut. These cuts, the viscosity of which is about 3 mm^2/s at 20°C, have a high aromatic content. They cost slightly less than pure light aromatic hydrocarbons such as benzene, toluene or xylene.

3 - COMBINED INJECTION OF STEAM AND SOLVENT

Laboratory experiments have been carried out to study the combined injection of steam and solvent in porous mediums in a one-way configuration.

The effect of various parameters was studied and in particular the influence of the type of crude, the type of solvent and the size of the plug of solvent.

A prototype cell 50 centimetres long was built to study displacement by steam in porous media containing highly viscous oils capable of causing very high head losses.

This cell (figure 3) consists of an external chamber capable of withstanding a service pressure of 125 bars, and an internal testing tube with a very high mechanical strength and an internal diameter of 7.3 cm. The annular space between the two chambers is filled with an insulator with very low heat conductivity and air at reduced pressure so as to reduce heat losses despite the absence of heating collars along the testing tube. The steam is generated at the inlet of the cell by a pair of resistors each servo-controlled by a regulating device. Modifications were made to this part of the equipment during the project in order to improve the thermal efficiency of the steam generating device.

The following are the main results obtained (references 1 and 2):
- injection of a plug of solvent before injecting the steam considerably reduces the head loss through a porous medium containing very heavy oils. This is observed immediately on injecting a plug of solvent representing 3 to 4% of the pore volume. This influence on the injectivity is the most spectacular result of injection of solvent,
- if the solvent injection speed is too high, highly marked preferential paths form in the medium. For certain tests, this results in the hot fluid completely flooding the medium,
- with certain volatile solvents injected together with the steam, an increase in the rate of production of oil and a decrease in the residual oil saturation compared to the saturation obtained with steam alone is observed (figure 4). This drop in saturation amounted to 10% of the pore volume with certain light solvents compatible with the crude. The most clear-cut results were obtained with simultaneous injection of steam and solvent,
- of all the solvents tested, xylene gave the best performance,
- it would appear necessary to inject a quantity of solvent representing over 5% of the pore volume to induce any favourable effect on the residual oil saturation,
- three methods of injecting the solvent were examined:
  . injection of all the solvent before injecting the steam,
  . injection of a small starting plug to improve the permeability to fluids followed by simultaneous injection of a small plug of solvent and steam,
  . simultaneous injection of the solvent and steam without a starting plug,

The latter method of injection enabled a reduction in the residual saturation to be obtained both with xylene and n-heptane, despite the fact that the latter is a precipitating agent of the asphaltenes in the crude. This operating method hence appears capable of reducing the problems of incompatibility between the crude and the solvent.

The viscosity of the oil produced by combined injection of solvent and steam is considerably less than that of the "virgin" oil. This will facilitate transport of crude extracted from very heavy oil reservoirs.

4 - IN-SITU COMBUSTION COMBINED WITH INJECTION OF SOLVENT

The effect of injecting solvent together with in-situ combustion was studied in a horizontal adiabatic cell available to IFP when the project started (reference 3). The combustion tube was 210 cm long with an internal diameter of 20 cm. The purpose of the experiment was to determine whether addition of solven resulted in modification to the essential parameters of the process.

In particular, it was expected that addition of solvent would reduce the quantity of coke available for maintaining the combustion front, whilst at the same time reducing the quantity of air needed.

The main results obtained during the project can be summarized as follows (reference 2):
- with the very heavy crudes, very long transient conditions are observed. This is probably caused by the times needed for a stable saturation profile to be formed ahead of the combustion front,
- addition of the solvent to the very heavy crudes results in a reduction in the maximum head loss observed during the tests,
- the favourable effect of the solvent on the propagation of the combustion assumes good compatibility between the crude and the solvent,
- for all the crudes and solvents studied, no significant effect is to be observed of the injection of solvent on the speed of the combustion front and on the quantity of air necessary when operating at high pressure, whether in limestone or sand media,
- the behaviour of the crudes on adding a solvent varies considerably with the composition of the oil:
  . for certain crudes used at moderate pressures in combination with compatible light solvents, the speed of the combustion front and the quantity of air needed fall considerably (by about 1/3 in one of the tests) (Table 1). This is generally accompanied by better recovery rates.

TABLE 1

INJECTION OF SOLVENT
BEFORE IN-SITU COMBUSTION
WEST CANADA CRUDE (11.3° API)

SOLVENT	-	-	XYLENE 0.085 VP	XYLENE 0.042 VP	XYLENE 0.083 VP	nC_{16} 0.088 VP
p (MPa)	1	10	1	1	10	1
AIR NEEDED (Nm^3/m^3 porous medium)	375	445	242	267	440	279

. for other crudes, no favourable effect on the solvent on the propagation of the combustion front is to be observed, even at low pressure,

. in a few tests with very heavy oils, carried out in particular with solvents of mediocre compatibility with the crude, the coke in the zone swept by the combustion front is not completely burnt,

- where the solvent has a favourable effect on the in-situ combustion parameters, this influence is all the greater, the larger the size of the plug ; the range studied in the project corresponds to solvent quantities of from 7 to 20% of the volume of oil in place,

- the density of the oil produced in the case of very heavy crudes is lower than its initial value ; the difference is greater than with conventional oils. The difference between the oil produced and the oil in place is yet more where in-situ combustion is combined with the injection of solvent.

CONCLUSIONS

- The first problem arising from the use of solvents for recovery from very heavy oil reservoirs concerns the risks of certain heavy constituents in the oil precipitating.

- Correlations of the same type have been demonstrated between the viscosity and the temperature on the one hand, and the viscosity and the diluting agent content on the other.

- Injection of solvent is an effective way of increasing the injectivity of a drive fluid such as steam or air.

- Addition of solvent in certain cases yields an appreciable gain in the displacement efficiency with steam injection. Injection of solvent combined with in-situ combustion reduced the quantity of air needed in only very few cases.

- The range of application of injection of solvents for producing very heavy oils is restricted by the cost of the solvents, which is of the same order of magnitude as that of the crudes. The light cycle oils produced by catalytic cracking could offer certain advantages with a view to using them in combination with thermal methods.

ACKNOWLEDGEMENTS

The author wishes to express his earnest thanks to C. GADELL, M. ROBIN, P. MIKITENKO and J.L. ZIRITT for their contributions to the experimental research reported here.

REFERENCES

(1) ZIRITT J.L. and BURGER J. - Combined steam and solvent injection. Second Internation. Conf. The future of heavy crude and tar sands. Mc Graw Hill, New York, p. 760-772 (1984).

(2) BURGER J. and ROBIN M. - Combinations of injection of solvent and thermal methods for producing very heavy oils. Eleventh World Petroleum Congr., Proc., John Wiley, Chichester, vol. 3, p. 251-260 (1984).

(3) BURGER J. and SAHUQUET B. - Laboratory research on wet combustion. J. Petroleum Technol., vol. 25, p. 1137-1146 (1973).

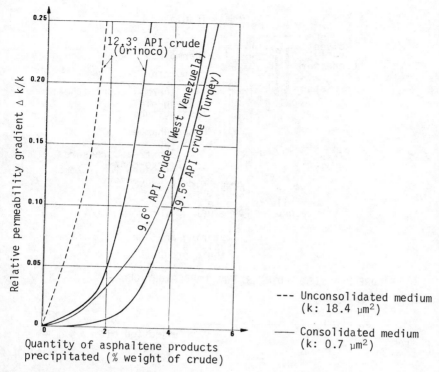

FIGURE 1 - RELATIVE LOSS OF PERMEABILITY OF POROUS MEDIA FOR 3 CRUDES
DILUTED IN BENZENE-N HEPTANE SOLUTIONS CONTAINING LESS THAN 60% BENZENE

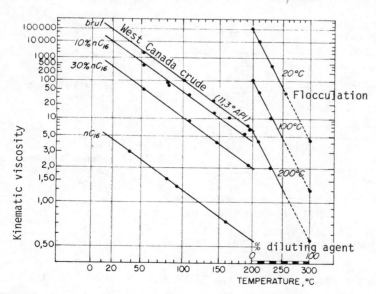

FIGURE 2 - VISCOSITY-TEMPERATURE-DILUTION DIAGRAM - ASTM scale

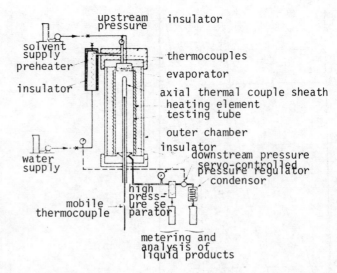

FIGURE 3 - SINGLE-DIMENSIONAL LABORATORY CELL

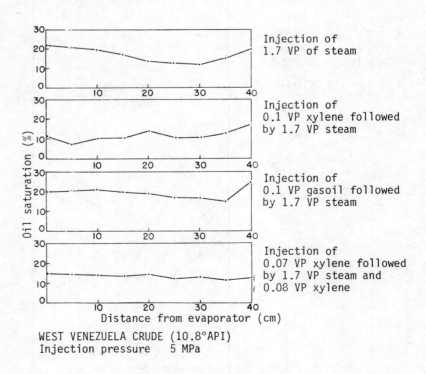

WEST VENEZUELA CRUDE (10.8°API)
Injection pressure   5 MPa

FIGURE 4 - PROFILES OF RESIDUAL SATURATION IN LABORATORY CELL

(05.38/82)

ENHANCEMENT OF DISTILLATION EFFECTS DURING
STEAMFLOODING OF HEAVY OIL RESERVOIRS

J. Bruining, D.N. Dietz, W.H.P.M. Heijnen,
G. Metselaar, J.W. Scholten and A. Emke
Delft University of Technology
Petroleum Engineering and Technical Geophysics

Abstract

Distillation effects give an important contribution to the recovery efficiency. This has been attributed to the formation of a distillable oil bank near the steam condensation front.
The aim of the work described here is to show that the beneficial effect of distillation can be enhanced by the addition of distillables to the injected steam.
Laboratory tube experiments and visual studies have led to new insight and drawn our attention to the following:
First, small amounts of distillable components added to the steam are capable of enhancing the oil recovery.
Second, the distillable oil bank, built up by the condensing distillables, is also broken down. This is mainly due to bypassing of oil patches by the steam at the upstream side and unstable displacement of heavy oil by distillable components at the downstream side of the oil bank.
Third, and most importantly, the low oil saturations in the steam zone must not only be attributed to these distillation effects. Also the spreading of the oil between the water that envelops the (water wet) sand grains and the steam leads to very low residual oil saturations. These effects must be incorporated in the available models when a meaningul extrapolation to field conditions can be made.

1. Introduction

    Steamflooding is routinely applied to heavy oil reservoirs, with a low primary and secondary recovery potential, where a high extra oil production can be expected.
    The high efficiency of steamdrive assisted recovery has been ascribed to distillation effects (1,2) and a relatively high sweep efficiency, which is assisted by these distillation effects (3). In other words distillation effects give an important contribution to the recovery efficiency (4).
    In the steam zone a larger part is removed from the heavy oil than anticipated from the distribution of boiling points of the components comprising the oil. This has been attributed to the formation of a distillable oil bank near the steam condensation front. Still a finite oil saturation is found in the steam zone (5,6,7) and distillation effects are less pronounced in oil that contains a small percentage of distillable components (2).

Therefore Dietz proposed to enhance these distillation effects in heavy oil by the addition of distillables to the injected steam.

In order to evaluate whether the addition of distillable components to the steam is economically justified with respect to pure steam injection, a better understanding is needed of the effect of steam distillation efficiency on oil recovery. This efficiency is related to the maintenance of an adequate solvent bank and thus to the factors which help and work against the building up of this bank.

However, in the simulation models that have been described in the literature (8,9,10) no terms have been included that describe the breakdown of a distillable oil bank.

Therefore it is necessary to carry out laboratory tube experiments and visual studies to find the form of the terms that describe the breakdown of the oil bank. When these terms are incorporated in simulation models a meaningful prediction of the enhancement of distillation effects on oil recovery by the proposed method can be made.

2. Experimental methods

The experiments can be broadly classified in two categories.
a) First, experiments have been carried out to measure production rates and cumulative productions (recovery efficiencies), b) second, visual studies have been carried out to observe the microscopic displacement mechanism.

a) Fluid production studies

Experiments have been carried out in a rig shown schematically in Figure 1. This figure also serves to explain the visual studies. An item list has been given in table I. The numbers of the item list are underlined to distinguish them from reference numbers.

Heart of the set up is a stainless steel foil (0.5 mm) reactor (7) with a diameter of 5 cm and a length of 1 meter. This reactor is mounted in a vessel (8). The N_2 pressure (supplied by a cylinder (24)) in the annulus, filled with cotton wool, exceeds the pressure in the reactor, which is essentially atmospheric. Graduated measuring cylinders 9 (crude oil), 10 (alkane) and 11 (water) serve for filling and injection of fluids. The water/alkane mixture vaporized in the steam generator (18) is injected from above. The temperature of the injection mixture exceeds the boiling point at the prevailing pressure.

Table I	Item list of figure 1		
1 =	Endoscope	11 =	Measuring cylinder water
2 =	(Video) camera	12 =	Oil pump
3 =	Pulley	13 =	Alkane pump
4 =	Video recording	14 =	Water pump
5 =	Photocamera control	15 =	Thermocouple connector
6 =	Glass tube	17 =	Temperature data acquisition
7 =	Reactor tube	18 =	Steam generator
8 =	Vessel	22 =	Cooling tube
9 =	Measuring cylinder oil	23 =	Fraction collector/cylinder
10 =	Measuring cylinder alkane	24 =	CO_2 and N_2 cylinders

However, some time is needed to heat the lid of the reactor after the time of the experiment. Produced fluids are cooled (22) and collected in a fraction collector or measuring cylinder (23). Temperature data from thermocouples mounted in the centre are acquired by a microcomputer.

b) Visual studies

When visual studies are carried out, a glass tube (6) ϕ = 30 mm is placed in the centre of the reactor tube. In this glass tube an endoscope is lowered. This endoscope (ϕ = 15 mm) is essentially a periscope with a glass fibre cable to transmit the light generated by a xenon lamp. A photocamera or videocamera (2) was used to record the visual data, taken at the lower end of the endoscope. The advantage of this set up is that heat loss problems are avoided.

ad a,b Preparation of the oil sand

The reactor was filled with 350 Darcy sand by the method of the seven sieves (11), and a more or less reproducible sand pack with a porosity of 35% is formed. The sand pack is flushed with carbondioxide to remove the air from the pores. Subsequently water is injected from below. In spite of the flushing with CO_2, visual studies show that some gas bubbles remain trapped in the pores. Finally oil (Clavus 32(is injected from above until this oil is produced below. The properties of the fluids have been summarized in table II. Clavus 32 represents the non-distillable crude and the alkanes represent the distillable oil components. The reason for the use of alkanes is that it eases the determination of the oil composition.

	M (g/mol)	ρ g cm^{-3}	μ (20°C) (cP)	μ (100°C) cP
Table II Fluid properties				
H_2O	18	1.00	1.001	0.282
Clavus 32	364	0.884	84.4	5.156
hexane	86.18	0.6603	0.326	0.173
heptane	100.21	0.6838	0.409	0.197
octane	114.23	0.7025	0.542	0.255
decane	142.29	0.7300	0.92	0.357

M = mol weight, ρ = density and μ = viscosity

Collection of fluid samples

Fluid samples are collected in graduated tubes put in the fraction collector. The tubes are centrifuged to break the emulsion and the oil and water level can be determined. The composition of the oil sample is determined with a refractometer. In other experiments cumulative productions are measured with the graduated measuring cylinder.

3. Results

i) Cumulative productions

Cumulative productions have been measured in a number of experiments. Steam injection rates varied between 5 grams and 20 grams per minute. Various amounts (0-17 v/v% liquid alkane/liquid water) pure alkane (i.e. n-hexane, n-heptane, n-octane and n-decane) were added to the steam. The results have been summarized in figure 2. In this figure the recovery effifiency is plotted versus the time (t_s) that steam was injected divided by the time (t_{sd}) required for steam break through at the outlet. The numbers near the symbols indicate the v/v% alkane added to the steam. For small steam injection times $(t_s/t_{sd}) = 1.4$ we observe that the recovery increases from 82% for pure steam to 90% when 2% distillable components were added to the steam.

On the other hand, the recovery efficiency with pure steam increases from 82% for small steam injection times $(t_s/t_{sd} \sim 1.4)$ to 90% when longer steam injection times $(t_s/t_{sd} \sim 3)$ are applied.

For these longer injection times, the recovery increase, owing to distillable components is less because already a high recovery has been achieved when pure steam is applied (e.g. 88%).

ii) Fluid production rates

The production rates of water, oil and alkane have been measured in a number of experiments. Two of these measurements have been presented graphically in figure 3a and figure 3b.

In these figures the fluid production rates have been plotted versus the time from the start of the experiment. The arrow indicates the time required for steam break through to occur at the outlet. The fluid production rates increase from the start of the experiment due to the fact that part of the steam is used to heat up the lid of the reactor.

The experimental conditions have been indicated in the figures. In figure 3a a small hexane bank is formed indicated by the dots, whereas in figure 3b a larger octane bank is formed owing to the higher octane/steam injection ratio. We observe an early production of alkanes, which has also been observed in the other 7 experiments.

iii) Visual studies (12)

a) Steam by passing of oil clots

This phenomenon can be discerned on photograph I made during an experiment where 9% octane was added to the steam (16.1 ml liquid H_2O/min). On this photo the sand grains can be observed as white-grey irregularly shaped bodies. The oil can be recognized as black (purple) patches clearly visible in the middle of the photo. Steam bubbles (injected from above) penetrate from above in the left hand part, but leave the oil zone in the right hand part untouched. Only some time later steam also penetrated in the right hand part.

b) Film flow

This phenomenon discerned on any photo where steam, water and oil are present is shown here on photograph II. This photo was made during an experiment where 4% heptane was added to the steam (20.1 ml/min H_2O (1)).

The sand can be recognized as white-grey irregularly shaped bodies. In the centre of the photo films of oil formed between the water (capilarily trapped between the glass tube and the sand) and the steam are clearly visible. These films can be observed in the steam zone of all our experiments and also during gravity drainage of oil (i.e. when connate water, oil and air are present).

4. Discussion

i) Conceptual basis for experiments

The underlying concept of our experimental configuration is that the oil recovery is caused by an expanding steam zone (1). This limits the applicability of our results to situations where this occurs in practice (6).

At the time that we started the project we adopted the generally accepted view, that the low values of the residual oil saturation in the steam zone could be only attributed to the formation of a distillable oil bank, which is formed near the steam condensation front. The distillable oil bank displaces the oil ahead of the front. The residual oil capilarily trapped by the condensing water is rich of light ends, which evaporates when the steam front approaches, thus leaving less oil (1).

These concepts of an expanding steam zone and a distillable oil bank have been translated to the experimental configuration shown schematically in figure 4. In a vertical tube (originally filled with sand, connate water and oil) steam and distillable oil (alkane for the ease of interpretation) are injected from above. In the same way as described above a distillable oil bank will be formed near the steam condensation front. Fluids will be produced at the bottom. The experiments showed us that our concept (figure 4) had to be adapted in two ways. First, the low saturations in the steam zone can also be partly attributed to film flow of oil. This will be discussed in paragraph (ii). Second, the distillable oil bank is unstable and there will be an (non-zero) amount of heavy oil in the steam zone. This will be discussed in paragraph (iii).

ii) Also film flow of oil causes low oil saturations in the steam zone

The low oil saturations in the steam zone also occur when pure steam is injected, i.e. no distillable oil bank is formed. Indeed for long injection times (see figure 2) recovery efficiencies of 87% have been attained. This corresponds to an average oil saturation of $S_O \approx 0.11$ (connate water saturation $S_{wc} = 0.12$). These low oil saturations can be attributed to the spreading of oil between the water that envelops the sandgrains and the steam, as illustrated by photo II. This spreading effect is known to occur during gravity drainage (13,14) where the oil spreads between the water that envelops the sandgrains and the air. This effect is known to lead to surprisingly high recoveries (13).

iii) Unstable behaviour of the distillable oil bank

The oil bank formed by the condensing light components near the steam condensation front does not uniformly cover the whole cross section. At the upstream side of the oil bank steam bypasses part of the oil blobs. This has been observed visually (photo I).
At the downstream side the miscible displacement of heavy oil by distillable oil is unstable. This can be inferred from the production rate data

presented in figure 3: distillable components are produced soon after the start of the experiments.

Also theoretically an unstable displacement is predicted. This statement warrants some further explanation. In the vertical tube experiment the miscible displacement is stabilized by gravity effects as the density of the distillable oil is less than the density of the heavy oil. The viscous forces, however, tend to destabilize the displacement as the viscosity of distillable oil is less than the viscosity of heavy oil. We assume a transition zone in which the oil composition varies from 70% distillable components near the steamcondensation front to 100% non distillable oil downstream. It follows, from the proceeding of the pressure gradients, calculated with the help of Darcy's Law, the experimental oil production rates and a viscosity/density mixture rule, that the displacement is unstable (15). In other words, the distillable oil bank shows strongly unstable behaviour.

iv) Practical consequences

The experiments have drawn our attention to two important aspects of steamflooding of oil, which were not incorporated in the available models (8,9,10). The relative importance of these effects can only be appreciated when these models are extrapolated to field dimensions. Indeed the negative effect of an unstable oil bank on the recovery efficiency exhibits itself more emphatically in tube experiments where there is insufficient time to generate an adequate solvent bank. Still also in field conditions there is a finite oil saturation in the steam zone (4,5,6).

The film flow of oil can be incorporated easily in the available simulation models, as a different form of the relative permeability curves (13,16).

Conclusion

Small amounts of distillable oil components added to the steam are capable to enhance the distillation contribution to the recovery. A distillable oil bank is formed near the steam condensation front.

Experiments have drawn our attention to two important aspects of steamflooding which were not incorporated in the presently available models.

First, the distillable oil bank exhibits strongly unstable behaviour. Upstream steam bypasses oil patches in the bank. Downstream there is unstable displacement of heavy oil by the less viscous distillable oil. Even though there is a substantial recovery increase, the distillable oil bank is still in its building up stage in the experiments.

Second, the low residual oil saturations must not only be attributed to distillation effects. Also the spreading of oil between water and steam leads to low oil saturations. This effect is analogous to the spreading effect that occurs during gravity drainage and is known to lead to high recovery efficiencies (13).

These experimental results can form a basis for the development of appropriate physical models. The relative importance of the two aspects can only be appreciated when these models are extrapolated to field conditions.

Only then a meaningful estimate of the distillable oil requirement (added to the steam) for the maintenance of an adequate distillable oil bank can be made.

Acknowledgements:

We are indebted to Prof.Ir. H.J. de Haan for many practical suggestions.
We thank Dr.Ir. J. Hagoort for many useful discussions. We also thank
Mr. K. Kamps for his photographic advices and Mr. A.P.E. Maljaars for his
technical assistence.

REFERENCES

1. M. Prats, "Thermal recovery", Monograph volume 7, SPE, Henry L. Doherty
   Series, New York, 1982, Dallas, chapter 7, ISBN 089520-314-6.
2. B.T. Willman, V.V. Valeroy, G.W. Runberg, A.J. Cornelius and L.W.
   Powers, "Laboratory studies of Oil Recovery by Steam Injection", J.Pet.
   Tech. (July, 1961), 681-690.
3. J. Hagoort, A. Leijnse and F. Poelgeest, "Steam Strip Drive: A potential Tertiary Recovery Process", J. Pet. Tech. (December, 1976), 1409-
   1419.
4. T.R. Blevins, J.H. Duerksen and J.W. Ault, "Light-Oil Steamflooding -
   An Emerging Technology", J. Pet. Tech. (July, 1984), 1115-1122.
5. T.R. Blevins and R.H. Billingsley, "The Ten-Pattern Steamflood, Kern
   River Field, California", J. Pet. Tech. (December, 1975), 1505-1514
   (fig. 7).
6. C. van Dijk, "Steam Drive Project in the Schoonebeek Field, The
   Netherlands", J. Pet. Tech. (March, 1968), 295-302.
7. T.R. Blevins, R.J. Aseltine and R.S. Kirk, "Analysis of a Steam Drive
   Project, Inglewood Field, California", J. Pet. Tech. (September, 1969),
   1141-1150, (figs. 8 and 9).
8. H.G. Weinstein, J.A. Wheeler and E.G. Woods, "Numerical Model for
   Steam Stimulation", Society of Petroleum Engineers of AIME, paper
   number SPE 4759 (1974).
9. K.H. Coats, "Simulation of steamflooding with Distillation and Solution
   Gas", SPEJ (Oct. 1976), 235-246.
10. N.D. Shutler, "Numerical Three-Phase Model of the Linear Steam Flood
    Process", SPEJ (June, 1969), 232-246.
11. R.J. Wygal, "Construction of Models that simulate oil reservoirs",
    SPEJ (December, 1963), 281.
12. G. Metselaar, "Oil displacement by mixtures of steam and distillable
    oil; Photographic and mathematical description on a microscopic scale",
    Delft University of Technology, Report No. 1984-13.
13. J. Hagoort, "Oil recovery by gravity drainage", SPEJ (June, 1980),
    139-150.
14. J.M. Dumoré and R.S. Schols, "Drainage Capillary-Pressure Functions
    and the influence of Connate Water", SPEJ (October, 1974), 437-444.
15. J.M. Dumoré, "Stability considerations in downward miscible displace-
    ment", SPEJ (December, 1964), 356-362.
16. F.J. Fayers and J.D. Matthews, "Evaluation of Normalized Stone's
    Methods for Estimating Three-Phase Relative Permeabilities", SPEJ
    (April, 1984), 224-232.

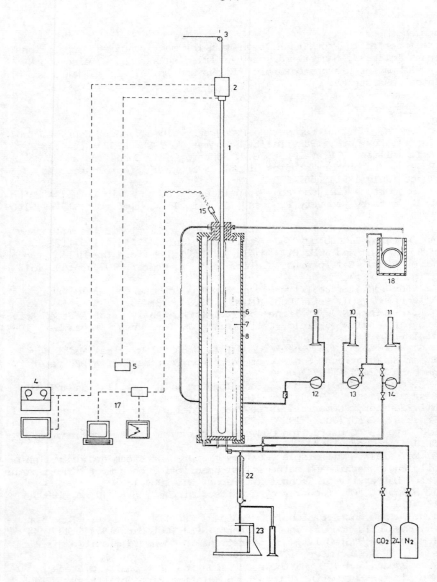

Figure 1 Equipment for steamflood experiments and visual studies.

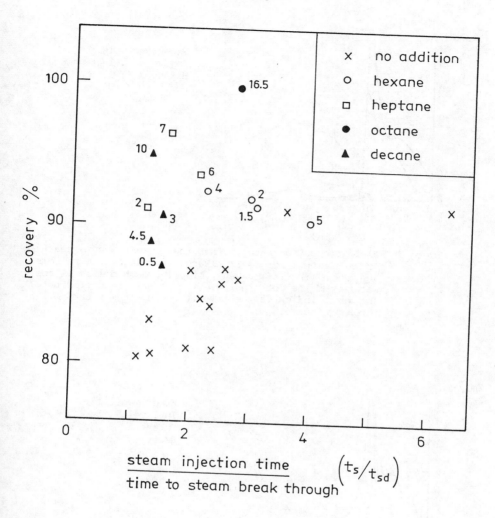

Figure 2 Recovery efficiency versus the steam injection time (t_s) divided by the time to steam break through (t_{sd}). The numbers near to the symbols (explained in the right upper corner) indicate the (liquid alkane/ liquid water) % alkane added to the steam.

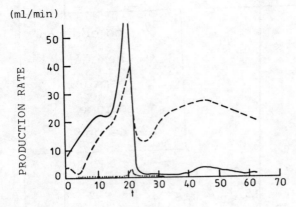

Figure 3a Fluid production rates versus the time (t = 0 is start of experiment) for small alkane/steam injection ratio's. The drawn line is the heavy oil, the dashed line is the water- and the dots indicate the alkane production rate. The arrow indicates the time at which steam break through occured.

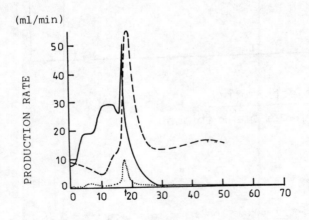

Figure 3b Fluid production rates versus the time for high alkane/ steam injection ratio's. The drawn line is the heavy oil-, the dashed line is the water- and the dots indi- cate the alkane production rate. The arrow indicates the time at which steam break through occured.

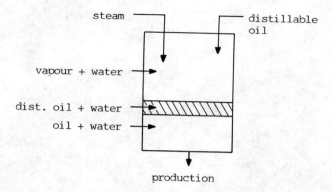

<u>Figure 4</u>   Experimental configuration and original concept.
Steam and distillable oil are injected from above.
A distillable oil bank is formed near the steam
condensation front. Oil is produced below.

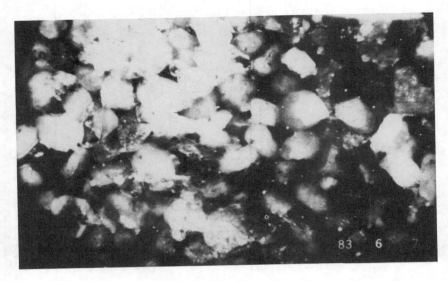

Photo I   Photo of steam displacement process near steam condensation front. Sandgrains are irregularly shaped white-grey bodies. Steam bubbles penetrate in the left hand part, but bypasses the oil zone in the right hand part.

Photo II   Photo of steam displacement process in steam zone. Sandgrains are irregularly shaped white-grey bodies. Films of oil are formed between the water that envelops the sandgrains and the steam.

PIPELINES

J-configuration laying - Electron beam welding (09.07/77 and 09.19/80)

Use of mechanical joints for laying pipelines in very deep waters (09.17/79)

Electron beam welding of horizontal pipelines (15.38/82)

Development of high pressure flexible piping (10.16/78)

Development of an ultrasonic riser pig (09.18/80)

Self-destroying vehicle for the inspection of pipelines (10.26/81)

Rip Rap protection of underwater pipelines by Velpo vehicle (09.21/79)

New offshore pipeline laying technology (09.06/76)

Thermobloc waste heat recovery system - (Pipeline Booster Station) (10.12/77)

(09.07/77 and 09.19/80)

J-CONFIGURATION LAYING - ELECTRON BEAM WELDING

B. de SIVRY, Compagnie Française des Pétroles

Summary

Laying large diameter pipelines in very deep waters requires laying
supports the form of which can already be conceived today. Use of
the J-curve laying method where the pipeline is lowered vertically
from the laying ship can be envisaged from vessels that exist today.
It calls for the use of a single-weld junction station, for which
the electron beam welding method appears to be the best solution.
The work, which started in 1977, resulted at the end of 1981 in a
series of trials of a prototype machine enabling pipeline sections
24" in diameter in X 65 or X 100 grade steel to be welded at a
simulated rate of one weld every 15 minutes. These trials at the
actual production rate are carried out on an oscillating test bench
enabling the equipment to be subjected to roll and pitch movements
during the welding operation. It was observed that the movements of
the ship in no way impaired the excellent quality of the welds.

INTRODUCTION

The maximum depth at which pipelines can be laid at sea by the
"conventional" method is 400 to 600 metres, owing to the anchoring
capabilities of the support vessels. For greater depths, use of dynamically
positioned superbarges would require considerable installed capacities
(in the range of 50,000 to 60,000 HP).

The J-curve laying method appears to be highly interesting for great
depths (300 to 3000 metres). Previous studies have shown that the innate
laying problems could easily be solved by technological solutions
approaching those used for a drilling support: the possibility of dynamic
positioning thanks to the low horizontal tensions, a vertical or low
inclination laying ramp and the handling and storage of the tubes. However,
application of the method requires the development of a rapid welding
process at a single welding station enabling a length equivalent to that
enabling the 4 or 5 wells carried out simultaneously on a conventional
laying barge to be welded without taking any more time.

Electron beam welding, a method well adapted to high grade steels in large thicknesses would appear to be the best solution for welding at a single station.

The feasibility study performed in 1977 and 1978 under EEC contract TH 09.07/77 has enabled some of the key points to be solved, for instance the positioning of the tubes and the vacuum seal for the welding chamber, following verification on a full-scale test bench. Likewise, the metallurgical study has enabled welds with good characteristics to be obtained on certain high grade API steels.

The technological development, which is the subject of EEC contract TH 09.19/80. consisted in building a complete prototype machine for electron beam welding of pipelines and testing it in the workshop on a test bench simulating the movements of the laying barge in order to prepare the method for homologation by the certification bodies.

FEASIBILITY STUDY - LIMITS OF APPLICATION

In order to determine the limits of J-curve laying, the preliminary work concentrated on the method of laying itself. These studies enabled the thicknesses of the pipelines to be calculated in terms of the grade of steel and the depth of water. The tension on the ramp, the horizontal tension and the laying ramp angle of inclination can then be deduced by studying the deformation of the tube during laying. The results obtained can be summarized as follows:
- a ramp tension in the range of 400 tons enables a 32" pipeline to be laid in a depth of 1500 metres, or a 20" pipeline in a depth of 2500 metres,
- the horizontal tensions are low (below 40 tons) and are compatible with the dynamic positioning of the support,
- the angle of inclination of the laying ramp is close to the vertical (maximum 10 to 15°) enabling laying equipment similar to that used during drilling to be used,
- the possibility of using high grade steels (X 100) becomes the predominant factor as the depth increases.

However, rational application of this method requires that a method of welding rapidly at a single welding station be developed.

In view of the nature of the tubes to be assembled (large thicknesses and high grades of steel) and the advantage to be gained in keeping the laying equipment and laying support similar to those normally used for drilling, the assembly must take place in length of about 24 metres. The overall dimensions of such a ramp installed on a semisubmersible or a ship allow of only one working station, whence the need to carry out rapid welding at a single station.

This holds particularly true for assembling large diameter tubes (16 to 36") and thick tubes (1 to 2"), since the economic imperative of a high laying rate (one joint every 15 minutes) is dependent on very fast welding (less than 5 minutes), which can not be achieved by a conventional method.

CHOICE OF THE METHOD

A survey was carried out in 1977 to compare the various methods that were capable of being applied for rapid welding of large diameter pipelines. They can be classified into the following two categories:

- Methods acting by melting the base metal:
  . electron beam (E.B),
  . laser.
- Forging methods:
  . friction,
  . spark-welding,
  . rotary arc,
  . high frequency induction.

In view of the requirements with respect to the welding time and the advantage of using high grade steels (X 100), communicating a much faster heat cycle to the part, appeared more suited than forging methods. The electron beam method offers the advantage compared with the laser of that of a technique that is already applied on industrial scale in the aviation, nuclear and automobile industries. Accordingly, it was adopted in 1978 as the best conceivable procedure.

There are three advantages to the method:
- suitability for large thicknesses,
- high speed and automated welding,
- welding requiring no heat treatment.

In 1978 and 1979, about 500 test welds were made on 6 types of steels from a variety of sources and of various grades. The welds were verified in accordance with an acceptance-testing procedure evolved jointly with the testing organizations, in order to homologate the method for welding pipelines.

The study enabled welding parameters to be defined ensuring the widest possible tolerances with respect to the positioning of the tubes to be welded and the "aiming" of the E.B gun onto the plane of the weld. The outcome is that the usual tolerances of the electron beam have been considerably broadened, making the method easier to adapt to welding pipelines laid by the J-curve method.

The verification of the various welding positions (full weld, lap weld, and tailing-out) by means of non-destructive examinations and mechanical tests enabled the soundness of the seams to be assessed, particularly with X 100 grade steels. Lastly, the reproducibility of the tests and the performances obtained have enabled the cycle times of the machine to be determined. These cycle times undercut the specification requirements by 5 minutes.

ENGINEERING STUDY OF THE LAYING INSTALLATION

The engineering study of the laying installation, which was carried out with the participation of A.C.B, Nantes, covered:
- the architecture of the laying ramp, consisting of a vertical or inclined mast maintaining the tube to be welded to the pipeline in the appropriate position close to the vertical (see figure 1),
- the specific equipment, and in particular the loading arm enabling the tube to be installed in the mast in the vertical position, the hoist for lowering the pipeline after welding and the centering and anchoring clamps holding the pipeline in position during the welding operation.

The engineering file will enable a request for tender to be sent out for building and industrial scale ramp.

ENGINEERING OF THE WELDING MACHINE

The engineering design of the welding machine concerned a number of components specific to the J-curve laying method and to electron beam welding:
- A docking system enabling the two ends of the tubes to be welded to be brought into contact with each other and held together firmly during the welding operations by means of a clamp designed by Société PRONAL, known as a "soft" clamp owing to the design of its jaws, consisting of shoes resting against the tube without damaging its lining (see figure 2).
- A prototype welding machine designed by S.A.F.
  This forms a chamber around the ends of the tubes to be welded, inside which is raised a vacuum needed for propagating the electron beams. The welding takes place by rotating the welding set generators around the tubes inside the chamber, the dimensions of which are determined to accommodate the equipment it contains (electron beam generators, vacuum devices, radiation protection devices, cable paths, etc..).
- A retractable vacuum joint capable of wide play, designed by Société PRONAL. The purpose of this joint, when in the closed position on the bare or lined surface of the tubes, is to provide a seal enabling the vacuum inside the welding chamber to be maintained, and when in the open position, to allow the pipeline to pass during the laying stage, together with its collars, protection anodes, "buckle arrestors", etc.. (see figure 3).
- A device known as the "internal clamp", designed by AC.B and S.A.F. This device is installed inside the tubes to be welded and fulfils several functions. First, it enables the ends of the tubes to be guided. Next, it provides the seal inside the tubes so that the vacuum needed for propagation of the electron beam is not lost, and lastly, it sets into position a heavy part known as the fire barrier, the purpose of which is to absorb the excess electrons which pass through the weld.
- A weld testing device was also studied. It is based on the use of X-ray sources situated outside the weld, and an internal display device. Tests performed on existing equipment turned out to be insufficient and showed the need to develop specific equipment.

QUALIFICATION OF THE WELDING METHOD

The study ended with the "preliminary qualification of the welding method" carried out by S.A.F and the INSTITUT DE SOUDURE DE PARIS (Paris Welding Institute), with the approval of the Laboratoire National d'Essais Français (French National Welding Laboratory).

The electron beam welding method was selected as offering characteristics compatible with use aboard ship, and in particular a high energy density ensuring good metal characteristics in the welding zone, even without heat treatment.

To check the values of the resilience in the welding zone, tests were performed on 16 different qualities of API type high strength steels in accordance with the specifications elaborated during the study. These steels came from European, USA and Japanese origins.

The tests showed that welding seams can always be obtained by electron beam welding that are perfectly sound, whatever the quality of the steel of the tubes to be connected, their method of elaboration, their thickness and their geometrical characteristics.

Nevertheless, if high values of the tenacity in the molten zone are to be obtained, specifications are required for the chemical composition of the steels, which, whilst stricter than API 1104, remain perfectly compatible with the present possibilities of the steelmaking industry.

Lastly, the study has enabled a method of controlling the welding parameters to be developed making full allowance for the variations in the chemical and geometrical characteristics and the various positions of the tubes to be welded that are encountered during welding.

These parameters have been defined exactly for X 65 and X 100 grade steels, and for welding a section of pipeline in the vertical position and inclined up to an angle of 10° to the vertical.

Additional tests would be needed for other qualities of steels that may be envisaged for J-configuration laying.

A new study would be needed to be able to use this method for welding pipelines in positions other than thé vertical, to counter the risk of collapse of the molten bath.

CONSTRUCTION AND TESTING OF THE J-CURVE LAYING SYSTEM COMPONENTS

The construction work took place during 1980 and the beginning of 1981.

The welding chamber was built by Société S.A.F in its workshops at Parthenay (Deux-Sèvres). It was equipped with two 45 kW standard electron beam generators and the associated pumping sets. The seal on the pipeline was ensured by means of two inflatable joints providing wide movement, built by Société PRONAL, at Roubaix.

A specific docking system was made for the shore tests. To cut down its size, this system is of smaller capacity and dimensions than the system already studied in the form in which it would be installed on a laying barge.

The jaws of the "mild" clamp system were built by PRONAL and installed inside the star-form docking system designed and built by S.A.B at Nantes.

The internal clamp was also built by S.A.B at Nantes.

The left hand part of the control console contains all the controls for the welding chamber, whilst the right hand part contains the docking device control system. These two systems are themselves controlled by interconnected microprocessors built by S.A.F and A.C.B for their respective equipment.

TESTING THE J-CURVE LAYING SYSTEM IN A OSCILLATING TEST BENCH (see figure 4)

The equipment was tested in the workshops under conditions approaching the real-life conditions on a pipeline laying barge, in an oscillating test platform designed specifically for this purpose and assembled in their workshops by A.C.B, Nantes.

This platform was located in a pit 9 metres deep. It weighs 140 tons, with a total height of 20 metres, of which 13 metres rise above the floor of the workshop. The mobile part of the platform can be made to roll through ± 6° at a variable period and at a fixed inclination of up to 15° in a direction perpendicular to the roll.

The platform is capable of assembling 4.3 m length of 24" tube. After assembly, each length of tube is lowered and then cut below the machine and removed rearwards by a discharge device.

During the last three months of 1981, all this equipment was subjected to a complete series of tests in the vertical position and subsequently the inclined position, with or without roll, during which numerous welds were made in X 65 and X 100 grade steel 24" tubes.

These tests were first controlled manually, after which control was gradually taken over by automatic devices, finally leading to fully automatic welding, even with the test platform inclined or undergoing roll motion.

The tests were performed with series of three tubes enabling two welds to be made without interrupting the roll motion.

The components of the laying ramp (hoist, loading arm, etc..) were not tried out. However, assuming they operate correctly, one can conclude that it would be easy to achieve the production rate envisaged of one weld every 15 minutes.

Throughout the tests, incidents occurred at the various components and equipment units. Note was taken of these incidents, laying the ground work for study of the overall reliability of the machine built.

This reliability study, conducted by Société SIMMAGE, showed that the probability of carrying out an incident-free welding cycle was 99.5% (this value is comparable to that of a ship's radar system).

CONCLUSION

A maintenance study was carried out in 1982. Combined with the results of the reliability study, it enabled the operational availability of the system to be determined, thus confirming the validity of the techniques used.

To determine the reliability of the method, a sea trial on a dynamic-ally positioned drilling ship is envisaged, which will be the subject of a subsequent project, and the prototype material developed has been set aside, with this in view.

The satisfactory operation of electron beam welding for this application has led the participating companies in the project to launch a new research programme covering the use of electron beam welding for laying pipelines by the conventional method.

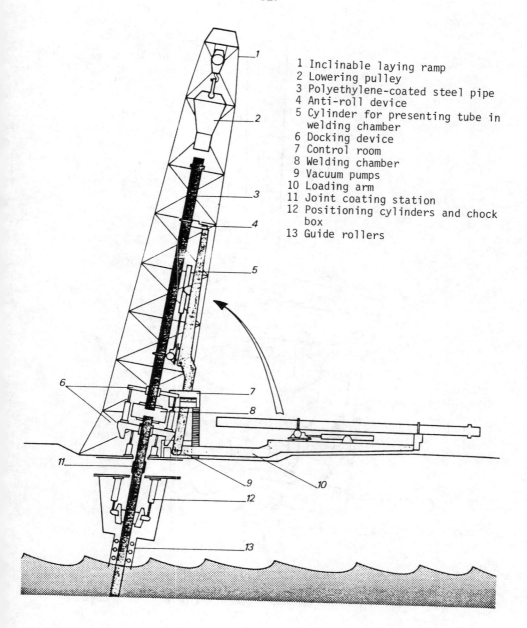

1 Inclinable laying ramp
2 Lowering pulley
3 Polyethylene-coated steel pipe
4 Anti-roll device
5 Cylinder for presenting tube in welding chamber
6 Docking device
7 Control room
8 Welding chamber
9 Vacuum pumps
10 Loading arm
11 Joint coating station
12 Positioning cylinders and chock box
13 Guide rollers

FIGURE 1 - J-CONFIGURATION LAYING METHOD - Laying ramp
- Application of the loading arm

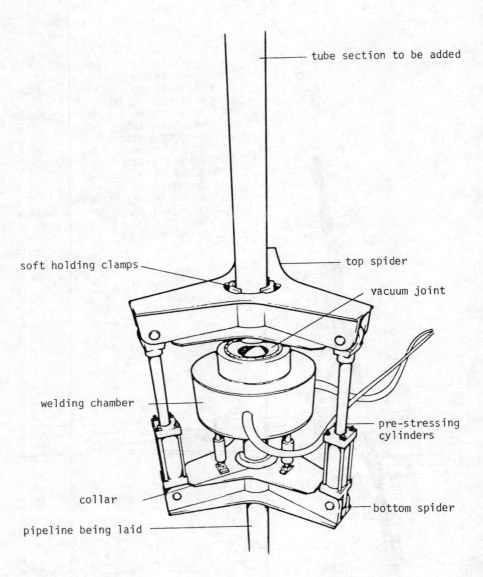

FIGURE 2 - J-CURVE LAYING METHOD - Operating principle of docking device

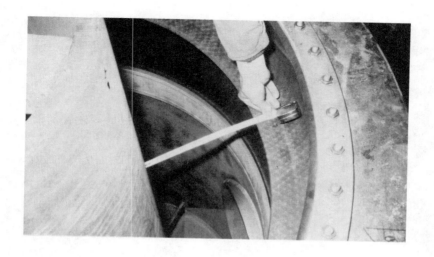

FIGURE 3 - TESTING OF VACUUM WELD PROVIDING WIDE PLAY

FIGURE 4 - J-CURVE LAYING - Testing complete system at maximum angle
of inclination (15°)

(09.17/79)

USE OF MECHANICAL JOINTS FOR LAYING PIPELINES IN VERY DEEP WATERS

E. PALIA
TECNOMARE SpA

Summary

Tecnomare has been engaged since 1980 on a research project on the development of new methods for laying pipelines in very deep waters. The study, which is expected to be completed by the end of 1984, concentrates in three main concepts, J-laying plant, new methods for mechanical connection at a single station, automatic welding methods used for pipeline junctions at a single welding station. The project is now within a few months of its conclusion and the following have sofar been completed : basic design of a pipe-laying system, design and construction of a new type of mechanical connector, design (construction is underway) of an automatic welding set with several orbital welding heads. The results of the function and qualification tests for the connector, and the acceptance and qualification tests for the welding unit, will pave the way for the possibility of development on an industrial scale.

1. INTRODUCTION

It is envisaged that within the next ten years approximately 50 % of world hydrocarbon production will come from offshore fields, many of them in deep waters. One of the major technological problems posed by this is that of laying pipelines in very deep water, and in particular rapid welding of the lines at a single welding station to make more efficient use of the J-laying system. Research into methods of welding at a single station has so far resulted in no satisfactory solution to the problem. On the other hand, the petroleum industry has not until now considered mechanical connectors, often used in drilling engineering, reliable for the connection of pipelines. The aim of the project is a basic study of the J-laying system and the definition and development of methods for joining pipelines at a single station in a system of this type : both welding and mechanical methods are examined. The project, commenced in December 1980, will be completed by the end of 1984. It is partly financed by the EEC and the IMI.

2. MAIN FEATURES OF THE PROJECT

2.1 Phase 1 : Configuration of the system

- Determination of the alternative configurations for the laying system;
- Calculation of the basic parameters for vertical laying;

- Feasibility study and preliminary design of a TIG multihead orbital welding system;
- Preliminary description of mechanical connectors.

The field of application equipment is designed for operation in the following conditions :
- **air:** capable of functioning in a marine environment at a temperature of between – 25°C and + 45°C;
- **seawater:** maximum operating depth 1 450 m, water temperature from – 2°C to + 32°C, salinity: 4 %;
- **fluid in the pipes:** oil or natural gas, maximum pressure 29.4 Mpa, temperature between – 10°C and + 90°C.

2.2 Phase 2 : Design, construction and qualification test on the mechanical joints and welding set

This phase includes the design, construction and acceptance and qualification testing of both subsystems.
In particular :
- the welding set must be capable of performing the automatic welding of lines with an external diameter of 508 mm (20") and a wall-thickness of 25.4 mm (1") in vertical position in under 18 minutes;
- it must be possible to make the mechanical correction quickly at a single station, and its performance must be in line with the current regulations for subsea lines.

2.3 Phase 3 : Basic study of the pipelaying system and operating methods

The aim of this phase is the general design of the plant and qualitative and quantitative definitions of the methods suitable for J-laying.

3. TIG MULTIHEAD WELDING SYSTEM

3.1 Preliminary design

Welding wide-diameter, thick-walled lines must be carried out rapidly and produce geometrically and metallurgically flawless joints. These requirements are more difficult to meet in the case of the J-laying of sealines, which not only requires a high degree of reliability and speed, but must also be carried out at a single station. In the case of lines with an external diameter of 508 mm (20"), and a wall-thickness of 25.4 mm (1"), the maximum limit for a complete welding cycle was set at 18 minutes. It was decided to use narrow gap welding geometry, together with the TIG Multipass process, TIG welding being adopted in view of its geometric and metallurgical reliability, and the multipass method, with several welding heads working simultaneously on the same joint, to compensate for the relative slowness of the TIG process. Theoretical assessments of the welding times referred to above have shown that between 6 and 8 welding heads are needed. The plant comprises a mechanical assembly, a power generating assembly and a control system. The configuration of the system is shown in Fig. 1.

3.1.1 Configuration of the mechanical assembly

The requirements of lightness, mechanical and thermal resistance and ease of maintenance were taken into account in the design of the apparatus. Several welding heads (between 1 and 8) can be mounted on the same tractor which travels around a track clamped hydraulically onto the pipe to be welded. The connecting cables from each welding head are connected to a winding drum which also rotates around the pipe. All welding heads are connected to a process unit and a generator supplying the welding current and performing and monitoring the programmed work cycle.

3.1.2 Configuration of the control system

The control system for the welding system comprises :
a) Central computer
b) Process units (between 1 and 8)
c) Programming unit

The computer formulates and monitors the welding programme to be carried out by the programming unit and used by the process units (one for each welding head) to control the mechanical functions.

3.2 Detailed design, construction and tests

To assess the actual potential of a TIG Multipass system with several welding heads, it was decided at the end of the first phase to construct a prototype with three welding heads and to carry out acceptance and qualification tests. These will be undertaken by the Italian Welding Institute and will enable the process and the welding equipment to be classified in accordance with the API standards. The detailed design of the entire prototype is now complete, and construction is practically finished. The main features of the equipment are the fact that the individual components are modular and the flexibility due to the fact that operations are controlled by the microprocessor and that the range of parameterrs is wide (see table below). A diagram of the welding set is given in Fig. 2.

Parametres	Ranges	Parametres	Ranges
Radial positioning of heads	0 – 50 m	Down slope	0 – 30 sec
Axial positioning of heads	20 mm	Arc suppression current	5 – 30 A
Arc voltage	0 – 30 V	Final current time	0 – 10 sec
Amperage	0 – 300 A	Direction of rotation	Clockwise and anti-clockwise
Speed of wire feed	0 –3.5 m/min	Axial oscillation	
Speed of rotation	50–1000 mm/min	a) amplitude	0 $\pm$ 15 mm
Angular frequency of arc	0 – 1 Hz	b) right and left position stops	0 – 5 sec
Synchronism of wire feed	0 – 1 Hz	d) velocity	0 – 200 mm/min
Synchronism of welding head speeds	0 – 1 Hz	Welding position	horizontal sloping vertical
Forming	5 – 300 A		
Immersion	0 – 10 sec	Radial space requirement of the welding head (with Midget reels)	75 mm
Up slope	0 – 10 sec		

3.2.1 Main mechanical groups

a) Welding head

The welding head is very compact. It contains the motors which control the radial and axial movements of the torch, supporting the tungsten electrode. The torch is cooled by water and designed for currents of 300 A. The filler wire units are at the sides of the central section of the head and comprise a feed reel, a back-geared motor and the drive mechanism.

b) Welding head tractor

This structure comprises two separable half sections. Apart from carrying the welding heads, it has three back-geared motors for the orbital traverse of the system. The tractor travels around a circular track which is clamped onto the surface of the pipe; it carries the rack to which the orbital traverse reducers are coupled.

c) Cable tractor

Its configuration is similar to that of the welding head tractor, but it is not motorized. It is connected by two flexible couplings to the welding head tractor, which drives it round. The two tractors are on the same track.

3.2.2 Principal electrical groups

a) Current generators

Each head is supplied by a current generator controlled from the relevant process unit. Commercial generators with a current range between 0 and 300 A were chosen.

b) Central computer

The computer carries out the following :
1. stores the adjusted programmes on one or more process units via the programming unit;
2. transmits programmes stored on the central computer to the various process units;
3. starts the programmes loaded on each individual process unit simultaneously or one by one;
4. monitors the operations (welding phases) of all process units;
5. monitors alarm conditions arising on any process unit.

c) Process unit

The process unit carries out the welding programme stored in it. The programmes are loaded by the central computer or are adjusted by the programming unit. While the programme is being carried out, the process unit monitors all the mechanical functions and operates the current generators and alarms.

d) Programming unit

The automatic welding system has a programming unit which can be connected to the process unit to adjust the welding programmes. The programming unit permits :
1. definition of the welding "sectors";
2. definition of the basic parameters sector by sector and for the initial and final phases;
3. variation (during welding) of the parameters selected for the sector being worked on as required.

4. MECHANICAL JOINT

4.1 Selection of the basic configuration

The solution adopted from the preliminary selection seemed very promising : the main junction was carried out by a threading, while the seal was provided by a "lip" preloaded axially to yield point. A further more detailed analysis using finite element methodologies has however shown that the springback of the lip would not be sufficient to keep the surface sealed under the envisaged load conditions (see table). An alternative was proposed based on a surface sealed radially as well as axially. The elasticity required to maintain sufficient pressure on the sealing surfaces is provided by the radial rigidity of the joint. The finite element analysis confirmed the effectiveness of this second configuration, and on the basis of this positive result a basic design was drawn up for a joint for sealines of a diameter of 508 mm (20") and thickness of 19.05 mm (3/4") (Fig. 3).

4.2 Detailed design and test

The second phase of the project involved the detailed design of the joint. This has now been constructed and a test programme has been drawn up.

The principal design conditions are listed in the table below :

SEALINE JOINTS — DESIGN CONDITIONS

Materials	API X 65
Line	Diameter 508 mm (20"), thickness 19.05 mm (3/4")
Seal	PLASTIC
Maximum bending moment	1 550 kNm
Maximum traction	13 130 kN
Maximum external pressure	14 500 kPa
Internal operating pressure	19 600 kPa
Maximum torque	1 850 kNm

As the most critical part of the entire welding system was felt to be the seal, three pairs of simplified full-scale models were constructed, as well as the three pairs of actual joints, so that the behaviour of the surfaces, in particular the surface undergoing plastic deformation, could be evaluated individually. It is hoped to carry out the following tests by September 1984:
- Dimensional checks
- Non-destructive tests
- Torque wrench measurements
- Tightness test at low air pressure
- Tightness test at high air pressure
- Water-tightness test
- Tests under various conditions of load acting simultaneously.

Extensimetric surveys will be carried out during the tests and at the end of each test the residual deformation will be checked and non-destructive tests carried out on the structural integrity.

5. LAY SYSTEM

5.1 Analysis of configuration

The first phase of the project involved the consideration of two alternative configurations. It was decided to use two dynamic positioning units, one a two-hulled semisubmersible and the other single-hulled. A preliminary assessment was carried out on the operating limits for wind, maximum wave and current action in different directions of laying, and potential down times were estimated by a simulation programme. A simplified computer programme was drawn up to demonstrate the feasibility of the J-system for laying wide-diameter pipes in very deep waters.

5.2 Development of the project

The third phase of the project comprised :
- preliminary design of the laying installation;
- formulation of a laying procedure and assessment of the time required for each phase;
- complete programme to determine the level of tension to be exerted on the pipe as it is lowered.

The configuration chosen for this first development uses the semi-submersible for laying. This type of craft has good stability and a wide deck for installation of the actual pipe-laying equipment and the ancillary equipment (Fig. 4). It is designed for a pipe diameter of 508 mm (20"), with a wall thickness of 24.5 mm (1"). The laying ramp is a reticular steel structure which supports the tensioners : the conventional track-type tensioner is replaced by a dual system of hydraulic cylinders and four hydraulic clamps (two for each group of cylinders) support the pipe during its descent.

It is planned to join the sections of pipe by the two systems being studied (see points 3 and 4) alternatively. The pipe sections are fed by a subsidiary ramp which is moved by a winch system (Fig. 4). The "laying procedure" comprises a list of the main phases of the operation and an assessment of the time needed to execute them. This has enabled the critical phases to be identified and preliminary optimization to be made. The computer programme is an improved version of the programme carried out in phase 1 : from the depth of the water, the tube diameter and the horizontal pull, the programme can calculate the thickness of the tube, the thickness of the ballast coverage and the total tension to be applied to the pipe.

6. RESULTS

Although the research is not yet complete, three main results have so far come out of it :
- A welding system is nearing completion which, although based on conventional technologies, is very advanced in terms of speed of execution, flexibility, lightness of the welding equipment, design of the control system and setting of the parameters. Mechanical connectors have been built designed for the first time specifically for sealines of wide diameter.

- It was shown that it is theoretically possible to adapt conventional and commercial equipment to construct a J-laying system suitable for wide diameter pipelines and this will prove to be a valuable factor if it is to be developed in the future on an industrial scale.

7. CONCLUSIONS

The project, which is expected to be completed by the end of 1984, will provide valuable information on :
- J-laying procedures and equipment
- Systems and methods for weldign at a single station
- Systems and methods for mechanical connection of wide-diameter sea-lines.

The results so far, together with the results which should come out during the tests, will be a good basis for the development of new equipment and new methods for laying pipelines in very deep water for widespread use on an industrial scale in the future.

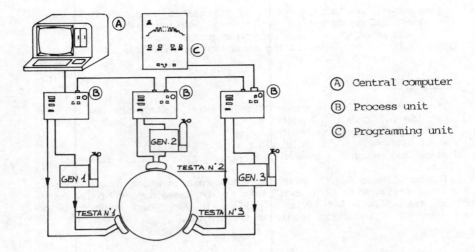

FIGURE 1 – WELDING INSTALLATION – Configuration of the system

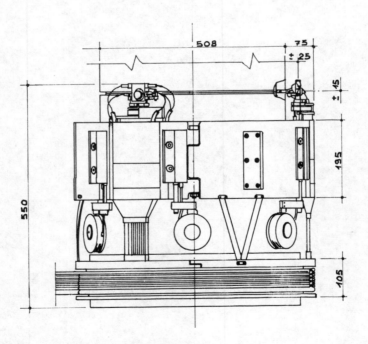

FIGURE 2 – WELDING INSTALLATION – Welding unit

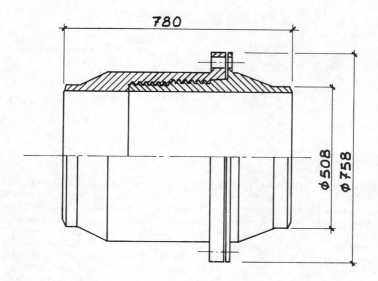

FIGURE 3 – MECHANICAL CONNECTOR – General view

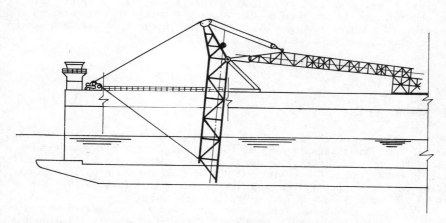

FIGURE 4 – LAYING SYSTEM – General view of the installation

(15.38/82)

ELECTRON BEAM WELDING OF HORIZONTAL PIPELINES

X. PUISAIS, Entrepose GTM pour les travaux pétroliers maritimes
M. JECOUSSE, Alsthom-Atlantique (Offshore Division)
J.P. HAMON, Soudure Autogène Française
M. KALUSZYNSKI, Compagnie Française des Pétroles

Summary

A feasibility study of the use of electron beam welding was made to
determine the savings it could bring in laying offshore pipelines
from conventional laying barges. Most of these savings are linked to
the possibility of welding steels with high characteristics whilst
avoiding the need for heat treatment that is generally involved in
conventional welding methods, thus enabling tubes to be laid in
greater depths of water. Equally, these gains are linked to the
possibility of increasing laying rates, thus improving the
profitability of the laying barges. Economic studies have enabled
these savings to be determined, whilst technical studies have enabled
the specific problems involved in electron beam welding to be
overcome, and in particular those set by the collapse of the molten
welding bath when the thickness of the tubes to be welded is greater
than 16 mm. A welding technique in two passes, one external, the
other internal, has been developed, and drawings of the internal and
external welding machines draughted. The layout of these machines on
the laying barge has also been determined.

1 - INTRODUCTION

Following the technological success obtained by the use of electron
beam welding when testing a machine for laying pipelines in the J-curve
method, it occurred to CFP, ACB, SAF and ETPM that it would be worthwhile
adapting this method to conventional S laying of subsea pipelines. This
welding method without filler metal should be competitive with other
methods now under study: friction and spark-welding.
    Welding on the barge is at present achieved very well with manual or
semi-automatic methods (CRC, SATURNE,...), all of which use filler metal
in accordance with a variety of techniques.
    The reliability is satisfactory and a fast laying rate is achieved
today by increasing the number of welding sets and using "double jointing"
(welding in 24 metre length of tubes instead of 12 metre length).
Accordingly, any new welding methods must tackle a market that is well
defended. They must yield high gains in productivity and reliability whilst
providing solutions to the technological difficulties that are solved only

untatisfactorily at present (welding of steels with high level characteristics).

All this must take place under worksite conditions combining the difficulties of public works with offshore work. Within the context of this technological and commercial research, what interest can be offered by electron beam (EB) welding ?

When the electrons strike the metal, they raise its temperature to the melting point: it suffices to keep the two parts in contact with one another to merge the metals together thus achieving the weld after cooling ; this method is increasingly used in the mechanical assembly industries. The welding rate is independent of the thickness: it suffices to vary the intensity of the beam. In this way, the method appears particularly attractive for thick tubes, i.e. large diameter tubes (over 24").

Another advantage of the method is that since the beam is very thin, the area affected by the heat remains very small, and cools rapidly. This enables the mechanical qualities of the steel to be left intact. Accordingly, one can assemble tubes in high performance steel without any procedural constraints.

There are however limitations on EB welding: the welding zone must be kept in a vacuum, and the end faces of the tubes must be positioned precisely with relation to one another.

Solutions were found to these limitations during the "J-curve laying" project. However, these results can not be transposed directly to conventional laying, i.e. "S configuration" laying.

The fact is, that the working conditions on a conventional laying barge differ from those for J-curve laying, and mainly:
- The joints to be welded lie in an almost vertical plane.
The welding plane in J-curve laying is horizontal and the welding position unique: an angle well. With conventional laying, however, the welding plane is vertical: four welding positions occur in succession: flat weld, vertical down-hand weld, ceiling weld and up-hand vertical weld. Tests have shown that with a through beam, thicknesses greater than 16 mm can not be flat-welded or ceiling-welded ; the molten metal falls under its own weight.
- The tubes are coated with concrete, which is particular hindrance for maintaining the vacuum needed to propagate the beam of electrons.

A solution therefore had to be conceived to the problems mentioned, and all the elements of the method optimized (metallurgy, machines, laying bench, handling, productivity, market study and layout on the barges) in order to render the method competitive compared with its rivals.

2 - WELDING MACHINE

To solve the problem of maximum thickness, welding in two passes was chosen: one internal, the other external, requiring two types of machines, an internal machine and an external machine, to be designed.

To reduce its overall dimensions, the gun of the internal machine is parallel to the axis of the tube. Accordingly, the beam must be turned through 90° to attain the wall of the tube. This is achieved by magnetic deviation (see figure 1). Several prisms are arranged so that welding can be carried out at several different points with a single beam deviated in succession at a fast rate. Accordingly, in addition to these guns and prisms, the machine comprises a sealing device and a device for raising a vacuum, together with a mechanism for displacing the machine to change to the plane of the next welding joint.

The external welding machine involves the difficulty of maintaining the vacuum in an annular space (see figure 2). In the J-curve laying project, inflatable O-rings proved entirely satisfactory. However, with conventional laying, the fatigue behaviour and presence of the concrete made it impossible to use these joints. Several methods have therefore been envisaged to reconcile the requirements of a high production rate. However, at the present stage of the project, no solution has been adopted and further studies at the drawing board and testing of prototypes will be needed. The weld is made by a ring of fixed guns. This configuration offers the advantage of reducing the overall dimensions, reducing the volume requiring a vacuum and limiting the number of mechanical parts. The beams scan by means of magnetic deflection.

In the case of tubes less than 16 mm thick, a single machine will be used, namely the internal machine. Where the tubes are over 16 mm thick, the internal machine makes the first weld and the external machine completes this weld.

3 - DESIGN OF THE LAYING BENCH

A docking clamp is associated to each machine designed externally for the internal machine and internally for the external machine. In addition to its function of holding the part, it must provide the seal between the welding zone and the fire barrier function.

The internal clamp adopts a layout similar to the clamps used for conventional welding.

The external clamp involves the same difficulties as the external machine, namely maintaining the vacuum. Naturally, a common method is adopted for these two devices.

Lastly, to assist in docking and relieve the loads on the clamp, a compensating support carries part of the weight of the tube.

4 - STUDY OF HANDLING THE MACHINES

During operation, the internal machine lies inside the tube ; when a new tube is presented into position for welding, this internal machine emerges completely, whilst remaining connected to its power supply. The machine is then housed in a receptacle. The power supply cable is wound and unwound with each operation.

The external machine is simply mounted on a carriage: it is hence linked to the tube and independent of any of the bucking movements of the barge.

5 - STUDY OF THE LAYOUT ON THE BARGE AND COST STUDIES

The layout of all this equipment on the barge was studied. Several existing barges were considered (see figure 3).

The fact that a thick tube is welded in only two stations enables the length of the laying bench to be reduced. This leads to two possible layouts:
- either a short barge, the cost of which is low, but the possibilities of which are meagre (tension in tube, seaworthiness),
- a long barge, the cost of which is high, but capable of satisfying a wider market ; it then becomes possible to weld in lengths of 24, 36 or 48 metres, to make best use of the available surface.

On the basis of these facts, several types of layout were designed.
Next, the profitabilities of these theoretical barges was assessed starting
from study of the operating rate of the machines: each operation was timed
and optimized and the overall production rate of the system deduced from
it: the working cycle of the internal and external machines is 8 minutes
for a 24" tube.

Next, the working rate of the laying sites was calculated for
several diameters (24 to 48") and several pipeline thicknesses. The
electron beam welding rates were compared against rates currently
achieved by conventional methods. The savings can amount to 10 to 50%,
depending on the type of barge involved.

However, it is of first order importance to demonstrate the economic
gains of the method by embodying all the investment costs in the
calculation.

The economic study covered two ETPM BARGES (Nos 801 and 1601) 110 and
190 metres long, revealing the interest of the electron beam method.

In addition, a more representative study was made by comparing them
to a real-life case, namely the STATPIPE laying site 250 kilometres long
in the North Sea, where ETPM barge N° 1601 laid a 36" tube 23 mm thick.
All the investment and operating costs have been allowed for, on the basis
of a tube assembled in 24 metre length.

The savings calculated concerning the duration of the laying work
amount to 25% and 10% of the cost. Whilst attractive, this result can be
improved yet further by assembling 36 or 48 metre lengths, which would be
possible if major modifications were made to the barge.

Now, the market study conducted by the BATTELLE Institute reveals that
the market involves first and foremost zones where the weather conditions
are difficult. For this reason, it seems clear that a large dimension
barge should be fitted out, capable of working in these zones, provided
the productivity can be improved by assembling 36 or 48 metre length of
tube.

6 - METALLURGICAL STUDY

At the same time as these mechanical studies were being made,
experimental testing of the welding method was going on.

An already existing vacuum chamber was fitted out to simulate S
curve laying conditions (see figure 4).

The tube is represented by a fixed cylinder ; this cylinder, held
from the outside, is contained within a vacuum chamber carrying the axial
gun (the internal machine) and the radial gun (the external machine).

The welds are not assembly welds, but work by melting the metal,
producing the same metallurgical effects as ordinary wells.

Since the welding positions change constantly, the adjustment
parameters of the beams must evolve. The purpose of the tests was to
develop a variation programme and to verify the quality of the seams by
inspection and mechanical tests. The initial stage consisted in adjusting
the parameters point by point, after which a continuous programme was
established for the inner seam and the outer seam.

With a complete procedure available, welds were made in various
grades of steel, followed by X-ray tests and mechanical tests.

However, most of these were high elastic limit steels (X70 and X100),
not representative of the steels used today for laying pipelines.

For certain cases with these steels, the hardness and resilience
tests did not prove satisfactory ; this tends to prove that the method can
not be accommodated to just any type of steel.

What has to be demonstrated is that the method is capable of welding steels currently used in pipeline projects.

7 - CONCLUSION

The studies and tests enabled the feasibility of the method to be demonstrated. However, certain points require additional investigation to decide whether to enter into further detail and invest in expensive equipment and modification of existing barges. This study must in particular cover the outer seal on the clamp tube and the external machine, and the weldability of steels at present used.

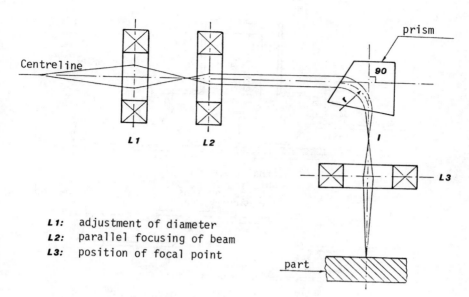

L1: adjustment of diameter
L2: parallel focusing of beam
L3: position of focal point

Operating diagram of electron beam gun

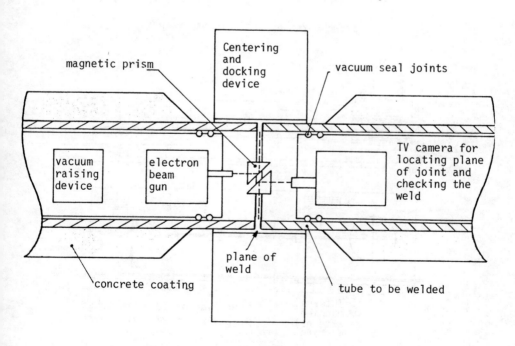

Operating principle of internal machine

FIGURE 1 - ELECTRON BEAM WELDING

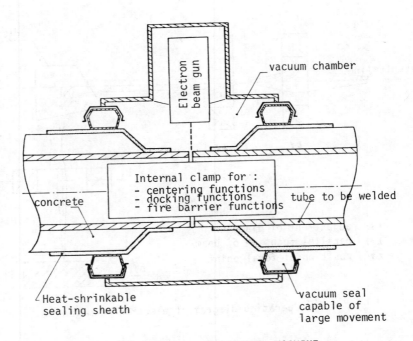

FIGURE 2 - OPERATING PRINCIPLE OF EXTERNAL MACHINE

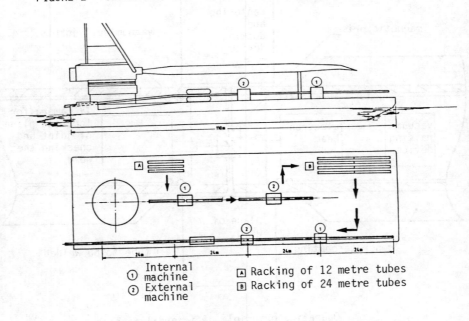

① Internal machine
② External machine

Ⓐ Racking of 12 metre tubes
Ⓑ Racking of 24 metre tubes

FIGURE 3 - LAYOUT OF WELDING MACHINES ON 110 METRE BARGE

VIEW TEST OF MACHINE WHEN OPEN
The cylinder support plate and axial gun prism can be seen in the background

FRONT VIEW OF TEST MACHINE
Centre : axial gun

FIGURE 4

(10.16/78)

DEVELOPMENT OF HIGH PRESSURE FLEXIBLE PIPING

A.D. GRIFFITHS
Project Manager, High Pressure Flexible Pipe
Dunlop Limited, Oil and Marine Division

Summary

An extensive research and development programme is in progress to
develop and produce a range of high pressure flexible pipes for use
in exploration for oil in subsea environments. Where commercial
'strikes' are made also to provide high pressure flexible links in
rigid flowlines and risers. Finally long length flexible pipes will
be developed for use as complete riser and flowline systems.

A wide range of materials have been evaluated under laboratory
conditions and selected materials have been incorporated into
prototype pipes which have been tested on a specially developed
series of test rigs. Test results and field trials indicate that
the pipes are superior to pipes incorporating nylon liners or
convoluted bellow liners. A complete facility to produce 150 ft.
pipe assemblies is now completed and a continuous pipe production
plant is planned.

1. INTRODUCTION

The development of marginal oilfields is dependent on reliable,
high pressure, flexible pipes. The environment in which these flexible
pipes are required to operate is hostile. The transference of crude,
diaphasic oil, especially where hydrogen sulphide, carbon dioxide, and
water is present at temperatures above 90°C creates severe problems
for flexible polymeric materials. Gases permeate into the polymeric
material and become dissolved at high pressure. If rapid depressurisation
occurs in the pipeline, explosive decompression of the polymeric material
and the pipe structure can occur.

When carbon dioixde or hydrogen sulphide permeate into the pipe,
in the presence of water, corrosion of most common metals will occur. High
tensile carbon steel will suffer from hydrogen embrittlement, whereas
austenitic and duplex stainless steels will suffer from chloride induced
stress corrosion.

Where the crude oil contains solid materials, or where through
flowline tooling will be used, the lining must be resistant to abrasion.

The cover of the pipe must be resistant to marine growth, ozone,
ultra violet, heat transmitted from the oil travelling within the pipe
and fire.

In operation the pipe must be capable of withstanding mud slides
and other seabed movements. The pipe configuration must be carefully
calculated and controlled to reduce resonance induced by vortex shedding
and tension variations, due to heave and pitch of the processing vessel.

Flexible pipes produced from nylon do not have adequate resistance
to the deleterious effects of the present operating environment.

2. TEST REGIME

A unique test regime has been developed, capable of simulating many of the operating conditions encountered. The expansion of this test regime is continuing and at present the following equipment is in continuous use ;

Liner compatibility bombs
Permeability, solubility and decompression bombs
Wire corrosion cells
Burst and proof pressure rig - 60,000 psi maximum
Tensile test rig 200 tonnes maximum
Hydrostatic collapse rig - 2,000 psi maximum
Pipe ageing and permeation apparatus - 5,000 psi max.
Flexibility - Kink tester
Waveform flex fatigue tester - 10,000 psi at 20 cycles per minute
Rotaflex tester - 10,000 psi at 10 rpm
Fire test - 700°C, 40,000 psi maximum

3. PRODUCT CONSTRUCTION

'Armalink' risers & 'Armaline' flowlines, will have a similar product construction. They will differ in their production methods, in that short lengths can be produced on purpose built lathes, where the pipe rotates and reinforcment materials are applied from static creels. For long and continuous production the reinforcing materials must revolve continuously around the pipe, which is traversed at a controlled speed through the reinforcing machinery. The construction of 'Armalink' - 'Armaline' is shown in Figure 1. The pipe bore consists of a stainless steel, interlock strip wound tube. This tube provides an abrasion resistant lining which is deliberately intended to slowly allow the passage of oil and gas through it. If this pipe were sealed it would take part in the pressure functions of the pipe and as it lacks sufficient hoop strength, it would burst before loads were transferred to the reinforcing cables. Over the interlocked tube is extruded a layer of Dunlop, proprietary, modified polyolefin, 'Duralon.' 'Duralon' is chemically stable within the temperature range -40°C to + 130°C. Duralon is resistant to permeability and the effects of rapid depressurisation and is believed to be superior to all, presently existing, moderately priced, polymeric materials. Anti-extrusion textile layers are wound over the Duralon, prior to applying paired layers of corrosion resistant, high tensile, aluminium clad, steel strands. The angle of application of the steel strand and design of the strand are carefully calculated to ensure maximum load sharing between each layer of steel strand. It is possible to achieve burst pressures up to 30,000 psi utilising two twin layers of steel strand reinforcement. An abrasion resistant, outer cover is extruded onto the pipe and during subsequent pipe vulcanisation, consolidation and complete bonding of all components in the pipe is achieved. At present time, pipe end fittings are build into the assembly and the whole assembly is bonded during vulcanisation.

4. DESIGN CONSIDERATIONS

If Duralon were immersed freely in crude oil, it would absorb oil and swell. Dependent upon the composition of the crude oil, the volume increase could be at least 20% and possibly as much as 100%. Aromatic

components of the oil will tend to swell more than paraffinic components.
In the Armalink and Armaline structure, crude oil and gas will leak
through the labyrinths of the interlock tube, at each seam. Oil and gas
will be absorbed by the Duralon. The Duralon will tend to swell, but
because it is contained and constrained between the interlock tube and
the textile cord reinforcement layer, the Duralon can only swell by a
small and limited amount. When the available space is completely occupied,
osmotic pressure will build up in the Duralon layer. This osmotic pressure
will reduce permeation of gases through the Duralon layer. Because the
Duralon layer is contained and constrained, the quantity gas, which it can
dissolve is reduced. If the gas/oil mixture in the bore of the pipe
depressurises rapidly the gas absorbed in the Duralon layer can not
rapidly depressurise because of the containment and common problems with
explosion of elastomeric materials on rapid depressurisation do not
occur. Additionally with pressures up to 10,000 psi in the bore of the
pipe, gas and oil migration through the strip wound structure will lift
the Duralon wedge, which fill the concave crevice on the outside of the
strip wound tube. This is effectively a non return valve, which allows
free passage of oil and gas into the pipe structure from the bore of the
pipe. After approximately six hours the pressure in the Duralon layer
equals, and may exceed marginally, the pressure in the bore of the pipe.If
the pressure reduced rapidly in the bore of the pipe, the pressure within
the Duralon layer will exceed the pressure in the bore of the pipe and
the wedge shape in the outside crevice of the strip wound tube will be
forced into the concave crevices in the tube, forming an effective,
sealed seat. With this effective seal formed the decompression of the
Duralon can not occur rapidly. Porosity and explosive expansion of gases
dissolved in the Duralon is eliminated. Pipes have been built incorporating these principles and with a 3" bore pipe, pressurised with crude
oil saturated with carbon dioxide at 100°C, over a period of 48 hours,
results in a pressure drop less than 5 psi. This pipe has been subjected
to multiple pressurisation and depressurisation cycles followed by
measurement of permeation rate through the pipe and near to zero
permeation is recorded during the nine months duration of this test.

The permeation of hydrogen sulphide, carbon dioxide, and water
is consequently significantly reduced. The quantities of these materials
contacting the steel reinforcment is also reduced. It can be concluded
therefore that with the reduced permeability there is reduced tendency to
stress corrosion or hydrogen embrittlement in the steel reinforcing layers.
Many design considerations intended to reduce corrosion of the reinforcement material, have been evaluated. In the Armalink / Armaline construction the high tensile steel wires are clad, under extreme pressure, with
pure aluminium. The aluminium is effectively pressure welded to the
steel substrate. The aluminium cladding provides an impermeable barrier
around the high tensile steel reinforcing member and in the event that
the cladding becomes damaged, the aluminium provides effective cathodic
protection. Traces of hydrogen sulphide and carbon dioxide, which permeate
through the Duralon layer will penetrate quickly through the outer layers
of the pipe and will be dissipated evenly through the outer cover of the
pipe. No external 'pricking' is required and consequently ingress of
seawater is minimised .

Where high burst strength is required it is necessary to use cords or strands of highly drawn steel wire. The pressure welded aluminium clad material is highly drawn and the cords utilised have high tensile strength. In order to provide a pressure balanced construction, the layers of strand are arranged in pairs. The first layer of the pair is applied at an angle slightly lower than the neutral angle for that structure and the second layer of the pair is applied at an angle slightly steeper than the neutral angle. When the pressure is applied, both layers assume their neutral angle and the result is that the first layer increases slightly in diameter and the second layer reduces slightly in diameter. Effective load sharing is obtained using this technique. Burst pressures achieved using one pair of strands indicate that approximately 99% efficiency is achieved. If a second pair of cords is applied it is not possible to obtain effective load sharing and only approximately 50% of the theoretical additional load from a second pair of wires is achieved. If a third pair of cords is applied then very little increase in burst pressure is realised. Armaline and Armalink construc- tions will incorporate either one pair or two pairs of cords. The structure is stable under pressure. Because high tensile steel has been used in an efficient manner the pipe is light in weight, dimensionally stable under pressure and is highly flexible. During vulcanisation of the pipe all layers become consolidated and bonded firmly together and the structure is not only flexible but very resistant to flex fatigue.

The stable structure also functions well if external hydrostatic pressure is applied. Verification testing has just commenced to confirm the capability of Armalink / Armaline to withstand hydrostatic pressures of 2,000 p.s.i. Similarly, the pressure balanced structure will also allow end tensile loads upto 100 tonnes with no collapse of the pipe bore.

5.   PRODUCTION CAPABILITY

The high pressure flexible pipe plant at Grimsby, costing £2.2 million, is now fully commissioned. With the exception of the cold feed rubber extruder, all plant and equipment was designed by Dunlop Limited. The equipment is manufactured to machine tool standards and the accuracy and reproducability obtainable is outstanding. The quality control test facility is also commissioned and further development testing at Dunlop Technology Division is fully utilised in the continuing development programme.

6.   PRODUCT RANGE

As the first phase of the development programme, a full range of Rotary Drill hoses, Choke and Kill hoses and Vibrator hoses were developed. American Petroleum Institute approval was obtained, January 14th, 1983. The development of non-oil carrying hoses is complete and several hundred pipes are now in service.

The Armalink, live crude oil high pressure flexible pipes are also in full production. The facility will allow production of pipes from 2" upto 12" inside diameter, 150 ft. long and mandrels and tooling are available at present for most sizes. The present burst pressure capability is as follows :

BORE	CONSTRUCTION	CARBON STEEL (PSI)	ALUMINIUM CLAD STEEL (PSI)	NITRONIC 50 STAINLESS (PSI)
2"	4 x 4.5 mm	32534	22256	28952
2½"	4 x 4.5 mm	28459	19469	25326
3"	4 x 4.5 mm	24659	16869	21944
3½"	4 x 5 mm	22885	17432	22639
4"	4 x 5.5 mm	23707	16323	21248
6"	4 x 6 mm	19073	13158	17097
8"	4 x 6 mm	15300	11009	14304
10"	4 x 6 mm	12718	8774	11401
12"	2 x 6 mm	8092	5582	7254

The above table shows current pressure capability for three styles of steel cord. It is expected that as the product construction is refined that the pressure capability will increase. Aluminium clad cables provide the best available option for sour service, but because the aluminium occupies space but does not contribute to the load sharing capability, then Nitronic 50 is available at present, if higher pressures are required.

The present range of pipes can operate with fluids in the temperature range -40°C to +130°C and the upper limit can be extended to 150°C but consultation on design features will be necessary. Armalink pipes are suitable for carrying all types of crude oil, sweet or sour, with or without gas, water or enhancement fluid contamination. As optional extra fire proofing of the cover and additional armouring can be provided.

Slimline fittings are presently built into the pipe during manufacture. Coupling systems which can be incorporated into the fittings include :

1. API Line Pipe Thread – upto 7,500 psi working pressure

2. Cameron Hub and Clamps

3. Grayloc Hubs

4. Hammer Unions – such as those made by Weco

5. Flanges to suit particular requirements

6. Cryogenic couplings

It is possible to attach other styles of couplings, including quick release, self sealing couplings.

7. TYPICAL APPLICATIONS FOR DUNLOP ARMALINK PIPES

Dunlop Armalink high pressure flexible pipes can be used for the following applications :

1. Jumper Connections : Short lengths of Armalink used to connect rigid piping in topside equipment on offshore production platforms

2. Spool Pieces : Short lengths of Armalink between sections of rigid piping in subsea equipment

3. Flexible Risers : For handling live or processed crude oil in floating production systems

4. Flowlines : Short length flowlines carrying live crude oil between elements of seafloor production piping

5. Transfer Pipes : For transferring live and processed crude oil from early production systems

When used as a link between well head and testing equipment in well testing procedures Armalink provides a, flexible, corrosion and abrasion resistant connector.
Armalink can also be used to inject fluids into an oil well in cementation and acidising operations providing a safe and reliable high pressure flexible pipe, which reduces time needed for setting up and dismantling. Armalink can also handle most oilfield enhancement fluids.
Armalink can be used with through flowline tooling, various pigs and down well tools, without damage to the stainless steel liner.

Eight typical applications for Armaline pipes

Dunlop Armaline high pressure flexible pipes can be used for the following applications :

1. Flexible Risers : For handling live or processed crude oil in floating production systems

2. Flowlines : Long length flowlines carrying live crude oil between oil well and seabed manifolds

ACKNOWLEDGEMENTS

1. Kokoku steel Wire Limited
2. National Standard Wire Company
3. Sandvik Svenskaforsaljnings AB
4. Armco Corporation

DUNLOP LIMITED - OIL AND MARINE DIVISION

'ARMALINE' - 'ARMALINK' PRODUCT CONSTRUCTION

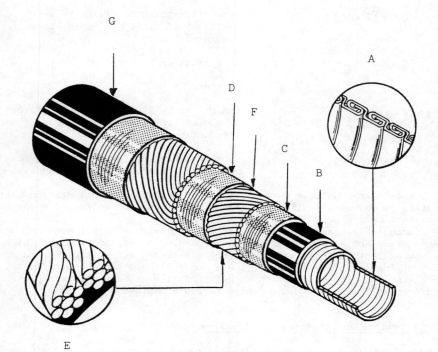

A.   Stainless steel, strip wound, interlock tube.

B.   Duralon elatomeric liner.

C.   Textile, anti-extrusion layer.

D.   Hydraulic transfer layer.

E.   Clad steel reinforcement.

F.   Anti-chaffing layer.

G.   Abrasion resistant cover.

FIGURE 1

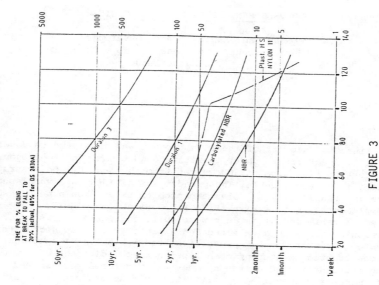

APPARENT LIFETIMES FOR 1mm THICK SAMPLES
IMMERSED IN WET SOUR(0.6±0.2 wt% H₂S ON OIL)
N.SEA CRUDE OIL UNDER LABORATORY CONDITIONS
AT VARIOUS TEMPERATURES.

TIME FOR % ELONG
AT BREAK TO FALL TO
20% (actual, 40% for DS 2030A)

FIGURE 3

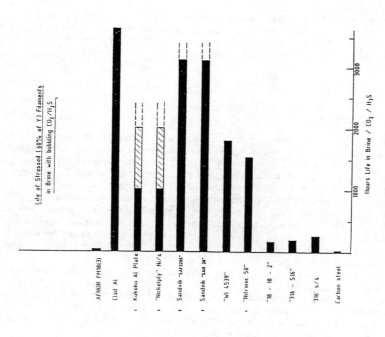

Life of Stressed (60% of Y) Filaments
in Brine with bubbling CO₂/H₂S

FIGURE 2

(09.18/80)

DEVELOPMENT OF AN ULTRASONIC RISER PIG

J.A. de Raad Ing. Manager Dept. Development & Systems
Röntgen Technische Dienst - RTD - Rotterdam
The Netherlands

Summary

With financial support of the EC a project was initiated to develop a
tool to inspect risers under operating conditions. The tool should be able
to enter the riserpipe to 300 metres from the launch/receive trap and use
an umbilical/rescue cable. The project was divided in several phases in
which the feasibility to solve the problem areas should be proven prior
to implementation to a Riser Pipe Inspection Tool - RPIT.
A special cable and winch unit were built and worked satisfactorily.
Ultrasonic probes were selected and a prototype ultrasonic system constructed. Prior to a high pressure system a Pipeline Inspection Tool(PIT)
was constructed to inspect open pipelines to a distance of 500 metres
using the constructed ultrasonic device. The field test showed its good
performance and technical justification of the project. Also attention
was paid to high pressure free swimming tools, as such the RTD-Caliper
was designed.
Problems anticipated with cable operated tools and new views of riser
operators do require free swimming inspection tools. A second project
is in progress to build RPIT's without umbilical for inspection of oil
and gas risers.

1. INTRODUCTION.

There is an increasing demand to establish the integrity of buried
and sub-sea pipelines which have been in service for considerable time.
Offshore pipelines are connected to platforms by risers. These risers can
be embedded, as usual with concrete platforms. On steel platforms they
are externally mounted against the vertical legs.

To avoid corrosion, pipelines and risers are protected by the use of
cathodic protection, inhibitors and selected materials which nevertheless
do not exclude the possibility of corrosion. Wind and water act heavily
on external risers, this causes ever changing temperature gradients over
the pipewall in the splash zone. This might despite protection increase
the possibility or speed of corrosion.

If diving conditions permit external risers can be inspected from the
outside.

Due to the harsh weather and tide conditions on parts of the North
Sea and Atlantic, diving operations are very restricted and hard to plan.
Even if possible the available timewindows do not allow for painstaking
inspections particularly at larger depths.

Embedded risers cannot be approached from the outside and thus require internal inspection. Platform operators consider it therefore
attractive to use an inspection tool which can operate inside the risers

of both types. Preferably such a tool should perform its job with minor influence on pipeline operating conditions. Already in the early seventies attempts have been deployed to inspect risers from the inside. Due to the limited technology available at that time the tools were slow and required open risers. This was unattractive because the long shut down times interfered heavily with the operating conditions of the pipeline. These cable operated tools could only pass a few bends whilst some of the riser systems show many bends. These limitations and additional unsolved safety requirements prevented application of these tools on risers. For several years not much progress could be observed.

The ever existing demand of platform operators and technological progress in the last decade allowed RTD in 1980 to propose a project to the EC to develop a cable operated Riser Pipe Inspection Tool (RPIT) which could operate in an oil riser under operating conditions.

The project incorporated several feasibility studies as well as the use of free swimming tools prior to development of the proposed RPIT.

2. RISER PIPE INSPECTION TOOL RPIT.

A basic set of equipment to perform riser inspection consists of:

° the actual crawler propelled by the oil flow, cable or powered by a drive motor, or a combination of those.
° umbilical with the electric and rescue cable.
° the winch unit housed in the instrumentation trap to handle the umbilical.
° the read out unit to analyse the data collected.

The crawler with electronics often indicated as "intelligent pig" consists of:

° several articulated high pressure containers to house the front end electronics, data treatment, data storage and power supply. In general lower power consumption components are used to safe scarce energy.
° The battery power source
° A vehicle with the sensors distributed around its circumference.
° Wheels and or discs to propel and guide the pig through the riser pipe carefully designed to enable passage of diameter restrictions, T-joints, valves etc.
° Odometer for distance information and several periphery electronics to enable remote and safe operation.

In principle the pig enters the riser from the instrumentation trap by its own traction or propelled by the oil flow. To retrieve the tool the winch unit located in the trap can be used. During its measuring run the data collected are stored with their respective locations as reference for retrieval use.

The read out unit is used to retrieve the data from the pig and to analyse them by diagnostics and routine formats for which small computers are to be used.

To operate the tool some remote control has to be developed also enabling the essential safe operation on platforms. During actual inspection but certainly during the preparation the pipeline operator is heavily involved for many reasons but particularly while the inspection interferes with the operating and safety conditions of the pipeline system.

3. THE INSTRUMENTATION TRAP.

To launch and retrieve instrumentation tools instrumentation traps
are present on almost all risers. They allow to enter the riser with minor
interference of pressure and flow. Their general lay-out is shown in
figure 1.

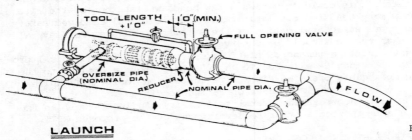

LAUNCH   Figure 1.

Despite a tendency to standardise several trap types exist. Their variations in dimensions have a high impact on universal tool design.
Particularly the nett trap length is of major importance. The trap should
in case of a cable operated tool provide housing for the inspection tool
and winch unit.

4. CONSIDERATIONS AND PROJECT PHASES.

It was considered useful to identify the problem areas and divide the
project in several phases.

- ° Measuring technique: Ultrasonic wall thickness measurement is
  considered the best method to quantify corrosion.
- ° Probe selection: For an oil riser the stand-off technique is most
  appropriate. The probe to be selected should be able to quantify
  in- and external corrosion. The probe must be capable to resist the
  high pressures as exist in risers.
- ° Compact ultrasonic equipment: To operate a large number of probes
  which cover the circumference of the pipe multichannel ultrasonic
  equipment is necessary. The feasibility of constructing compact
  equipment should be studied.
- ° Umbilical and winch unit: Lengths up to 300 metres with integrated
  rescue cable are necessary. If no precautions are taken a cable
  can only pass 4 bends (360°) due to the high friction forces generated by the Capstan effect. The design of a winch and umbilical
  allowing to pass many bends (800°) was considered necessary.
- ° Proof of principle: Prior to implementation in a high pressure resistant RPIT it was considered practical to check the above mentioned achievements. For this reason the "Pipeline Inspection Tool (PIT)
  was planned to be constructed, also to build a full scale simulation
  of a riser with bends.
  Dependent on the results of these tests the scope of the project
  should then be reconsidered.
- ° Free swimming tools: to study the capability and behaviour of a free
  swimming pig the "Caliper" was incorporated in this project. This
  relative simple tool uses electromechanical sensors to establish
  internal pipeline condition. The Caliper shows a number of similarities with the RPIT.

5. SUMMARY OF PROJECT ACHIEVEMENTS.

5.1. Probe selection.
An extensive study and experiments were performed to select appropriate ultrasonic probe systems for both oil and gas risers. The study incorporated following techniques:

° stand-off probes
° ultrasonic wheel probes
° Electro Magnetic Acoustic Transducers (EMAT's)

As a result of this study the proper probe to be used in an oil riser could be selected. This probe is suitable to detect in- and outside corrosion. Numerous probes as figure 2 shows were incorporated.

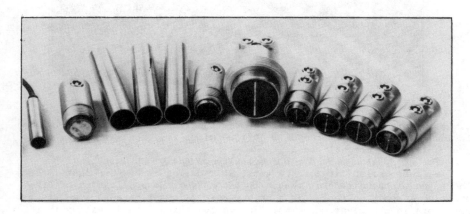

Figure 2. -   Selection of ultrasonic probes

Some basic studies were initiated on wheel- and EMAT ultrasonic probes to judge their alternative possibilities. At this moment the stand-off technique for oil risers is superior.

5.2. Compact ultrasonic equipment
A feasibility study with design for printed circuit boards proved the possibility to build multi channel ultrasonic equipment with a very dense packing of electronics.
According to these principles a stand alone system "SONOLOG 83" was built.

5.3. Umbilical and winch unit
To negotiate bends and eliminate or reduce the capstan effect several solutions were considered. One of the most promising concepts was checked for feasibility. To reduce cable friction in the bends a steel cable was provided with numerous rollers. To operate the cable a special winch unit as shown in figure 3 has been designed and optimised.

Figure 3. -   winch for cable with rollers

Cable and winch unit showed their proper functioning on a full scale simulation of a riser with 6 bends (540°). - Minor friction was observed of the cable under nominal tension.

The study also showed the complexity of the winch unit.

Because reliability of the total RPIT design is of paramount importance the winch unit forms reason to reconsider the application of free swimming inspection tools. This became more relevant because some riser operators in the meantime do not like cables through their main valves for safety reasons.

5.4. The pipeline inspection tool "PIT".

5.4.1. General.

Prior to the construction of a high pressure cable operated tool a low pressure Pipeline Inspection Tool - PIT - was developed. A sub-sea off loading line in Canada was considered a proper situation to perform field tests. Full Cooperation of the pipeline operator could be achieved to carry out extensive tests.

5.4.2. Concept of the PIT.

The PIT constructed for this inspection was equipped with 12 specially designed immersion probes as shown in figure 4. The probes are equidistantly mounted with a separation of approx. 125 mm. Six probes are springloaded to maintain a constant distance from the inner pipe surface. The other six are fixed to the probe-carrying ring to determine inside corrosion pits, dents and buckles and measure their profile on the basis of the variations in waterpath. During the inspection run the probe array is in a fixed position and inspection is carried out along 12 straight traces.

The ring carrying the probes can be rotated by remote control but is restricted to plus or minus 25°, sufficient to ensure detailed scanning in selected areas where 100% mapping would be desirable.

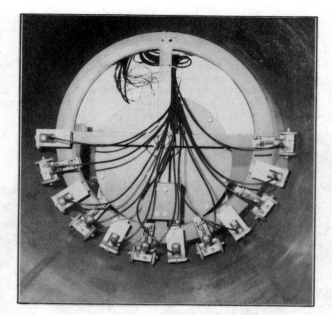

Figure 4. -   Multiprobe ultrasonic system of the PIT

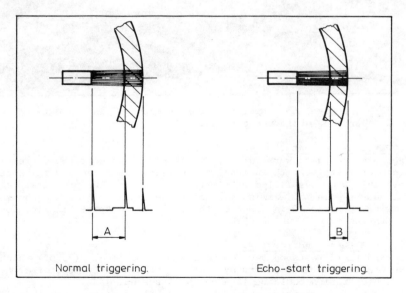

Figure 5. -   Principle of measurements

The principle of the measurements is shown in figure 5. Inside corrosion (profile) and outside corrosion as well as the remaining wall thickness could be measured simultaneously. To operate the 12 probes the Sonolog 83, see chapter 4.2, was used. The self-propelled tool as shown in figure 6, contains the probe multiplexer and batteries for the traction. This part of the PIT can be operated to pressures of 10 BAR and is connected to the electronics onshore by a buoyant cable with a length of 575 metres. The results stored on magnetic tape can for interpretation be recorded in an analog presentation of any scale.

Figure 6. - The pipeline inspection tool in front of inclined testpipe, and reel with buoyant cable.

5.4.3. Performance tests.

Performance tests were carried out at RTD on an appr. 50 metres long 36" diameter testpipe with internal and external cavities and thickness changes. This pipe was filled with water to simulate the practical conditions under which the examination had to take place. Figure 6 shows the PIT crawler, positioned in front of the inclined testpipe (inclination as found with the actual sub-sea line).

5.4.4. Field tests and results.

During field tests on the 540 metres of pipe length (36"Ø") which took 90 minutes, on line recording could be continuously monitored. Large areas with corrosion could be located and assessed.

The practical exercise proved the feasibility of an ultrasonic inspection tool for detection, location and quantification of inner- and outer corrosion.

5.5. The CALIPER

5.5.1. The Shell CALIPER

This system developed by Shell - KSLA was selected for further development in the frame of the EC contract. This Caliper was designed to detect and measure the pit depth, uniform internal corrosion and variations in diameter. The prototype available at that time was field tested in an 8" pipeline with a maximum length of 2 km.

The Caliper consisted of a liquid driven propulsion unit, a battery pack, a sensing unit, equipped with mechanical feelers with appropriate transducers to convert the movements of the 12 individual feelers into electrical variations, and a data storage unit. The latter collected all measuring data together with information on locations etc. in a solid state memory.

The information could be presented in digital form by a "read out unit" in the field, once the Caliper had been retrieved. The maximum and average diameter could be displayed as well as the number of pits per metre of pipelength found in 3 different classes of severity.

5.5.2. The RTD CALIPER.

Improvements.

During the EC contract several improvements have been implemented by RTD, e.g.:

° an extension of the range
° operation at working pressures up to 75 bar;
° improved application of microprocessor technology.

The range has been appreciably extended by the use of a more appropriate data reduction system and extension of the memory capability by the application of a bubble memory.

Data storage in the solid state memory now covers the information over a range of 50 km.
These data - stored per unit of length - include:

° 2 classes of pit depths
° minimum diameter
° maximum diameter
° average diameter

Figure 7. -   The Caliper with its different units

Concept of the RTD Caliper System.

  The Caliper electronic system consists of three major modules for:
1. Main control and odometer (distance information);
2. Data acquisition and data reduction;
3. Data storage.

The main control module is a micro processor system which is keeping track
of time and distances, giving commands to the data acquisition module, receiving data from it and compressing the data before storing them in the
data storage module. The data acquisition module is a second micro processor system receiving its information from the array of mechanical feelers. Their movements are converted into electric potential variations by
means of strain gages. The output of each is gated to an instrumentation
amplifier by analog multiplexers. The amplification is 60 dB. After that
a programmable gain (0 - 20 dB) and offset is applied to compensate for the
different straingage characteristics. The output is digitised. Then via a
programmable algorithm the pits are counted per unit of length, e.g.
1 metre. The data memory is 1 Mbit solid state bubble memory, from which
the power is automatically switched off when not in use, to reduce power
consumption. For the same reason all electronic components in the Caliper
are of CMOS technology.

Caliper "read out" unit.
  The read out unit is a small micro computer system with floppy discs,
keyboard, VDU (Video Display Unit) and a graphic printer as shown in
figure 8.

Figure 8. -   Caliper read out unit.

It is connected to the Caliper before and after a run, to program the
equipment for the measurements and in turn retrieve the data. Besides job
information, the read out unit presents the collected measuring data in a
numerical or graphical way on the VDU and on the printer by push button
request.

Several options are available on how the data can be presented:

  °numerical
  °graphical, see figure 9
  °histogram,

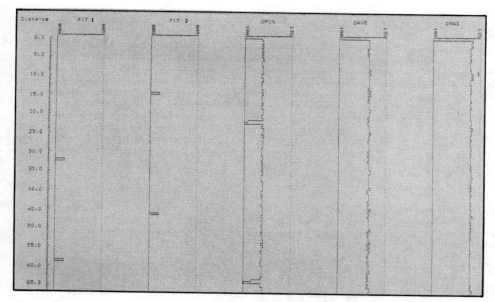

Figure 9.- Example of plot of pit numbers and diameters against distance.

For instance a histogram can be plotted to show the distribution of pits of different severity over a certain range. Bad spots and severity check programs are available. The data on the floppy discs can also be used for later comparison with the findings during subsequent tests.

Field tests.

At present a Caliper is almost ready to be field tested on a 10" two phase flow line. (Maui - New Zealand). Final laboratory tests prior to commissioning are in full progress. RTD will perform these tests by the end of 1984.

6. FUTURE WORK.

The results and achievements of the first project were reason to convert the attention to a free swimming RPIT. A second project is in progress aiming for tools without umbilicals to inspect oil and gas risers in which Royal Dutch Shell is heavily involved.
Also the work on the RTD Caliper is continued.

7. ACKNOWLEDGEMENT.

This work could only be performed at the level and extent as described with the financial support of the EC, Directorate General for Energy.

8. LITERATURE.

Systems to inspect buried and sub-sea pipelines
by R.van Agthoven, F.J. Nolet, J.A. de Raad, A. de Sterke
Eurotest Technical Bulletin E 52, March 1984.

A Caliper for Corrosion Inspection of Pipelines
By H.Bosselaar
Eurotest Symposium 24-26 March 1982, "New Trends in NDT".

(10.26/81)

SELF-DESTROYING VEHICLE FOR THE INSPECTION OF PIPELINES

M. FIOCCHI
SYMINEX Project Manager

SUMMARY :

The inspection of hydrocarbon pipelines by means of instrumented,
autonomous vehicles may be difficult from a technical point of
view, but this enables preparations to be made against the risk
of accidents whose ecological and financial consequences could
be considerable.
The types of equipment available at present are heavy, cumbersome,
expensive and the risk of jamming in the pipe cannot be put
aside. Therefore their use is generally limited to punctual
operations.
SYMINEX is developing a complementary tool designed to be used
periodically to eliminate the risk of blocking in the pipeline.
Its main characteristics are as follows :
- light and easy to handle vehicle,
- low cost operation permitting an extensive use of the system,
- global corrosion measurement in a section of a pipeline
  enabling regular monitoring of the evolution of the state of
  the casing,
- automatic dislocation of the vehicle if it becomes jammed in
  the pipeline.
This vehicle is made of composite materials with an electro-
magnetic measurement system controlled by a micro-processor
whose data are stored in a memory.

1. INTRODUCTION :

    The ecological and financial consequences of accidents with
hydrocarbon pipelines at sea are such that it is necessary to
carry out systematic and regular controls of the pipelines.
In this respect, the autonomous instrumented vehicles that can
carry out local measurements inside the pipeline present certain
advantages, for example, continuous measurement, localization of
defects while mining continues.

However, the equipment available on the market at present has the disadvantages of being heavy and cumbersome.
Moreover, the cost of an intervention is often an obstacle to periodical use. The risk of jamming of the vehicle in the pipeline cannot be excluded and the financial consequences, particularly offshore, are such that many operators prefer not to take the risk. To fulfil this need, SYMINEX is currently developing an instrumented scraper with the following characteristics :
- the vehicle is light and easy to handle ;
- the measurement made is a global estimate of (internal and external) corrosion in a pipeline section ;
- its manageability and low cost enables regular monitoring and detection of the evolution of the state of the casing ;
- if jamming occurs in a pipe, the system can be dislocated and the pipeline operation is not interrupted thus avoiding the cost of offshore operations.

2. SELF-DESTROYING VEHICLE :

After studying the various restraint linked to the self-destruction of a vehicle, such as the size of residual elements, their evacuation by fluids, their behaviour regarding the pipeline, the use of syntactic foams seemed to be the best suggestion for building the frame.
These composite materials consisting of an organic binder and preformed lightening cells are buoyant and resistant and have often been of great use in subea systems.
In our particular case, the foam retained is composed of hollow glass micro spheres associated with an epoxyde matrix.
This mouldable and workable foam enables a structure to be manufactured at a low cost.

2.1. Fragmentation mechanism :

The fragmentation of the vehicle is ensured by a pyromechanical transmission fuse embedded in the body of the vehicle.
It is released by an initialization block which is subject both to the pressure inside the pipeline (the mechanism cannot be triggered off if the scraper is not in a pressurized pipe) and a clock.
Most of the energy released by the explosion is absorbed by the fragmenting foam ; tests in progress have shown that no important over pressure occurs in the pipeline.

2.2. Propulsion :

The scraper proposed is propelled by the fluid and therefore does not prevent the continuation of working and does not require an embarked power supply for propulsion which is ensured by a set of obturator rings.

3. THICKNESS MEASUREMENT :

The chosen thickness measurement provides information on the quantity of metal in a tube section. It enables detection of the variations in volume of a few percent, which, in terms of thickness, represents a damage of 3 % for generalized corrosion or localized corrosion on the generator of less than 20 % of the thickness.

The sensitivity of the system is in the process of being evaluated with accuracy.

The operating principle is based on the deformation undergone by an alternative magnetic field going through a conducting casing. An "emission" coil generates a sinusoidal magnetic field. The current induced in a receiving coil, presents a phase-shift on the induced field proportional to the thickness of the casing.

A micro-processor system integrated in the equipment ensures the measurement of this phase-shift at each period as well as memory storage of up to 100 000 measurements.

The sensor and acquisition system require a small amount of energy to operate, which allows for total autonomy over a distance of several kilometres.

4. CONCLUSION

In safe conditions without any risks of jamming, this instrumented vehicle will ensure the monitoring of the evolution of corrosion phenomena in short pipelines (approx. 10 km which represents 50 % of the cumulated length of offshore pipelines).

This uncumbersome light weight, (dia. 12" ; length 40") autonomous and automatic equipment can be used by the operator himself.

The study currently in progress will result in the construction of a laboratory-tested prototype.

Tests on various types of pipelines will then have to be carried out to complete this preliminary phase.

(09.21/79)

RIP RAP PROTECTION OF UNDERWATER PIPELINES BY VELPO VEHICLE

J. MARTIN
Head of General Engineering Department
C.G. DORIS - PARIS (FRANCE)

Summary

The Rip Rap protection is one of the most efficient means of ensuring a good protection of the submerged pipelines which cannot be buried. In order to reduce the cost of such Rip Rap protection a group of french companies composed of :

- Commissariat à l'Energie Atomique (C.E.A)
- Omnium Technique des Transports par Pipe-lines (O.T.P)
- Chantiers de France-Dunkerque (C.F.D)
- Compagnie Générale DORIS

has developed an underwater vehicle named VELPO which can be described as a mobile hopper, able to travel and be accurately positioned over the submerged pipeline.

The system is rated for operation down to 300 meter water depth. The design of the vehicle, including all the positioning system has been completed, and a prototype has been built.

The prototype has been successfully tested underwater and is now ready for full scale operation.

1 - INTRODUCTION AND CONCLUSION

Protecting the pipelines has always been a concern for the operator of the offshore oil field. Although the pipelines can usually be buried on the sea bed, there is still a number of case where for example a hard sea bed, which makes burying or trenching impossible. In such cases, the rip-rap protection is the only solution in order to

- support free spans
- eliminate risk of under scouring
- provide protection against anchors, trawl boards and falling objects

The Rip-Rap protection is usually achieved by dumping stone from the surface or through a fall pipe. The efficiency of the system is usually low because most of the dumped stones do not reach the target. Sometimes, it is necessary to dump ten times as much as the quantity strictly required to correctly protect a line ; In such case, the cost of Rip-Rap protection increases tremendously.

The Velpo system has been developed to ensure an efficiency close to one in any circumstance.

A prototype has been built and tested, and the system is now ready for full scale operation.

2 - DESCRIPTION OF THE SYSTEM (See Figures I and II)

The system consists in an underwater vehicle which can be described as a mobile hopper, able to travel over the submerged pipeline. The Rip-Rap is dumped from a surface vessel, a dynamically positioned bulk carrier, through a feeding pipe. The hopper is large enough to accommodate the movements of the mouth of the feeding pipe.

As the Velpo vehicle can be accurately positioned above the pipeline, all the stones dumped from the vessel are collected and end up nicely laid above the pipeline to be covered.

3 - GENERAL CHARACTERISTICS OF THE SYSTEM

3.1. Main Data

- Operating conditions :

  - Operating depth   30 to 300 m
  - Current at the surface   3 knots
  - Current at the bottom   1 knot maximum
  - Wave characteristics (8 Beaufort) :
    - significant height   3.22 m
    - maximum wave height   6.00 m
    - period   8"
  - Wind characteristics (force 8)   30 to 40 knots
  - Sea-floor profile :
    - average gradient   10 %
    - cross slope   5 %
    - maximum gradient   35 %

- Basic rip-rap criteria :

  - Diameter of pipeline to be protected   8" to 48"
  - Bank section   3.5 to 10.5 cu.m/linear meter
  - Bank thickness above pipeline   1.20 m

- Operational characteristics to protect a 32" pipeline :

  - Dumping speed   50 to 72 m/hr
  - Daily rip-rap length   600 m
  - Dumping cubic capacity   500 cu.m/hr
  - Rock granulometry   3" to 8"

3.2. Surface Equipment

A dynamically positioned bulk carrier used as a surface support carrying the feeding pipe and the control room in which all operations relative to the sea bottom vehicle are centralized.

3.3. VELPO vehicle

3.3.1. Main characteristics

- Length overall   15.00 m
- Width overall   14.00 m
- Height :
  - maximum   10.50 m
  - minimum   9.00 m

- Surface of hopper :
  - upper aperture   8.0 x 8.0 m
  - lower aperture   1.4 x 2.2 m

Oscillations transmitted to the base of the casing pipe by the swell, the currents and the own movements of the surface ship determine the width of the hopper.

3.3.2. Propulsion system

The VELPO is propelled on the sea floor by four independent caterpillars (length overall 4.3 m - width overall 0.61 m) which are activated by hydraulic motors.

These motors are fed by two hydraulic units (power 30 HP each one).

3.3.3. Operational characteristics

Adjustment of pressure on sea floor and thrust for vehicle positioning through :

- 4 ballast tanks of   17 cu. m each
- 4 compressed air bottles
  (p = 30 bars)   6 cu. m each

For a mass of 124 tons (basic prototype), one will have :

- apparent buoyancy   50 t max.
- thrust   20 t max.

Velocity on sea-floor :

- in empty conditions   from 0 to 5 cm/s
- in operating conditions
  at 2 cm/s, that is   from 0 to 72 m/h

3.3.4. Power and control

The VELPO vehicle is linked to the surface vessel with an umbilical delivering electrical power, compressed air and driving instructions.

4 - OPERATION OF THE VEHICLE

4.1. The transmitter fitted on the casing pipe and the 4 receiving antennas installed in the upper corners of the hopper make it possible to control the position of the lower part of the feeding pipe in relation with the hopper.

The relative positioning of the transmitter in relation with the 4 antennas is thus measured.

Operating frequency of this electro-magnetic system : 1 kHz.

4.2. Two acoustic cameras are fitted on the vehicle ; a 500 kHz acoustic camera located on the fore part of the VELPO makes it possible to get a visual control of the pipe under the vehicle, and a 1 MHz aft acoustic camera checks the rip-rap bank and controls permanently the laying accuracy and the shape of this rip-rap bank.

Each camera is able to scan over an angle of 128°.

4.3. A detection system with acoustic sonars is located on the fore part of the VELPO to position and to centre the vehicle over the pipe to be buried.

This system consisting of 4 sonars (frequency 20 kHz each) visualizes on a video screen the relative positioning of the VELPO and of the pipe, even if the pipe is already naturally embedded.

5 - MODEL TESTS

Trials on small scale models were carried out in model tanks in the course of 1980 -1981.

- On a 1/10th scale installation, hydraulic flow tests were conducted to determine the size of the installations involved in the transport of rock materials and to establish the conditions of their transfer to the casing pipe.

This series of tests was carried out with different :

. rock granulometry
. water injection making it possible to increase the quantity of rock materials in the casing
. and feeding pipe lengths.

- Tests regarding the feeding pipe's mechanical behaviour were conducted considering the oscillations transmitted to the base of this feeding pipe by its own movements and the movements of the surface ship, residual motions entailed by the positioning of the ship.

Swinging amplitude at the base of the feeding pipe fully justify the use of a VELPO vehicle carrying a well adapted hopper to avoid the inconvenience of scattering, even in shallow waters.

- Trials on the rip-rap bank behaviour in wave and sea currents were carried out considering the gravitary destabilization on the bank gradients and the overspeeds induced by the bank shape.

These trials make it possible to confirm :

- the size of rip-rap granulometry to ensure a good stability
- the size of the riser in relative with granulometry and water flow output
- the size of the upper part of the hopper, considering the accuracy of the ship's positioning and its platform motions.

6 - FULL SCALE VELPO TRIALS

The first full scale tests of the VELPO took place at the end of 1981 in an open basin in the port of Dunkerque. These acceptance tests made it possible to carry out the inclining experiment of the vehicle, to check the hydraulic and electrical circuits.

New tests were conducted in March 1982, in a deep open basin in Dunkerque. The following operations have been carried out :

- testing the magnetic sensors for positioning the feeding pipe in respect of the hopper
- testing the acoustic cameras and the pipeline detection systems
- overall operating simulation : moving and driving the vehicle on the sea floor, above a dummy pipeline.

The system is now ready for full scale operation.

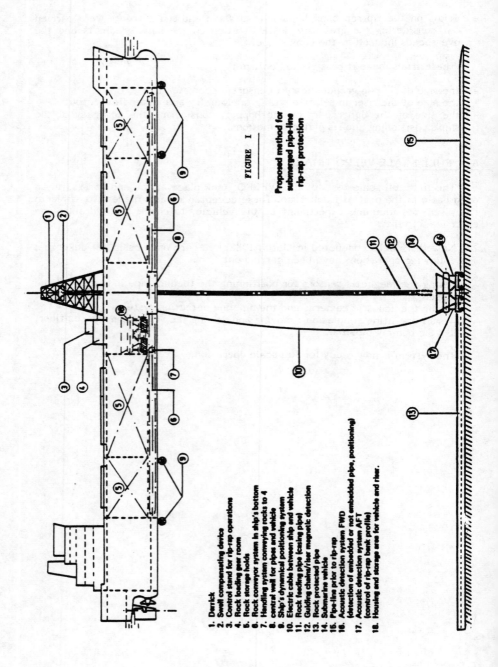

FIGURE I

Proposed method for
submerged pipe-line
rip-rap protection

1. Derrick
2. Swell compensating device
3. Control stand for rip-rap operations
4. Rock loading gear room
5. Rock storage holds
6. Rock conveyor system in ship's bottom
7. Handling system conveying rocks to 4
8. central well for pipes and vehicle
9. Ship's dynamical positioning system
10. Electric cable between ship and vehicle
11. Rock feeding pipe (casing pipe)
12. Guiding channel/riser magnetic detection
13. Rock protected pipe
14. Submarine vehicle
15. Pipe-line prior to rip-rap
16. Acoustic detection system FWD
    (detection of embedded or not embedded pipe, positioning)
17. Acoustic detection system AFT
    (control of rip-rap bank profile)
18. Housing and storage area for vehicle and riser.

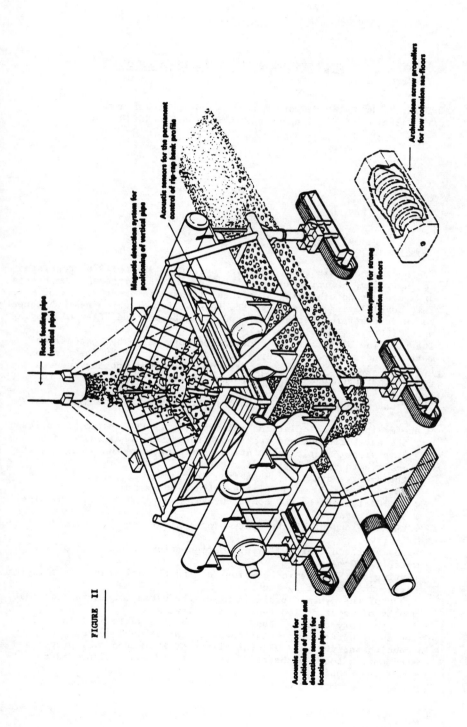

FIGURE II

Archimedean screw propellers for low cohesion sea-floors

Acoustic sensor for the permanent control of rip-rap bank profile

Magnetic detection system for positioning of vertical pipe

Caterpillars for strong cohesion sea floors

Rock feeding pipe (vertical pipe)

Acoustic sensors for positioning of vehicle and detection sensors for locating the pipe-line

(09.06/76)

NEW OFFSHORE PIPELINE LAYING TECHNOLOGY

L. AUPERIN, Technical Director of Société BOUYGUES Offshore,
Y. ROSE, Director of Pipelines Department of B.G. ENGINEERING

Summary

The purpose of this study, carried out from 1976 to 1978, consisted
in analyzing, defining and possibly testing the different components
of a new method of laying pipelines offshore in great depths of water.
This novel laying technology, the feasibility of which was subsequently
to be demonstrated during the study, was termed the "flexible laying
unit" (FLU).
The pipeline being laid is constructed on a barge by contentional
methods and accompanied over a considerable length by the flexible
beam providing it with the support and additional buoyancy needed to
ensure that when under tension, the S-curve is approximately in
equilibrium.
The study proper wad divided into several subjects, the main ones
being:
- the design and calculations relevant to the FLU system, comprising:
  . the establishment and testing of two-dimensional and three-
    dimensional static computation programmes for the design of the
    S-curve in great depths of water,
  . the S-curve dynamic computation programme,
  . the design and testing of the ballasting system,
  . the design of the FLU constituents, and in particular the pipeline
    support systems,
  . the development of the system for controlling the geometry of the
    S-curve,
- study of the different laying configurations in great depths of
  water (2500 metres),
- testing of the dynamic behaviour of the FLU and that of the laying
  barge, in the testing tank,
- study of the dynamic positioning of the barge for laying in great
  depths of water at low tensile load levels,
- a preparatory testing phase for laying at a scale of 1/10 in 250
  metres of water and simulation of laying at 2500 metres.
In conclusion, the general feasibility of the method was proved and
the computing tools elaborated. Laying tests would be needed to
confirm the studies.

1 - FLU (flexible laying unit) DESIGN AND CALCULATIONS

1.1 - S-curve static design calculation programmes

This study started with general consideration of the behaviour of the combination of tube plus FLU, in order to retain the essential parameters for laying in analysis of electronic computing programmes.

The following points were examined in particular:
- Influence of the tube link on the FLU. Can one consider in the calculations that these two elements lie along the same axis, even when they are not concentrical ?
- Do the frictional actions between the tube and the FLU affect the overall form ?
- Do the currents and rate of advance of the barge have any appreciable effects on an S-curve and how can these effects be expressed in figures ?
- How can one symbolize the contact of the tube with the bottom, the very nature of this contact influencing the complete S-curve ?

The initial analyses led us to adopt the following working assumptions:
- In most cases, the distance between the centrelines of the tube and the FLU can be neglected.
- Making this assumption, the nature of the links between the tube and FLU do not affect the form of the S-curve.
- On the other hand, the currents and the rate of advance of the barge have a non-negligible influence on the tensile forces in the tube and the shape of the S-curve.

Accordingly, it was decided to allow for these actions in the EDP programmes in the form of a hydrodynamic drag.

The relative speed between the pipe and the water is determined by the programme and added vectorially to the speed of the current.

Having defined the parameters of the problem, a method of solving the problem then had to be chosen. Three methods were envisaged:
- the method of finite differences: resolution of the differential equation of the elastic curve of the pipe by finite differences,
- the method of deformations applied to finite elements of bars with iterations in order to consider the major distortions and movements,
- the method of variation of the initial conditions: this method consists in studying the forms of a cantilever loaded at one of its ends by loads that can be varied within a given range until the shape sought is obtained, by successive approximations.

After different tests in which the power of the order required and the objective to be reached were considered, the third method turned out to be the most efficient and it is this method that was selected.

Despite the relative simplicity of this method, the analysis was complicated by the need to determine the intrinsic equations of a bar element that are valid for large distortions with compressive or tensile forces, leading to computerized calculations ensuring a very high mathematical accuracy, which is not the case of the DTANH, DATAN, etc... functions of the FORTRAN language for very small arguments.

Accordingly, in the intrinsic equations, all the expressions containing circular and hyperbolic functions had to be replaced by their serial development.

Lastly, the essential difficulty of programming resided in finding a satisfactory method of variation of the convergence parameters.

After several attempts, a dichotomy process followed by interpolation was adopted. This method transpired to be both fast and accurate. In addition, it requires only little central storage capacity.

Two computing programmes were hence written:
- PIPLAN 1 is a programme for determining the length of the stinger needed for a given laying configuration,
- PIPLAN 2 is a programme for verifying the shape of the S-curve and the stresses in the tube and stinger, for a stinger of fixed length.

Having completed this initial part of the mathematical study, the work was continued by study of a three-dimensional programme, based on the same principles as PIPLAN 1 and PIPLAN 2, namely using a method of varying initial conditions generalized to three-dimensional space.

Various programmes were hence established, expressing different stages of evolution and capable of being used to greater or lesser effectiveness, depending on the exact nature of the problem set:

. PØP 1 is a test programme of the computing block common to the other programmes. It turned out to be necessary for finalization, particularly in order to master the problems of mathematical precision,

. TRIP 3 is a programme for determining the shape of the S-curve for fixed conditions:
  - total length of S-curve fixed,
  - angle of entry onto barge of 0 (horizontal tangent),
  - soil horizontal,
  - bottom tension in tube fixed,
  - moment of axis horizontal to attachment aboard barge fixed,
  - zero torsion in the barge, by approximation.

The solution is obtained by double dichotomy on the vertical and horizontal shear efforts at the barge and double interpolation of these values.

. TRIP 4 is an extension of TRIP 3 with double dichotomy and triple interpolation on the two shear efforts at the entry aboard the barge and over the length of the S-curve.

It provides a solution for a fixed depth H.

. TRIP 5 is also an extension of TRIP 3 with double dichotomy and triple interpolation between the two shear loads at the entry to the barge and the tension in the bottom tube.

A solution can also be found for a fixed depth H.

. TRIP 6 is an improvement on TRIP 4 and TRIP 5, combining the advantages of these two programmes. However, the computing times are longer.

1.2 - Dynamic studies on a mathematical model

We analyzed and wrote a simplified programme of the dynamic behaviour of the S-curve by dividing the pipe/FLU and time into discrete intervals. However, this programme was limited by the following assumptions:
- the programme was planar,
- it involved no bending stiffness,
- the external hydrodynamic actions along the S-curve were not considered,
- the pipe was fixed to the bottom,
- no account was taken of the laying speed.

1.3 - Design and testing of the ballasting system

The geometrical optimization of the S-curve and the laying tension, depending on the type of pipe to be laid and the depth of water, requires fine adjustment of the buoyancy of the FLU, and hence the application of an adjustable ballasting system.

To do so, the FLU is divided into independent compartments equipped with buoyancy devices with a variable air space that can be controlled through a rubber bladder exactly matching the contour of the internal volume when inflated.

This device was tested on a 762 mm diameter tube 12.7 mm thick. The operations consisted in filling and emptying the buoyancy devices with water by inflating and disinflating the compressed air bladder.

The test enabled the initial design concept to be refined whilst confirming the satisfactory operation of the principle adopted.

1.4 - Design of the FLU constituents, particularly the systems for supporting the pipe during laying

In its basic concept, the long and flexible stinger (FLU) offers many advantages for laying in great depths of water:

- less external tension, enabling a dynamically positioned barge to be used,
- the possibility of geometrical control of laying throughout all or part of the S-curve,
- adaptation to variable depths,
- the possibility of carrying out preservation manoeuvres,
- the good behaviour in the presence of the waves at the top.

However, use of the FLU for laying pipelines in depths of water of up to 2000 metres and more raised new problems that had to be examined, namely:

- resistance to the external pressures,
- deballasting gases at very high pressure,
- the problem of friction over a considerable length,
- the finalization of the support systems.

Accordingly, systematic study of the phenomena involved in this application had to be made, which was done for the typical case of a 14" pipe 3/4" thick in API 5LX Gr X 80 grade steel, for a depth of water of 2500 metres.

At the same time, an initial study of the repair procedures was made, for dry or wet repair.

This study yielded many results, but it transpired that tests would be needed to reach a definite conclusion concerning certain aspects of the FLU concept.

As an example, not being able to give the detail of the results of the study here, the laying case previously defined leads to using a concentrical FLU in aluminium alloy, deballasted with helium, featuring magnetic lift support, with an external diameter of 1300 mm and an internal diameter of 400 mm. The FLU accompanies the pipe down to a depth of 600 metres, and the external laying tension is restricted to 40 tons.

1.5 - Development of the method of controlling the geometry of the S-curve

The complexity of the laying operations in great depths of water means that a laying control system has to be set up.

One of the basic elements of this system is control of the shape of the S-curve in the vertical plane.

Knowledge of this shape enables comparisons to be made against the forms calculated and corrective action thus to be taken, to ensure the safety of the pipe.

Use of a long "FLU" stinger makes it possible to install pressure sensors at a sufficient number of points to be able to know the shape of the curve. These analog signals are sent to the laying barge via electric cables. They are then digitized and form part of the computer input data. The computer is also responsible for collection of the information (scanner).

We analyzed and developed a programme for polynomial approximation of the curve capable of being applied to a microcomputer and easy to set up on a laying barge. We also analyze the structure of the overall management programme of this control system.

To conclude, this type of control is in our view applicable both from the standpoint of the available hardware and the computing procedures.

2 - STUDY OF THE VARIOUS LAYING CONFIGURATIONS IN GREAT DEPTHS OF WATER

For the standard case of the 14" diameter, 3/4" thick tube in X 80 grade steel laid at a depth of 2500 metres, systematic study was made of the laying curve by scanning the FLU buoyancy and tension parameters.

It appears that, for this depth, the incidence of the stiffness of the pipe and the FLU on the geometry of the S-curve is small and in general below 2%. A low variation in the internal load at the tensioner, oscillating within the range of 160 to 190 tons, is also to be observed. For an external tension of 8 tons, the length of the FLU is 900 metres, and for a tension of 36 tons, it is 310 metres.

3 - TANK TESTS OF THE DYNAMIC BEHAVIOUR OF THE FLU AND THE BEHAVIOUR OF THE LAYING BARGE

These tests were carried out in the TRONDHEIM laboratory of the Ship Research Institute of Norway, with the assistance of the Fluid Techniques Department of ALSTHOM, in order to develop similitudes and make 1/40 scale models of the pipe simulating the stiffness, geometry and weight, and equipped with distortion sensors.

The purpose of these tests was:
- First, to reveal the flexion and displacement problems that may occur during the different pipe laying phases using the FLU. The tests were carried out using an AKER H3 semisubmersible barge specially fitted out for a submerged laying module.
- Second, if possible to determine figures relating to the phenomena observed in order to define the maximum wave conditions for ordinary laying operations and for connecting and disconnecting the FLU from the barge for preservation of the pipe.

Eight complete tests were carried out in accordance with table I, under the following conditions:
- depth 240 m,
- length of FLU 385 m,
- pipe diameter 30",
- thickness 16.31 mm,
- concrete covering: density 2.3 tons/m³, thickness 74 mm.

Analysis of the results obtained highlighted the following essential points:
- for this type of pipe, making allowance for the barge used, the FLU probably makes it possible to lay in waves with significant amplitudes of 4 metres,
- the nature of the link from the pipe and FLU to the barge requires special study. It appears that the maximum working conditions can be improved by making this length sufficiently flexible.

TABLE 1

TEST N°	WAVE		BARGE LINK	
			PIPE	FLU
2	8.59s	5.78m – IRREGULAR	ARTICULATED	ARTICULATED
4	8.81s	5.72m – IRREGULAR	ARTICULATED	ARTICULATED
6	7.04s	3.79m – IRREGULAR	FIXED	FIXED
7	3.01s	3.26m – REGULAR	FIXED	FIXED
9	9.23s	5.62m – REGULAR	FIXED	FIXED
10	7s	3.55m – IRREGULAR	FIXED	FIXED
11	9.01s	7.56m – REGULAR	FIXED	FIXED
12	8.98s	3.36m – REGULAR	FIXED	FIXED

4 – STUDY OF THE DYNAMIC POSITIONING OF THE BARGE FOR LAYING IN GREAT DEPTHS AT LOW TENSION LEVELS

Laying of pipes in great depths necessarily sets the problem of dynamic positioning, which is required to ensure the laying tension on the combination of FLU and pipe and to follow the laying path. The FLU method enables the laying tensions to be considerably reduced and can consequently be envisaged with a dynamically positioned barge, which has so far been used for drlling in great depths. The wave, current and wind sensors are installed on the barge and operate continuously in conjunction with a computing unit. This computing unit sends commands to the system of thrusters and tensioners. The FLU control system is also connected to the central unit and ensures pipe safety during laying.

A third type of sensor is also needed to calculate and control the geographic position of the barge.

The installed power on the barge depends in particular on its geometrical characteristics. The study, which was entrusted to the French Petroleum Institute (FPI) led to a power of 9600 HP, divided between four swivellable thrusters, each capable of 2400 HP.

These values were calculated for a conventional barge with a length of 100 metres and a beam of 30 metres. This power enables severe conditions to be withstood:
– wind velocity 36 knots,
– laying tension 50 tons,
– significant wave amplitude 4.60 m, period 9 s or 3.00 m period 10 s,
– current 3 knots.

Safety was also quantized by full analysis of the allowable excursions of the barge around its nominal position.

5 - PREPARATORY PHASE FOR 1/10 SCALE LAYING TESTS IN 250 METRES OF WATER

The aim of these tests was to simulate laying a pipe in 2500 metres of water, i.e. 250 metres at a geometrical similitude scale of 1/10. The laying case corresponded to the standard case of a diameter of 14" and a thickness of 3/4", equipped with an FLU 600 metres long, balanced by a laying tension of 30 tons. The complete preparation of these tests was hence performed, though not the tests proper.

The Centre Nationale d'Exploitation des Océans (CNEXO) at Brest was appointed to analyze the problems of similitude, construction of mock-ups and for making the measurements covered by the test programme.

The new offshore pipeline laying technology therefore appears to be well suited for laying operations in great depths of water. In addition to the advantages already expounded, particularly its compatibility with dynamic positioning, this method enables conventional means to be used for end-connection of the sections of pipe.

(10.12/77)

THERMOBLOC WASTE HEAT RECOVERY SYSTEM
(Pipeline Booster Station)

H. MATTES, W. MALEWSKI
Borsig GmbH, 1000 Berlin - 27
Federal Republic of Germany

Summary

A plant to recover waste heat from the exhaust gases of gas turbines has been installed and operated within a natural gas compressor station in Alberta Province, Canada. The closed-loop, Rankine Process system uses supercritical ammonia as the working fluid during evaporation and converts the recovered heat into mechanical energy to power a separate compressor run in parallel to the main compressors. This has allowed the pumping capacity of the entire station to be raised without additional fuel consumption since the thermal efficiency of the coupled gas turbine - Rankine Process plant system increased from 34 % to 42 %.

The construction and operation of this plant (which is monitored and operated completely automatically from a distant control room) are described together with the operational experience gained. Standard construction components were used. Ammonia proved to be a suitable working fluid for areas with Arctic climates where outside temperatures reach -50°C.

1. OVERVIEW

The Rankine process makes it possible to recover mechanical energy from waste heat. There are a number of criteria for selecting the working fluid for this closed loop system, depending on the kind of waste heat, external conditions and the use to which the energy is to be put.

Preliminary studies (1) indicated the basic relationships between variables and it was decided to test these in a commercial-scale pilot plant. Interest in such a plant was shown both by a company operating gas turbine-powered natural gas compressor stations (Nova Corporation, Canada, formerly Alberta Gas Trunk Line Ltd) and by a mechanical engineering company with experience in construction components (Borsig GmbH, Berlin). Arrangements were made to use the waste gases of a compressor station in Alberta Province, Canada, equipped with a GT 61/GE LM 2500 gas turbine producing 16 MW shaft horsepower at 5 250 r.p.m. and yielding 34 % thermal efficiency at 450°C. The pilot plant was designed to recover mechanical energy from the waste gases of the gas turbine and to use it to drive an additional gas compressor. If standard construction components were to be used, the most suitable closed loop working fluid was ammonia since it met thermo-dynamic and thermic requirements (see 1) to the greatest possible degree as well as coping with environmental influences (external temperatures reaching −40°C). A heat recovery plant suitable for these conditions was planned, built and run in order to test its feasibility, to gain experience in the running of such a plant and to reveal weak points.

2. WASTE-HEAT RECOVERY PLANT

The waste-heat recovery plant was incorporated into a gas compressor station, commissioned in 1972, containing two gas-turbine-powered turbo compressors. It is connected to one of these units and uses ammonia as the working fluid in the following closed 100 P (cf. diagram 1).

Heat given off by the exhaust gases from the gas turbine vaporizes the ammonia in the evaporator (E) or, if it is in a supercritical condition, raises its temperature. After the ammonia vapour has expanded in the expander (ET), generating mechanical energy to drive the twin compressor (V3), it is then liquified in the air-cooled condensor (C). The feed pump (P) then raises the pressure of the recuperated ammonia fluid and returns it to the evaporator.

The pressure ratio of the pumping station is determined by the main compressor using its end pressure as the regulating variable for the revolution setting of the gas turbines and the main compressor. The impeller settings on both the main compressor (V2) and the second compressor (V3) are such as to avoid any mutual disruption to pumping when they are working in parallel. References (2), (3) and (4) give further details of the procedure used to match pumping parameters and of the additional possibility of linking the compressors in series.

Figure 2 shows the entire plant including engine room and condensor, while Figures 3, 4 and 5 show the individual components, the expander, the compressor run in parallel, the condensor and the feed pump.

The original intention was to equip the plant with its own generator using the recovered heat to supply all the necessary energy for the ammonia feed pump, the condensor ventilators and the oil systems. Because the optimal settings for the compressor and the generator proved irreconcilable outside sources were used to power the auxiliary equipment described above.

The waste heat recovery plant was designed according to the following specifications :

Evaporator E	– mass flow rate of exhaust gas	57.4 kg/s
	– inlet/outlet temperature	486/70 °C
	– fall in pressure	10 mbar
	– mass flow rate of NH_3	16.6 kg/s
	– inlet/outlet temperature	30/240 °C
	– heat generation	25.55 MW
Expander ET	– inlet status	132/240 bar/°C
	– back pressure	10.5 bar
	– rpm	8 970
	– design (shaft) power	5.2 MW

Condenser C	– heat generation	20.75 MW
	– mass flow rate of air	1 575 kg/s
	– air at inlet/outlet	4.5/17.5 °C
	– fan rating	200 KW

Pump P	– mass flow rate of NH$_3$	16.6 kg/s
	– feed pressure	10/139 bar
	– power consumption	500 kW

The following design features are to be found in the main components :

Evaporator : The ammonia makes a single passage from top to bottom, in the opposite direction to the gas turbine exhaust, through a group of pipes with collectors at both the inlet and outlet.

Expander : A horizontally-split axial turbine with one regulating step and eleven reaction stages.

Second compressor : Single-step radial turbo compressor

Feed pump : Vertically-mounted 18-stage centrifugal pump.

Condenser : A forced draught is applied to the bundle of finned tubes which are horizontally mounted in two trains each with eight fans (50 % with variable speed controls).

3. RESULTS

3.1 Operating experience

Since little gas had to be pumped during the test period, the plant was run for only about 1 500 hours in the period 1982 to spring 1984. Problems occurred during the start-up phase because :
– Weather conditions (external temperatures below freezing) meant that it was not possible to properly clean and dry the plant after assembly and pressure trials.
– The ammonia used contained not only water but also dissolved oxygen and traces of cleaning fluid from the shipping tank.
This led to oxygen corrosion in the evaporator and to the formation of a thin black magnetite layer which normally acts as a protective layer stopping further corrosion. Trials carried out in a small pilot plant to select materials before the project was executed had shown that significant quantities of the ammonia would separate out into hydrogen and nitrogen above 200°C although up to 450°C decomposition is so slight that extrapolation suggested less than 5 % of the charge would be lost in the course of 8 000 hours of operation. The materials used, comprising unalloyed or low-alloy carbon steels (similar to the materials chosen for absorption cooling plants using ammonia and water as working fluids) produced a thin fissureless nitrid layer at temperatures above 300°C. By contrast, hydrogen did not produce any surface decarburization.

In the full scale plant, the nitride layer formed under the magnetite layer was five times as thick at temperatures in excess of 350°C as it was at temperatures under 300°C. Since the growth of these nitride layers is associated with on increse in volume, they shatter the magnetite layer on top of them as soon as a certain thickness of nitride layer has been reached. In consequence, nitride and magnetite are constantly being produced if :
– it is not possible to keep wall temperatures below 300°C
– additional oxygen cannot be prevented from entering the loop.

Such conditions existed during the first operational phase of the plant and resulted, due to the presence in the closed loop of impurities in the form of fine splinters of the nitride and magnetite layers, in errosion to parts of the feed pump, the control devices and the impeller blades of the expander.

During trials, it became apparent that there was an uneven distribution of temperature within the evaporator in the case of both the exhaust gases and the ammonia. Studies showed the presence within the exhaust gases of flows, differing in speed and temperature, that were carryover effects from the gas turbine and appeared as temperature peaks in the ammonia.

The following measures must therefore be routinely adopted to keep the formation, described above, of nitride and magnetite layers from exceeding critical limits :
– the plant should be built in such a way that greater evenness in temperature distribution is obtained on the exhaust gas side so that on the ammonia side temperatures in excess of 300°C do not occur;
– the ammonia used should be exclusively water and gas-free and of the standard used for cooling fluid;
– before the plant is filled, and after any section of it has undergone maintenance, it should be carefully dried and emptied;
– care should be taken to ensure a constant flow of inert gas from the loop.

3.2 Performance

The waste heat recovery plant achieved approximately 90 % of the planned generation of mechanical energy. This shortfall was due to a lower than expected quantity of vapour (ammonia) being generated by the waste heat in the evaporator. This in turn was due to flow variations within the exhaust gas flow from the gas turbine and the associated reductions in the efficiency of heat transfer.

The thermal efficiency of the gas turbine was raised from 34 % to 42 % as indicated in the following energy balance :

– shaft horsepower of the gas turbine	16 000 kW
– additional mechanical energy recovered from waste heat	4 600 kW
– energy consumed in heat-recovery	700 kW
– total power rating of the gas turbine including waste heat recovery with no increase in fuel consumption	19 000 kW

Rectifying measures in both the exhaust gas flue and the ammonia distribution permitted a more even temperature profile and thus prevented the formation of nitride layers and solid particles within the loop. The plant was run both during winter, with outside temperatures

falling to −40°C, and in summer at up to +25°C. Input, output and monitoring were all completely automatic with remote controls being operated
both from within the plant and by radio at long distance.

REFERENCES

1. W.F. Malewski, G.M. Holldorff
   Combined Cycles for Pipeline Compressor Drives
   ASME Publication 79-GT-162 (1979)

2. K. Lüdtke
   Gasturbinen-Abwärme trübt Pipeline Verdichter
   VDI Bericht Nr. 377, (1980)

3. G.M. Holldorff, A.R. Hladum, S.A. Dunn
   Pipeline Compressor Station with Waste Heat Recovery
   ASME paper presented at the ASME Gas Turbine Conference in New
   Orleans, Louisiana in 1980

4. G.M. Vogt
   Operation Experience with a Waste Heat Recovery System installed on a
   Pipeline Compressor Station
   ASME Publication 83-GT-226 (1983)

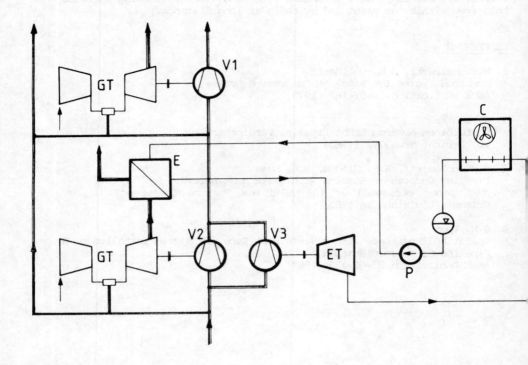

Figure 1 - Operation of the Rankine Cycle within the natural-gas pumping station

V1, V2	-	Main compressors
GT	-	Gas turbines
V3	-	Compressor operating in parallel
E	-	Evaporator
ET	-	Expander
C	-	Condensor
P	-	Feed pump

Figure 2 – General view of the waste-heat recovery plant

Figure 3 – Expander-gas compression unit

Figure 4 – Ammonia condensor, engine room and the evaporator into which the exhaust gases are fed

Figure 5
Ammonia feed pump

MARINE TECHNOLOGY OPERATIONS – MONITORING OF STRUCTURES – SUBMARINE VEHICLES – ROBOTS AND ENERGY SUPPLY

High density underwater energy sources – A radical transformation of subsea operations (13.05/78)

Development of a constant mass subsea power generation unit using high pressure methanol/oxygen combustion and a Rankine cycle (15.30/82)

TM 308 Deep water offshore structure inspection vehicle (07.48/82)

Development of a deepwater tethered manned submersible (07.46/82)

Study of remote–controlled workover and maintenance system for subsea petroleum installations (07.35/80)

The long range submarine (07.34/80)

DAVID – A versatile multipurpose submersible support system for remote control or diver assisted performance (07.22/78; 07.33/79; 07.43/81)

Prototype construction and prototype testing of the UW-work and pipeline repair system "SUPRA" (10.37/82)

Repairs in deep water ("RPM") (09.19/79; 10.42/83)

Repair of subsea pipelines by mechanical coupling (10.21/79)

Deepwater pipeline repair system (10.20/79)

Echomechanical diagnostic method for offshore structures (15.33.82)

Diagnostic methods for offshore structures (15.06/78)

Development of a reliability analysis system for offshore structures (RASOS) (15.03/77)

Development of a total structural monitoring system for offshore platforms (15.24/81)

Development of a measuring system for the optimization of programmes for calculating and monitoring - the dynamic behaviour of offshore platforms (15.08/78)

Meteoceanographical-structural measurement system for the safety of the Nilde single anchor leg storage - Sicily channel (15.11/80)

Meteoceanographical and structural data acquisition to improve platform design (15.07/78)

Automated processing of recordings obtained by side-looking sonar (09.18/79)

Deep water steel pipeline - Pipeline and connection techniques (18/75; 09.07/77)

The TM 402 deep water trenching machine (07.13/77)

Feasibility study of a wave dampener (06.05/76)

Personnel transfer device (07.36/80)

(13.05/78)

<u>HIGH DENSITY UNDERWATER ENERGY SOURCES</u>
<u>A RADICAL TRANSFORMATION OF SUBSEA OPERATIONS</u>
H.NILSSON, Managing Director, SubPower AB, SWEDEN
Y.DURAND, Managing Director, Comex Industries, FRANCE

Summary

The un-availability of powerful and long-endurance civilian underwater energy source has dramatically limited the development of surface-independent solution for subsea operations : manned submarines, remote-controlled vehicles or subsea installations are either limited in their capabilities or highly dependent from the surface for their energy requirements. Founded in 1968, United Stirling of Sweden has been dedicated to the development of the Stirling Engine technology, an external combustion engine. In 1979 Comex and United Stirling joined their efforts, with the support of EEC, to apply this technology to an air-independent underwater power system using the 4-95 Stirling Engine. This project, successfully conducted, has opened the way and its results allow to proceed with the next stage, i.e a real size demonstration of the capabilities of a large-size, manned diver lock-out submarine, the **SAGA 1**. This 500 tons submerged displacement prototype, presently developed in France jointly by COMEX and IFREMER (ex CNEXO), is to be completed by August 1986 and includes two V4-275R Stirling Engines of 80 Kw power each, and a liquid-oxygen storage system. This sub will be able of surface-independent operations for up to 25 days, and will have a maximum waterdepth capability of 600 meters. It will allow to test the "all-underwater" solution for subsea works in actual conditions, thus preparing for the next generation of underwater developments.

1. INTRODUCTION

Underwater operations, military as well as offshore, depend significantly on supply of large amounts of energy. Surface independent system has during recent years been subject to considerable development efforts, resulting e.g in nuclear power systems for large naval submarines.

Direct conversion of chemical energy into electrical power will ultimately be the most efficient way for utilizing energy, inherent in the fuel. This attractive potential is embodied by the fuel cell, but these systems assume, however, quite advanced technology and are hence still not cost competitive.

The lack of compact air-independent energy sources in the medium power range, commercially available, has prevented any significant evolution of underwater autonomous offshore operations using submersibles and habitats or when aiming for an extended underwater cruising range for naval submarines of medium size. The near term solution for supply of large quantities of energy to underwater systems might be based on closed cycle heat engines among which the Stirling engine embodies several unique features, recognized since many years.

2. BASIC STIRLING TECHNOLOGY

The Stirling engine, an externally heated engine, with a closed operating cycle, was invented in 1816 by the Reverend Robert Stirling.

The engine has many similarities to a conventional internal combustion engine (e.g a diesel engine). However, it differs in the sense that heat is supplied externally and continuously to a working gas (e.g helium or hydrogen), which is contained in a completely closed system.

Sensible heat energy is reclaimed through a packed-screen heat storage, called a regenerator, where a large portion of the heat of the working gas after expansion is stored and supplied back into the cycle, as the gas reverses direction.

The pistons in a double-acting Stirling engine have dual functions and each piston operates simultaneously in two cycles : the hot upper surface of one piston works within the same cycle as the cold undersurface of the adjacent piston.

Theoretically, the Stirling cycle thermodynamic efficiency is close to that of the ideal heat engine, i.e the Carnot engine. This efficiency level requires, however, quite high operating temperature of the working gas together with a high working gas pressure. Hence availability of advanced materials technology is necessary for high performing Stirling Engines. This status was established in the late 1930's after significant development work carried out by N.V. Philips in the Netherlands.

3. UNITED STIRLING AND CURRENT PROGRAMS FOR UNDERWATER ENGINE TECHNOLOGY

United Stirling AB (USAB), located in Malmoe, Sweden, was founded in 1968, starting its development work as a licensee of N.V. Philips. The company is now a subsidiary of the FFV group, being a large Swedish industrial group, owned by the government, for development and manufacture of advanced defence systems.

To adapt the standard engines to fit within underwater systems the following modifications essentially have been made :

3.1 Pressurized combustion

Energy support to underwater engineering activities employing saturated divers has been a near-term application assumed for the engine systems, why the operating depth for the systems is currently been set to 300 metres. The combustion pressure is hence chosen slightly above 30 bar. A corresponding engine design is shown in figure 1.

3.2 Combustion gas recirculation (CGR)

The combustion of a hydrocarbon fuel, e.g diesel fuel, with **pure oxygen** means an adiabatic flame temperature of approx. 4000° C. A thermic ballast for heat absorption is hence needed, and is provided using the USAB CGR-concept, similar in effect to a separate external supply of an inert gas.

The CGR-system is based upon oxygen supply through a set of ejector tubes of simple design, having no moving parts, creating a back-flow of combustion gases inside the combustion chamber thus reducing the combustion temperature to the design level (2000° C) for a standard engine heater. A basic design of the CGR-system is shown in figure 2.

4. THE 4-95 PROJECT

The current Stirling underwater technology at USAB originates from the program that started 1979-80 for modification of the 4-95 engine, in a joint project with COMEX INDUSTRIES, aiming for a power source for manned submersibles.

During the first phase of the 4-95 project certain "key functions" were identified as being crucial for a successful Underwater Modification of the standard engine, including a satisfactory system arrangement. These were hence to be initially qualified, with satisfactory results.
- Overpressure combustion
- Combustion gas recirculation
- Oxygen control system
- Exhaust valve design
- Arrangement analysis versus pre-set energy density objectives.

In addition to the extensive laboratory testing of the 4-95 sub prototype engine, the system was installed in a pod and tested submerged in the diving pool of COMEX in Marseilles, at a combustion pressure of 30 bar, thus corresponding to an operating depth of 300 m.

The 4-95 system was originally developed to serve as an energy source for air-independent underwater operations, using diver lockout submersibles (figure 3). The combustion pressure was hence chosen to 30 bar, which mates with the surrounding water depth at current limits for operational offshore diving (300 m).

This power module has a nominal electric output of 25 KWe and, additionally, supply of water of about 50° C, to be used for e.g divers' and compartment heating.

5. THE SAGA 1 PROJECT

5.1 Project's motivations

A revolutionnary concept in 1967, when initiated by Cdt COUSTEAU, the **ARGYRONETE** project is still revolutionnary today, more than 15 years later, under another name : **SAGA**. For indeed, it is the only approach, one which is still new, that enables divers to work on the bottom of the sea from a self-contained, independent supply base freed from surface conditions.

In 1982, COMEX decided to re-initiate this project to respond to the new challenges facing not only the offshore oil industry, but also the entire subsea operations world.

The requirements to perform deeper operations, in rougher or ice-covered seas, far from safe or sheltered, areas justified to investigate a "surface-independent", or "all-underwater" solution. Such a solution could be the only answer to the dramatic growth in size and cost of surface vessels (either monohull or semi-submersible type) able to cope with such conditions. The growing demand for "light" subsea works - like observation, cleaning, non destructive testing tasks or intervention on subsea installations - to be performed either with divers or with robots, will not continue for a very long time to be served cost-effectively with such heavy spreads.

On the other hand, a submarine fitted with a sufficiently large energy reserve to sustain a long-distance submerged trip to its working site, then work there with divers and/or robots for a significant period, and then come back to its base without surfacing, could be an ideal solution. A dream still a few years ago, such a concept can be put at work today, thanks to significant industrial and reliability advances in several key technologies and especially :

(a) Energy sources with the Stirling Engines, (b) computerized navigation and instrumentation systems, (c) composite materials for high pressure gas storage vessels and low-weight components, (d) closed-circuit breathing systems for divers, (e) underwater robotics and artificial intelligence.

The submarine will be able to carry out a variety of subsea tasks in oil and oil-related activities : geophysics, geotechnics, survey, etc... and especially :

- Maintenance of wellhead and subsea production systems
- High-resolution seismic operations on possible drill sites
- Unreworked sample-taking by rotary coring
- Static or dynamic penetrometry
- Construction assistance in hyperbaric welding operations
- Pipeline surveys and bathymetric surveys
- Operation of robot vehicles and systems : ROV's, remote télémanipulators, module's replacement or repair etc...

5.2 SAGA basic specifications

All these new - but already proven technologies - will be incorporated into the SAGA project, a true "house under the sea" composed of a hydrostatic pressure compartment for the divers and of an atmospheric pressure compartment for the pilots, crew and engines(see figures 4 & 5)
The major SAGA specifications are the following :

Main hull :	Diameter	3.7 meters
	Length	16.3 meters
	Max operating depth	600 meters
Divers' compartment	Diameter	2.5 meters
	Length	5.0 meters
	Max operating depth	
	– with divers	450 meters
	– without divers	600 meters
Submarine	Length overall	28.0 meters
	Breadth overall	7.4 meters
	Height	8.5 meters
	Surface displacement	300 tons
	Submerged displacement	545 tons
	Operational depth	600 meters
	Collapse depth	1200 meters
	Drop weight	5 tons
Crew	Skeleton crew	6 men
	Additional crew	1 man
	Divers (max)	6 men

Speed   Normal, submerged ....   4 knots
                         Maximum, submerged ....   6 knots
                         Surface ....   7 knots

Life support   Atmospheric compartment .... 25 days

5.3 SAGA program milestones

The major challenges of the **SAGA** program are not exclusively technical. They are also psychological and "cultural" since this project intend to demonstrate the operational feasibility of a concept, that appears as impossible as deep human diving in the late 60's - now an industrial reality down to 450 meters - or the space shuttle ...

This is why a large portion of the program will be devoted to operational demonstrations, in actual conditions and in varied areas, of the major tasks listed above.

The **SAGA** program will cost approximatively 135 Million French Francs and extends over 4 years distributed as follows :

- late 1982   : Launching of the program
- 83 - Mid 84   : Detailed Engineering studies
- Mid 84 - Late 85   : Fabrication, construction, assembly
- First half 86   : Sub-systems testing and quay testing
- Second half 86   : Sea trials and final testing
- 1987   : Operational demonstration
- 1988   : Starting of commercial operations.

6. THE SAGA 1 ENERGY SYSTEM

The energy system for the **SAGA** 1 is nearly identical to the system that has been given priority from the Royal Swedish Navy, and includes two 4-275R Stirling engines in the power range of 100 KWe and two liquid oxygen tanks giving an energy storage capability of 12000 KWh allowing to perform missions of up to 3 weeks submerged depending on their profile.

Within the 4-275 project, currently built for the Royal Swedish Navy, a similar qualification procedure has been performed with satisfactory test results and subsequent selection of system components.

The 4-275 engine and the LOX system will be integrated within 1984 and also undergo subsequent operational system testing, installed in a full scale submarine section.

The 4-275R engine means an engine with four cylinders and a swept volume of 275 cm3. The cylinders are arranged in a V which together with two balancing shfts gives a noise and vibration level of a very low magnitude, meeting the severe demands from the Navy. A silent environment is also of great importance in a civilian submarine, i.e for the crew's comfort and for the realiability of different hydro acoustic systems. The letter R stands for annular regenerators and gas coolers, an advanced development design compared to the former Stirling engine concepts, which decreases the engine width and simplifies the engine block configuration.

The combustion chamber is surrounded by a pressure vessel where the combustion pressure during operations is always kept slightly above 30 bar, independent of the diving depth of the submarine. A back-pressure valve in the exhaust line prevents the surrounding water pressure to influence the combustion chamber pressure. According to the combustion pressure the engine can operate submerged down to 300 metres with no need for either an exhaust gas compression or a dissolving system.

Deeper operating depths, to say 600 metres, will be achieved by adding exhaust gas compression of only marginal pressure ratio and hence insignificant power consumption.

With respect to the invidual alternators 3,000 rpm (4-95) and 2,200 rpm (4-275) have been chosen together with 13 MPa as the working gas mean pressure. The corresponding measured engine overall efficiency is now :

4-95 : 26% or 1220g. 02/KWh
4-275 : 38% or 820g. 02/KWh

Power performance for the 4-95 and 4-275 energy supply systems is given in the following table :

Engine	Continuous operation			Maximum output		
	Speed	Shaft Power	Heat	Speed	Shaft Power	Heat
4-95	3000 rpm	25 KW	40 KW	4000 rpm	40 KW	65 KW
4-275	2200 rpm	95 KW	110 KW	2600 rpm	120 KW	200 KW

7. CONCLUSION

The 4-95 Stirling Engine program, partly supported by EEC, is a good demonstration of the efficiency of R and D long time planning. A relatively small program at the origin, its results in a significantly larger project whose consequences could dramatically transform the subsea operations and procedures in a very near future, leading to a new generation of underwater developments.

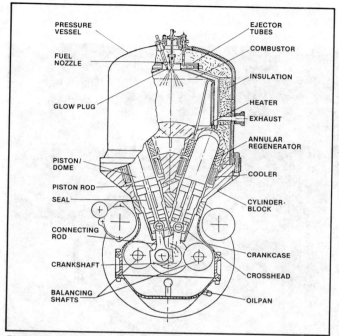

Figure 1. The 4-275 underwater engine

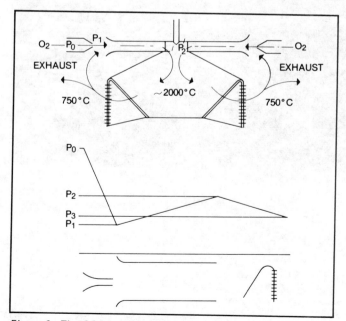

Figure 2. The CGR basic principle

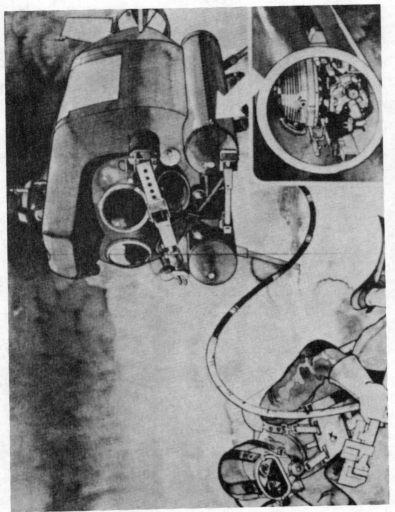

Figure 3. A Stirling powered submersible

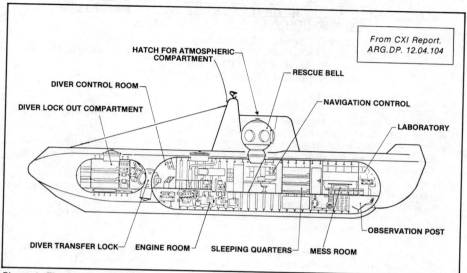

Figure 4. The Argyronète submarine

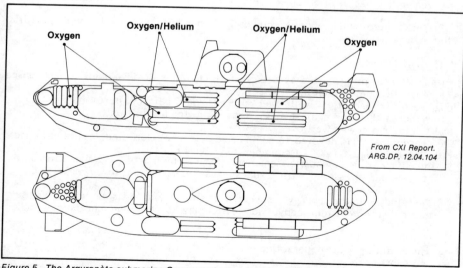

Figure 5. The Argyronète submarine Oxygene storage arrangement

(15.30/82)

DEVELOPMENT OF A CONSTANT MASS SUBSEA POWER GENERATION UNIT USING HIGH PRESSURE METHANOL/OXYGEN COMBUSTION AND A RANKINE CYCLE

J.L. BOY-MARCOTTE, M. BLONDEAU, S. GALANT, D. GROUSET,
M. JANNOT, M. LEFRANT, J.C. MULET, D. REBUFFAT, A. VERNEAU
Société BERTIN & Cie
B.P. 3, 78373 PLAISIR Cedex FRANCE

Summary

Account is given of the main results obtained by Société BERTIN & Cie during the first step of a development program dealing with a subsea electrical power generation module whose concept is based upon high pressure combustion and a Rankine cycle configuration. To date, the following results are obtained :
i) module specifications concern operational capabilities at depth ranging from 600m (short term) down to 3000m (long term) because of total exhaust gas condensation ; the performances are :
   - more than 1000 kWhe energy range capabilities with energy per unit mass for the ready-to-use module over 100 Whe/kg at - 600m depth,
   - minimum peak electric power at 30 kWe, with no upper limit, by using an organic Rankine cycle, as an energy conversion unit,
ii) the combustion chamber is a sized-down jet engine configuration using a pressurized (60 bar) methanol-oxygen mixture : exhaust gas recirculation is necessary to keep the combustion chamber wall temperature within acceptable ranges for metallic materials,
iii) extensive experimental testing of the exhaust gas condensation unit : feasibility has been obtained yielding very compact heat exchanger designs.

The next step will concern combustion chamber tests and coupling with a full size Rankine cycle.

1. INTRODUCTION

The technical program led by Société BERTIN aims at developing a new subsea power generation unit which allows :
- long range capabilities well over present day battery performances, i.e., a few thousands kWhe against a few tens kWhe for battery units,
- peak electric power well over 50 kWe.

The present paper is devoted to summarize technical work and results obtained over the period August 1982 - December 1983, viz. :
   . the study of a combustion chamber burning methanol-oxygen mixtures at 60 bar pressure, with exhaust gas recirculation to lower flame temperature,
   . the study of the energy converter : an organic Rankine cycle whose thermal to electrical conversion efficiency at the nominal point reaches 0.2,
   . the study and the experimental tests of a condensation heat exchanger which allows full liquefaction of the exhaust gases (a mixture of H_2O and CO_2),
   . the system study of the subsea power generation unit with potential designs of the control system to make it suitable for subsea operations.

2. TECHNICAL SPECIFICATIONS

Recent prospective studies concerning subsea power generation units [1] show that, at the end of the 80's, needs for more appropriate systems will appear. Applications will cover inspection work for off-shore oil and gas production systems and exploration travels, especially within the Arctic Seas.

The main technical specifications will then strongly differ from current power generation units (battery systems), viz. :
- peak power capabilities well over 50 kWe with large variations below the nominal point, at acceptable efficiencies.
- power distribution (thermal, mechanical, electrical) for submarine operations such that :

- mission range capabilities well over 1000 kWhe without refueling.
- overall energy constant per unit mass of ready-to-use unit reaching 100 wHe/kg at - 600m depth.
- no a priori depth limitations which implies constant mass working conditions, i.e. no rejection of exhaust gases to the sea.
- life time of the power generation unit around 2 000 hour MTBF and simple overhauling every 200 hours.
- full safety and reliability for intensive subsea operations, which yield adequate redundances in the unit control system to allow for possible emergency operations.

3. THE UNIT CHARACTERISTICS

The above technical specifications lead to the following basic technical choices :
- energy storage : methanol to minimize soot and unburnt hydrocarbon production since exhaust gases must be liquefied,
- oxidizer : pure oxygen to minimize storage volume,
- combustion chamber design of the aeronautic type to allow highly pressurized working conditions (viz. 60 bar for liquefaction purposes at 20°C) and exhaust gas recirculation to minimize flame temperature : the CO_2 recirculation rate is around 7.
- thermal to mechanical conversion unit chosen to be an Organic Rankine cycle : overall conversion efficiency reaches 0.2 at the nominal design point with maximum hot gas temperatures within the heat exchangers not above 900°C and exit gas temperatures around 150°C at the liquefaction unit inlet.
- full liquefaction of the exhaust gases, using a new heat exchanger design to condensate both CO_2 and H_2O before recirculating CO_2 only to the combustion chamber.
- an overall mechanical design which separates the combustion chamber and the conversion unit compartment to insure the best safety and reliability characteristics for subsea operations.

In the following, each of the main system components is described .

3.1. The combustion chamber
A jet engine configuration was chosen because :
- combustion efficiencies will reach 0.99, hence yielding low concentrations levels of unburnt hydrocarbons and CO,
- very compact designs are obtained due to very high pressure combustion conditions,
- stable working modes are obtained over a wide range of thermal output (10 to 100 percent) and equivalence ratios (10 to 100 percent of stoichiometric conditions) with extended reignition capabilities,

- very high life times compatible with present subsea specifications.

However it must be emphasized that subsea operations yield quite different operating conditions when pressurized methanol and oxygen are used
- on the difficult side, elevated pressures (60 bar) exhaust gas composition (a mixture of CO_2 and H_2O) do change very much the overall thermal balance of the combustion chamber. Moreover higher gas densities may change drastically the usual vibratory modes of classical combustion chambers,
- on the easy side, the weight and volume constraints are much less stringent than for jet engine configurations : furthermore exhaust gas (CO_2) recirculation allows a better control of both maximum chamber temperature and gas emissivities (only due to the molecular contribution of CO_2 and H_2O).

Société BERTIN used its present day knowledge of aeronautic combustion chambers and computer codes to optimize the combustion chamber design. Fig.1 is a schematic of the final design : film cooling of the chamber walls is extensively used to maintain wall metallic materials within acceptable ranges (i.e. less than 900°C) . The chamber volume stands around 300 cm^3 for thermal outputs around 200 kW. It is the lower acceptable thermal output due to metallic material limitations (elevated life time, no corrosion, no mechanical failure). A specific study of the ignition system led to two basic systems which will be experimentally tested : a spark plug and a torch design.

3.2. The thermal to electrical conversion cycle

Two designs have been studied for the chosen Rankine cycle [2] (see its thermodynamic diagram on Fig.2) :
- a water vapor cycle because it is a safe (i.e. non flammable) well known thermodynamic fluid. The overall conversion efficiency could reach 0.25 at a nominal 50 kWe power. However very difficult mechanical problems must be solved, due to the high number of turbine stages, the very tormented water vapor flow configurations and potential corrosion problems,
- an organic fluid cycle (Fluorinol 85) where Société BERTIN is acquiring a lot of industrial experience (two 50 kWe systems installed on a thermodynamic solar power plant, one 1 MWe system to recover energy out of Diesel exhaust gases on a power plant, one 3 kWm and one 30 kWm systems under development to recover energy from exhaust gases of automotive power plants).The overall conversion efficiency will reach 0.20 at a nominal 50 kWe power.

It is the Organic Rankine Cycle which is chosen. The combustion chamber and the organic fluid evaporating heat exchanger are located in two different compartments : the connection is carried through a neutral thermodynamic fluid (CO_2) which prevents the thermodynamic fluid to reach critical dissociation temperatures (above 400°C for Fluorinol 85). Fig3 is a schematic of the system configuration, giving overall sizes for the hot heat exchanger between exhaust gases and the CO_2 closed circuit. The CO_2-Fluorinol heat exchanger is very small sized (less than 5 liters) with very low power requirements to recirculate the CO_2 gas (about 20 Watts shaft power).

3.3. Design optimization and tests of the exhaust gas liquefaction system

The constant mass working conditions for the subsea energy generation unit are insured by liquefying all the exhaust gases. Total condensation is obtained by working around 60 bar, 20°C thermodynamic conditions.
- a mathematical model of the heat exchanger (radial internal fins) was set up to optimize the heat exchanger as a function of the inlet thermodynamic conditions (temperature and chemical composition to take care both of variable recirculation rates and possibly non condensable species). Heat exchange surfaces are calculated via the Chilton-Colburn analogy,
- a thorough experimental study was carried out to verify the heat exchanger computational scheme. Fig.4 illustrates the laboratory apparatus which was

used : the combustion gases were made by mixing appropriate amounts of CO_2 and water vapor at 60 bar. Reheating of the mixture was necessary to simulate properly hot exhaust gases and to prevent from any water vapor condensation prior to the heat exchanger. The condensation efficiency was measured by comparing mass and energy balances at various points on the system. The maximum mass flow rate tested was equal to about 1/30 of the total mass flow rate of 50 kWe full size systems.

More than 15 successful experimental runs were carried out. They showed that

i) it is actually possible to totally condensate such exhaust gases at 60 bars, 20°C or less on the cold side,

ii) the heat exchanger calculation methodology is very conservative : it is possible to halve the heat exchanger length for the same overall efficiency since the water droplets on the condensation side do increase the actual heat exchange surface quite considerably,

iii) the main parameters which must be considered when sizing the heat exchangers are : the flow rate and exhaust gas inlet temperature, the cold side temperature and the fraction of non condensable species.

4. CONCLUSIONS

All the present results have been integrated within an overall system analysis of the subsea power generation module. Cruising capabilities at depth ranging from - 600 down to - 3000m are depending mainly upon methanol/oxygen storage modes (cryogenic or standard, titanium or new resin type reservoirs, etc...). Yet it is clear that standard existing storage configurations allow the overall system to reach more than 100 Whe/kg at - 600m whereas even advanced battery systems cannot go well over 50 Whe/kg. Furthermore energy capabilities can reach well over 1000 kWhe, whereas battery units stand around a few hundreds kWhe.

Hence Société BERTIN is now ready to continue on the two next program steps :

- first, the combustion chamber must be tested on an appropriate bench scale at a nominal 300 to 400 kWthermal (a 36 months program): special attention must be paid to control and safety systems to make sure that a 2000 hours MTBF and full reliability during emergency operations are obtained,
- next integration of the Rankine cycle and the methanol/oxygen storage systems must be studied within a complete package to be tested both at ground level and within quasi marine environment (a 36 months program).

Successful completion of both steps would demonstrate the capabilities of a subsea power generation unit tor submarine applications at the end of the 80's.

ACKNOWLEDGEMENTS

Thanks are due to Mr. DEGEN (CNEXO), Mr. DURAND (COMEX) and Mr. GUILLON (ASTEO) for many helpful discussions during this phase of the development program.

REFERENCES

[1] - M. DEGEN "A market analysis of present and future small submarine applications" presented at the ASTEO Meeting, November 1983.

[2] - J.L. BOY-MARCOTTE et al., BERTIN Technical Note, 1983.

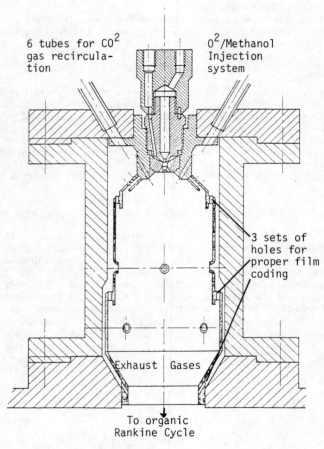

FIGURE 1

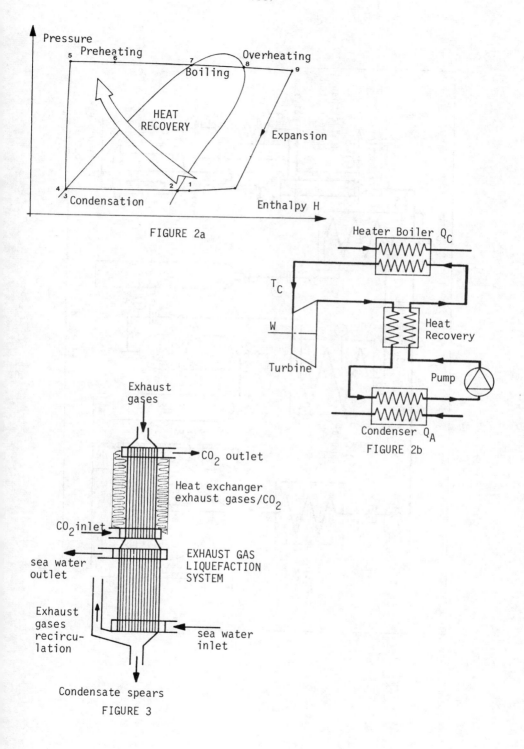

FIGURE 2a

FIGURE 2b

FIGURE 3

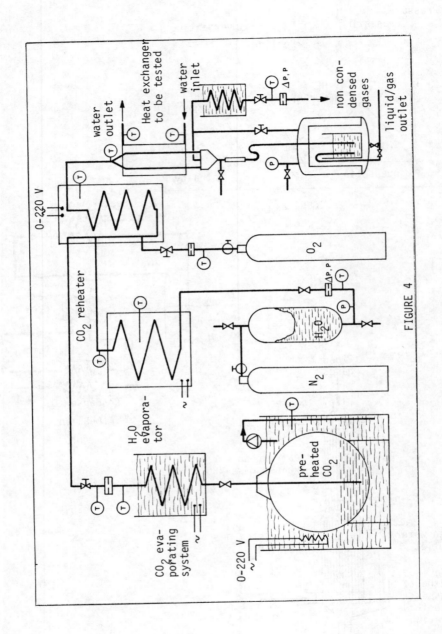

FIGURE 4

(07.48/82)

TM 308 DEEP WATER OFFSHORE STRUCTURE
INSPECTION VEHICLE

M. MAZZON
TECNOMARE SpA
(Società per lo Sviluppo delle Tecnologie Marine)

Summary

Underwater inspections of offshore platforms are conducted now-adays by saturation divers, generally operating between 40 and 150-200 metres below the surface. The high costs of such operations, the personal risk involved, the dependence on weather conditions and the prospect of installations at much greater depths in the future have created a need for technology relating to ROVs (Remotely Operated Vehicles) to replace saturation divers gradually.

This aim cannot be achieved by the present generation of ROVs, and new concepts must be developed for vehicles specially designed for such working environments, featuring for example, complex tubular steel structures which restrict the manoeuvrability of umbilical-cable vehicles.

A description is given of research in progress at present to design and construct a prototype of the TM 308, a ROV generating power for captive use, which would be employed for inspecting offshore structures and would have a daily performance comparable with that of a team of saturation divers.

The form considered most promising at present uses a closed cycle diesel engine, with oxygen reserves providing 10 to 12 hours' endurance under the anticipated operating conditions.

The control system, designed with the idea that there would be a low-capacity underwater communications channel, makes provision for remote-controlled operation of the vehicle with the aid of two computer systems, one on board and the other at the surface.

1. INTRODUCTION

In recent years the use of ROVs (Remotely-Operated Vehicles) in the offshore industry has been on the increase, and one of the fields of application on which attention is focusing is the inspection of platforms.

Up to the present, experience has been gained mainly in the North Sea, where regulations, which are applied in other areas also, require all installations to be thoroughly inspected every five years by means of a series of annual checks.

A typical inspection, which normally lasts 10 to 20 days at sea, involves the following types of checks :
- green : a general visual inspection (with TV), aimed at revealing obvious damage. Sometimes measurements of the cathodic potential are required;

- blue : a local inspection of approximately 20 % of the structural nodes to reveal any hidden defects. This involves cleaning the welded area to be inspected and conducting a detailed visual or photographic examination;
- red : where necessary, a "blue" inspection combined withnon-destructive tests (e.g. magnetoscopy or ultrasonic inspections).

Saturation divers operating at depths greater than 40 or 50 metres are mainly employed to carry out these inspections.

Reasons connected with the high costs and risks associated with diving operations and considerations relating to the development of offshore fields at increasingly greater depths beyond the practical limits of direct human intervention clearly make it necessary to replace man progressively with remote-controlled vehicles, particularly for repetitive routine tasks, such as structural inspections.

Another consideration is cost. The potential savings which may be realized on present offshore installations by using ROVs only, rather than divers, to conduct annual inspections may amount to a substantial percentage of the cost of inspections. Savings may be made by :
- reducing the cost of the craft;
- reducing manpower costs;
- eliminating expenditure due to decompression time (which may amount to five working days).

Existing vehicles are not capable at present of replacing man for these tasks, although ROVs are widely used for "green" inspections (generally, however, these involve the exterior of the structure only).

It is widely recognized that the limitations on the use of ROVs stem from problems such as :
- the handling of the umbilical cable, which is normally of considerable thickness and constitutes a major encumbrance when, for instance, it is necessary to bring it into and out of the structure;
- the difficulty of locating the vehicle and the cable, and memorizing the path followed;
- ineffective remote-controlled cleaning and handling methods;
- the need to develop satisfactory systems for mooring the vehicleto the structure or the node to be inspected, in order to carry out cleaning and inspection operations;
- the man-machine interface;
- reliability.

The special nature of problems related to structural inspections has also shown (this emerged from other applications of ROVs offshore also) that general-purpose craft are not suited to such work.

Consequently, ROVs must be purpose-designed to undertake such tasks. To this end Tecnomare decided to commence research, at present in progress, on the development of a prototype craft designed for platform inspections. Interim findings are described below.

2. THE OBJECTIVES OF RESEARCH ON THE TM 308 VEHICLE

The research aims to develop a vehicle meeting the following basic requirements :

- capacity to replace divers executing saturation dives during inspections of offshore platforms;
- energy endurance.

The former requirement follows on the considerations set out above; the latter takes account of the desirability of eliminating unwieldy umbilical cables, which place one of the basic limitations on the performance of present vehicles utilized for this purpose, and the possibilities afforded by recent developments in the field of subsea generators.

In addition to EEC contributions, the research makes use of funds provided by the Istituto Mobiliare Italiano and is sponsored by Agip and Micoperi.

It is divided into four phases:

Phase 1 : analysis of the vehicle's mission, definition of operating and system requirements, choice of design and system specifications;

Phase 2 : design of system and tests on critical components or
(in progress) subsystems;

Phase 3 : construction of subsystems, tests and assembly of prototype;

Phase 4 : sea trials of prototype.

Some of the findings obtained, subject to changes following the results of present work, are given below.

3. DESCRIPTION OF RESEARCH WORK TO DATE

Phase 1 of the research was aimed at establishing the system configuration which is most capable of achieving the project's objectives.

A survey was carried out of the major technological developments relating to the construction of a highly-sophisticated vehicle of this type, viz :

- power generation systems;
- underwater communications;
- navigation and steering;
- remote manipulation;
- cleaning systems;
- inspection systems;
- video image processing;
- materials;
- man-machine interface.

Work was carried out in conjunction with specialized organizations on some of the major aspect of the technology, with a view to developing applications specific to the TM 308.

Information obtained was used to work out an initial design, to use as a basis for defining the requirements and analysing performance.

3.1 Definition of requirements

A market survey and study of current operating procedure, involving contacts with oil-rig operators and diving companies, were conducted enabling the operating requirements of the system to be defined.

The major ones are given below :
- energy endurance : 10-12 hours;
- daily performance equivalent to that of a team of saturation divers;
- operation from a support ship or platform;
- launching and recovery in Force five seas;
- capacity to operate down to at least 400 metres.

The operation was analysed in detail bearing in mind current regulations and practice, so that specific requirements relating to power, energy, control and communications for the craft could be defined. The operation was subdivided into three main phases :

- launching : comprising testing of functions on board the craft, launching to release depth and travel to the inspection area;
- inspection : comprising the performance of all work on the structure, such as :
  . travel along pre-established routes, with television filming ("green" inspection);
  . location of a specific structural node;
  . mooring to component and removal of any debris;
  . cleaning the welds to be inspected;
  . close-up television filming;
  measurement of cathodic potential, thickness, location of any defects;
  . cleaning of anodes, etc.;
  . departure from inspected area;
- recovery : includes exiting from structure, connection to hoisting device and recharging of generator.

Each phase was subdivided into elementary tasks, with the conditions under which they can be performed and the estimated duration and power requirements defined in respect of each.

An analysis of the operation was thus obtained, permitting simulation trials of the various tasks to be carried out on the computer, and providing amongst other things a good estimate of the energy requirements of the craft, the basic components of which are described below.

3.2 Generation system

The generation of power on board is the most critical problem, as regards both capacity to perform pratical work and the dimensions and shape of the vehicle.

The power required on board depends on the requirements of the cleaning and propulsion systems; endurance depends on the trade-off between the length of the inspection operations, time lost due to recovery of the TM 308 for recharging, and the craft's volume, which in turn determines the power requirement for steering the craft.

The attention paid to this problem is consequently understandable, particularly in view of the demanding requirements revealed by simulation trials.

It became clear very quickly that primary and secondary batteries were inadequate on account of their dimensions or cost per work cycle; interesting possibilities such as fuel cells or chemical reactors (e.g. lithium plus seawater) were found to be too costly or not sufficiently developed. The choice therefore went to an airless subsea generator based on a closed-cycle diesel engine drawing oxygen stored in cylinders.

The design worked out provides the vehicle with the required endurance, with a peak power of approximately 50 kW.

3.3 Vehicle design

Once the energy requirements and type of tasks to be carried out were defined, a thorough analysis was made of the arrangement of off-shore steel structures and foreseeable environmental conditions in which work must be undertaken.

Conditions for travel, mooring to the component and weld cleaning and inspection operations were analysed with reference to several types of vehicles, using models of structures and the considerable experience accumulated by Tecnomare in designing all types of platforms.

It was found that a more or less cylindrical shape along a vertical axis was most suitable for this type of application, in view of the fact that :
- no specific direction in relation to the current requiring the choice of a cross-section of minimum resistance needs to be maintained for the craft;
- as regards mooring, a vertical cylindrical shape reduces to a minimum the distance between points on the craft and the resultant of the forces acting;
- the interface area with the node to be inspected is increased to a maximum;
- the stability of the craft is improved.

On the basis of the above, an initial outline was drawn up of the vehicle, featuring :
- an upright body housing the diesel engine, fuel and equalizing tanks, oxygen cylinders, propellers and hydraulic units for the cleaning and drive system;
- a front-end structure in the shape of an upright rectangle, with suction caps fitted to four gripping arms at the extremities and two manupulators inside.

The vehicle is served by equipment at the surface, in modules, comprising control room, oxygen generation and charging facilities, launching system and store-room. It is anticipated that the vehicle will be protected by a housing during launching.

To perform inspection and cleaning operations, the vehicle will be equipped with television cameras, sensors of various types (for measure-ment of cathode potential, thickness, etc.), a water-jet cleaning system and various devices, such as a grinder, brush, etc. It will be possible to change these devices on the manipulators using hydraulic arms.

3.4 Control

The control system of the TM 308 is based on the assumption that the remote system is an extension of the persons operating it. Thought was given to automatic systems, with the aim of providing greater help for the operator rather than replacing him.

It was decided that the operator should be given ample assistance by computers in the performance of his duties; this decision was based on general efforts to simplify operation of the ROV and the need to make best use of the modest capacities of the communications channel, which is in fact the basic problem in designing the control system and defining the operator's tasks.

The absence of an umbilical means that an underwater transmission system must be used, and in the light of present technology it is planned to use an acoustic system with the following features :
- low capacity, of the order of 5-10 Kbit/sec;
- delay in emission, varying with the position of the vehicle;
- intermittence, on account of possible masking effects between hydrophones owing to structural components, or because of specific conditions temporarily affecting the signal/noise ratio.

In this situation, the system must be capable independently of acting to ensure its own safety should there be a temporary, unforeseen break in communications : for example, the vehicle must stop and remain in position should this occur during navigation.

It is therefore vital that the craft should have on board a considerable degree of intelligence capable of steering the vehicle using limited information supplied by the operator. Conversely, the operator must be provided with sufficient information on the system and the environment in which it is operating, using limited information transmitted by the vehicle, one of the major sources being television images.

An outline of the control system was therefore based on the following three items :
- onboard controller;
- communications channel;
- surface controller.

The breakdown of tasks between the two computers gives the surface controller the role of interface with the operator, involving a display of the system's status and high-level input of commands to be carried out. The onboard controller deals with the corresponding initiation of control loops, managed locally, and of transmission to the surface of as much information as possible.

The on-board controller's most important functions include :
- control of navigation and trim, which implies a knowledge of the craft's position in relation to the structure, and the ability to overcome environmental disturbance (current) automatically while tracking a velocity vector stipulated by the operator;
- local control of manipulator movements, on the basis of desired position data transmitted from the surface;
- processing of signals for transmission to surface, so as to maximize the information per bit: this involves in particular the TV signal, which is transmitted in various ways depending on the phase of the mission, and basically involves varying the resolution and number of frames per second, while special procedural steps deal with the image's "compression";
- control of auxiliary functions (status instruments, control of motor, etc.);
- local monitoring and high-level communications with operator.

In addition to managing the system's communications, the surface controller allows the operator to "observe" the system and the environment in various ways, in order to be able effectively to carry out his planning, command, monitoring and intervention functions.

Methods were examined for displaying the platform and the vehicle, the route, the node and handling operations.

The procedures for the command of the vehicle when travelling and for remote handling were designed with a view to enhancing the system's efficiency through the automation of basic or repetitive functions such as the cleaning of the area being inspected.

As far as the latter is concerned, research is being undertaken on the possibility of carrying out operations under operator control, such as :
- collection of data on the configuration of the welded area;
- determination of trajectories to be followed by the cleaning tools;
- automatic execution of trajectories.

Similar operations would be carried out to perform other inspection operations.

As regards the communications channel, a study was made of the arrangement of hydrophones on the vehicle and around the structure. A signal/multipath ratio enabling commercial hydroacoustic transmission systems of suitable capacity (5-10 kb/sec) to be used can be obtained by apt positioning of the hydrophones.

However, in view of the vital importance of communications and the novelty of the application, thought is also being given to equipping the prototype with an optical fibre cable, 1-2 mm in diameter, which will be paid out and retrieved by a winch placed on board the vehicle. In addition to providing greater security during sea trials, the redundancy of the communications channel will also permit a full-scale trial to be conducted of hybrid solution for communications in this type of work situation.

4. CONCLUSIONS

Progress is being made on the TM 308 project though much work remains to be done. As a preliminary, guidelines have been worked out for all aspects of the research; results to date indicate the complexity of the project, with technological aspects covering numerous disciplines.

Further work during subsequent phases of research should provide worthwhile results for the corresponding practical applications.

(07.46/82)

DEVELOPMENT OF A DEEPWATER TETHERED MANNED SUBMERSIBLE

D. J. HAMPSON and F. C. BOARDER
Chairman and Manager, Engineering and Design
OSEL OFFSHORE SYSTEMS ENGINEERING LIMITED

SUMMARY

Hawk is the name given to this design of a one atmosphere diving
system currently being developed as an advanced, general purpose,
deep diving work system. It is a unique concept, and has been
evaluated with respect to three roles. Firstly, as a
commercially orientated vehicle optimised for deep water use.
Secondly, as a powerful scientific tool for deep ocean research
and thirdly, a development platform for the next series in this
design, which will have a capability of descending to 20,000 ft.
(6,000 metres) and ultimately 38,000 ft. (12,000 metres).

The first stage of the project has been to design a working
prototype submersible capable of transporting man safely to a
depth of 5,000 ft. (1,500 m). Throughout the design programme
materials, specifications and design have been specifically
chosen to allow integration to stages 2 and 3 of the next series.

1.   INTRODUCTION

The Hawk was conceived during 1981 to fill a growing need for
deep manned intervention and oceanographic study. At the design study
stage of the concept it was thought, to achieve our alternative goal
of reaching the deepest oceans, a hull of titanium with small view
ports would be required. This immediately highlighted two problems to
avoid, a) limited amount of available lift from the hull, and, b)
limited viewing through small ports by the pilot. Ideally all round
vision was required, and a need to take full advantage of using low
density materials in respect to lift from the hull. The use of two
acrylic hemispheres were then considered and decided upon as the right
choice of material for the 1500 m first stage of the project (a
separate long term programme will be pursued to ascertain the
applicability of glass as a transparent structural material for the
6,000 m and 12,000 m versions.)

With the knowledge and expertise inhouse of the interfacing of
acrylic hemispheres to aluminium rings readily available. No major
design problems existed and by mid 1983. The pressure hull and
associated penetrators had been assembled and await pressure tests to
be carried out.

See fig. 1 overleaf

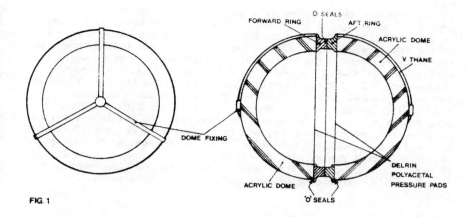

FIG. 1

2.0 BRIEF DESCRIPTION OF BASIC VEHICLE LAYOUT

The pressure hull is supported by a 'C' frame mounted to the aft ring and in turn supported by a tubular aluminium framework on the vehicles underside - this is known as the supporting framework. The framework supports the submersible while on deck or seabed, it also provides fixing points for battery pods, contactor pods, life support O2 bottles, cameras, lights, manipulators, etc. See fig. 2

2.1 'C' Frame

The 'C' Frame is of fabricated aluminium and is the main load bearing component of the submersible, capable of taking all non-pressure related loads away from the pressure hull. It provides the lifting points and mounting for fixed buoyancy units.

2.2 Fixed Buoyancy

Syntatic foam is added to enable the submersible to carry a 90 kg pilot and associated equipment + 150 kg of supplementary equipment.

2.3 Drop Weights

10 x 2.5 kg lead weights, jettisonable by the pilot, are provided to achieve a neutral state on reaching the desired work area. In addition a 22.5 kg emergency drop weight is fitted to the 'C' frame and can be released hydraulically by the pilot.

The 'C' frame, exostructure and all associated equipment to the above have been completed and are ready for assembly. See fig. 4.

2.4 Propulsion

A number of propulsion types were considered, i.e. cycloidal, hydraulic, electrical, 3 phase A.C. and D.C. A full design and test programme was conducted on the various alternatives. The cycloidal propellers were chosen first because of their attractive 360° thrust directional control. It was possible too, using only two propellers, to achieve forward, aft, side slip, turning, descend and ascend. A gear box has been designed and produced using 4 x 24 volt printed circuit motor driving a common output shaft. Tests on various propellor lengths were conducted but with limited results, our best result being 150 lbs static thrust. As 150 lbs represented only 1/4 of our total requirement, the cycloidal concept was then dropped from the test programme. The hydraulic thrusters at this stage were considered but due to the power to weight ratio, and the need to accommodate a D.C. drive for emergency situations, the weight penalty was unacceptable. Our final design then was to use a combination of 3-phase A.C. units, and D.C. units. The A.C. and D.C. units can be used in any combination producing an excess of 600 lbs thrust whilst running from the mains supply, and in the event of mains failure the D.C. units using onboard batteries at 60 volts produce 100 lbs of emergency thrust for 60 minutes. A total of 12 thrusters are fitted. 10 off A.C. and 2 off D.C. thruster trials have now been completed together with our production requirements of 12 thrusters. See fig. 4.

2.5 Seawater Pump and Manipulators

Manipulator development has progressed with the development of a new double ended seawater hydraulic pump unit. The pump situated either end of a 660 V 3 phase motor. Significantly reducing the size and complexity of our previous manipulator control pods.

Specification

Size:	117 mm dia. x 453 mm long
Motor:	660 V 3 phase 1 H.P.
	Oil filled and externally compensated
Pump output:	30 L.P.M. @ 100 P.S.I.
Weight in Air:	13 kg weight in seawater 9 kg

Material selected for the motor housing is HE30TF Aluminium; Pump Unit, stainless steel.

The manipulator arms are constructed from inert neutrally buoyant polymer and are operated by seawater at 80 p.s.i. above ambient. The requirement was to produce a robust manipulator with the same versatility as a spatially correspondent arm. By removing many of the complicated proportionally controlled valves found in other systems, and with less moving parts, an arm with exceptional freedom of movement and realiability has been achieved.

Specification

Weight in Air:	35 kg
Weight in Water:	5 kg
Dimensions:	See fig. 3
Working Envelope:	See fig. 3
Movement Function:	See fig. 3
Operator Control:	7 function joystick.

Prototype arm and pump units have been built and undergone extensive trials with excellent results. Manufacture of arms and pump units due for completion late 1984.

2.6 Vehicle Power and Distribution

a) Vehicle Power Pod

This unit contains all switching and protection for the vehicle power systems. This is to ensure that the penetrations in the pressure hull carry low energy control signals only. The items housed in the power pod are as follows:-

- i) Fuses and contactors for the 8.3 phase 660 V 3 phase thruster motors.
- ii) 4.5 KVA transformer to supply 120 V lights, 24 V for control circuits and 30 V to instrument battery charger.
- iii) All relays required for control of lights, manipulators, pumps, tools, etc., as required.
- iv) Fuses and contactors to control the 660 V supply to manipulators.
- v) Fuses and contactors for emergency D.C.
- vi) Battery contactors and summing diodes.
- vii) Instruments battery charger.

b) Main Standby Batteries

Two sets of standby batteries are used, these are housed in pods mounted on the exostructure lower frame. Each pod contains 24 off 2 V 25AH sealed lead acid batteries connected in series. Taps are provided at 60 V and 24 V. The 60 V line is used to drive the emergency D.C. thrusters and the 24 V supply is used for control circuits. All output lines are double pole fused in the battery pod.

No provision is made for charging these batteries during vehicle operation. This is achieved by using a charger unit mounted in the control cabin and a deck cable. Access to the charing connectors is obtained by removal of the pod end cap, this is to ensure that the pod is fully ventilated during charging. Another benefit is that the charging connector is not exposed during operation and need not be a waterproof type. A further pair of fuses are fitted to protect the charging connectors.

c) Instrumentation Batteries

In order to minimise interference from the other vehicle electrical systems a separate set of batteries are used to power all instrumentation. These batteries are mounted in a separate pod and connected via double pole fusing and screened cables to the pressure hull penetrators. Also, wherever possible, in the hull, screened wiring is used to distribute the instrument supplies. Two sets of batteries are provided, each is 24 V 25 AH. On the pilots switch panel a five position rotary switch controls use of these batteries.

All above equipment is under manufacture and are scheduled for wiring during September, 1984.

2.7 Winch

Taking into consideration the limitation of deck space it was necessary to limit the overall dimensions of the winch and make it as compact as possible. Having defined the umbilical diameter and minimum bend radius a first layer diameter of 1500 mm. was established. In consideration of the test load which will be applied to the full drum a total of eight layers is the maximum acceptable. With a width of 1,828 mm. between flanges it is calculated that the required storage capacity can be achieved in eight layers.

Whilst it is appreciated that recovery speed of a manned vehicle is of prime importance calculations of power requirements have been made to provide an 'in-air' hoisting speed of 15 m./minute for the test load. Assuming constant power from the power pack then hoisting speeds will vary according to load being raised. Therefore the test load can be lifted at 15 m./minute for a constant rotational drum speed of 8.3 R.P.M. An average layer circumference between 2nd and 7th layers of 5.5 metres would result in the above hoisting speed. Recovery time for the vehicle through 1,500 metres would be approximately 33.5 minutes

Design Criteria

Recovery time from 1,500 metres:	30 - 40 minutes
Cable storage:	1,600 metres
Cable diameter:	36.5 mm.
Cable weight:	2.04 kg/metre
Hoisting speed in air (test load):	15 metres/minute
Test load:	1.5 x vehicle weight
Drive:	hydraulic motor and spur gears

Umbilical and winch design finalised. Scheduled completion date January, 1985.

3.0 CONCLUSIONS

Predictions suggest that the impact of the oil industry will diminish insofar as the need for deeper vehicles is concerned, although it will continue to foster the development of subsystems that will broaden vehicle capabilities for task peformance. The need to develop vehicles with 2,000 m plus depth rating will come from scientific needs and ocean mining industry commencing in the late 80's with full commercialisation realised by the early 90's.

In view of the potential capacity for development offered by the current revolution in microcomputers it is perhaps presumptious to attempt to predict what deep ocean vehicles will be like in the next decade. It is probably safe to assume that the major developments will be in the evolution of subsystems such as data acquisition, controls, navigation and manipulators. Many of these will however have been developed as requirements in shallower water exploitation.

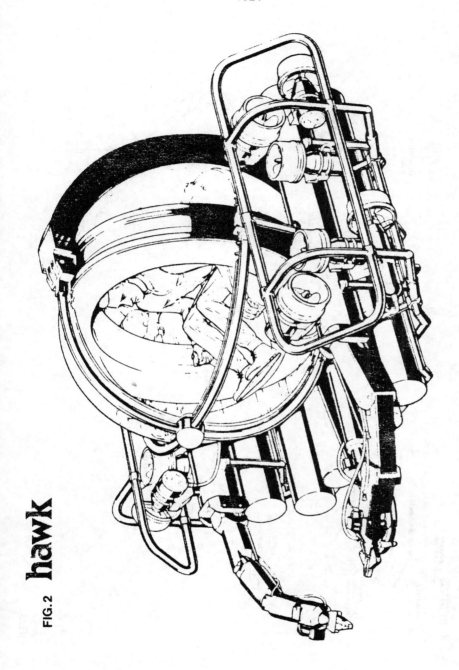

FIG.2 hawk

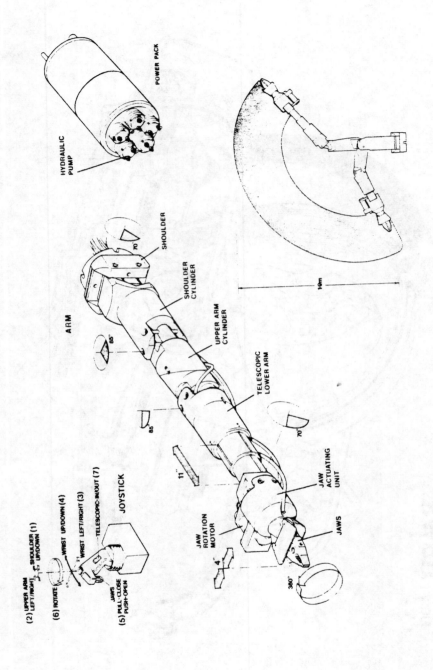

FIG. 3

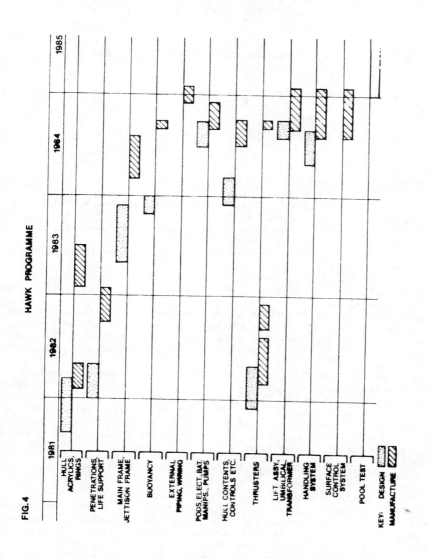

FIG. 4   HAWK PROGRAMME

(07.35/80)

STUDY OF REMOTE-CONTROLLED WORKOVER AND MAINTENANCE SYSTEM FOR SUBSEA PETROLEUM INSTALLATIONS

C. de VAULX, R & D Manager, ALSTHOM ATLANTIQUE, Nantes

Summary

ACB and the Commission of the European Communities have set up a work programme, entrused to ACB, covering the following technological project: "Remote-controlled workover and maintenance system for subsea petroleum installations".
The main technical aspects of this study are summarized in this article, namely:
- definition of the tasks to be accomplished,
- evaluation of the tools and equipment needed to carry these tasks into effect,
- drafting specifications of the systems of navigation, location and propulsion required for the missions entrusted to the system,
- design of the man/machine interfaces in order to cut the routine tasks of the operator and enable him to concentrate on the essential tasks,
- drafting of specifications of an offshore handling system,
- definition of the power production system and the specifications of its main elements.
This article gives the conclusions of the study with relation to the assumptions whereby a remote-controlled system for maintenance of offshore petroleum installations can be implemented and the requisite operating conditions.

1 - PREAMBLE

This study, conducted by Ateliers and Chantiers de Bretagne in 1981 and 1982, was aimed at defining the constituents of a remote-controlled system the purpose of which was to replace divers, together with the associated tools, in order to intervene on hydrocarbon production equipment installed in depths of water exceeding 300 metres.
In 1980, when this study was first proposed, its starting point was the fact that, whilst remote-controlled equipment was already widely used for observation and monitoring tasks, it was practically never applied for operations of installing or repairing subsea hydrocarbon production equipment. Four years later, this statement still applies and most oil companies prefer envisaging platforms of exorbitant cost in order to place a minimum of equipment in the water, thus avoiding problems of working under water, in particular when the equipment is not accessible to divers. What are the obstacles against the development of working by remote-control

underwater ? Are they technical, economic or other ? The study proposes to answer these questions and to stimulate discussion.

2 - PRESENTATION OF THE STUDY

When the study was first launched, we thought that the obstacles to development of remote-controlled work underwater were technical and technological, and in particular that there were major gaps concerning the following points:
- lack of tools specialized by task,
- inadequate ergonomics of the piloting facilities,
- fragility of the umbilicals resulting from unduly rough-and-ready handling at sea,
- lack of propulsive power.

We also consider that technical and economic analysis at system level (tools + underwater vehicle + surface support) would enable us to identify the factors limiting remote-controlled work.

The results of the study have led us to conclusions that differ from those that we expected ; this is why we lay more stress on the conclusions than on the technical development of the report.

The following is the structure applied for the study:

2.1 - Preliminary studies
- studies of the tasks on an offshore petroleum site,
- definition of adequate tools and facilities,
- preliminary study of the architectures of the adequate systems,
- definition of the main constituents and critical subsystems.

2.2 - Studies of the tools
- powered manipulator,
- energy generating source (electrical and hydraulic),
- handling tools (lifters, etc...),
- connection tools,
- repair tools.

2.3 - Studies of the piloting and ergonomics of the system
- analysis of the positioning constraints of the system and the operating modes,
- study of dynamic positioning (sensors - propulsion systems - servo-control),
- study of the man-machine interface (display - control),
- simulation of model.

2.4 - Study of the means of implementation
- definition of the underwater and surface environment, taking into consideration the surface support,
- studies of the handling facilities of the device and the umbilicals,
- simulation on a computer of the interactions between the surface support - cable - device),
- optimization of operating costs by means of an econometric model combining all costs (surface support location - mobilization - personnel cost - reliability - etc...).

2.5 - Studies of the energy aspects
- analysis of the mission profiles from the standpoint of energy and power,
- analysis of the available energy sources (umbilicals - independent sources),

- definition of an energy system necessitating the optimum operating cost,
- detailed specification of the necessary equipment.

3 - ANALYSIS OF THE TASKS AT AN UNDERWATER PETROLEUM SITE

The tasks are classified in accordance with a systematic list arranged in accordance with the following main groups:
a)   Oil equipment category: satellite wells - clusters - manifolds - flowlines - risers - platforms - anchoring systems - main pipelines).
b)   Type of work: observation/monitoring - installation - repair following foreseable failures - emergency repairs following incidents.
c)   The deterministic (example in a workshop) or non-deterministic (example on a site) nature of the task is a character of first-order importance.

4 - DEFINITION OF THE NECESSARY TOOLS AND FACILITIES

For each task, we have attempted to define the necessary and sufficient specialized tooling, by characterizing it in terms of its essential interfaces with the underwater vehicle that is to carry it:
- weight and overall dimensions)
- power   )   for each tool
- information flow   )
    The following conclusions emerge:
a)   Tools for light tasks are practically never of the right dimensions for the vehicle.
b)   Tools for heavy tasks can not be multipurpose, but must form a tool box with mechanical, electrical and DP interfaces with the vehicle that must be standardized for rapid and reliable connection, made either on the surface or at the bottom.
c)   Heavy handling operations (> 200 kg) require an underwater hoisting device resting on the sea floor provided this has sufficient bearing capacity or a surface vessel equipped with a handling beam.
d)   The tool/oil equipment interface varies considerably, depending on the non-standardization of the production equipment ; for the future, it is indispensable that the underwater station and the working tools be designed simultaneously.

5 - ARCHITECTURE OF THE SYSTEM OF INTERVENTION

Agreed that:
- the platforms are not to undergo heavy remote-controlled interventions,
- the repair of pipelines and flowlines deserve specific tools,
- the risers are recoverable on the surface for repair,
it is the wellhead production and subsea manifold units that comprise the most valuable elements and call for special consideration.
    The study covered a remote-control system specifically designed to ensure the installation and maintenance of production and subsea manifold units. The system must be capable of:
- carrying the above-defined specific tools,
- ensuring transfer between surface and the bottom (vertical) as well as horizontal movement of these tools,
- an adequate supply of energy to these tools,
- acting as handling beam for the heavy loads,
- operating in the least favourable environment (strong currents - poor

visibility).

These constraints as a whole have led to the definition of a device with the following leading characteristics:
- 3500 kg self-propelled device,
- ballasting capacity of 300 kg for handling heavy loads,
- electric power supply bearer umbilical: load 15 tons - power capacity 50 kW,
- standard mechanical interface used either for hooking on tools, or for berthing and locking into a subsea structure,
- high definition forward-looking sonar navigation with short base and high frequency.

The device is completed by:
- a set of interchangeable tools, including a position control handling arm,
- a subsea power generating set that can be installed on the bottom capable of outputs of over 50 kW (up to 500 kW),
- a dynamic positioning surface support designed for handling loads of 20 tons in the air, through a central diving shaft.

Figure 1, appended, gives a rough illustration of the architectural configuration of the device.

This provisional configuration is intended merely to provide a framework for the specific studies to follow.

6 - STUDY OF THE PILOTING SYSTEM

Considering that the device is designed to evolve in a restricted and perfectly defined space delineated by the surface support and the subsea production system on the bottom, both being perfectly positioned with relation to each other, one can define a quasi-automatic piloting system obliging the device to follow predetermined trajectories conceived in the light of the possible obstacles (risers) and the direction of the current. This enables two objectives to be attained:
- relieve the operator, who no longer has to pilot by sight or by instruments ; he merely has to check that the predefined trajectory is being followed correctly,
- avoid the contingencies involved in manned piloting, and in particular the risks of collision between the device and the production equipment,
- provide "soft" docking of the subsea structures.

The only specific equipment is a high frequency short base high definition sonar. The subsea structures are equipped with acoustic reflectors to beacon the field of action of the device.

No technological barriers were revealed by the study ; intensive use of microcomputers ensure great flexibility and safety in the piloting, for instance making it possible to use self-adapting servo-control systems.

7 - STUDY OF THE SURFACE FACILITIES

The availability of the system under difficult sea conditions depends to a considerable extent on the combination of surface support and handling device. The reliability of the umbilical, which is a fragile device heavily stressed in particular by heave, calls for a heave compensator, and the vessel should remain of reasonable dimensions if excessively high operating costs are to be avoided.

The study has led us to define a handling system that is original in two aspects:
- preference has been given to a linear tyred winch, rather than a drum winch,

- an umbilical ballast has been used to compensate for the drag of the current on the vertical part of the umbilical.

The study has revealed the need to design the handling device by adapting it specifically to suit the underwater device, though without requiring a special technology.

8 - STUDY OF THE POWER SUPPLY

Lacking high performance independent power sources, one has to accept the need to bring in the energy through an umbilical cable carrying the electric power line.

When navigating in the open water, the device makes use of almost all the installed capacity (50 kW) for propulsion purposes ; when supplying heavy tools, since it is stopped and locked to the subsea structures, this power is then available for the tools.

The umbilical selected features the following specific characteristics:
- Kevlar-reinforced electric power bearer cable,
- monoaxial cable feeding the medium voltage and multiplexed signals via the same power conductor,
- lightening of the bottom part of the cable to endow it with sufficient buoyancy to float.

Tools requiring over 50 kW of energy are connected to a subsea generating set made on the bottom, with the device making the underwater connection between the generating set and the tool.

Clearly, the development of an independent energy source would enable the performances of the intervention system to be enhanced by simplifying the umbilical, which would be less fragile and less expensive ; one can even envisage doing away with it entirely, by replacing it with short range acoustic links between the device and a cage situated above it. However, this goes beyond the scope of the study.

9 - ECONOMIC ASPECTS

The conclusion of the study is that, whilst the capital investment can be envisaged by a services company, it is well within the means of an oil operator seeking for operating safety.
- first, the cost of the vehicle and the tools is marginal compared to that of the production equipment,
- second, there is no need for a specialized surface support, since the intervention equipment can be applied from the production platform or from one of the service ships operating permanently on the field.

At all events, the cost of the combination of the subsea production system plus the intervention equipment is much lower than that of the platforms that would be required to do away with the need for subsea equipment. It must be considered as the "tool box" assigned to each production system.

10 - CONCLUSIONS

The replies to the questions asked in the preamble concerning the obstacles to the development of remote-controlled underwater intervention are hence the following, as we see it:
- the obstacles are not technical or technological ; the state-of-the-art is adequate,
- nor are they economic ; the cost of the platforms installed in great depths of water give the scope of the possible budgets.

The main problem is essentially linked to the fact that oil companies, save for a few exceptions such as ELF or SHELL have not yet integrated underwater intervention techniques in the design of their production systems: the latter must be easily capable of disassembly, in other words be modular, both at the level of the circuits and the subsea structures ; the connectors or connecting elements between the modules must be capable of being disassembled by simple tools. Integrated teams must be formed made up of oil production specialists and intervention specialists, both in the design and operating phases ; this is the requirement for subsea production systems to develop together with the associated remote-controlled means of intervention.

(07.34/80)

THE LONG RANGE SUBMARINE

G.C. SANTI

SSOS - SUB SEA OIL SERVICES SpA - Italy

Summary

The first ever Long Range Submarine to be built, the CEE - 22, is being briefly described. The Submarine, of 22,5 tons of submerged displacement, is 9,5 o.a. mt long, has a maximum diameter of 2.18 mt and an underwater endurance of over 500 nautical miles at 5 knots while cruising at maximum diving depth of 650 mt. The submarine can support saturation diving operations, by a team of four divers, or can carry out surveys of 100 n. miles per day.

This time, after the paper we presented here in 1980, I would like to start, as a good classical author, in "medias res", showing you the photograph of the PH-1350, plate (I), under steam.

This is, to my knowledge, the first operational submarine in the world powered solely by an efficient closed circuit diesel engine, i.e. an anaerobic diesel with a stoichiometric oxygen consumption. And when I say the first, I mean that the submarine can indifferently snorkel or dive on the same engine and stay submerged at maximum operational depth on diesel power, for the total duration of the mission. This is, in round figures, an 80 tons submerged displacement craft, which can carry a maximum of 2,2 cu mt of pressure vessels, or 1100 kg of oxygen, to feed a 100 HP closed circuit diesel, to steam underwater at 8.5 knots. The pressure hull is of conventional design, cylindrical plating reinforced by ribs, and the oxygen is stored, gaseous, in four h.p. bottles placed outside the main body.

And this was the starting point for the long range submarine.

Let's first define what we mean by "Long Range".

Definition

"The Long Range submarine is a vehicle capable of steaming continuously underwater, at maximum operational depth, at five knots true speed over a distance equal to its submerged displacement multiplied by 20".

This is an arbitrary definition, like all definitions are, but it contains all elements necessary to illustrate the operational capabilities of any vehicle falling within the definition itself. A fifteen tons submerged displacement vehicle will carry enough energy to steam for three hundred nautical miles at five knots; this is a 60 hours mission, so it is easy to extrapolate that, if the submarine is engaged on a lock in-lock out saturation dive she will be able to carry out a mission of full ten hours bottom time over any twelve hours period.

Equally a 150 tons submerged displacement will steam for 3000 n. miles, or 600 hours; i.e. she will be capable of supporting a saturation team for fourteen days some 200 n. miles offshore.

At this point the objective was well defined and, with the full support of the EEC, whom I want to thank warmly for their confidence in our capabilities, we started to work.

We defined four mission profiles:
- a full saturation team of six divers working in 350 mt waterdepth for 14 days 200 nautical miles offshore.
- 2000 nautical miles of geophisical survey starting 200 n. miles offshore.
- Ten hours bottom time in every twelve hours period, for a saturated team of four divers in 350 mt waterdepth.
- 100 nautical miles of survey in 650 mt waterdepth, in any 24 hours period.

and we carried out the preliminary design of two submarines, each capable of carrying out two missions (Plate II and III); one of 250 tons submerged displacement and the other of 22 tons.

Then we got down to the detailed design and when we had the work completed, inclusive of research work in the compact closed circuit diesel, the pressure hull and the avionic outfitting, we found that the two projects were identical but for the dimensions. At this point we took the decision to fully develop the CEE 22 in order to operate her in the North Sea, and to complete the detailed construction design of the 250 tons.

The Long Range submarine, like all submarines, was divided in three major items: hull, power plant and outfitting. But in the LR the power plant is inherently "heavy" as, in addition to the fuel, we must carry the oxydant, which is roughly three and a half times more weight. And this create, for any given displacement, a problem of weight and volume, complicated by the fact that the weight is variable with the consumption, while the displacement must be kept constant.

The oxidant is available in three different forms: liquid

oxygen, gaseous oxygen and hydrogen peroxide. We selected for our project, the gaseous form. The reason why I am starting to talk about the oxidant, while I just mentioned the three main items into which the project was subdivided, is because gaseous storage is part of the pressure hull. If your pressure vessels are located outside the main hull, for any given net volume, there will be an increase of submerged displacement; and therefore any given speed will require more power than the power required solely by the net volume.

To give an example the PH 1350 has a net volume, i.e. volume available for crew and machinery, of some 20 cu mt, while the total submerged volume is of some 78 cu mt.

Therefore we carry oxygen to propell eighty tons while we use only twenty of them. Should we stock the pressure vessels inside it would be the same story.

The ideal solution would be to adopt the double hull configuration and to use the anulus between the outer and inner hull. This is how it was born the hull of the CEE 22; twenty cubic meters of net volume, 2,15 cu mt of pressure vessels for the oxygen, a submerged volume of 22,45 cubic meters.

The double hull, which allows oxygen to be stored in the anulus, and withstands the bottom pressure, in the toroidal hull (Plate IV).

The toroidal hull has three main advantages over the conventional hull:
- it has the best ratio net volume/total volume;
- it can be shaped into the tear drop form;
- it has a greater diving depth.

In order to prove these points we built a module of 2,4 mt X 7 mt and we carried out an extensive series of pressure tests, twelve, fully strain gauged, toghether with a full mathematical analysis by the finite elements method.

The toruses of the module had a pipe diameter of eight inches, a wall thickness of 7.5 mm and a yield of 24 kg/mm^2; collapse depth was found at 1200 meters in line with the mathematical prediction.

At the same time two working groups had been fully engaged in the development of the compact closed circuit diesel and in the avionic outfitting. It would be tedious and outside the scope of this paper get involved in to too many details.

The objectives were: for the engine, a reliable unit, depth indipendent and totally unmanned; for the outfitting, two tons less than the PH 1350.

Both the objectives were achieved. With these results in hand the CEE 22, the Long Range submarine, took shape.

The pressure hull is made of X-60 grade toruses of 90 X 7,5

mm with a maximum OD of 2200 mm and a pure tear drop form. Lenght over all is 9,5 mt, diving depth 650 mt, cruising speed submerged 8,5 knots. The submarine is divided in three compartments; cockpit for three, lock in - lock out for four and unmanned engine room.

The oxygen storage is of 2.15 cu mt at a maximum pressure of 350 Kg/cm^2; a 12 hours mission at maximum speed will require a charging pressure of 150 kg/cm^2. Each torus is equipped with a pencil pressure reducer, inserted inside, so no high pressure oxygen is ever coming outside the pressure vessel. The CEE 22 carries 22 Kwh of lead acid batteries to propell the submarine, in case of emergency, for 12 nautical miles at 3 knots.

The engine room houses the compact closed circuit diesel which drives the main propeller through an hydrostatic transmission. A brushless electric motor/generator is in line with the same transmission and can supply 14.7 Kw of stabilized current or develop 14 Kw as a motor.

The main propeller has a diameter of 1325 mm and turns at 240 rpm; four thrusters, of 8 HP each, are provided.

The CEE 22, whose hull is almost completed at the time of writing this paper, (Plate V) will be equipped with a "sail" housing a ROV and all ancillary equipment necessary to launch and recovery the vehicle. Another important feature will be the operation in rescue mode; 20 men at a time, either under hyperbaric or athmospheric conditions, will be shuttled in one mission, which includes the mating and unmating with a distressed submarine or a hyperbaric plant.

Thank to the support of the EEC a major break through in underwater technology has been possible and will be available to the oil industry by early '86. But in order to make full use of this avenue it is mandatory that the people involved in underwater engineering start to get acquainted with the new submarine partecipating to the demonstration program that we will carry out by mid '85.

PLATE I

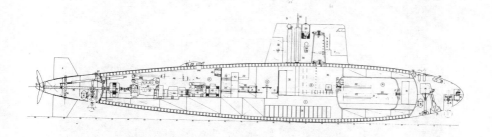

PLATE II

PLATE III

PLATE IV

PLATE V

(07.22/78; 07.33/79; 07.43/81)

D A V I D

A VERSATILE MULTIPURPOSE SUBMERSIBLE SUPPORT SYSTEM FOR REMOTE CONTROL OR DIVER ASSISTED PERFORMANCE

G.E. MARSLAND and K. WIEMER
ZF-Herion-Systemtechnik

Summary

The expansion of the offshore industry associated in particular with the development of North Sea oil and gas fields has produced the requirement for inspection, maintenance and repair work on a large number of submersed and semi-submersed offshore installations. As such installations become older, and as governments and classification organisations insist on ever more stringent regulations for safety and environmental protection, the demand for inspection, maintenance and repair must increase.

The tasks involved in this work require special equipment and the system described in this project report is designed to provide a submersible vehicle which is firstly capable of performing certain tasks under remote control, secondly has sufficient power and stability to transport tools and equipment to the subsea worksite and thirdly to provide the diver with all the tools and facilities which he may need to complete his work program.

The submersible system prototype, designed and manufactured by ZF-Herion Systemtechnik with the support of the Commission of the European Communities, has successfully completed a series of sea-trials both with Det norske Veritas in Bergen Norway and with Shell Expro working in the North Sea out of Aberdeen.

1. INTRODUCTION

The inspection, maintenance and repair of offshore installations and equipment essentially means going underwater to carry out usually pre-defined work programs.

The growth of the offshore industry in recent times has led to corresponding demands on the diving industry and as a result diving tech-niques have been developed to a degree where the task of putting a free swimming working diver at a water depth of, for example, 300 meters has become largely a matter of routine.

The costs involved in operating a Diving Support Vessel (DSV) equipped with the necessary saturation diving spread are however high and this fac-tor, together with the ever present risk of accident, injury or loss of life have created a requirement for submersible machines which can perform the various work programs under remote control from the surface.

The development of such machines has been continuous since about the mid 1970's and although a number of remote controlled and manned submer-sibles are already available and doing useful work, for many tasks there is as yet no real substitute for the diver working on site.

This situation was recognized by ZF-Herion Systemtechnik and the company set out to design and manufacture a new kind of multi-purpose submersible which could be employed either as a conventional Remotely Operated Vehicle (ROV) or as a Diver Assistance Vehicle.

2. COSTS ASSOCIATED WITH DIVING OPERATIONS

A typical example of the equipment required to maintain a diver working in deep water would begin with a specially constructed Diving Support Vessel (DSV) of approximately 2000 tons with accomodation for about 100 personnel. The vessel would carry a saturation diving chamber spread, a diving bell with appropriate handling arrangement and the various diving life support systems such as gas and hot water supplies. During diving operations it is essential that the DSV should maintain accurate position and heading above the work site and the vessel would therefore be equipped with a sophisticated dynamic positioning control and drive system.

The costs to operate such a vessel at present in the North Sea amount to approximately thirty thousand pounds sterling each day that the vessel is on contract. Expressed another way, to transport and maintain a man at a subsea site for the purpose of carrying out manual work tasks costs, on average, 1250 pounds sterling per hour.

Depending upon the particular contract, the diving bottom time (i.e. the time when the diver is available at the subsea work site to perform useful work), can be between 30 % and 70 % of the total port to port contract time. Figure 1 shows the breakdown of a typical contract in terms of time required for the various parts of the operation.

From this analysis, two points are immediately obvious. Firstly, because of the high rates for DSV operations, any reductions in the times required to complete contracts automatically result in significant savings with regard to costs involved. Secondly, the breakdown of the total contract time clearly shows that the area where effective savings are most likely to be realised is the diver bottom time.

3. CONDITIONS EXPERIENCED IN OFFSHORE DIVING

The environment in which the diver has to work is difficult and dangerous. The water temperature is generally about 4° C and natural illumination is either poor or non-existant. Tidal flow can be strong enough to prohibit diving completely and visibility is often drastically reduced by the particle content in the water. North Sea divers can be often working at depth of 150 meters where they experience a pressure of sixteen times atmospheric pressure.

The 30 kg or more of life support equipment which the diver carries or wears is cumbersome and makes movement difficult and strenuous. The umbilical which tethers the diver to the bell usually consists of a hot water hose; two gas hoses and two communication cables, making a bundle of about 50 mm diameter. Decompression sickness can occur as a result of an uncontrolled movement upwards through the water and to prevent this the diver usually has a slightly negative buoyancy in the water. Swimfins are not always worn when working in deep water at the sea-bed or on structures, the divers preferring to use weighted boots for extra stability.

In order to perform useful work tasks, additional tools and equipment must be made available to the diver and this is done for example by lowering a container from the ship above. Apart from the first problem of locating the container, the diver then has the task of transporting tools and equipment from container to work site, which means rigging lines to a suitable structure so that the weight of the equipment can be supported. This preparation procedure is essentially non-productive but because of the high risk to the diver's life it can only be carried out with extreme caution and is therefore time consuming. Any piece of equipment which is suspended from the ship will of course follow the movement of the ship and delays in communication between diver, diver supervisor and handling system operator make locating and positioning of such equipment difficult.

4. THE DAVID SPECIFICATION

Using reputable diving companies as sources of information, engineers of ZF-HERION Systemtechnik GmbH made a comprehensive study of the problems associated with diving operations. The findings indicated that there was a requirement for a submersible which could be operated by the diver locally and which could provide tools and facilities as and when required by the diver.

The specification for the submersible was greatly influenced by the demand for diving operations associated with the maintenance of jacket type structures where work is carried out by saturation divers working from a diving bell.

It was seen to be of particular importance that the vehicle should be already at the subsea work site when the diver emerged from the diving bell and a remote control facility was therefore considered to be essential.

Facilities which were seen as first requirements were:
o A submersible having a full ROV (Remotely Operated Vehicle) capability.
o A claw, adjustable in diameter, for clamping the vehicle to tubular structures.
o A moveable platform to provide a safe stable support for the diver.
o A source of power for hydraulic power tools.
o A range of underwater power and hand tools.
o A power winch, attached to the vehicle and with controls available for use by the diver.
o Adequate lighting
o TV equipment.

5. SAVINGS IN DIVING TIME USING THE DAVID

With the basic specification complete, the next step was to study actual work programs in order to assess the effectiveness of the proposed DAVID system. Independant diving specialists were contracted to define and analyse various subsea tasks and to compare task completion times using conventional methods with those achievable using the DAVID.

Seven examples, all considered to be normal everyday tasks, were chosen and results were as follows:

Task	Diving Time Saved
M.P.I. structural weld (Saturation)	56 %
M.P.I. structural weld (air)	47 %
M.P.I. structural weld (seabed)	34 %
Anode installation	51 %
Installing riser clamp	58 %
Pipeline repair	38 %
Riser repair	37 %

Figure 2 shows a detailed explanation using the anode installation as an example. A general assessment based on these results alone must obviously be treated with care as contracts and conditions can vary considerably. Items not taken into account are the lost time due to bad weather and the aspect of diver safety. If tasks are completed in shorter times then lost time due to bad weather will also be reduced. These are bonus points which can only be really evaluated in practice but can only increase savings.

Taking an average of 45 % saving in diving time and relating this to the contract example previously described, it can be seen that using the DAVID reduces the contract time from 500 to 392 hours. This results in a cost reduction of 135 thousand pounds sterling on the contract and clearly provides commercial justification for the project.

6. THE DAVID

The prototype system was manufactured according to the specification outlined in section 4 and starting in the Summer of 1983 went through first commissioning in Lake Constance in Germany, through pilot and diver familiarization trials at NUTEC in Bergen, Norway and finally to operation in the North Sea for trials with Shell Expro during 1984.

The design engineers responsible for the development of the system were actively involved in all stages of the trials so that both the concept and the design of the final product were subject to continual updating based upon operational experience.

An outline drawing of the final design is shown in figure 3 and this illustrates how the specification was realized.

6.1 Control and Navigation System

A fundamental requirement for the concept is that the surface operator must be able to pilot the submersible to the subsea worksite and there to dock onto a structure at any chosen position. The control system was designed to provide the degree of stability required for this phase of the operation.

An on-board computer receives actual value input signals from sensors indicating depth, roll, pitch and heading, and set value signals from the surface control station. The sensor inputs are used to calculate the actual vehicle orientation and this is then compared with the required orientation derived from the control station inputs. The difference is evaluated and outputs to the thruster speed regulators effect the required correction. From the surface operator's point of view, the vehicle remains stable in attitude, depth and heading, and commands from the steering joystick are superimposed. Commands to travel vertically upwards or downwards are derived from push-buttons and movement in this direction is made at a constant speed irrespective of load on the vehicle.

6.2 Claw arrangement

The claw is designed to attach to the tubular members of offshore structures and can be aligned either by rotating or tilting as the vehicle approaches the chosen docking position. The claw pressure pads grip the tube at points which are approximately equally spaced and the mechanism functions using a single hydraulic cylinder. In order to achieve a compact design, diameter setting is provided in two ranges, the range adjustment being made by the surface operator as required.

6.3 Platform design

Because the diver is virtually weightless in water it is difficult for him to work with equipment which is not itself weightless. Without some kind of support there is no way that he can exert pressure when using tools.

The platform is a ladder type arrangement and is dimensioned so that the diver can obtain a firm anchorage using only the legs. Both hands are then free to work with equipment.

6.4 Tools and equipment

A set of power tools and a range of small hand tools are carried on the vehicle as standard equipment. The hydraulic power supply for these tools is separated from the main system. Additional heavy equipment units can be mounted on the rear of the vehicle and each of these units is trimmed to be neutrally buoyant. Connection is made into the main hydraulic supply before the vehicle is launched.

6.5 Power Winch

In order that the diver can use the winch, the submersible must first be clamped onto the structure by means of the claw. The line of action is then from the winch through the claw and into the structure itself. The obvious mounting point for the winch therefore is on the claw frame. The winch/claw arrangement has been tested with loads up to 3 tons.

6.6 Lighting and TV equipment

The submersible is equipped with five separate lighting circuits. The TV cameras are mounted on pan and tilt units and cameras can be removed by the diver for close-up observation. The pan and tilt units are controlled from the surface.

7. PROJECT STATUS AND CONCLUSIONS

Trials with the prototype vehicle can now be regarded as completed. Results show that the system has fulfilled the specified requirements and that technical quality and reliability have been achieved at a price acceptable to industry.

Trials have also shown that the system is commercially viable and that the estimates for savings in diving times are realistic. The contribution offered by the DAVID to the offshore industry is threefold. Costs of maintenance and repair work are significantly reduced for the oil companies, operators and service companies achieve greater productivity and the diver receives a powerful and flexible tool which reduces physical exertion and improves conditions for safety at the worksite.

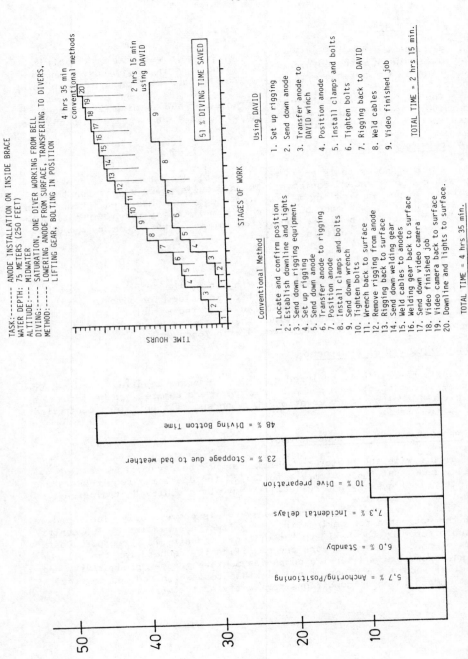

TASK:------- ANODE INSTALLATION ON INSIDE BRACE
WATER DEPTH: 75 METERS (250 FEET)
ALTITUDE:--- MIDWATER
DIVING:----- SATURATION, ONE DIVER WORKING FROM BELL
METHOD:----- LOWERING ANODE FROM SURFACE, TRANSFERRING TO DIVERS,
             LIFTING GEAR, BOLTING IN POSITION

4 hrs 35 min
conventional methods

2 hrs 15 min
using DAVID

51 % DIVING TIME SAVED

STAGES OF WORK

TIME HOURS

Conventional Method

1. Locate and confirm position
2. Establish downline and Lights
3. Send down rigging equipment
4. Set up rigging
5. Send down anode
6. Transfer anode to rigging
7. Position anode
8. Install clamps and bolts
9. Send down wrench
10. Tighten bolts
11. Wrench back to surface
12. Remove rigging from anode
13. Rigging back to surface
14. Send down welding gear
15. Weld cables to anodes
16. Welding gear back to surface
17. Send down video camera
18. Video finished job
19. Video camera back to surface
20. Downline and lights back to surface.

TOTAL TIME = 4 hrs 35 min.

Using DAVID

1. Set up rigging
2. Send down anode
3. Transfer anode to
   DAVID winch
4. Position anode
5. Install clamps and bolts
6. Tighten bolts
7. Rigging back to DAVID
8. Weld cables
9. Video finished job

TOTAL TIME = 2 hrs 15 min.

FIGURE 2 – EXAMPLE OF AN OFFSHORE OPERATION WITH A DRIVER

48 % = Diving Bottom Time

23 % = Stoppage due to bad weather

10 % = Dive preparation

7.3 % = Incidental delays

6.0 % = Standby

5.7 % = Anchoring/Positioning

% OF TOTAL CONTRACT TIME

FIGURE 1 – EXAMPLE OF DSV OPERATION
Total Contract Time = 500 hours

Figure 3 – <u>The DAVID Diver Assistance Vehicle</u>

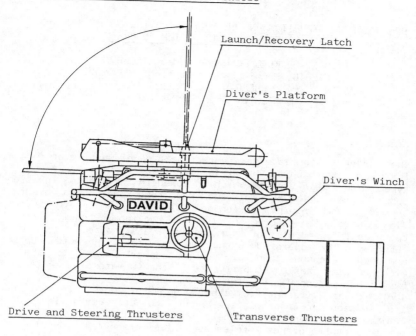

Launch/Recovery Latch

Diver's Platform

Diver's Winch

Drive and Steering Thrusters

Transverse Thrusters

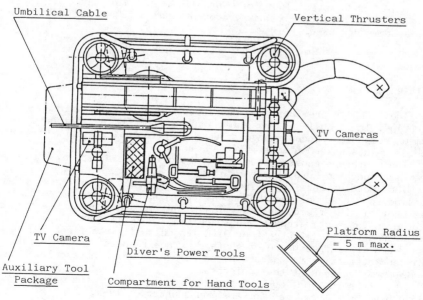

Umbilical Cable

Vertical Thrusters

TV Cameras

TV Camera

Diver's Power Tools

Platform Radius = 5 m max.

Auxiliary Tool Package

Compartment for Hand Tools

(10.37/82)

PROTOTYPE CONSTRUCTION AND PROTOTYPE TESTING OF THE UW-WORK AND PIPELINE REPAIR SYSTEM "SUPRA"

R.-D. KLAEKE, Ocean Consult GmbH
H. FIEBIG, Howaldtswerke-Deutsche Werft AG
P.V. LUECHAU, Ferrostaal AG
Members of ArGe SUPRA, FRG

Summary

The SUPRA system is a new, versatile type of underwater work system for use at depth up to 420 m. It is integrated into a catamaran-type submersible vessel which is towed to the working location by a support vessel. The SUPRA is submerged and positioned on the sea bottom by its own propulsion and ballast system. Energy is supplied via a cable which ensures an unlimited working autonomy. The SUPRA operates independently of surface crane assistance.

SUPRA can be operated either manned or by remote control and unmanned. Personnel transfer from the support vessel to the inside of the atmospheric compartments of the submerged SUPRA is achieved by a 1-bar personnel transfer system.

The SUPRA is different from other underwater work systems, due to its design characteristics and to its versatile underwater equipment: dry hyperbaric habitat / 4 hydraulic alignment clamps / up to 2 gantry cranes / telescope swivel crane / 4 hydraulic support legs / underwater plugs for hydraulic supply for hand-held tools.

For diver assisted uw-works, conventional offshore saturation diving systems are used.

1. INTRODUCTION

The SUPRA system has been developed since 1979 and is now under construction. It is built by the joint venture ARGE SUPRA, composed of the German companies Ocean Consult GmbH, Ferrostaal AG, Haux-Life-Support GmbH, Howaldtswerke-Deutsche Werft AG, and SCHIFFKO GmbH.

The prototype construction and testing is subsidized by the German government and the European Communities.

The development objective was to construct an uw- work system for pipeline repair as well as the necessary preparatory work and subsequent treatment in the field, or heavy load handling e.g. for subsea completion or other maintenance, repair and inspection tasks (Figure 1).

SUPRAs versatility together with its capability to float and to submerge and its independence of surface barges renders it particularly economic and attractive to operators and companies which have only small diving support vessels at their disposal.

The construction now is nearing completion and the harbour and sea trials have been scheduled for the 2nd quarter of 1985.

The "Submersible Underwater Pipeline Repair Apparatus" SUPRA is characterized by its great flexibility, independence of heavy surface ship support, and by easy handling and mobility.

The SUPRA system is designed mainly for pipeline repair operations down to 420 m water depth. SUPRA is well suited for operations without or to minimize diver assistance by integrated remote controls.

SUPRA is fitted out with basic equipment such as alignment frames and hyperbaric welding habitat, and additional equipment such as e.g. one or two gantry cranes as required, plugs for hydraulically driven uw-hand-held tools. Other optional equipment such as e.g. a swivel crane, a storage platform or 1-bar systems can be installed into the existing structure.

2. THE SUPRA CONCEPTION

The complete SUPRA system consists of the SUPRA vessel itself, the surface support equipment to operate the system, and an 1-bar transfer sy-stem for access to the interior 1-bar compartments or crew exchange of the SUPRA vessel.

The uw-work and pipeline repair system is integrated into the rigid steel structure of the catamaran SUPRA vessel, which can be towed from the harbour to the offshore working location by means of a supply or di-ving support vessel or a tug. Model towing tests have shown that a towing speed of 5 knots is possible up to sea state 5 to 6 which corresponds to wave heights of approximately 4 m.

Once at the working location submerging can start without delay since the umbilical is linked to the SUPRA vessel and is immediately rea-dy for energy transfer. UW-manoeuvring to the working position on the sea bottom is accomplished by means of SUPRAs own navigation, propulsion drives, and ballast water systems.

Consequently SUPRA is highly independent of bad weather and heavy sea conditions and can operate in unfavourable areas such as the North Sea during 90 % of the year. Once SUPRA is positioned on the sea bottom, the working limits are mainly dictated by the capability to keep the diving support vessel in position, and in diver-assisted operations by the safety criteria of the diving system.

For diver-assisted uw-works, conventional saturation diving systems are used which are not part of the SUPRA system.

The SUPRA vessel can be operated either manned from the 1-bar com-partment or remotely controlled and unmanned. All the auxiliary systems and components installed to operate and drive the vessel and its external uw-work systems are contained in the 1-bar compartment.

The necessary surface support equipment to operate the SUPRA system is installed in containers or directly on deck (umbilical winch, person-nel transfer system) of the support vessel.

3. GENERAL DESCRIPTION OF THE SUPRA SYSTEM

The complete SUPRA system consists of the main components mentioned below.

3.1 SUPRA Vessel
The equipment carrying SUPRA vessel essentially consists of two cy-lindrical floats with flat bottoms, two cylindrical longitudinal girders, 4 cylindrical and 4 square vertical connecting supports and 2 by 2 cylin-drical and square cross connecting girders. The diving tanks are situated at the sterns and bows of the floats.

The catamaran type steel structure consists of pressure-proof and pressure-equalized sections based on stress calculation criteria. All cy-lindrical sections which contain the 1-bar compartment and 4 ballast and trim control tanks are pressure-proof.

All systems required to operate the vessel are contained in the atmospheric compartment. All system components are accessible when SUPRA is submerged.

The pressure-equalized section consists mainly of the 4 diving tanks and 2 U-shaped portals to absorb external stresses.

3.1.1 Existing UW-Work Systems

Alignment System. The alignment system consists of 4 equally shaped frames with pipe clamps which are fixed in vertical guides to the steel structure. Movements in the vertical and horizontal directions and the operation of the claws are controlled hydraulically from the uw-control desks by a diver or remotely from the inside of SUPRA.

- Momentum (between 2 frames)		180 mkN
- Pipe diameter (for claws)		305 to 1067 mm
		(12 to 42 inches)
- Forces	vertical up	450 kN
	vertical down	200 kN
	horizontal	200 kN

Hyperbaric Welding Habitat. A habitat is integrated into the midsection of SUPRAs structure. Vertical adjustment is controlled hydraulically. The habitat is equipped with TV, lights, electricity supply, plugs for the hydraulic supply, spare gas volume, pressure resistant dry box, chain hoist, and an adapter flange for docking a modular supply unit. The habitat design can be modified to customer requirements.

Gantry Crane. One or two gantry cranes as required can travel on the rail running over the total length SUPRA. Drive and lift are controlled hydraulically from an uw-control desk. The crane is equipped with a pressure equalized hydraulic aggregate and electrical energy supply via a flexible cable.

The drum has a capacity of 50 m lifting rope which allows the gantry crane to be used as an uw-winch.

- Lifting capacity	200 kN
- Pull when operated as a winch	100 kN

Telescopic Swivel Crane. A hydraulically operated swivel crane can be installed as an option on top of SUPRA. Foundation, external hydraulic supply lines and the control valves inside of SUPRA are already provided. The crane is controlled from an uw-control desk. A TV-camera with lamps and uw-plugs for electrical and hydraulical energy supply are provided on the telescopic arm.

- Outreach	appr.	18	m
- Reach over bottom	appr.	26.5	m
- Slewing angle		360	degrees
- Lifting capacity	at 6 m: appr.	35	kN
	at 18 m: appr.	7	kN

UW-Hydraulic Plugs. Quick couplings at the fore, mid and aft section of the SUPRA vessel provide hydraulic power for hand-held tools etc. Hydraulic pressure and flow are remotely controllable.

3.1.2 UW-Observation Systems

UW-TV Systems. The SUPRA vessel is provided with several uw-TV-cameras. In both of the control stations on the surface support vessel and

inside SUPRA 2 monitors are provided for simultaneous display from 2 cameras. The following cameras are provided:
- Hyperbaric welding habitat, inside
- Alignment system: on top of each clamp
- Telescopic swivel crane or alternatively a diver guided uw-TV
- Front of SUPRA.

Obstacle Avoidance Sonar. Observation and obstacle detection is achieved by means of a scanning sonar. The sound dome is positioned at the bow of SUPRA. The acoustic signals are displayed on monitors simultaneously in both control stations (surface and inside SUPRA).

3.1.3 Optional UW-Work Systems

The SUPRA vessel is prepared for outfitting with additional or exchangeable equipment and working systems, for example:

Transfer Platform. For transport of heavy loads up to 200 kN from the surface to the sea bottom to be installed instead of the habitat.

2nd Gantry Crane. The crane rails are designed for the simultaneous operation of 2 similar cranes.

1-bar Welding Chamber. For 1-bar atmospheric pipeline welding a pressure-resistant welding chamber can be installed instead of the hyperbaric habitat.

Handling of Loads, Water Jetting or Suction Systems. The handling of loads can be provided with the telescopic swivel-crane, as a manipulator arm, which also may handle a water jet nozzle, a suction nozzle, grippers, hydraulic tools or a diver working platform, etc.

3.2 Operation Systems

SUPRAs major operation systems are as follows:

3.2.1 Electrical Power Supply System

The required power is either supplied by the support vessel or generated by a separate diesel-generator or in a combination of both (440 V, 60 Hz, 3 Phs).

The power is fed in to the main switch board of the remote control system container. The voltage is transformed to 3000 V and supplied to the SUPRA consumers via the umbilical. Inside SUPRA the voltage is retransformed to 440 V and fed to the switch and distribution boards.

Black Out or Circuit Break Down. In case of emergency the electrical supply for consumers of particular importance is ensured by battery systems.

For the electrical safety, especially for the external consumers and the part of the umbilical which is underwater, an earth leakage circuit breaker, permanent insulation monitoring and isolation transformers are provided.

Umbilical. Energy supply, control data and video signal transmissions are ensured by a steel armoured umbilical which is designed for an operational traction of 150 kN and in case of emergency of max. 400 kN. It can also be used as a lifting wire for a salvage operation of the SUPRA vessel.

3.2.2 Propulsion System
SUPRA is equipped with 10 independent hydraulic propulsion units, which consist of the following external components
- Thruster with hydro-motor
- Hydraulical aggregate in a pressure equalized box
- Connecting pipework
and in SUPRAs 1-bar compartment
- Electrical equipment and switch board.

Steering and manoeuvring is performed by the following thruster units:
- 2 vertical thrusters
- 4 longitudinal thrusters
- 4 transverse thrusters
- The speeds predicted for SUPRA (design data):   longitudinal   2.5 kn
                                                  transversal   1.0 kn
                                                  vertical   0.5 kn

3.2.3 Trim and Ballast Water System
The system includes 2 piston pumps (2 x 100 %), the pipework and valves, 4 pressure-proof combined compensation and trim tanks, arranged in the fore and aft section of the floats. Normally the ballast water intake and discharge is performed by means of a logical circuit controlled by the remote control system.
The systems main functions are:
- Buoyancy control
- Trim and heel balance
- Water ballasting to produce pressure on the sea bottom
- Bilge drainage
- Blowing out the dive tanks for surfacing.

The piston pumps are normally used for ballasting, discharging and water shifting for trim and heel balance. A compressed air system ensures the trim and heel function in case of emergency as a back-up system.

3.2.4 Remote Control System
The components, aggregates, engines and operational functions of the SUPRA system are controlled and surveyed by the remote control system. For necessary and important functions alternatively manual operation modes are provided.
The remote control system consists of:
- 1 central control station on the support vessel (integrated into the container) including the central computer with priority circuit.
- 1 similar unit installed inside the 1-bar compartment of SUPRA.
- 1 signal transmission unit connecting the surface and SUPRAs internal central computers via the umbilical.
- 10 subcomputers decentralized sited inside of SUPRAs engine rooms.

3.2.5 Hydraulic System
The hydraulic energy supply of the external uw- working systems and the valve drives of the trim and ballast water system is ensured by a pump station which includes all solenoid control valves. The 2 x 100 % pump capacity and an additional 100 % to serve the valve drives are installed inside the 1-bar compartment. The control valves for the external uw-working systems are activated remotely by means of the diver operated external or SUPRAs internal control desks.

3.3 Surface Support Equipment

The following components of the SUPRA system are installed on deck of the support vessel.

3.3.1 Remote Control System Container

The central control station contains the complete surface control station with control panels, switch boards, remote observation systems (TV, sonar, navigation) and the electric power supply main connection with transformer.

3.3.2 Diesel Generator Container

This is an option and only used if the support vessel cannot supply the total electric power required for the SUPRA system.

3.3.3 Umbilical Winch

The winch with variable constant pull for the umbilical is designed for a maximum pull of 400 kN in case of an emergency for a SUPRA vessel salvage operation.

3.3.4 Personnel Transfer System

The system designed for SUPRA consists of the 1-bar bell with positive buoyancy for dry docking on top of the SUPRA vessel. The SUPRA docking hatch is integrated with a pull down winch. On the support vessel a winch including a heave compensation system and a launch and recovery equipment is installed . The system functions are in standard cases: crew transfer from the support vessel down to SUPRA and back to the surface, or crew rescue in an emergency.

The personnel transfer system can be adapted to customer requirements.

4. CONTROL AND MODES OF SAILING

Underwater manoeuvring, sailing and positioning at the working location on the sea bottom is possible either by remote control or by manual joystick control. An uw-navigation system is installed which is used particularly during the positioning procedures at the working location.

For safety reasons any uw-manoeuvre either submerging or surfacing are performed with a residual positive buoyancy.

For submerging and surfacing operations particular programs will be defined. They include a series of predetermined steps for certain water depths so that a sensitive slowing down, balancing and uw-maoeuvring of the SUPRA vessel can be performed.

4.1 Automatic Controls

The automatic controls for course and depth are normally locked.

Automatic course control is used during all uw-manoeuvres. Values for the control are determined from the gyro-compass.

The Automatic Depth Control gets the actual values from pressure sensors and provides the factors "Diving Depth" and "Sinking Speed".

4.2 UW-Navigation System

SUPRA is fitted with a high frequency short baseline acoustic uw-positioning system to perform the positioning procedure on the sea bottom with the required accuracy. This system consists of a triangle transponder system. The calculated data are displayed on a graphic screen.

4.3 Sailing Conception

Manoeuvring and sailing of the SUPRA vessel is subdivided in the following phases:

Surface. This means SUPRA is in standby to the support vessel. Manoeuvres are possible with the transverse thrusters arranged below the water line in the floats.

Underwater. During submerging and underwater sailing between the surface and a certain distance of approximately 30 m above the sea bottom SUPRA and its position relative to the support vessel is observed by the DSV-sonar system during all uw-manoeuvres.

Positioning. Approximately 30 m above the sea bottom the uw-navigation system is used for the touch-down procedure on the sea bottom.

5. SYSTEM SAFETY CONCEPTION

SUPRA is classified by the classificaton society German Lloyd. The essentials of the safety conception are as follows:

5.1 Hull Safety

The hull of the SUPRA is designed and pressure tested in accordance with submarine standards. Stability calculations are in accordance with the American rules for the dimensioning of submarines as well as the rules of the German Navy. For stress calculations the program EASE 2 was used.

5.2 Operation Systems

Important functions or operation systems, components, gauges and measuring instruments are secured by redundancies under the view of safety.

The redundancy for the umbilical in a case of severance or power black-out is realized by means of an ultrasonic emergency control system on board of SUPRA.

5.3 Manoeuvring Concept

In this concept are the controls and modes of sailing established, particularly that the submerging and surfacing operation will be performed in any case with a residual positive buoyancy.

5.4 Manning Concept, Life Support, Rescue

The following definitions are established:
- SUPRA is a tethered, cable supplied uw-vehicle, which operates in pre-determined water depths up to max. 420 m.
- Qualifications required for persons with access to the submerged SUPRA.
- Provisions for emergency cases and the amount of life support material.
- In an access-plan the occasions of access and different compartment categories with certain required room conditions.
- The possibilities of crew rescue by means of:
  - The umbilical used as a lifting system.
  - A 1-bar personnel transfer system if SUPRA is attached to a pipeline.
  - The emergency battery capacity or deballasting by pumps.
  - Deballasting by the compressed air system.

6. SUPRA APPLICATIONS

SUPRAs general versatility makes it easy to employ the system for a number of different construction, maintenance and repair works. Combined with additional equipment, SUPRA can easily be adapted for different work requirements.

The various external work systems of SUPRA can be operated either manually by divers or by remote control from the inside of SUPRA. Additionally, surface-operated remote controls can be provided, if required. Thus it is possible to execute a specific underwater work job either in a manned or unmanned operation. SUPRA is able to execute the following underwater jobs:

- Pipeline alignment by means of the 4 integrated hydraulically operated alignment frames.
- Hyperbaric welding in the integrated dry gasfilled habitat which is equipped with TV, hoist, electric and hydraulic power supply. For midline connections and riser-tie-ins, two different habitat positions are provided.
- Supply of hydraulical energy for hand-held tools etc. by means of uw-plug connectors arranged in the fore, mid and aft section of the SUPRA vessel.
- Heavy load handling by means of 1 or 2 gantry cranes, each having a lifting capacity of 200 kN. The cranes can travel on rails along the total length of SUPRA.
- Uw-winching with a pull of 100 kN using the gantry cranes as uw-winches for positioning of pipe sections etc.
- Heavy load transport from the surface to the sea bottom with an additional transfer platform, installed instead of the habitat.
- Load handling with a hydraulically operated telescope swivel-crane, which can also be used like a large manipulator to handle water jets or suction nozzles, hydraulic tools or a diver working platform.
- Pipeline connections by means of mechanical pipe couplings.
- Atmospheric pipeline welding in an optional 1-bar welding chamber, to be installed instead of the hyperbaric habitat.
- UW-inspection and testing works using the dry habitat as an uw-work-shop.

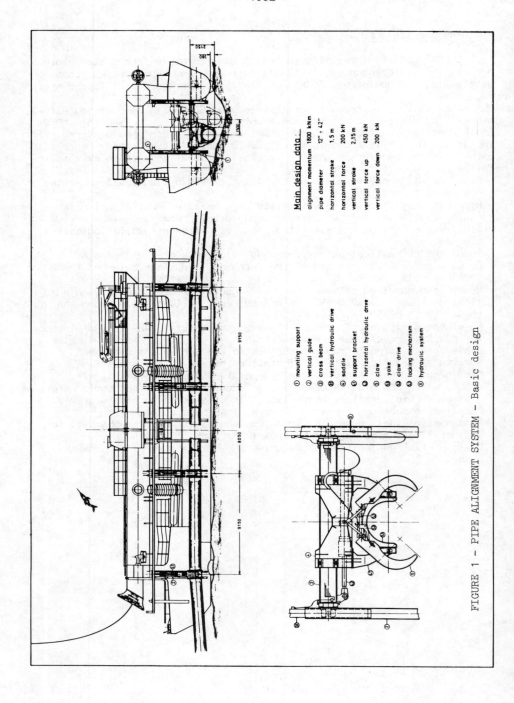

Main design data:

alignment momentum	1800 kNm
pipe diameter	12" ÷ 42"
horizontal stroke	1.5 m
horizontal force	200 kN
vertical stroke	2.15 m
vertical force up	450 kN
vertical force down	200 kN

1. mounting support
2. vertical guide
3. cross beam
4. vertical hydraulic drive
5. saddle
6. support bracket
7. horizontal hydraulic drive
8. claw
9. yoke
10. claw drive
11. locking mechanism
12. hydraulic system

FIGURE 1 – PIPE ALIGNMENT SYSTEM – Basic design

(09.19/79; 10.42/83)

REPAIRS IN DEEP WATER ("RMP")

J.A. TEYSSEDRE, Compagnie Française des Pétroles
P. THIBERGE, Société Nationale Elf Aquitaine (Production)
G. HERVE, Alsthom Atlantique – ACB – Offshore Division
M. BAYLOT, Comex S.A

Summary

Whereas deep water subsea connection and pipeline laying techniques have already undergone development work, there is no proven method in existence today for making repairs on a damaged section of pipeline in great depths of water.
Amongst the various stages in a repair operation, from removing the damaged section to replacing it with a new spool and connections, a gap exists as regards the preparation work required on the ends of the pipeline on the bottom.
Accordingly, this project consists in designing, building and testing in the workshop and in shallow water a prototype robot for preparation of the ends. This robot is of the modular type.
It is fully remote-controlled from the surface and does not require any divers. Handling takes place without guide-lines using a self-propelled vehicle.
The robot can carry out the following tasks:
- concrete stripping,
- removal of the anticorrosion lining,
- a clean-edge cut,
- cleaning of the inside of the tube,
- levelling of the longitudinal weld of the tube.
The equipment was tested in seawater at shallow depths in the port of St. Nazaire in July and August 1984. The sea trials are to take place in October 1984.

1 - INTRODUCTION

Use of subsea pipelines in great depths of water means that both the techniques for laying the pipelines and for maintaining and repairing them must be mastered.
The development of these laying techniques has gained a considerable lead over their repair, particularly through the following projects:
- J-configuration laying with electron beam welding (1980),
- laying and connecting test of flowlines between a wellhead and a manifold (1981).

The repair method developed for deep water lines consists in removing the damaged section and replacing it with a new spool as compared to certain local repair methods already applied in shallow waters based particularly on the use of collars bolted around the tube at the point of failure.

Amongst other techniques, the method adopted applies deep water subsea connection techniques that have already been covered by development work:
- connection by atmospheric welding (WELDAP),
- connection by mechanical coupling,
- extension of the connection limits by hyperbar welding (tests already performed in a depth of 300 metres).

On the other hand, before connecting the new section of pipeline, a number of preparatory tasks must be performed, in particular preparation of the ends of the pipeline on the sea bottom. This is the objective of the system developed within the framework of this project.

2 - PRINCIPLE OF COMPLETE REPAIR

Since this is a technique developed for deep waters - the target being 1000 metres - the repair principle was to do without divers owing to the inherent limits to saturation diving.

Likewise, the guide-lines commonly used at normal depths become a handicap at great depths. The packages of the system are therefore handled with a self-propelled transfer vehicle hung from the hoisting system of the operation support ship.

Accordingly, the system is modular, essentially in order to simplify handling operations, by reducing the weight and size of the packages, compared with a compact system laid on the sea bottom for the entire duration of the operation, which is necessarily more complex.

The complete repair would hence comprise the following phases:

1) Location of the failure, notably by routine inspections of the line, carried out by manned or remote-controlled submarine. These are now conventional operations commonly carried out in the North Sea and elsewhere.

2) De-trenching of the line, if necessary: the necessary de-trenching machines already exist. However, they will have to be adapted for an operation remote-controlled from the transfer vehicle, though this would not appear to set insuperable problems.

3) Cutting of the section by explosive collars installed by submarine into one or several pieces. Such collars have been developed and tested under this project.

4) Recovery of the pieces on the surface by means of the transfer vehicle equipped with special tongs.

5) Preparation of the ends of the pipeline on the sea bottom, this being the actual objective of the RMP project.

6) Measurement of the length of the replacement spool, for example by means of measurements by existing acoustic systems.

7) Insertion of the replacement spool, for instance by means of the transfer vehicle together with appropriate slinging. Positioning on the bottom by means of lifting devices operated by the transfer vehicle.

8) Connections between the replacement spool and the ends of the line, if necessary between several replacement spools, by the above-mentioned methods.

Phase 5 turns out to be the most delicate, since it requires stripping the concrete from the line together with its anticorrosion coating, making a clean cut of the end, cleaning the inside of the tube and if necessary levelling its longitudinal weld, without human intervention.

The work performed under the RMP project was accordingly concentrated essentially on this phase, and on cutting the damaged section with explosives.

3 - MAIN BASIC CHOICES

Making an explosive cut and then a fine cut by the end preparation robot is one of these basic choices. By removing the damaged section first, one can then work on the ends without being hampered by residual stresses.

Aside from the choices listed above: a system without divers, guide lines and of modular design, several other options were selected:
- Power generation: electric power on the surface and submerged hydraulic power plants for reasons of efficiency and umbilical diameter in order to reduce drag by the current,
- System architecture: a self-propelled transfer vehicle, or active beam, is the key component of the system, a veritable remote-controlled subsea device,
  a lifting device holding the end of the pipeline clear from the bottom,
  a work table supporting the robot on the end of the pipeline,
  tool modules, one for each specific task and one per range of pipeline diameters,
  lastly, a module support beam secured beneath the transfer vehicle and set down on the work table enabling the modules to move longitudinally.
- RMP system assisted by manned or remote-controlled submarine,
- Neither electrical nor hydraulic links between the packages remaining on the sea bottom and those raised to the surface for changing modules,
- Standby power at the bottom in hydraulic form, applied either by remote-control from the surface, or by "manual" valves operated by the assistance submarine,
- Location of the packages by acoustic system,
- Precise positioning by video system.

4 - GENERAL DESCRIPTION OF SYSTEM (see figure 1)

4.1 - Transfer vehicle or beam
This rectangular-shaped device consists of a tubular lattice structure equipped with 4 pushers (screws driven by hydraulic motors), the purpose of which is to fulfil the following functions:
- positioning of the packages (in X, Y and Heading),
- supply of the main hydraulic power on the bottom by means of 2 motor pump sets (MSP),
- hoisting function:
  . of the lifting device with the handling connector (HM), disconnectable on the bottom,
  . the combination of beam-module and table by means of 2 pins not disconnectable on the bottom.

This vehicle carries various containers at atmospheric pressure or maintained at equipressure by a bladder accommodating the bottom transformer, electrodistributor blocks, relays and telemetering systems and various connection boxes.

4.2 - Lifting device
This has been designed on the basis of existing lifting devices. Nonetheless, it comprises a number of specific characteristics, namely:

- it can travel parallel to the axis of the pipeline so as to pick it up at the desired place,
- it is equipped with specific connectors:
  . a handling connector (HC) for setting it down by means of the transfer vehicle,
  . a "hydromechanical" connector (HMC) capable of transmitting the power and orders either from the transfer vehicle or the assistance submarine. The HMC is patented. In principle, the lifting device carries the complete weight of the pipeline.

4.3 - Modules support beam
This beam is secured beneath the transfer vehicle and fulfils the following functions:
- guidance of the work modules: the beam carries two rails on which slides a carriage carrying the module,
- lifting of the work table, the beam then acting as lifting beam. It can be disconnected from the table on the bottom by means of cylinders,
- transmission of power to the table via an HMC, just like the interface between the transfer vehicle and the lifting device,
- holding the pipeline by means of hydraulic tongs.

The combination of transfer vehicle, beam and module is indissociable on the bottom (owing to the many electrical and hydraulic links) and hence has to be raised to the surface each time the module is changed.

4.4 - Work table
This consists of a rectangular frame resting on 4 telescopic feet. Its functions are:
- to act as shock absorber for the packages, thanks to 4 pneumatic wrenches,
- to carry the combination of transfer vehicle, beam and module,
- to maintain the pipeline by means of hydraulic tongs used in conjunction with those of the beam.

4.5 - Work modules
In all, six modules have been developed:

1) Concrete stripping module: the tools (milling cutters) capable of cutting the concrete and its steel reinforcement have been successfully tested on various samples of tubes and patented. The module, the architecture of which is similar to the following two modules, has been designed (production drawings), but not built, because it was considered as having less priority, since in principle deep water pipelines (1000 metres) would not have to be ballasted by concrete.

2) Module for removing the anticorrosion coating: this consists of a frame, an opening circular rail with a cage rotating in it, which itself opens. The cage carries three hydraulic motors driving three brushes. The latter have already undergone tests successfully and have been patented on various types of coating: bituminous pitch, epoxy and polyethylene.

The module advances in successive steps by means of the carriage of the beam. With each step, the coating is removed by brushing it away with the cage turning through one third of a revolution.

3) Cutting module: the architecture and principle are identical to that of the above module. It comprises in addition a pair of hydraulic tongs holding the cut end and two chocks maintaining the tube in the closed position. Instead of brushes, three milling cutters are mounted consisting of standard tools similar to those of milling cutters (example WACHS) used by divers.

4) Internal cleaning module: this is designed to remove the deposits inside the tube and consists of a horizontal arm comprising two brushes mounted on the ends identical to those of module N°2, together with a scraper to remove the debris. The tube is cleaned from the inner end towards the outer end.

5) Weld machining module: the purpose of this is to level off the longitudinal welding bead of the tube. This module has been designed (production drawings) but has not been built since it was considered as having lower priority. It forms part of the system for connecting by atmospheric welding (the WELDAP process), which requires a perfect seal where the pipeline penetrates into the working chamber. In addition, recent developments concerning elastomers have shown that the excess thickness of the longitudinal welding bead can be overcome.

6) Ovalization measuring module: this is designed to measure the ovalization of the tube so as to select the best place for repairing it. The sensor for making the measurement through the concrete has been successfully tested. The module was not built, since it was considered as having lower priority. The sensor has been patented.

5 - SCHEDULING OF THE WORK

The general studies during which the main basic choices were decided started in 1979. These were followed by the detail studies and lastly drafting of the production files, which were completed at the end of 1981. The control and monitoring system was covered by a separate study. Definition of the system was completed at the end of 1982.

Construction of the equipment started in mid-1982 and ended in October 1983 (see figure 2). This work included building the structure of each device and assembly of the hydraulic systems. The acceptance tests with and without load were performed separately on each device.

Workshop tests of the system were then resumed until June 1984.

They enabled all the devices to be assembled and run together on a testing pipeline about 100 metres long built specifically for this purpose.

These tests break down into the following three phases:
- tests in the control and monitoring system (CMS) until January 1984,
- integration on the CMS devices during February and March 1984,
- tests with the CMS, ending in June 1984.

The workshop tests were completed with final acceptance of the complete system, controlled from the control cabin with the modules working underwater in a tank (see figure 3).

The shallow depth tests, which are the ultimate phase of the present project, took place during July and August 1984 in a former shipbuilding dock.

6 - ASSOCIATED DEVELOPMENT WORK

6.1 - Coarse explosive cut

A prototype explosive ring designed for making a coarse cut through the section of damaged pipeline was successfully developed and tested ashore and at shallow depth. In addition, study of the damage to a tube cut by explosives and the influence on the subsequent welding conditions was carried out and showed the compatibility of the two operations, provided that the end of the pipeline removed after the fine cut is sufficiently long, which is possible with the RMP system.

6.2 - Obturators

These can be used in the case of connections made by atmospheric or hyperbar welding, and two prototype obturators inside the pipeline were successfully built and tested:
- obturator for a pressure of 15 bars in an 810 mm diameter tube,
- obturator for a pressure 150 bars in a 160 mm diameter tube.

7 - CONCLUSIONS

Development of the robot for preparation of pipeline ends has filled the gap that existed in the stages of a repair operation associated with connection methods that have already been developed. A coherent combination of proven techniques are hence now available for working on a deep-water pipeline that has been damaged accidentally.

In addition, all or part of the RMP robot may well soon find applications at more commonplace depths.

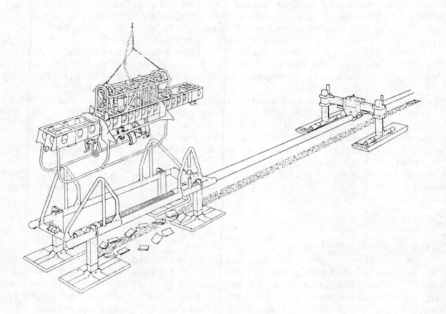

FIGURE 1

DEEP-SEA REPAIR

Active beam

Work table plus beam assembly

DEEP-SEA REPAIR

Background : fine cutting module
Foreground : external brushing module

Internal brushing module

DEEP SEA REPAIR
Shallow depth tests

Lowering the combination of lifting beam + main beam + cutting module
into the water

(10.21/79)

REPAIR OF SUBSEA PIPELINES BY MECHANICAL COUPLING

P.L.H. THIBERGE, Société Nationale ELF AQUITAINE (Production)

Summary

It was in 1979 that Société Nationale ELF AQUITAINE (Production) started a series of tests and preliminary trials to verify the possibility of repairing a pipeline in great depths of water, i.e. without being able to use divers. The repair principle is based on a metal-to-metal seal between a collet formed by stamping the end of the pipeline and a hydraulically actuated mechanical coupler. After preparing the pipe, de-trenching, concrete stripping and cutting the pipe, the collets are formed by a hydraulically remote-controlled gauge and their relative positions and dimensions measured. The replacement section constructed on the basis of the measurements made by the gauge essentially consists of a deformable S-piece fitted with mechanical couplers.

1 - INTRODUCTION

Large diameter petroleum pipelines have already been laid in depths of water of up to 600 metres, where divers can no longer intervene in the present state-of-the-art. This is why Société Nationale Elf Aquitaine (Production) has carried out research on possible repair techniques. To avoid the need for any human presence on the bottom, the research work was directed towards a mechanical type connection achieving a metal-to-metal seal. Preliminary studies and tests started in 1979 leading to adoption of a repair principle consisting in forming a collet at the end of the pipeline onto which a coupler can be secured, forming a seal. Since all this equipment can be remote-controlled and actuated electrohydraulically, in the end, it was this solution that was adopted.

2 - COLLET FORMING MODULE

Since forming by stamping gives better metallurgical characteristics, a hydraulically actuated die is used to form a collet about 100 millimetres long. A considerable number of tests have been carried out on pipes 10" in diameter and 1/2" thick and 20" in diameter and 1" thick in API 5L X.52 and X.65 grade steel. Despite the increase in hardness resulting from cold-working, the elongation and toughness remain highly satisfactory at the edge of the collet. Since 20" pipelines are the most commonly used, it is this large diameter that was adopted for designing the forming module (Figure 1), which is capable of developing a force in the range of 10 metric tons. The forming die is actuated by an annular jack. Although more complicated, this solution provides better guidance of the die and absorption of the forming loads by a device secured to the central shaft. This device consists of a striated flexible ring which slides on conical ramps mounted on the central shaft until it becomes secured to the internal wall of the pipe. Using appropriate profiles, tests performed with a hydraulic press have shown the excellent behaviour of this system up to a force of 1200 tons. On the other hand, the radial loads resulting from the flexible ring securing system have necessitated mounting a reinforcement system on the outside of the pipe based on the same principle, secured to a forming die with a gearing around its perimeter. This provides the anchor point for the mechanical coupler during the repair phase. To facilitate installation of the flexible rings inside the pipe, the forming module has an active self-centering device at the front end. The system is applied by means of a carriage (figure 2) in which the forming module is hung from a jack, the supply pressure of which is adjusted in order to balance the weight. The combination is electrohydraulically remote-controlled from the surface via an umbilical, with monitoring by pressure sensors, proximity detectors and TV cameras capable of pan and tilt motion.

3 - MECHANICAL COUPLERS

These must ensure a perfect metal-to-metal seal on the rough-formed collets, maintain the clamping effort and lend themselves to automated application. The solution adopted is direct actuation by hydraulic jack and tie-bars for maintaining the clamping tension. The coupler consists (figure 3) of a body fitted at the front with a gear ring which comes up against the back of the die flange by a claw action in the longitudinal direction. Hydraulic actuation of the coupler results in plastic deformation of the seal and exerts a pull on the tie-bars, which thus undergo elongation. After the lock nut has again been driven up against the rear flange by the motor drive, the hydraulic pressure is released ; the stretched tie-bars maintain the clamping action. The shape and nature of the seal ring, the nose angle of the coupler, the clamping force and the analysis of the metallurgical stresses after the clamping force is applied were examined by many tests so as to optimize the parameters and obtain the best possible results. Since this is a new operating principle, a 12" diameter prototype coupler was built. Following qualification tests enabling the technical choices made to be verified, this prototype was installed on the production pipeline of an onshore well in order to carry out endurance tests. It has now been in position for 2 years and no incident has occurred.

4 - REPAIR S-PIECE

By installing ball-joints and telescopic joints in the repair spool, any inaccuracy in measuring the distance between the two ends of the pipeline to be repaired is thus compensated. On the other end, these ball joints and telescopic joints necessitate further seals. To avoid this drawback, a deformable S-piece (figures 4 and 5) has been adopted, despite its cumbersome size. If the ends of the pipe to be repaired are fairly well aligned, the piece retains its "S" shape. It is equipped at each end with a mechanical coupler and mounted in a supporting frame positioning it with relation to the collets by means of two guide columns carried on end supports. By making use of the flexibility of the S-piece, a hydraulic jack enables its overall length to be decreased, enabling it to pass between the two collets formed on the ends of the pipe to be repaired. After positioning the S-piece on the ends supports, it is expanded by releasing the pressure in the contraction jack, bringing the nosepieces of the couplers fitted with their metal seal rings into contact with the collets. Application of the mechanical couplers described in paragraph 3 completes the repair sequence. Thanks to its flexibility, the S-piece can be built to measurements with very wide measurement accuracy tolerances. These measurements are made by means of a gauge.

5 - <u>METROLOGY BEAM</u>

The S-piece is built up from bended sections, requiring the appropriate information concerning the position of the collets with relation to one another in space, i.e. the gap and the angle between their centrelines. The 1980 review of methods and equipment for carrying out underwater metrology concluded that the accuracy of acoustic systems was inherently inadequate and that other methods such as laser and inertial navigation systems had not reached a sufficient level of industrialization to be usable.

The extremities of the pipe to be repaired are carried on end supports enabling the pipe to be lifted from the bottom. This configuration has been used to measure the relative positions of the collets by means of a gauging principle. The metrology beam (figure 6) therefore consists of an articulated installation remote-controlled from the surface that is re-entered on the guide columns with which the end supports are equipped and then takes a print of the collets. Accordingly, it is the guide columns that provide the common reference datum for the metrology beam and repair S-piece. The metrology beam consists essentially of a central section carrying the electrical and hydraulic containers, the length of which can be modified hydraulically, with two identical heads on either side that can be swivelled and adjusted in length. All the operations can be remote-controlled and actuated electrohydraulically from the surface.

After setting the gauge, the following work is carried out on the surface:
. positioning of the funnel sections on the frame carrying the S-piece, in accordance with the measurements made,
. positioning and welding of the couplers on the S-piece, after the latter has been built up from the necessary bended sections.

6 - <u>HANDLING</u>

Before the end of 1984, repair of subsea pipelines by mechanical coupling will be subjected to testing in shallow depths in order to demonstrate the technical feasibility of the concept studied by Société Nationale Elf Aquitaine. To carry out these tests, the handling methods have been simplified, though all the equipment is nonetheless remote-controlled from a control cabin on the surface and monitoring is ensured by a set of sensors and black and white TV cameras. At great depths and with a view to application without requiring divers, the equipment designed and built under contract 09.19/79 "Deep-sea repair" will be used to carry out the operations. In this context, the forming operation will be carried out following the pipe preparation work: de-trenching, concrete stripping, internal and external brushing and cutting. The plan is to apply the forming module complete with all the equipment of contract 09.19/79, namely the work table on the bottom straddling the pipe and lowering the forming module together with the active beam and main beam. The transfers between the surface and the bottom and the bottom to the surface are carried out by handling with a single wire-line, with the active beam directing the package as it approaches the bottom so that the forming module can be correctly inserted into the table.

After forming the collets, the following operations are required, again with the active beam carrying out the handling:
- installation of the end supports equipped with guide columns,
- lowering and setting of the gauge with the metrology beam,
- construction of the repair spool on the surface, fitted with the two mechanical couplers,
- lowering and connection of the repair S-piece,
- testing the tightness of the repair.

7 - PROGRESS OF THE WORK

All the engineering studies of the equipment needed to carry out the shallow depth tests have now been completed and all the equipment constructed and subjected to acceptance tests without load, as of 1st July 1984. The first functional tests are now taking place and the shallow depth tests are to start on 15th October 1984.

8 - MEANS EMPLOYED

A number of companies have taken part in this project, among which the following should be mentioned in particular:
- ALSTHOM-ATLANTIQUE (Offshore Division), who built the forming module, prototype couplers and piloting and control system,
- COMEX-INDUSTRIES, who built all the equipment needed for the underwater tests (baseplate, support gantry, metroloy beam and repair S-piece).

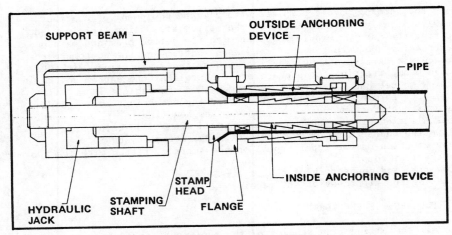

FIGURE 1 - DIAGRAM OF FORMING MODULE

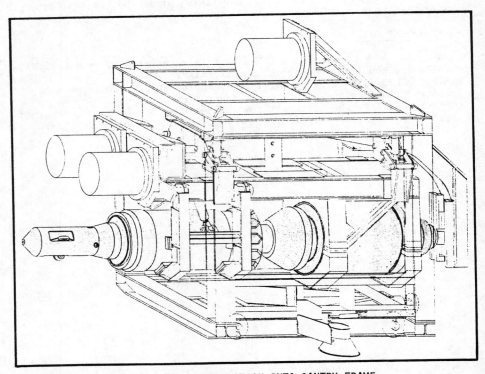

FIGURE 2 - FORMING MODULE - INTEGRATION INTO GANTRY FRAME

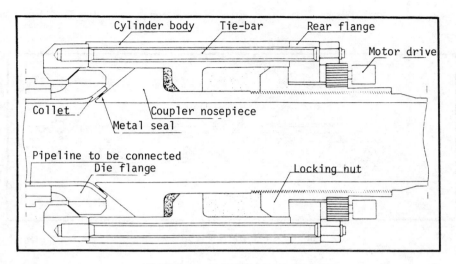

FIGURE 3 - MECHANICAL COUPLER

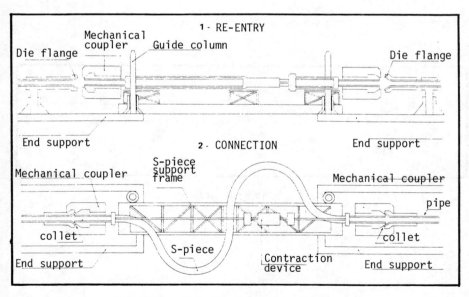

FIGURE 4 - DIAGRAM OF REPAIR S-PIECE

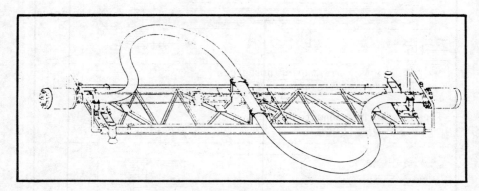

FIGURE 5 - REPAIR S-PIECE

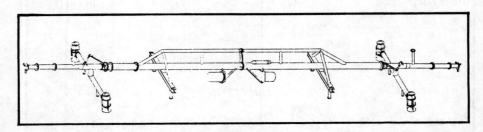

FIGURE 6 - METROLOGY BEAM

(10.20/79)

DEEPWATER PIPELINE REPAIR SYSTEM
(Progress report of SNAM's SAS project)

G. BONVECCHIATO
SNAM S.p.A. - Milan

Summary

The Transmed gas pipeline which joins Africa to Europe crossing the Mediterranean Sea in the Sicily Channel and in the Messina Strait includes two sections laid at a depth never achieved before by pipelines: 608 and 360 metres respectively. New technologies have been adopted for laying the pipeline and now others are being developed for maintenance and possible repairs.

This report describes a repair system which for its cost and availability of means is competitive with the alternative solution of laying a new section to reach shallow water in order to replace the damaged section.

This new system has a modular concept, i.e. it is made up of several modules, each having specific duties, the operative sequence of which is controlled by a surface computer. The heart of the system is a new type of joint that has already been effectively used. The system is now in the final construction stage and will be tested at sea in operating conditions, equivalent to real conditions, during 1985.

1). Introduction

Even before constructing the Transmed gas pipeline which connects Africa to Europe across the Mediterranean Sea, it was SNAM concern to study the possible methods to repair a pipeline laid in deep waters i.e. beyond the limits of human intervention and hyperbaric welding.

The only possible solution was cutting the pipeline before the damaged section, lifting it to the surface utilizing the lay vessel used for its construction, laying a new section to a point where such depths are reached that allow the use of the part of the pipeline remaining intact by means of either a surface connection (slack loop technique) or of hyperbaric welding, on the sea bottom. Estimates of the necessary time to mobilise a vessel with the necessary characteristics and the cost forecast for the above described operation where such that SNAM felt it necessary to initiate a research and development programme to realize a system of immediate availability and lower cost.

Considering the interest that this research work can present in the off-shore activities field, financial support was requested from the European Economic Community, which granted it in 1979. Other companies of the ENI group began to cooperate in this research programme, such as NUOVO PIGNONE, SAIPEM and SNAMPROGETTI.

The results achieved, i.e. the repair system in general, are applicable to the off-shore petroleum industry which continues to push towards increasingly deeper waters. Furthermore, the joint can be conveniently used also in shallow waters because of its safety and easy application.

2). Development of a new joint

The first step towards the realization of a new repair system was to obtain a joint between two pipeline sections:

- to be installed without human intervention;
- to compensate misalignment of the ends to be re-connected;
- to be adjustable lengthwise;
- to withstand axial loads and moments exerted on the pipeline;
- to adapt itself to manufacturing tolerances of the pipe, welded longitudinally;
- to allow, after installation, testing of the seal which must preferably be of the metal to metal type;
- to allow passage of electronic pigs;
- to be of a material compatible with that of the pipeline and have a technical working life equal to the duration of the pipeline itself.

By comparison with existing joints (connectors) it was found that:

- the connector-pipe joint tolerance were too tight;
- preparation of the pipe end, including flattening of the longitudinal welding seam, was too complex to be performed at a distance yet achieving the required characteristics;
- handling, centring and clamping of the connector and the spool piece required operations and degrees of freedom that were too complex for remote controlled operation.

For these reasons a research programme was started with NUOVO PIGNONE to develop a joint which could satisfy all of the conditions described above. Various concepts were developed and in the end the one that corresponded best was based on a new blocking principle, obtained by plastically deforming the material of the pipe and of the connector body under high pressure.

This principle, although conceptually very simple, presented many difficulties for its practical realization, due to the high pressures involved and the stresses caused by them on the metallic parts.

However, several tests, firstly on scaled down models and then on full size pieces, successively subjected to test cycles, have removed all doubts.

The joint, as a complex, consists of:

- two sleeves, to be applied on the outside of the pipeline to be repaired, each with a ball coupling to compensate misalignments (see fig. 1);
- a central telescopic body, to allow insertion between the positioned ball connectors, complete at each end with the female part of the spherical joint and equipped with a clamping systems (the detail of the sliding slip joint part is shown in fig. 2).

The assembly procedure is as follows:
- the first coupling is slipped over one end of the pipe;
- a hydraulic press is then inserted inside the pipe, that expands a hollow hard rubber cylinder against the pipe wall to deform it above the yield point and making it adhere to the inside grooved surface of the coupling.
- The press is then retracted and the other end is prepared in the same manner, the telescopic body is positioned between them, and, after being clamped onto the ball ends, it is fixed into position.

3). The repair system

The final result of the research programme is a system that has the following characteristics:

- easy mobilization, since it does not require a special surface means dedicated to it;
- human presence on the sea bottom is not required;
- versatile, since it is possible to carry out numerous operations.

The system components are illustrated in figure 3, all to the same scale, except for the support vessel.

The various components and their construction status at the end of July are:

- a module which supplies power to the others, controls them and positions them accurately on the sea bed (thruster module). Its construction has been completed and the control software is under development. Delivery is foreseen by October.
- a dredge module, in the event that the damaged section is covered by deposits. Detailed engineering is under way and delivery is foreseen by January 1985;

- two support frames with four feet to raise the pipeline off the sea bed;
- a module to prepare the pipe surface and cut the damaged section. Construction of the mechanical parts has been completed;
- a module to recover the cut section;
- an installation module for the joint system, consisting of three sub-assemblies:
  - a device to measure the distance between the two pipe ends to be connected;
  - a base structure;
  - a trolley to install the end couplings and the telescopic joint.

There is also a crane to handle all said modules, with system to compensate the oscillations due to the action of the waves. The entire system is foreseen to be available within this year.

4). Repair procedure

The sequence of the remplacement operations of a damaged section, after indentification of the point to be repaired can be summerized as follows:

The dynamic positioning type vessel, by means of a remote operated vehicle (ROV), positions a grid of transponders on the sea bed at the two sides of the pipeline to permit the thruster module to identify its position at any moment.

If necessary, the dredge module uncovers the pipeline that can then be raised off the sea bed by the two lifting frames. The preparation module is positioned astride the pipe to remove the concrete and the polyethylene and make the first cut of the pipeline. It is then positioned on the other side of the damaged part, repeats the preparation operations and the second cut, therefore allowing removal of the damaged section that may possibly be recovered for study.

The base frame, positioned over the pipe, blocks its ends allowing the trolley to install the ball type connector with the use of the hydraulic press. Successively the central telescopic body of the joint is positioned, clamped and blocked at the necessary lenght.

5). Programmes to be carried out

A full series of tests is foreseen at various stages, starting from those in the work-shop, passing to those in shallow waters and finally to the installation of a connector on a section of pipeline constructed with the same materials as the Transmed and laid for trials in sea-beds similar to those of the Sicily Channel. The last test is foreseen for 1985.

From what has been realized to-day and from the experiences that will be made during the tests a considerable increase in know-how has derived and will derive on the methods of performing interventions of various types in deep waters, on remote operated controls and operations, know-how that will have considerable repercussions on the operating techniques of the off-shore gas and oil industry.

Bibliography

- "Deep water pipeline operations and maintenance problems and their impacts on the design".
(A. Pedrazzini, G. Scurati - SNAM S.p.A.)
D.O.T. - Palma de Mallorca - SPAIN (Oct. '81)

- "Development of a deepwater pipeline repair system"
(G.P. Bonfiglioli - SNAM S.p.A.)
Italian - Norwegian Seminar - Milan - ITALY (Oct. '82)

- "Status of SNAM Deepwater automatic pipeline repair system"
(G. Bonvecchiato - SNAM S.p.A, A. CONTER - SNAMPROGETTI, F.P. - SAIPEM, G. Ferrari Aggradi - NUOVO PIGNONE)
D.O.T. - Malta (Oct. '83)

- "R.O.V.'s within multifunction deepwater spreads: concepts and operational results".
(G. Melegari, M. Brambilla - SAIPEM S.p.A.)
R.O.V. '84 - S. DIEGO - USA (May '84)

- "SEALINES: Maintenance and repair problems in deep water (experienced on the Transmed)
(G. Bonvecchiato, E. Migliavacca, A Colombo - SNAM S.p.A)
Seminar on off-shore natural gas technology, - UNITED NATIONS - Dubrovnik - YUGOSLAVIA (Oct. '84)

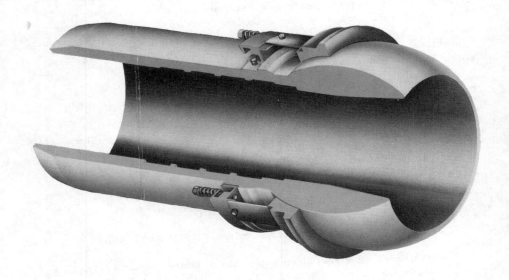

FIGURE 1

FIGURE 2

Snam

DEEPWATER REPAIR SYSTEM MODULES

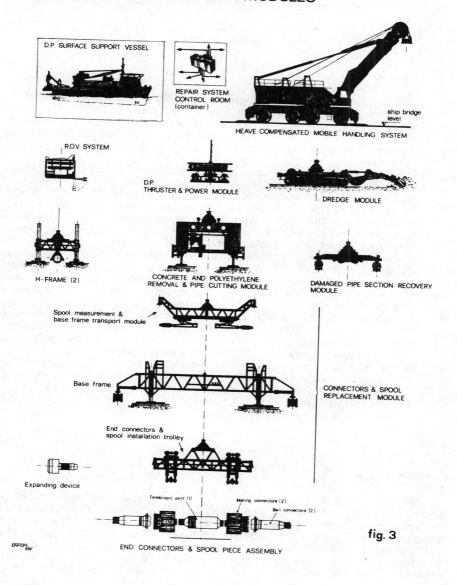

D.P. SURFACE SUPPORT VESSEL

REPAIR SYSTEM CONTROL ROOM (container)

ship bridge level

HEAVE COMPENSATED MOBILE HANDLING SYSTEM

R.O.V. SYSTEM

D.P. THRUSTER & POWER MODULE

DREDGE MODULE

H - FRAME (2)

CONCRETE AND POLYETHYLENE REMOVAL & PIPE CUTTING MODULE

DAMAGED PIPE SECTION RECOVERY MODULE

Spool measurement & base frame transport module

Base frame

CONNECTORS & SPOOL REPLACEMENT MODULE

End connectors & spool installation trolley

Expanding device

Telescopic joint (1)   Mating connectors (2)

Ball connectors (2)

END CONNECTORS & SPOOL PIECE ASSEMBLY

fig. 3

(15.33/82)

ECHOMECHANICAL DIAGNOSTIC METHOD FOR OFFSHORE STRUCTURES

L. DOSSI
Tecnomare SpA

Summary

The periodic controls required to check structural damage in offshore platforms entail high costs and considerable tasks which weigh heavily on the oil supply economy. The aim of the research project on the "echomechanical diagnostic" method is to achieve greater precision and reliability, with lower costs. The project is innovative.
The operating principle of the echomechanical method is to impress variable frequency elastic waves on the structure and to measure the frequency and energy of the elastic waves reflected (echoes) by structural discontinuities, standardising both according to the frequency and the energy of the impressed waves. The preliminary phases of the project include study activities, laboratory tests, construction of a portable prototype of a diagnostic system, sea tests and real size structures. At present (July 1984) the following activities have been completed: study of the elastic wave propagation, study of the analysis method, design and procurement of the laboratory test system and software for the forced response simulation and for the calculation of wave velocities. The signal analysis study made it possible to establish the system properties and to computer simulate its operation. The block diagram of the laboratory instrumentation system for the reflected waves signal conversion with synchronous sampling, and for the FFT or Maximum Entropy analysis has been described. The result of a test carried out during the system setup, which gives a very preliminary indication of the system sensitivity, has been reported. This result seems promising in order to continue the project and arrive at a successful conclusion.

1. INTRODUCTION

No less than 20 % of the world's oil supply is derived from off-shore resources. These are exploited mainly by steel platforms which are subject to damage caused by sea waves and collision. The passibe safety rules and the periodic checks of damage absence entail high costs during platform operations. The integrity check is carried out, mainly, by divers who clean the area to be checked and carry out the visual inspection and the magnetic particle and ultrasonic inspection. The time involved and the costs of this operation greatly limit the possibility of a complete structure check, with the consequent risk of leaving undiscovered damage which may compromise the safety of the structure and of the men working on it. Much research is being carried out on instrumental methods capable of detecting damage without the need of an extensive use of divers. Several methods have been proposed; e.g. acoustic emission, flection analysis, strain control. Each instrumental method has its own merits. However, none of them has yet reached a confidence level sufficient to allow a substantial reduction of the usual diver inspections. Take, for example, the dynamic analysis method. This method is one of the most promising for the detection and measurement of the damage in each tubular element and for the complete structure analysis. However, according to the experimental data presently known, the method sensitivity is limited to cracks of not less than 25 % of the circumference or, in other cases, not less than 50 %. Moreover, this method requires the use of an extremely weighty vibrator, in the order of 1 metric ton.

The aim of the echomechanical method is to use, in a different way, the dynamic properties of the structural elements to detect cracks of almost 1/8 th of the circumference, thus achieving twice the sensitivity of the dynamic method. Another aim is the possibility of using less weighty vibrators which are easier to handle. The research also aims at a system capable of locating and measuring the damage without taking as reference previous experimental data of the undamaged state. We believe that achievement of the first two objectives, and if possible the third, would lead to a better control of the offshore steel platform state of safety, with lower costs. This would contribute to the economy of the oil supply. Presently the theoretical study of the method has been completed and the laboratory tests are about to start. The principles of the new method, the results of the theoretical activities performed, an experimental result obtained in the course of the experimental controls carried out to set-up the laboratory system are described here below. During the course of the research activities many laboratory tests are foreseen and field tests on real scale structural elements in the sea. The research has been financed for the period from 1.12.82 to 30.11.85 (project number TH 15.33/82) and it is divided into two main phases of 18 months each. Other sponsors of the project are the Istituto Mobiliare Italiano and the AGIP, SAIPEM and SNAMPROGETTI companies.

2. MAIN CHARACTERISTICS OF THE PROJECT

The main characteristics of the project are :
a) The innovation in the field of structure diagnostics, with respect both to the working principles and the technologies employed by the new method.
b) Great scientific and technological contents, because the method development phases entail the full acquisition and some progress of the state of the art in the following fields : signal analysis, software for structural modelling, instrumentation to generate and analyse structural vibrations, devices for underwater structural diagnostics.
c) The cooperation, during the course of the project, with companies and University Institutes fully qualified in their respective fields.
d) The completeness of the project which is carried out from the method conception to its practical application.

3. DESCRIPTION OF THE RESEARCH AND OF ITS MAIN RESULTS

3.1 Description of the Principle

The method principle is, basically, to impress elastic waves with variable frequency on the structure and to measure the frequency and the energy of the waves reflected (echoes) by the mechanical discontinuities. In greater detail, a vibrator impresses a sinusoidal oscillation on a point of a structural element. The oscillation frequency varies with the time, as indicated in fig. 1.

The elastic wave propagates along the structural element, and when it encounters a discontinuity, e.g. a crack, it divides its energy into a transmitted wave and a reflected wave. An accelerometer senses both the impressed elastic wave and the reflected one.

In every time instant the reflected wave has a frequency different from the impressed wave frequency, as shown in fig. 1, because of the different path travelled by the impressed and the reflected wave from the emission time instant. As shown in fig. 2, in general there are many reflected waves and the energy of each wave is a function of the number of the reflections encountered and of the corresponding reflection factors.

By means of discrimination of the different waves according to the different instant frequencies they have, it is possible, in principle, to locate each reflecting discontinuity and to measure the corresponding reflection coefficient which is a function of the discontinuity magnitude. This operation principle resembles, somewhat, other system operations, e.g. the operation of continuous transmission, frequency modulated sonar. However, contrary to the sonar case, the elastic wave velocity in the steel structure is an order of magnitude greater; moreover these elastic waves can be longitudinal, flexural, torsional and each of them has a different velocity range.

These and other differences require a different approach to the analysis.

3.2 The Method Development

The method development requires the theoretical and experimental study of many problems, concerning mainly :
- the elastic wave propagation in the structural elements
- the signal analysis method
- the design and procurement of the laboratory instrument system and software
- the laboratory testing
- the design and procurement of the field system and software
- the field testing.

The study of the **wave propagation** in the structure requires the realization of a mathematical model to allow a computer simulation. The study of the model has been given to the "Istituto di Meccanica Applicata" of the University of Padua.

The study resulted in three computer programmes. One programme simulates the structure as a series of resonnant elements and it gives the forced time response in the axial-symmetric propagation case, with or without damages. A second programme gives the elastic wave velocities in the tubular structural elements of different diameter and thickness, according to the tridimensional elastic wave theory of

Prochhammer and Cree. A third programme, of finite elements type, gives the complete time response of the damaged or undamaged structural elements to different types of elastic vibration and, in particular, to the frequency variable vibration adopted in this method.

The **signal analysis study,** carried out at the same time as the design and procurement of the instrument system, has been made in cooperation with the "Istituto di Elettronica" of the University of Padua. The study carried out by the "Istituto di Electtronica" gave the definition of the analytical expression and properties of the signals, detected by the system accelerometers and frequency converted according to the procedure which will be explained in the following. By means of the analytic expression it has been possible to computer simulate the operation of the system which generates the structural vibration and carries out the signal analysis. The simulation allowed the verification of the system properties.

The **instrument system design** resulted in the laboratory hardware setup according to the functional block diagram of fig. 3, briefly described in the following. The generator outputs a sinusoidal signal with a frequency logarithmically variable in a few seconds between the upper frequency F_{max} and the lower frequency F_{min}. The power amplifier drives a vibrator which impresses the longitudinal, flexural or torsional vibration on the structural element model. The accelerometer senses the impressed elastic waves and the elastic waves reflected by the structural discontinuities. The amplified signals are sent to a sampler. The sampling is synchronous with the generator frequency. It can be demonstrated that, because of the logarithmic frequency variation, the ratio between the reflected elastic wave frequency and the impressed elastic wave frequency is constant during the sampling period. In fact, if in the frequency/time domain the imparted wave frequency can be expressed by the following law

$$f(t) = F_O \, a^{-t}$$

the frequency of the wave reflected by a discontinuity is :

$$f_r(t) = F_O \, a^{-t(t-dt)}$$

where dt is the additive time interval due to the wave propagation. The ratio between $f_r(t)$ and $f(t)$ is a ^{dt}, constant in the interval T between the reflected wave onset and the impressed vibration end. The result of the synchronous sampling for each reflected wave is a sinusoidal component having a constant frequency in the interval T. The elastic energy distribution between the reflected waves is converted, by the spectrum analyser (FFT module) into an energy/frequency distribution. A reflecting discontinuity appears as a spectral peak at a frequency which is a function of the discontinuity location relative ot the vibration and the accelerometer location. The peak magnitude depends on the reflection multiplicity and on the discontinuity reflection factor.

The FFT analysis requires many samples, generally 1024 or 2048 samples. It can be demonstrated that the frequency resolution obtainable by the FFT analysis and, correspondingly, the spatial resolution of the discontinuity depends on the frequency difference between the F_{max} and F_{min} of the sweeping not on the sweeping time. The order of magnitude of the obtainable resolution is 0.3 meters when the sweeping band is 10 kHz. The specially made software allows the computer control of the whole system, the carrying out of a post-analysis (cepstrum, etc.), memorization and display of the results.

The software for a Maximum Entropy analysis has also been made.

At present (July 1984) the laboratory tests are about to start on steel models of tubular structural elements. The test equipment is almost completed. Some preliminary tests have been carried out to verify the system operation, using provisional equipment and a provisional model. The structural provisional model is a brass pipe, of 80 mm diameter, 1 mm thickness, 5.15 m length, with a circumferential welding cut at 1.5 m from the same end. The pipe was suspended at both ends.

Fig. 4 shows the result of a test carried out with a frequency sweeping from 20 kHz to 10 kHz and with a torsional vibration by a vibrator located at the opposite end from the welding. The signal was sensed with an accelerometer close to the vibrator. Fig. 4 graphic is the autospectrum of the frequency converted signal, in the zero to 100 Hz band. The first peak (double peak) close to the axis origin corresponds to the impressed wave. The main following peaks correspond to the reflection on the pipe ends. Other peaks, marked S, corresponding to the 1/8 th cut of the circumferences can be noted. The correspondence of the T peaks with the cut has been verified by inserting a wedge into the cut and by noting that this insertion gives amplitude and shape changes in the T peaks and, to a lesser degree, in the following peaks. The shape of the first peak (double) close to the origin and the oscillation in its descent correspond to the impulse response of the provisional coupling between the vibrator and the pipe.

This test gave a first indication of the possibility of detecting cracks of 1/8 th of circumference by the new method. In this case of damage of the structural nodes it is expected to find a significant difference between the damage node response and the undamaged one. The laboratory tests will also be carried out on pipes clamped at both ends and on elementary structures, with and without damage. Following those tests, it will be possible to design and procure the **final instrumentation and software** for a portable system to be used for **field tests.**

4. CONCLUSIONS

As the research project is still being developed, only partial conclusions are possible. The activities carried out so far led to the acquisition of the basic theoretical knowledge and the procurement of the laboratory instrumental system. The simulation programmes allowed the simulation of both the structural element behaviour, damaged or undamaged, and the system operation. This lays the premises for an effective comparison between the method theory and the experimental results. After some preliminary laboratory tests it was possible to verify experimentally the system operation and to obtain a first indication, as reported, of the possibility to locate the damage and to supply information concerning the sensitivity obtainable.

The partial results so far achieved are deemed positive and are a valid basis for the continuation and conclusion of the project.

5. REFERENCES

The main references are cited below. Other references can be found in
the literature referred to in the text.

1.   R.C. HEYSER, Acoustical Measurement by Time Delay Spectrometry,
     J. Audio Eng. Soc. 15, 370 (1967)
2.   University of Southampton, Institute of Sound and Vibration Re-
     search, Applied Digital Signal Processing, Course Notes, Dec. 1982
3.   P.J. HOLMES, The Experimental Characterization of Wave Propagation
     Systems : II°, Continuous Systems and the Effects of Dispersion,
     J. of Sound and Vibr., 35 (2), 1974
4.   S. LERY, J.P.D. WILKINSON, The Component Element Method in Dynam-
     ics, McGraw-Hill

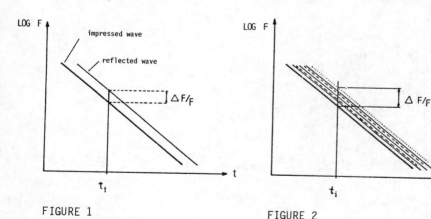

Representation of the behaviour of a
single reflection in frequency and
time (ideal case).

In the presence of numerous reflections :

FIGURE 1

FIGURE 2

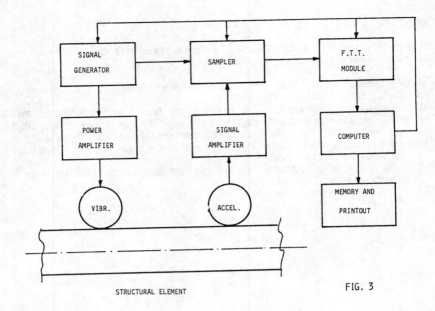

STRUCTURAL ELEMENT

FIG. 3

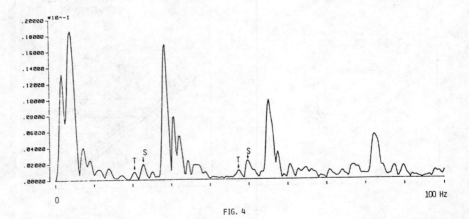

FIG. 4

(15.06/78)

DIAGNOSTIC METHODS FOR OFFSHORE STRUCTURES

V. BANZOLI
Tecnomare S.p.A.
Società per lo Sviluppo delle Tecnologie Marine

Summary

Throughout a platform's working life, to avoid risk to personnel and plant and to permit the planning and execution of the necessary maintenance and repair work, the operator needs to be kept informed about the state of repair of the platform.
The main aims of the research project "Diagnostic methods for offshore structures', carried out by Tecnomare between November 1979 and May 1983, were as follows :
- to detect damage in offshore structures by means of a portable instrumentation system and regular or random checks;
- to design a permanent instrumentation system for continuous monitoring;
- to design structures capable of withstanding damage and to integrate the design, monitoring and inspection phases.
The main results of the project were as follows :
- a portable instrumentation system was developed and tested, designed to carry out regular or random checks by monitoring the vibrations produced by artificial excitation of the structure.
The procedures and computer programs for optimizing the measurement programme and analysing the test results, so as to detect any damage, were drawn up.
- A permanent instrumentation system for continuous monitoring of offshore structures was developed.
The computer program for the calculation of cumulative fatigue and simulation of the propagation of fractures, based on the extensometer readings, was designed and tested on an actual platform.
- Different design methods, based on the "damage tolerance" approach, were examined to determine their applicability to offshore structure design.
The theory of fracture mechanics was analysed in depth to assess how far it could be applied to these methodologies. The criteria for an integrated approach to design/monitoring and inspection were defined.

1. INTRODUCTION

At present, the only way to determine the performance in real terms of an offshore structure, and hence its state of repair, is to carry out regular inspections. In view of the cost and risk involved in subsea inspections, there is growing interest in diagnostic methods for off-

shore structures, using instruments and data analysis methods to :
- reduce the need for and cost of subsea inspections;
- permit rapid assessment of the structure's state of repair;
- detect and pinpoint damage more reliably and at lower cost than with conventional inspection methods.

The research project "Diagnostic methods for offshore structures" approached the problem from two separate but complementary angles.
- design and construction of a portable instrumentation system for random or regular checks, and establishment of the procedures for analysing results;
- design of a permanent instrumentation system for continuous monitoring of offshore structures; development of the computer programmes for the calculation of cumulative fatigue and assessment of the speed of propagation of damage.

Since it is considerably easier to ascertain precisely a structure's state of repair if monitoring and inspection requirements are taken into account at the design stage, and since the "damage tolerance" approach to the design is a useful way of doing so, the research involved :
- examination of various design methods based on the "damage tolerance" approach to establish their applicability to the design of offshore structures;
- analysis of the theory of fracture mechanics to assess its applicability to these methodologies;
- definition of the criteria for an integrated approach to design, monitoring and inspection.

The aim of this paper is to outline the main elements and results of the research project carried out between November 1979 and May 1983.

2. MAIN FEATURES OF THE PROJECT

The research was divided into four phases, described briefly below.
- Study of the most suitable methods for monitoring the state of repair of offshore structures.
After analysis and classification of typical damage occuring on offshore structures, and examination of the alternative diagnostic methods, monitoring of the vibrations resulting from artificial excitation of the structure was selected as the most suitable method. After extensive laboratory tests on a reticular steel structure, the procedures for optimizing the measurement programme and for analysis of the results of the rests were developed.
- Design and construction of a portable instrumentation system for regular or random checks on offshore structures.
On the basis of the results of the laboratory tests carried out in the first phase, a portable system was designed and constructed to meet specific requirements such as portability, the possibility of being used on existing structures, ability to detect relatively minor damage.
- Study of a permanent design system for continuous monitoring of structures.
As well as the definition of the permanent monitoring system, computer programmes were written for the calculation of cumulative fatigue and simulation of the propagation of cracks.

- Analysis of the application of fracture mechanics to diagnostic methods for offshore structures and definition of new design methods, based on "damage tolerance" approaches, integrated with monitoring and inspection.

3. "VIBRACHECK" PORTABLE INSTRUMENTATION SYSTEM, DATA ANALYSIS PROCEDURES

3.1 General

The problem of structural diagnostics by means of regular or random checks was investigated with a view to developing a monitoring method and instrumentation system which would meet the following basic requirements :
- able to be used on existing structures
- easily transportable - low operating costs
- able to give an evaluation of the state of repair of the structure as a whole
- able to detect relatively minor damage.

The method adopted is based on subsea monitoring of vibrations induced by forced local excitation. The choice was based on theoretical and experimental analysis.

A systematic analysis of the various methods of monitoring vibrations (local and global behaviour of the structure in conditions of forced and natural excitation) and extensive laboratory tests on a scale-model jacket showed that, by monitoring local vibration modes, damage could be identified with a detection threshold equal to a crack of approximately a quarter of the circumference of the tubular element. When the global vibration modes are monitored, this detection threshold rises to a crack of half the circumference.

The instrumentation system essentially comprises three triaxial subsea accelerometers and the electronics for acquisition of the data, a subsea electro-hydraulic exciter, a surface container for the control and data recording systems.

Procedures were developed for :
- detecting damage on the basis of the response of the structure to the forced excitation;
- minimizing the number of excitation and recording points required for a full description of the behaviour of the structure.

The measurement programme comprises the following phases :
- selection of the best excitation and measurement points on the basis of a computer analysis of the structure, carried out using modal analysis methods;
- execution of the measurement programme, including attachment of the exciter and accelerometers at the selected points. The operation is repeated, with the exciter and accelerometers being repositioned, until the measurement programme is complete (see Fig. 1).

To monitor a platform of medium size, standing in approximately 100 metres of water, six sets of measurements, each with the accelerometers and exciter in different positions, are felt to be more than sufficient.

3.2 Description of the instrument action system

The system is subdivided into the following components :
a) Instrument subsystem comprising three triaxial accelerometers, a subsea module, a surface control unit, the communications cable.
b) Excitation subsystem, comprising the exciter, the hydraulic system,
   the power and signal cables, the control unit and the signal generator
c) Surface module comprising a container (3 x 2 x 2) for the power cable
   handling gear and the power distribution panel.

The triaxial accelerometers are connected to the steel structure by
permanent magnets. Each accelerometer is enclosed in a cylindrical
container and connected to the subsea module by an electric cable. The
subsea module contains a microcomputer controlling the strength of the
signals, the frequency of the filters and the frequency of sampling.

It also controls the acquisition, coding and transmission of data
to the surface unit.

The data are transmitted to the surface by means of a communications cable. The surface unit contains two microcomputers which
monitor the data and record them on tape.

This unit makes it easy for the operator to change the operating
parameters and ascertain the state of the system. A spectrum analyser
enables the operator to check the quality of the readings and gives him
some indication of the behaviour of the structure.

The electro-hydraulic exciter is designed to generate an excitation
force of approximately 20 kN and weighs virtually nothing in water. The
exciter is clamped to the structure at the preselected points and the
force is measured together with the acceleration.

The hydraulic gearbox is designed to operate underwater :this
obviates the need for a hydraulic umbilical, which, in view of the type
of excitation required (irregular frequency varying from 1 to 20 Hz)
would have meant using wide-diameter hoses.

The exciter is designed for use both in and out of water and can,
with an additional external mass, provide sinusoidal excitation of the
whole of the platform.

During transportation, all the elements of the system are encapsu-
lated in the container which serves as an operations centre during the
measurement programme. The subsea equipment is installed inside a metal
structure to facilitate the operation.

The procedure for the measurement programme is as follows :
a) the surface equipment is installed
b) the subsea module is lowered into the water to the pre selected
   depth.
   The support structure for the module contains the accelerometers, the
   exciter, the hydraulic plant, the electric cables and the hoses
   required for the operation;
c) the diver releases the exciter and clamps it to the platform. The
   operator commences the excitation of the structure;
d) the diver connects the accelerometers to the platform at the preselected points, after cleaning the surfaces for the attachment of the
   magnets;
e) the data are collected and recorded;
f) the operator checks the quality of the data, using a spectrum analyser to make sure that they are consistent, and takes any necessary
   corrective action;

g) if the results of the check are favorable, the accelerometers and exciter are replaced in the support structure and the operation is repeated at points a) or b) until completion of the measurement programme.

3.3 Description of the diagnostic methods

The method of optimization, aimed at minimizing the number of excitation and measurement points needed for a full description of the behaviour of the structure, and the method for detecting damage, based on the analysis of the response of the structure to the excitation, are as follows (see Fig. 2) :

Optimization
- Assessment of a sufficient number of modes of vibration. These vary according to the structural element being monitored.
- Selection of the best excitation points, i.e. those of the selected excitation points where the highest levels of acceleration can be achieved for the different modes of vibration to be monitored.
- Choice of the best measurement points, i.e. those where the levels of acceleration are highest.

Detection
- Comparison of the theoretical modal frequencies with the measured frequencies to establish which modes of vibration have changed.
- Damage probability evaluation for each structural element, using the components of the eigen vectors of the moments of the modes of vibration which have changed, on the basis of the measured variations in the modal frequencies.
- Evaluation of damage which could explain the variations in the modal frequencies, by assessment of the influence on the modal frequencies of the significant parameters of the structural elements selected (moment of inertia, mass etc.).

4. PERMANENT MONITORING SYSTEMS, CALCULATION AND DATA ANALYSIS PROCEDURES

4.1 General

A reliable permanent monitoring system must be based on a combination of methods, such as :
- monitoring vibrations.
  There are two possible methods : the first is based on analysis of the form of the first two or three global modes of vibration with natural excitation; the second (which involves the permanent version of the portable system described above) is based on analysis of the local vibration modes produced by forced excitation;
- monitoring the foundations;
- acoustic emission;
- survey of losses from and seepage into the structural elements;
- monitoring the tensions.
  This system is based on analysis of the tensions at various points of the structure, using extensometers, with subsequent calculation of the cumulative fatigue and simulation of the velocity of propagation of cracks.

4.2 Procedures for the calculation of cumulative fatigue and the simulation of propagation of cracks

A method has been formulated for updating cumulative fatigue of an offshore structure and simulating the propagation of cracks, using, in both cases the actual ambient forces acting on the structure.

The basic premise is that, for a fixed load, the distribution of the forces and tensions in a structure, can be obtained theoretically by structural analysis.

If the tension at a predetermined point of the structure is known (by in situ measurement with extensometers), it is possible to determine the actual tensions in all parts of the structure by theoretical calculations based on the same conditions of load.

Measurements need therefore to be taken on only a limited number of structural elements. The procedure also defines the method for selecting these elements.

The identification of the ambient forces is carried out directly in situ by processing the readings taken by the extensometers and using other instruments (for example, accelerometers on the deck of the platform) to determine the direction of the incident waves.

The sea conditions and tensions in the elements being monitored are correlated by a statistical method to improve the accuracy and reduce the volume of data for analysis.

When the tensions have been ascertained throughout the structure or, to be more precise, when the "history of the tensions" in the points subject to the greatest stress is known, the cumulative fatigue can be calculated using the SN curves.

Cumulative fatigue is ascertained for all the joints of the structure, thus allowing inspections to be planned and the state of the structure as regards fatigue to be assessed. Priority can therefore be given to inspections of those joints with the greatest cumulative fatigue. For junctions with cracks, the speed of propagation of the cracks can also be evaluated using the theory of fracture mechanics.

5. ANALYSIS OF VARIOUS "DAMAGE TOLERANCE" APPROACHES TO THE DESIGN OF OFFSHORE STRUCTURES, AND IMPLICATIONS FOR MONITORING AND INSPECTION

The following methodologies for design, both conventional and based on "damage tolerance" approaches, were examined :
a) Safe life: structure designed to remain completely free from damage in all parts for its entire operating life.
b) Fail safe : every part of the structure designed in such a way that the damage assumed as being initially present will not exceed predetermined critical dimensions for a limited period of time.
c) Slow crack growth : critical elements and those which cannot be inspected designed in such a way that the damage assumed as being initially present will not exceed predetermined critical dimensions during the entire operating life of the structure.
d) Intrinsic safe : structure designed to be able to withstand the complete failure of structural elements for the period of time between the discovery of the damage and its repair.

The methodologies for the designs listed in b) and c) use the theory of fracture mechanics, while the methodologies in a) and d) use conventional methods of calculation.

The possibility of using the theory of fracture mechanics for the design and diagnosis of offshore structures was analysed by using simplified programmes and finite element programmes to estimate the speed of propagation of the cracks in plates of small thicknesses and tubular joints.

It was found that the simplified programmes were very useful for diagnostic purposes (although the speed of propagation of the cracks was higher than in reality) but that their application to structural design gave excessively conservative results.

The finite element programmes cannot be used for the design of a whole structure in view of the computer time this would require.

From these results and the analysis of the design methodologies listed above, the following conclusions were drawn :

- the "safe life" methodology uses a conventional and proven method of calculation which, however, is not entirely reliable for very thick welded structures
- the "fail safe" methodology cannot be used for the overall design of welded stuctures in view of the present limits of the theory of fracture mechanics.
- the "slow crack growth" methodology is suitable only for the design of critical elements and/or elements which cannot be inspected, but is more reliable than the "safe life" methodology.
- the "intrinsic safe" methodology, which uses conventional methods of calculation, is more reliable than the "safe life" methodology despite the higher design costs and, in some cases, greater structural weight. This method makes the best use of existing monitoring systems and could result in a substantial reduction in the number of inspections.

6. <u>MAIN RESULTS</u>

The procedures for optimizing the monitoring programme using the vibration method, and for data analysis for the purposes of identifying damage, have been defined and successfully tested on, respectively :
- an actual platform
- a model of a jacket, on laboratory scale and at sea.

The portable instrumentation system, developed in the course of the project was tested at sea on a test structure and was found to be both practical and efficient.

The procedure for calculating cumulative fatigue and for estimating the propagation of damage by measuring the tensions on some of the structural elements, in conjunction with the relevant computer programmes, was successfully tested on an actual platform.

Various "damage tolerance" approaches to design were analysed with a view to the implications for monitoring and inspection.

A real platform, designed by the conventional method, was redesigned on the basis of the "controlled damage" (intrinsic safe) approach. Comparison of the two designs showed the advantage of a "damage tolerance" approach, were by the structure is designed to withstand the total failure of structural elements for the period of time between discovery of the damage and its repair.

7. CONCLUSIONS

The development of the research project "Diagnostic methods for offshore structures" has produced a number of technical results which will enhance the current technology of monitoring structures and in the "damage tolerance" approach to offshore structure design. The following systems and methodologies have been found to be particularly suitable for use in the field :
- a portable system for structural diagnosis, both global and local, based on monitoring the vibrations caused by forced excitation.
- a computerized procedure for the calculation of cumulative fatigue and assessment of the speed of propagation of flows, based on measurement of the tensions at a small number of points on the structure.
- definition of the applicability of the possible "damage tolerance" approaches to the structural calculation and establishment of the more promising calculation methodologies for monitoring and inspection.

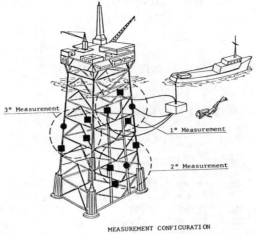

MEASUREMENT CONFIGURATI ON

● Exciter Position
■ Accelerometer Position

Fig. 1 Measurement programme using portable system (vibration monitoring)

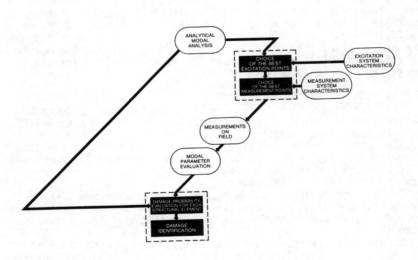

Fig. 2 Diagnostic procedures

(15.03/77)

DEVELOPMENT OF A RELIABILITY ANALYSIS SYSTEM FOR OFFSHORE STRUCTURES.RASOS

W. BRENNAN and P. ROONEY

Institute for Industrial Research and Standards (IIRS) Dublin.

Summary   (23.7.84 current project)

The structural monitoring system installed by IIRS under a previous
project (No. 15.03.77) was maintained to provide data for developing
a Reliability Analysis System for Offshore Structures (figure 1).
This paper outlines the areas concentrated on to date and does
not present an overall system for Reliability Analysis. Topics
covered include, data processing, forced v dynamic response,
transfer functions and characterisation of stochastic variables
for Irish waters.

1. INTRODUCTION
    The flow diagram for a Reliability Analysis system shown in
figure 1 provides a framework for incorporating the structural and
environmental data recorded on the Kinsale Alpha platform,in the
Celtic Sea, into an ongoing Reliability Analysis System. IIRS are the
Certification Authority for the Kinsale Head Gas field installations
and overall reliability is monitored with respect to planned maintenance,
inspection and operating procedures. Marathon Petroleum Ireland Ltd.,
operate the Kinsale installations.

2. DATA ACQUISITION AND PROCESSING
    The data acquisition system used on Kinsale Alpha is described
under project No. 15.03.77. Both projects were carried out by IIRS
on the Kinsale Alpha platform.
    The data processing and analysis carried out for this project (RASOS)
are outlined in figure 2. A detailed account of the programmes used is
included in the project status reports. Only interim results are
discussed in detail in this paper.
    In addition to acceleration records and wave records IIRS has
gathered data on wind and currents in the Celtic Sea area. IIRS have
also had access to Reports from similar data acquisition systems in
the North Sea.
    Maintenance inspection and operating procedures are major inputs
to a Reliability Analysis system. IIRS have compiled a comprehensive
dossier on these topics and the IIRS "Rules for periodic survey and
certification of offshore petroleum installations" incorporates many
of the results of this investigation.

3. CHARACTERISATION OF STOCHASTIC VARIABLES

Wave data collected between 8.10.81 and 14.1.84 has been processed to provide seasonal scatter diagrams for the Kinsale Head Gas field area of the Celtic Sea. This data has been compared with other environmental studies carried out for the Celtic Sea and an overall probabilistic model has been developed to simulate the Celtic Sea environment and to predict extremes.

Wind and current data was collected from previous surveys in the Celtic Sea and from meteorological sources.

Marine growth is monitored regularily on the jacket and is removed when it approaches design limits. Thus the actual thickness at any particular time can be simulated.

4. PLATFORM RESPONSE

Figure 3 summarises the response of the platform to the wave spectrum shown. These results are recorded simultaneously on the wave staff and accelerometers and it is evident that the response is predominantly a forced response.

Figure 4 shows a wave trace with the corresponding x and y deck displacement records. The incident waves were $+15°$ from the x direction and it can be seen that the response in the y direction is less than in the x direction and does not have a high correlation with the incident waves.

Figure 5 plots 25 data points of wave height versus deck displacement and a log/linear regression relationship is derived. This transfer function was found to be similar in both x and y directions (for incident wave in x and y directions respectively). The scatter on the data points is due mainly to ignoring wave period, wave direction ($\pm15°$) and current variability.

5. THEORETICAL STUDY

A theoretical study of the quasi-static response of the platform to waves, wind and current was carried out using the ASAS finite element suite. The elements marked in figure 6 were found to be most highly stressed and representative of the overall platform. From the analysis two aspects of the deck displacement to stress transfer function were studied:

- zero offsets due to wind, current and waves
- Quasi-static response/oscillations due to cyclic loading from the waves.

Thus relationships were developed to estimate the stresses in the jacket directly from the environmental parameters. These relationships were used to study the fatigue life of the platform. The recorded displacements were also used to calibrate the quasi-static model of the platform and to provide feedback on the level of safety inherent in current design practice.

All results indicated an acceptable safety factor on design.

6.   SUMMARY
Referring back to figure 1 the work outlined in this report can be
placed into an overall reliability analysis system. Much work has yet
to be done within this current project however it is evident that results
to date indicate that the complex loading and resistance solution systems
used by designers are acceptable and for a particular platform simplified
models can be used to study the overall reliability of such a multi-
variate system.

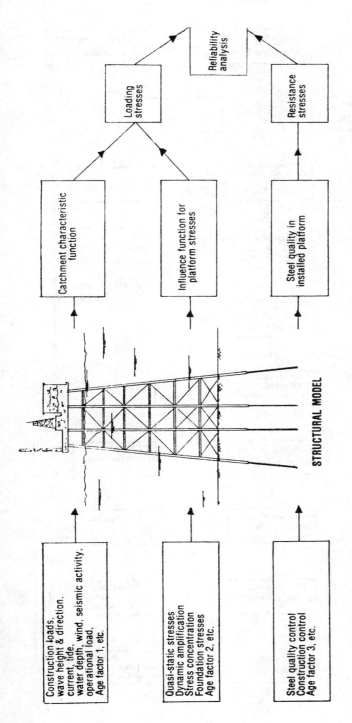

STRUCTURAL MODEL

Age factor 1 includes marine growth
Age factor 2 includes corrosion, structural damage
Age factor 3 includes fatigue

FIGURE 1

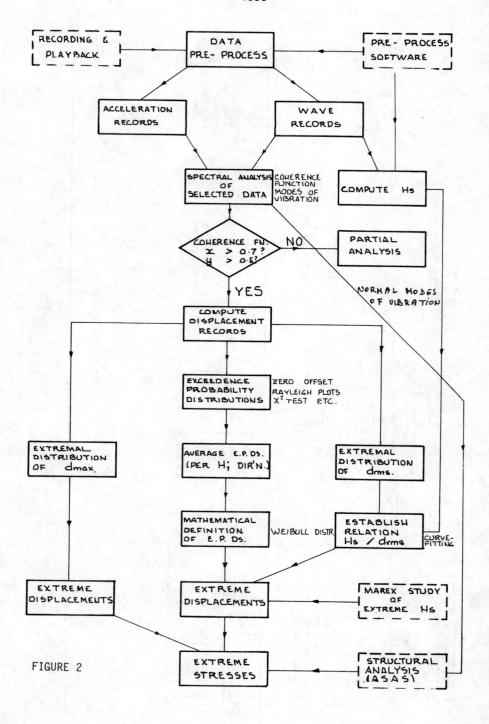

FIGURE 2

Wave Spectrum

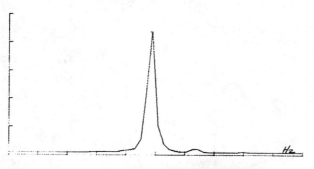

FREQUENCY RESPONSE FUNCTION

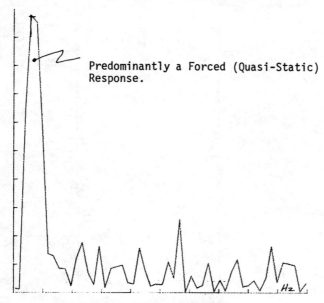

Predominantly a Forced (Quasi-Static) Response.

DISPLACEMENT SPECTRUM

FIGURE 3

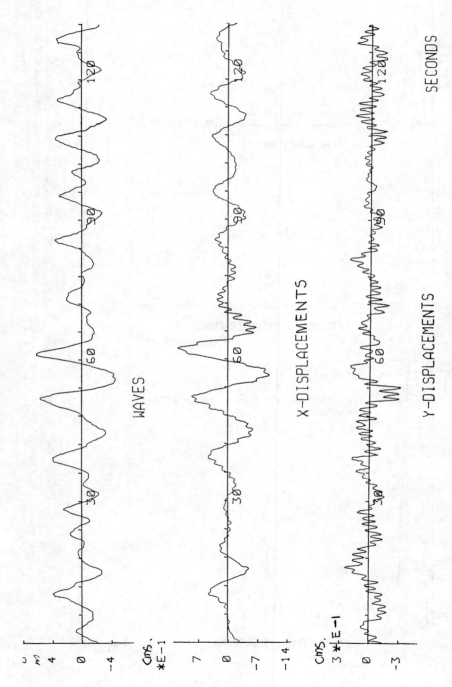

FIGURE 4   TYPICAL RECORD SHOWING DISPLACEMENTS DIRECTION W.   Hs = 7.2 m

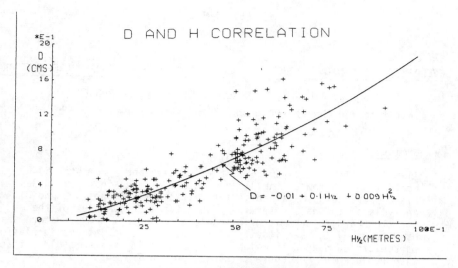

FIGURE 5   PLOT OF WAVE VERSUS DISPLACEMENT AMPLITUDES FOR 25 DATA SETS

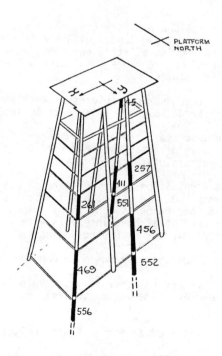

FIGURE 6   ELEMENTS SELECTED FOR EXTREME STRESS ANALYSIS

(15.24/81)

DEVELOPMENT OF A TOTAL STRUCTURAL MONITORING SYSTEM FOR OFFSHORE PLATFORMS

W. BRENNAN, E. HERLIHY, G. KEANE, P. ROONEY
Institute for Industrial Research and Standards (IIRS) Dublin.

Summary

IIRS installed accelerometers, a wave staff and a computer based data
acquisition system on the Alpha platform of the Kinsale Head Gas
Field operated by Marathon in the Celtic Sea (figures 1,2 and 3).
Analysis of the data determined the natural frequencies of the
structure and a 'dynamic' finite element model of the structure was
used to study the sensitivity of the system to variations in platform
parameters such as deck weight. The monitoring system could not act
as a replacement for current inspection requirements.

1. INTRODUCTION
    The Kinsale Alpha platform (figure 1) was installed in 1977. Attempts
were made to lower strain gauges with the insert piles however some became
disconnected during installation and those remaining did output strain
data for a period before gauge insulation decreased to an unacceptable
level for valid results.

Eight Sunstrand servo accelemeters were installed on the platform's
spider and cellar deck levels (figure 2) to monitor platform vibrations.

A Baylor wave staff installed by Marex (U.K.) was used to monitor
wave height. Wave direction was recorded by observation and was cross
checked with Meteorological wind data.

The data acquisition system (figure 3) was designed and installed by
Microconsultants Ltd. U.K. to IIRS specifications. The system records 9
channels at 16 samples per second for 17 minutes as one data set. Data
sets were recorded at 12 hour intervals with a storm trigger over-ride
to facilitate continuous recording during storms.

Figure 4 shows simultaneous wave and accelerometer time histories.

The data processing software was developed in-house and validated
against a commercial time series analysis package.

2. RESULTS
    The data presented in figure 4 was analysed to determine;

- Natural frequencies of the structure
- Modes of vibration of the structure
- Damping coefficient for 'ringing' responses.

Figure 5 summarises the results from the accelerometers. The natural frequencies were consistent for all the data sets and as shown here the cross correlation between accelerometers is very high except for torsional modes of vibration. (For further discussion of the structures response to waves see project no. Th/15.24/81 "Development of a Reliability Analysis System for Offshore Structures, RASOS". In RASOS the forced response of the structure is quantified and fatigue investigated).

The accelerometer record in figure 4 shows a ringing response by the platform to wave, 'slamming' on the horizontal bracing at spider deck level. This is due to impact loading and it was possible to determine the damping ratio for the excited mode of vibration using the log decrement method. The average damping ratio for the two fundamental modes of vibration is 0.036 (3.6%). It was not possible to positively identify a torsional ringing response.

3.   THEORETICAL STUDY
The values of the natural frequencies in figure 5 were used to tune a dynamic finite element model of the Kinsale Alpha structure. This work was carried out in conjunction with Mr. N. Starsmore of Atkins Research and Development using the ASAS suite.

The first ten eigenvalues and eigenvectors for the model (1,700 degrees of freedom) were solved.

Eigenvalue No.	Hz.	Mode Shape
1	.7633	x dir. translational
2	.8530	y dir. translational
3	.8820	Torsional
8	1.5120	x dir. 2nd translational
10	1.7260	y dir. 2nd translational

These results correspond to the recorded results in figure 5 and a sensitivity study was carried out to identify the most significant parameters in determining the theoretical natural frequencies. These parameters included:

-   Grout annulus in jacket leg
-   Point of inflection in pile
-   Deck mass
-   Coefficient of added mass
-   Stiffness of unbraced legs (spider to cellar)
-   Stiffness of bracing members.

The main factors affecting the natural frequencies were deck mass, point of inflection in the pile, the stiffness of the unbraced legs, and the stiffness of the bracing members. It is of interest also to note that the coefficient of added mass had a greater influence on higher order modes than on primary modes.

4. CONCLUSIONS
The requirements of a structural monitoring system are:

(1) Respond to the severance of a major load bearing member
or a foundation change.

(2) Be sufficiently stable and sensitive to verify no such
damage has occurred.

(3) Be free from false alarms.

Based on these requirements it was not possible to develop a total
structural monitoring system. Variations in deck mass, marine growth,
corrosion and system reliability would be difficult and very expensive
to incorporate into a monitoring system and authorities would not accept
such a system as a replacement for current inspection programmes.

However the information gathered does provide valuable feedback to
verify theoretical tools used in design and a monitoring system can be
used where parameters affecting the recorded results are controlled.

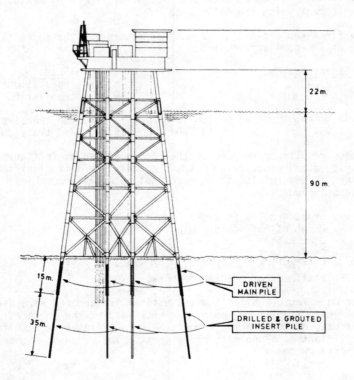

FIGURE 1   KINSALE "A" PLATFORM

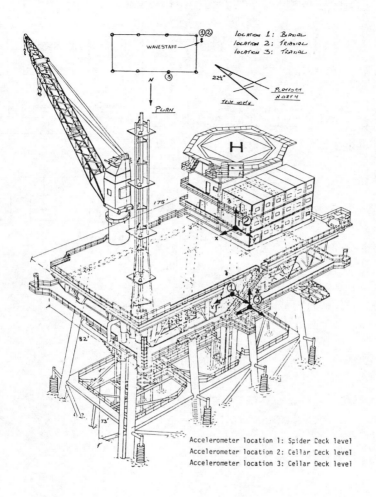

Location 1: Biaxial
Location 2: Triaxial
Location 3: Triaxial

WAVESTAFF

N

PLAN

22½°

PLATFORM NORTH

TRUE NORTH

Accelerometer location 1: Spider Deck level
Accelerometer location 2: Cellar Deck level
Accelerometer location 3: Cellar Deck level

FIG 2   LOCATION OF THE ACCELEROMETERS

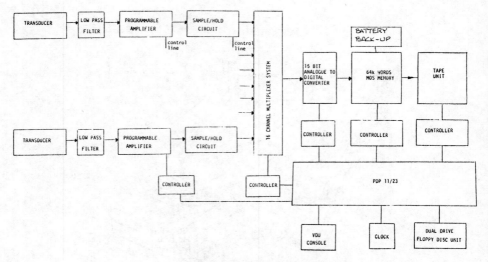

BLOCK DIAGRAM OF THE ACQUISITION SYSTEM

FIGURE 3

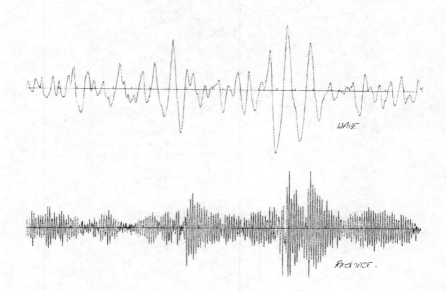

FIGURE 4

Cross Spectral Density Between Accelerometers	Frequency of Peak		Coherence at Peak Frequency		Phase Difference Between Accelerometers At Peak		Estimated Mode Shape
Accelerometers	Set 42	Set 70	Set 42	Set 70	Set 42	Set 70	
1X, 2X	.756	.756	.9945	.9965	0.0	0.0	Main Translation
2X, 3X	.756	.756	.93	.972	0.0	0.0	Main Translation
2X, 3X	-	1.73	-	.915	-	0.0	Higher Order
2X, 3X	1.03	1.03	.64	.89	180	180	Main Torsion
1Y, 2Y	.756	.756	.987	.998	0.0	0.0	Transverse
2Y, 3Y	.756	.756	.995	.998	0.0	0.0	Transverse
2Y, 3Y	-	1.98	-	.913	-	0.0	Higher Order
2Y, 3Y	1.03	1.03	.720	.913	180	180	Main Torsion

FIGURE 5

DEFINITION OF NATURAL MODES OF VIBRATION OF THE STRUCTURE

(15.08/78)

DEVELOPMENT OF A MEASURING SYSTEM
FOR THE OPTIMIZATION OF PROGRAMMES FOR CALCULATING AND MONITORING
THE DYNAMIC BEHAVIOUR OF OFFSHORE PLATFORMS

A. BAUDRY, SYMINEX, Director, Research and Development

Summary

The objective of the project was to develop a procedure and equipment
enabling the dynamic characteristics of offshore structures to be
measured in order to:
- optimize the theoretical models of structures,
- detect any cracks.
In the initial phase, after choosing a computation programme by finite
elements, a reduced-scale model platform was built and a numerical
model adjusted to this physical model. Various methods were then
studied in order to attain the objectives.
The choice of a hydraulic exciter to bring about the general
excitation of the platform (transfer functions method), together with
the local excitation (method of impedance functions or vibrodetection)
led to the construction of specific equipment. In the second phase,
this equipment and these methods were tested on the site.
The following conclusions emerge from this study:
Modal analysis provides interesting information on the dynamic
behaviour of the structures, enabling a numerical model to be roughly
set. It can enable major faults to be detected. However, there is no
question of detecting and locating partial cracks in elements.
Vibrodetection enables partial cracks to be detected and located
reliably in members, without first having to clean them. Application
is rapid and enabled 121 nodes to be inspected, i.e. about 600 welds
in 13 days. This is a fast, reliable and economic method of
inspection.

1 - INTRODUCTION

The objective of this project was to develop a procedure and the
equipment needed for measuring the dynamic characteristics of an offshore
structure, with a twofold aim:
- to optimize the theoretical models of the structure, by associating
  realistic dampening values to them in addition,
- on these bases, to detect any possible damage to the structures.

This project took place in two successive phases:
1)   The design and construction of the equipment and the preparation of a real-life sea trial (development of the methods, procedures and equipment).
2)   The trial proper and the processing of the results in order to develop definitive measurement procedures.

The first phase made it possible to define three measuring methods intended either to detect and locate partial damage to the structure, for instance a crack (method of transfer functions, method of impedance functions), or to adjust a theoretical model of a structure to experimental results (method of matrices), and also to develop the necessary equipment for application on a real-life site. To do so, calculations on a finite elements model, tests on a reduced-scale model and tests on the equipment developed were carried out.

In the second phase, a real-life test was successfully conducted in the Persian Gulf. In enabled the methods to be evaluated, the ones truly appropriate to be selected and conclusions to be drawn as to their real performance. The methods, procedures and equipment were then optimized so as to offer an industrial solution to the problem studied.

2 -   PHASE 1

2.1 - Choice of a programme
Since one of the objectives of the project was to optimize computing models of platforms, the studies required the application of a "finite elements" type computation programme.
Three programmes were compared by calculating the first twenty natural modes of a reduced-scale offshore structure model and on a simple tube model constrained at one end.
On completion of this preliminary phase, it was decided to use the ASAS programme, owing to the facility with which it can be accessed.

2.2 - Verification of the model by means of measurements
A 1/27 scale model of an offshore platform was used, both at experimental level (measurements) and theoretical level (calculations) to define the procedures and methods that were subsequently tested on a real-life offshore platform. This is why, using the ASAS programme, a model built with beam elements was validated, by comparison against the results of measurements and calculations, for the following two cases :
- statically, under different loads,
- dynamically, for several specific modes.

2.3 - Evaluation of performances
During this phase, the methods were evaluated by calculation and by laboratory measurements, on a reduced-scale model of the platform. The following were the conclusions:
- Whilst measurement of the natural frequencies of a structure is extremely accurate, on the other hand, measurement of the modal deformates involves greater uncertainty, particularly for the higher ranking modes.
- In addition, if a major fault affects the fundamental modes of a structure, on the contrary, a partial fault affects only the higher ranking modes, which are also those measured with the least stability.
- Consequently, whilst modal extraction is of certain interest for adjusting mechanical parameters of a numerical model, this method is unsuitable for detecting and locating the partial damage of a member.
Two new methods concentrating particularly on the detection and

location of partial damage were studied and tested on the model.
- The method of transfer functions consists in making modal analysis measurements of a structure in the conventional way (overall excitation at a point, measurement of the response at several points) and in comparing the transfer functions between responses and excitations directly, without extracting the modes from them. This is hence an adaptation of the initial method, eliminating the considerable processing that is a source of numerical errors and hence by no means negligible instability of the results.
- The method of impedance functions or vibrodetection consists in calculating the impedance function between the response of a member measured at its mid-point and a local excitation applied at the same point.

These two methods were experimented upon using the model, simulating a progressive fault. A horizontal face member was progressively cut. After each elongation of the cut, new measurements of transfer functions and impedance functions were made. The following conclusions were to be drawn:
- by the method of transfer functions, one can expect to detect an open through-crack equal in length to 60% of the circumference of the member, together with highly approximative location of the fault,
- vibrodetection justifies envisaging a detection threshold equal to 20% of the circumference and very accurate location of the fault.

Excitation methods

The performance of the vibrotary measurement methods depend to a very great extent on the means employed and particularly the means of excitation that were studied in succession, namely:
- "beating" by means of a device for imparting repetitive shocks,
- snapping of a taut cable between two elements of the platform,
- snapping of a taut cable between the platform and an external anchoring deck (ship),
- shock (semi-sinusoidal, inverted sine, explosion...).

Following this examination of the various possible modes of excitation, it was decided to direct the study towards the execution of a slaved hydraulic vibrator mainly consisting of the following four parts:
- an inertia mass the traversing movement of which is guided,
- a double acting cylinder, the piston of which is secured to the mass and the body of which is rigidly linked to the structure to be excited,
- a servo-control electronic system enabling the movements of the mass to be controlled perfectly,
- a hydraulic power plant supplying the piston with oil under pressure.

Local exciter

In addition, an underwater local exciter designed to excite each member individually within the context of the vibrodetection method (method of impedance functions) was developed and perfected. This consists in the example here of a portable system capable of imparting dampened shocks, the energy of which is concentrated at the low frequency end of the range (from 0 to 80 Hz). The impacting device is secured to the member to be tested by means of a clamping collar surrounding the member.

Accelerometer box

The underwater box built contains three accelerometers laid out in an orthogonal trihedral. In this way, the vibrations of the structure are measured simultaneously in each of these three directions. The measurement signals are sent to the data acquisition system on the surface via a cable specifically designed for this purpose. This box is secured to the members either directly by a collar around them, or by means of a fixed base, particularly when repetitive measurements have to be made.

Acquisition and control system

The equipment consists of an overall exciter and a mobile and submersible accelerometer box to apply the method of transfer functions (detection of damage) or to carry out modal analysis of the platform (setting of the numerical models), together with a local pulse exciter for testing the method of impedance functions (vibrodetection - detection of damage). This equipment is completed by an electronic system for acquisition of the measurements of servo-control of the overall and local vibrators.

To acquire and process the data in order to calculate at the site the transfer, impedance and coherence functions needed to validate measurements, a Fourier analysis system was used. The following are the operations this system is capable of:
- acquisition of the data (x", y", z", force F responses)
  . anti-fold filtering,
  . digitization,
  . recording onto magnetic tape,
- data processing
  . calculation of spectra,
  . calculation of transfers and associated coherences,
  . storage of these results on magnetic tapes,
  . extraction of vibration modes (off-site),
- exciter control system
  . numerical generation of excitation signal,
  . conversion of this signal into analog form to go to the servo-loop of the exciter.

This electronic circuitry has been built into an antivibration container designed to ensure transport of the equipment and its used at the site.

3 - PHASE 2 - ON-SITE OPERATION AND INTERPRETATION

3.1 - On-site operations

The platform selected to carry out the first tests was the jacket type structure carrying the living quarters of the ABU AL BU KOOSH field owned by Compagnie TOTAL, in the Persian Gulf. A horizontal member was added to one of the faces of the jacket between two diagonals of a K node. This member, together with the neighbouring members, were measured for transfer functions (with overall excitation) and impedance functions (with local excitation) as soon as they were installed. They were then flooded with water and a new series of measurements was made, similar to the previous ones, against which they were then compared. Next, one of the ends of the member was gradually sectioned. With each cut, the length of the crack thus simulated was increased by 10% of the circumference, until the member was almost completely cut through. After each extension, all the measurements made were compared against the results of the reference measurements.

In this way, the following results were determined:
- the method of transfer functions is sensitive to a crack of over 60% of the circumference and does not enable it to be located,
- the method of impedance functions or vibrodetection transpired in this test to be sensitive to a fault involving 20% of the circumference. However, without any prior knowledge of the fault, it would probably have

been detected at the 30% to 40% stage, when its influence on the impedance function of the member has become undeniable and distinctly amenable to interpretation. As for the ability of this method to locate the fault, this turned out to be highly interesting, since on the basis of the results of this test, it seemed possible to determine the cracked member without any ambiguity.

Various offshore tests (L7 platforms in the North Sea and Balik Papaan in Indonesia) enabled the equipment and methods to be tested and improved.

The local pulse exciter was widely reviewed to enable it to respond to the following criteria:
- sufficient power to excite 20 to 30 ton members,
- handlable beneath the water, namely:
  . as small as possible,
  . zero apparent weight in water,
  . equilibrium in all positions in the water,
  . readily transportable,
  . highly capable of withstanding shocks, vibrations, etc...

The economic advantages of vibrodetection thus came to the fore, since it proved possible to inspect 121 nodes, i.e. 600 wells in 13 days without first having to clean the structure (a long and expensive operation).

3.2 - Automatic setting of the model

A new automatic setting technique of the model was developed, by combining a direct orthogonalization method applied to the mass matrices and the stiffness of the model with a method based on sensitivity formulae.

The method developed was tested on a simple model with 20 degrees of freedom, with the solution to the problem known by other means. The result was disappointing, since even with extremely precise measurement results, a very high number of modal parameters were needed to hope to effectively correct a digital model, which initially differed only slightly from the exact model. Now, in the case of measurements at sea, high frequency modes are obtained only with a limited accuracy on deformates.

4 - CONCLUSIONS

The research project work led to two well defined techniques:
- modal analysis did not truly justify its original expectations. Nonetheless, there are two main applications for it:
  . in inspection, it can enable major faults to be detected (modifications to the foundations, failure of a leg or main element). It is out of the question to detect, and yet less so locate partial cracks in elements, as vibrodetection can do,
  . in design, it provides information on the behaviour of offshore structures and provides a coarse setting for a model of the structure with site measurements, once this structure is operating. However, today, it is not possible to carry out this setting automatically and hence systematically. This distinctly confines the interest of this technique.

It will be noted that the considerable personnel and equipment to be applied results in an expensive measuring operation, for results that are of no direct interest to operators.
- vibrodetection enables significant cracks (i.e. dangerous cracks) in a metal jacket to be reliably detected. It also enables the faults to be accurately located.

Particularly efficient and practical equipment was developed, permitting low-cost and reliable monitoring of the structures. Vibro-detection appears to be a highly promising inspection techniques that has retained the attention of those responsible for oil companies operating offshore.

(15.11/80)

METEOCEANOGRAPHICAL-STRUCTURAL MEASUREMENT SYSTEM FOR THE SAFETY OF THE NILDE SINGLE ANCHOR LEG STORAGE - SICILY CHANNEL

D. DE MARZO

Agip SpA offshore department, S.Donato Milanese Milano, Italy

Summary

In the last few years, with the increase of oil demand, the oil companies were induced to exploit "marginal field" considered not economically exploitable by conventional steel jackets. In this context Agip decided to develope the Nilde field (50 miles NW of Trapani) by means of a new more economical structure called SALS (single anchor leg storage) fig. 1, composed by a tanker anchored to a riser by means of an articulated system. In order to compare the behaviour of the structure at sea with the result of the model tests previously performed, Agip on 1980 developed an integrated system for monitoring the dynamic aspects and the structural integrity of the SALS. On December 1980, a storm caused the breakage of the SALS and of the transducers, cables and conduits on the riser and the yoke. On September 1982, after the replacement of the SALS, the instrumentation was ready to start. To date, the acquisition period is terminated. Data of good quality have been collected and their processing is in course. The results will be available toward early 1985.

1. INTRODUCTION

The meteoceanographical and structural measuring system, installed on Nilde SALS is essentially composed by two sub-systems: the first-one is mainly dedicated to the acquisition of the meteocean, dynamic and structural data; and the second-one is concerned with the on-board preanalysis of the data collected from the field.

The main task of the monitoring system consists in checking the dynamic behaviour of the components of the SALS (riser, yoke and tanker), as well as their structural integrity during one year operation.

In particular, by means of the elaboration of the recorded data it is possible to point-out the following:

a) to ascertain possible and significative failures rose-up in the structure, simply by analysing the spectral correlation between the structural response (stresses) and the meteoceanographical data.

b) to evaluate the fatigue damages account on the monitored members.

The acquisition procedure and the data elaboration are schematically resumed in fig.2.

This monitoring system has been designed by Tecnomare (Venice) taking into account that the area where are installed transducers and cables is safety restricted and trying to satisfy the reliability and suitability requirements of the apparatus to be installed in severe envir onmental conditions.

The lay-out of the transducers and of the data acquisition system (DAS) has been sketched in Fig. 3.

A brief description of the transducers which have been used for recovering the data is reported hereinafter.

2. METEOCEANOGRAPHICAL DATA ACQUISITION TRANSDUCERS

For the collection of the environmental factors, have been employed the following instruments:
- n° 1 anemometer installed on the deck of the tanker for measuring the wind speed and direction.

The principle relies on the relation-ship existing between the vortex formation frequency in the wake of a stationary rod and the speed of the air moving around it.
- n° 1 waverider moored about 1 miles from the tanker.

The waverider is a buoy which following the movements of the water surface measures waves by measuring the vertical acceleration of the buoy. These information (wave height and period up crossing) are transmitted to the receiver in the control room of the tanker by a radio link.
- n° 1 currentmeter for the data regarding the direction and intensity speed of the current.

It has been installed about 11 meters below the sea-surface in such a way to eliminate disturbance due to the tanker.

The underwater unit principle is based on the acoustic phase-shift phenomena which is proportional to the current speed. The direction is individuate by a magnetometer compass sensor.

3. DYNAMIC DATA ACQUISITION TRANSDUCERS

The dynamic data of the structure are collected by means of the following instruments:
-n° 2 inclinometers, type potentiometric pendulum, installed on the riser close to the bottom.
- n° 2 inclinometers on the joke for the measurement of the rolls and pitches angles.
- n° 1 inertial platform installed on the tanker for monitoring the roll, pitch and yaw.

The inertial platform consists of a cageable vertical gyroscope with wirewound potentiometer pick-offs in the pitch and roll axes. The yaw output is via precision wirewound potentiometer.

4. STRUCTURAL DATA ACQUISITION TRANSDUCERS

In order to measure the axial stress and the bending moment on the riser and on the yoke have been used the following transducers:
- n° 4 vibrating-wire strain gauges on the riser, installed around the theorical maximum stress circumference.
- n° 8 vibrating-wire strain gauges on four bracings of the yoke.

These transducers operates as follows:
a change in natural frequency of a stretched wire depends on the change of the tension in the wire. By means of two electromagnets, the exciter and the pick-up, it is possible to transform the frequency variation of the wire in a current signal proportional to the stress.

5. ACQUISITION AND PREELABORATION OF THE DATA

All the transducers, via 22 cable channels, transmit the signals to a data acquisition system (DAS) installed in the control room of the tanker together with the receivers. Only the wave data (2 channels) are transmitted to the DAS by a radio link.

In the fig. 4 is reported the DAS configuration which includes:
- signal conditioner apparatus,
- filters
- analogue digital converter
- minicalculator with its periphericals (magnetic tape recorder, video display and the key board).

The 32 K words minicomputer supervises the acquisition modality and the recording of the data.

Moreover, it provides the operator on the video display with the summarized parameters of the environmental condition and makes a preelaboration of the data to be recorded.

The 24 channels are handled in two different ways:
a) continuously handled signal:

The environmental data relevant to waves, wind and current are acquired continuously and grouped in 20 minutes records. For each record are calculated and displayed the statistical parameters relevant to wave, current and wind. These values labelled with their collection time and date are recorded on magnetic tape.
b) Periodically collected data

All channel data are collected for an acquisition period chosen by the operator in different sea-states in order to follow the SALS dynamic and structural behaviour, for comparison with the model tests which have been performed in the similar environmental condition. The data are recorded on magnetic tape. At the end of recording period data are processed and recorded in a file.

6. ON-SHORE PROCESSING

The tapes are processed on-shore and quality controlled. The "meteocean" tapes are produced for statistical purpose. The structural data, together with the meteoceanographical data collected in the "periodically way" are processed in order to compare the actual SALS behaviour with model tests and in order to get fatigue data of the structure in particular operativity condition.

7. CONCLUSIONS

To date, the programmed data acquisition period have been completed.

The instrumentation, excepted for the gap period between 1980-1982 following the SALS breakage, have been working without heavy problems.

However, the experience gained by this project about the acquisition of the data, has suggested to reduce the number and the length of cables running the signals to the DAS, by replacing them by a radio link. In fact, the transducers them-selves, are todate, working satisfactorily.

But, what we supposed failed, has been the electrical insulation of the cables and connectors specially those in presence of severe environmental conditions.

The final on-shore data processing, together with the dynamic and structural conclusive study, is about to start. The definitive and significative results are expected toward early 1985.

Fig. 1 NILDE SALS

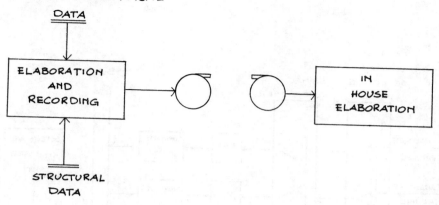

Fig. 2 <u>DATA RECOVERING AND ELABORATION</u>

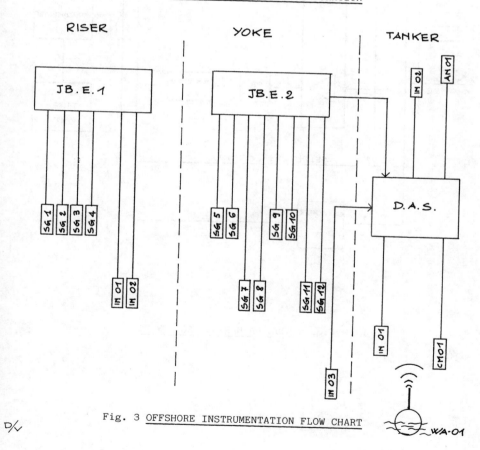

Fig. 3 <u>OFFSHORE INSTRUMENTATION FLOW CHART</u>

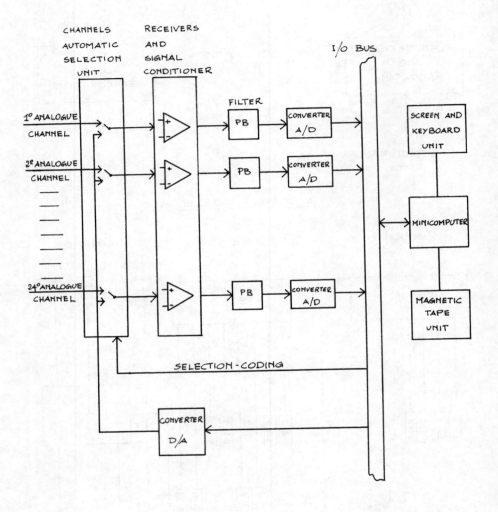

Fig. 4 DATA GATHERING FLOW CHART

D/√

(15.07/78)

METEOCEANOGRAPHICAL AND STRUCTURAL DATA ACQUISITION
TO IMPROVE PLATFORM DESIGN

D. DE MARZO
Agip S.p.a. offshore department
S. Donato Milanese Milano
Italy

Summary

In order to improve both future platform methodologies design and
new on-life inspection procedures, a meteoceanographical and structural data acquisition system has been developed and installed
on the Barbara platform (24 miles offshore Ancona Adriatic Sea).
The system, designed by Tecnomare (Venice), has been working suc-
cesfully since 1982.The project will be completed toward the end
of 1986. The main task to be accomplished are:
- to verify the scattering of the results of theoretical calculation from the actual measured value (specially for the dynamic
  behaviour of the platform),
- to calculate the fatigue damages of the structural members,
- to individuate irreversible failures of the structure,
- to plan the periodical inspection by means of a spectral analysis of the records.
To date, some results about the dynamic and the fatigue aspects of
the platform have been gained in addition to an interesting meteoceanographical report of the site.

1. INTRODUCTION

The Barbara "A" platform (fig.1) is a conventional steel jacket
platform with four piled legs, installed in 70 meters water depth 24
miles offshore Ancona,Adriatic Sea.

The meteoceanographical and structural data acquisition system,
designed by Tecnomare, is composed by two different parts (fig.2-3);
the first one, located on-board, includes the following:
a) the data acquisition sub-system, made up by transducers, signal
   conditionner and paper recorder,
b) the data trasmission sub-system, made up by multiplexer equipment, analogue-digital converter and a modem for data link with
   the onshore station,
the second one, located onshore, includes the following:
c) a modem for data link with the on-board station,
d) a 32 Kb computer with peripherical units including a recording

magnetic unit, a teleprinter, a reader and puncher of paper tape and a clock unit to govern the recording periods.

The system operates as follows:
a 20 minutes recording procedure is activated by the on shore computer every 6 hours by asking data from the on-board terminal in a master-slave configuration through the radio link. The computer records and elaborates all the data regarding the 20 minutes acquisition period, which in a reduced form, are printed in a real time for operative use and are in the meantime recorded on a magnetic support for in-house data processing.

2. THE INSTRUMENTATION

On board, the following transducers are installed:
- two accelerometers positioned on the deck of the platform for the measurement of the x-y axis accelerations induced by the waves and wind. The range is about $\pm$ 10 mg with and accuracy of 100 μg. The measuring principle is based on a seismic mass which moves under acceleration imput effect producing a capacity variation which is converted in a balancing current for a braking coil.

This current is proportional to the acceleration.
- Eight back-upped vibrating wire strain gauges installed on two beams of the jacket at 20 meters water depth. Range $\pm$ 200 $\mu\epsilon$; accuracy 5 $\mu\epsilon$.

Changes in displacement cause a change of the strain and hence of the natural frequency of a measuring wire stretched to vibrate inside the transducer. This wire vibrating within the magnetic field of an electromagnet induces, in the coil of the magnet, an electrical oscillation of the same frequency which is trasmitted by cable to a receiver for further processing to obtain the measured value. The same electromagnetic system serves to excite the measuring wire by the receiver.

- A wave staff measuring system. Range 0-7 m, accuracy 5 cm. The sensor is made-up by a wire rope plunged for its half part in the sea.

The wire is linked to a transducer whose output is an electric tension proportional to the emerged wire (and hence to the wave height).

- A vortex shedding anemometer mounted at helideck level for wind speed and direction measurement. Speed range 0 $\div$ 60 m/s; accuracy 0,5 m/s. Direction 0° $\div$ 360°; accuracy 5°.

In this kind of sensor the vortex formation frequency is proportional to the wind speed. The direction of the wind is on the contrary, converted in an electric tension by means of an accurate potentiometer.

- A tide meter, range 2.5 m; accuracy 1 cm based on a differential cell, type strain gauge, back-upped with a mechanical tide meter with floating and counter balance.

- A vibrating wire transducer for air temperature measurement. Range from -10°C + 40°C; accuracy 0.5°C. working on the same principle of the above reported strain gauges.

Moreover, on board we find:
the conditioning electronics which provide the transducers power supply,
the signal amplification and filtering, the signal multiplexing, digitizing and formatting; an eight-channel strip-chart recorder which
can monitor on board eight of the fifteen signal; a modem for data link
with the on shore station through radio telephone channel.

Finally, onshore are installed in Agip Ravenna base, a modem
for data link with the onboard station through radio telephone channel
and a 32 Kb. interdate model 70 computer with peripherical units
including a magnetic tape unit, a printer and a clock unit to govern
the recording periods.

3. OPERATION OF THE SYSTEM

The operation of the system is automatic and continuous.

The tape recording cycles start automatically every 6 hours as
a rule, and last 20 minutes. Extraordinary recording cycles can start
automatically or on the operator request, should any particular events
occur as for istance, the overcoming of predetermined thresholds by
some parameters to be registered under the minicomputer control. (Fig.4)

Between two successive registration, the minicomputer rereads
the recorded data and makes:
- a preliminary analysis of them in order to find out the significative
  values to be printed by teletype;
- a preelaboration in order to record in definite way the stress
  parameters so as to facilitate further elaborations.

The system provides, in particular, three types of real time
printouts:
1) A meteoceanographical data table for operative use containing
   wave significant and maximum and mean up-crossing period, wind mean
   and maximum velocity and mean direction.
2) A summary table of reduced data for the 20 minutes record of structural
   and environmental measurements (mean, maximum and standard deviation
   values) is normally printed every six hours.
3) A system failure alarm in case of instrumentation malfunctioning.
   The 20 minutes every 6 hours records in view of the further in-house
   elaboration, are packed onto 2 files, one for structural data and
   one for environmental data.
   By means of a set of specially developed computer programmes
   the data are used to calculate the fatigue suffered by a number of
   "hot spot" (280) on 23 knots of the structure fig. (5) and the propagation of possible microcracks, supposed the size of the cracks
   to be 5 mm lenght and 2 mm deep.
   Periodic account of the fatigue accumulated by the structure's hot
   sposts and of the crack propagation will give the warning to start
   with more accurate inspection procedures.

4. CONCLUSIONS

This project is the first one in Italy, and probably in Europe, in course of development.

From 1980 to 1982 the instrumentation has not worked very satisfactorily because of malfunctioning of some transducers. These difficulties, however, has been overcome and the system since 1982 has been working withouth serious problems and incovenient.

To date, Agip has collected and processed a large amount of data gaining good results and experience.

First of all for the offshore platform safety, in view of the new government rules and regulations regarding the structure monitoring, Agip, developing this project, has experienced the most suitable sensors, electronics and processing of data collected.

Second, Agip has recovered meteoceanographical information to get more accurate environmental data for designing of future platforms.

Moreover, the first results regarding the dynamic response of the structure have been gained.

In particular, comparing the calculated first vibration mode of the platform with that one measured by the accelerometers it has not been found a significative scattering of the results.

As regards to the fatigue analysis, two years of recording are not sufficient to verify the propagation of the microcracks supposed originally to exist in some particular hot spots of the structure.

We expect to conclude the project toward the end of 1986.

At that time we deem more data will be available to reach completely the goals of the study.

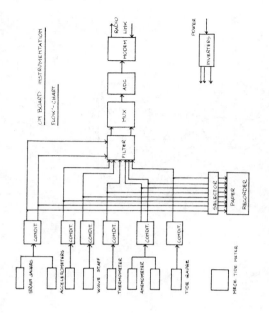

ON BOARD INSTRUMENTATION

FLOW-CHART

FIG. 2 OFFSHORE INSTRUMENTATION FLOW-CHART

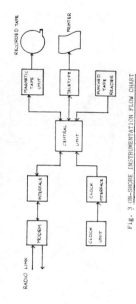

Fig. 3 ON-SHORE INSTRUMENTATION FLOW CHART

Fig. 1 BARBARA A PLATFORM

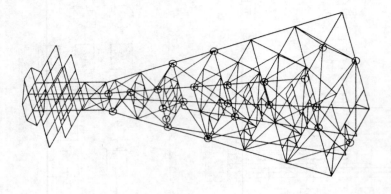

Fig. 5 FATIGUE CALCULATION HOT SPOTS

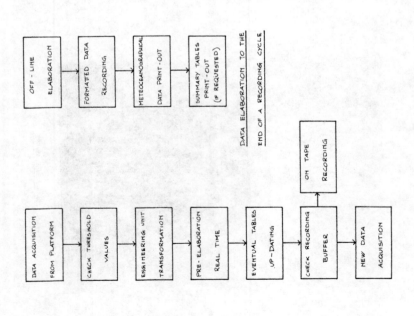

DATA ELABORATION TO THE
END OF A RECORDING CYCLE

Fig. 4 TYPICAL DATA PROCESSING COMING FROM
THE PLATFORM DURING A RECORDING CYCLE

(09.18/79)

AUTOMATED PROCESSING OF RECORDINGS
OBTAINED BY SIDE-LOOKING SONAR

Ph. GAUDILLERE, General Manager of SESAM
B. ROBERT, Consultant Engineer of SESAM

Summary

Two problems arise when studying SONAR films: the identification and
the interpretation of the forms recorded and their restoration
(resolution of the distortion caused by collection of the data at
the measuring instrument). Aside from considerable savings in time,
automatic processing of the image enables an image with excellent
definition to be obtained. The image is photographed and broken down
into points, namely 64 shades of grey that can be perceived by the
machine. It is hence possible to identify specific shapes (pipeline,
rocks, wrecks, etc...) or geological structures. The image can be
fully restored, which is impossible with manual processing, for
reasons of cost and time.

INTRODUCTION: FEASIBILITY OF AUTOMATIC PROCESSING

Before all else, automatic analysis of SONAR documents necessitates:
- elimination of the surface echo,
- indication of the bottom echo, the first incoming signal, that is the
basis for good restoration of the image,
- contrasting of the shapes to be interpreted,
- measurement of the "ranges" of these characteristics echos,
- complete restoration of the underwater image.
Accordingly, the feasibility study was concentrated towards an
understanding of the dynamics of the image in order to ensure good
detection of the only echos of interest to us, or the integrality of the
image. New equipment was used for these experiments.

1 - HARDWARE AND SOFTWARE FACILITIES

The digitization and image storage hardware comprises:
- 1 CCD 448 x 380 point matrix camera,
- 1 standard 50 mm object lens,
- 1 CCD video acquisition card capable of digitization to 64 shades of
grey,
- 1 video card enabling the digital image to be displayed on a standard

black and white monitor,
- 1 image memory,
- 1 Motorola 68000 fast processor.

Particular care was taken with the lighting, which was done by transparency with a neon tube light box housed in a dark cowl.

In view of the size of the documents and the precision of 488 x 380, a resolution in width of 0.7 mm was obtained.

The software used (the standard software of the makers or developed within the company in collaboration with Société BERTIN) consists of:
- a histogram of the image giving all the levels of grey perceptible on any line selected in the image,
- detection of contours (accentuation of two-dimensional intensity gradient) highlighting all the "breaks" in the intensity of the document, by digitizing the image stored,
- computation of the mean value and variance of the image, enabling a mean threshold to be selected and the intensity scatter to be expressed in figures in order to determine matched threshold processing, depending on whether the image is dark or light,
- threshold processing followed by concise digitization of the remaining levels (example: elimination of values below a certain threshold),
- two-dimensional filter enabling the isolated details of an image to be "erased" or emphasized.

2 - THE TESTS

The tests proved entirely satisfactory with respect to the definition of the image.

Measurement of the shades of grey yielded the following:

Position	Shade of grey	
BLACK	0	
WHITE of fluorescent tube	63	
Central band of document	56	
"Black" crest	20	25
Mean level	40	45
Maximum deviations	15	56

The dynamics of the image are hence more than enough to justify treatment such as threshold processing.

The echos were identified. However, there is still one problem to be solved, namely that of the interpretation.

The identification of the shapes resides on the image contrast, though also on geometrical identification (determination of the shape of the object by its contours). To do so, one must first know what one is seeking in the image. Certain information is of no interest and can easily be eliminated automatically: surface echos, excess or scale lines, various faults such as folds in the paper, etc... On the other hand, the image of a pipeline the alignment of which intersects the path of the ship frequently displays little contrast.

The same also applies to the "ripple marks", recognition of which is nonetheless of fundamental importance. Depending on the information that is to be extracted, it is therefore necessary to delineate the ranges of contrasts within which one is best advised to work.

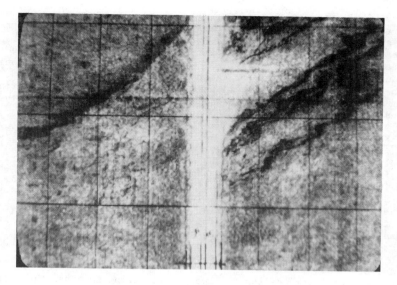

- A 0 -

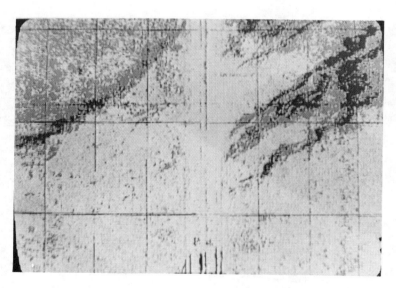

- A 2 -

The simplest technique is hence to run the SONAR recording through a TV camera, with the person responsible for interpretation following the progress of the images on a screen. Whenever a zone of interest is encountered, he can interrupt the advance, frame the zone, select the type of processing and request analysis of the image. The system is no longer fully automated, though no major information can be omitted.

On the other hand, in certain cases, it will be necessary to bring out a maximum amount of information and the image will then be completely restored point by point.

3 - <u>RESULTS</u>

Only three seconds are needed to scan an image. Relying on matrix processing of the data by means of a "submatrix" moving through the space of the image, one can estimate that from 5 to 30 seconds are needed for a 35 x 35 cm document, with a resolution of 0.7 mm.

These processing speeds render a paper advance rate of 1 to 6 cm/ second for automatic analysis. The following photographic documents result from analysis of the SONAR image: (AO and A2):

The echos marked stand out, which is of importance, for instance for defining the major geological structures.

Photo AO: original

Photo A2: threshold processing followed by filtering to erase the details.

Detection of the bottom and surface echos is more delicate, though simple detection of the gradient suffices.
Photo B0: original (on the right a fold in the paper is shown in white),
Photo B1: detection of the contours.

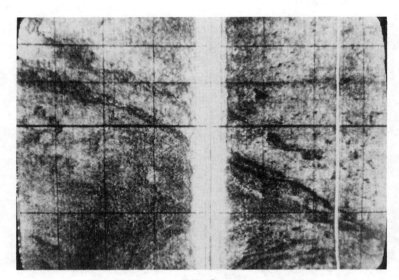

- B 0 -

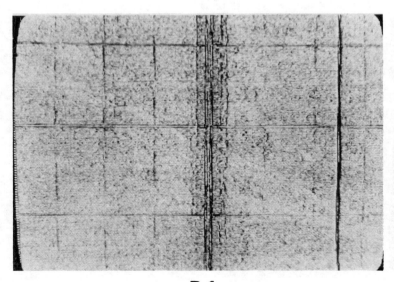

- B 1 -

The software was also tested on images where no notable echo was visible.

The pulse and scale lines are readily identifiable since they are known a priori (continuity test).

Photo C0: original

Photo C1: threshold processing.

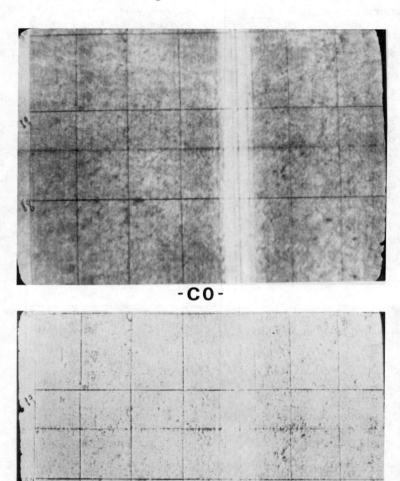

-C0-

-C1-

Automatic processing of a SONAR image hence yielded good results.

Automatic interpretation is particularly interesting for structural mapmaking, since the major lines of the relief and the accidents (faults, etc...) appear well-contrasted.

The sedimentological studies or the search for a precise and point phenomena are more delicate and in the present state of advance of the work, assistance by an operator would appear indispensable.

(18/75; 09.07/77)

DEEP WATER STEEL PIPELINE

PIPELINE AND CONNECTION TECHNIQUES

B. de SIVRY - Compagnie Française des Pétroles
P. THIBERGE - Société Nationale ELF AQUITAINE (PRODUCTION)

Summary

The purpose of the deep water steel pipeline was to solve all the problems set by laying, repair and use of subsea pipelines for diameters of up to 40", under difficult sea conditions and down to depths of up to 1000 metres.
The project had the following three main objectives:
. to improve existing techniques and means and extend the range of use of conventional laying barges. These limits have no been defined.
. depending on the operational or economic limits of these techniques, to study and develop new pipeline laying methods: the work essentially covered the RAT (French acronym indicating towing - end-connection - tensioning) method that underwent tests in the North Sea in 1977.
. to study pipeline repair methods: the work essentially concerned the hyperbar welding technique and the atmospheric pressure welding method known as WELDAP, tested in 1978 in a fjord in Norway.
The limits observed for application of the RAT method have led to examination of the possibilities of a new laying method known as the "J configuration curve" method. Likewise, the work performed on hyperbar welding and the WELDAP method have revealed the need for extensive work to prepare the ends of the pipelines and develop mechanical connectors. These three new subjects for study, on which work was started under contract N° 09.07/77 subsequently led to very considerable development and are dealt with in a separate part of this communication.

1 - INTRODUCTION

These projects started in 1974 at a time when the first major oil and gas pipelines serving the North Sea fields were nearing completion.
The difficulties encountered in laying these lines were to lead the oil companies to ask themselves three questions:

- what were the limits of the known pipeline laying methods, and could they be extended, where the depths most widely involved at the time

were from 100 to 150 metres ?
- what could be the new methods that might enable pipelines to be laid in depths of 500 to 1000 metres ?
- how could pipelines laid at such depths be repaired ?

These questions set the guidelines for the work started in 1974 and which are now nearing completion today.

2 - IMPROVEMENT OF EXISTING TECHNIQUES

Laying methods using a stinger that were applied in 1973 showed that this method could be improved for application to depths going as far as 150 metres, and a priori, even beyond this depth. The aim of the project was to seek to achieve possible improvements and to define the limits of this laying method, and then to undertake laying tests in a depth of 300 metres.

The initial phase of the work consisted in adapting a dynamic CAD programme simulating laying of a pipeline from a barge, based on the measurements made when laying from the barge ETPM 1601.

In actual practice, the development of this programme and its application turned out to be highly delicate and can not be considered as operational.

Elsewhere, studies performed on anchoring systems and barge laying equipment enabled the depth limits to be defined in terms of the tube types and sizes to be laid.

For large barges, tubes with large diameters (30") can be laid down to depths of 400 metres. This depth can be increased down to 600 metres by modifying the anchoring system.

For light barges capable of being anchored by dynamic positioning, the maximum depth for laying a 30" pipeline in X 65 grade steel would appear to be 300 metres, and 850 metres for a 12" pipeline in X 65 grade steel.

The logical followup of this project would be to perform a laying test in depths of from 300 to 600 metres, using conventional facilities. Thanks to progress achieved in this area in the Mediterranean (line across the Messina straits), it was decided not carry out this test, which would merely be to repeat the same exercise, and to concentrate the brunt of the effort on new laying techniques capable of higher performance as regards laying rates and depths.

Lastly, two systems have been designed to improve the safety of the pipelines during laying operations:

- an anti-heave system for handling bundles of tubes has been designed, though unfortunately no applications have been found for it, owing to its large size,
- a self-locking calibration system (CAB) the purpose of which is to restrict launching of the pipeline in the event of a laying incident occurring and to limit any crushing. However, progress achieved in the North Sea, both as regards laying techniques and repair techniques have considerably reduced the potential interest of such a system.

3 - NEW LAYING TECHNIQUES - THE RAT METHOD

This technique was essentially directed towards increasing the depths at which pipelines could be laid using the conventional method using the "stinger", namely a long articulated arm hung over the stern of the laying barge and carrying the pipeline being laid.

This new technique, known as the RAT (indicating towing, end-connection and tensioning) combined the following at one on the same time:

- a support with average dimensions,
- anchorage by dynamic positioning for laying in deep waters,
- short connecting-up times.

The basis of the technique resided on the end-connection of sections of pipeline from 1000 to 2000 metres long, constructed onshore and towed either on the surface or below the surface, with the assistance of buoys.

These sections were towed from the bowl of the barge, connected up by welding the new section on the barge to the pipeline being laid, and then laid from the stern of the barge. It was expected that by applying this method, the laying time would be speeded up considerably.

The studies covered:

- the diameters and apparent weights of the tubes for the various depths,
- the possibilities of manufacturing tubes and the short-term evolution of the tube products,
- the ability of the tube sections to be towed over great distances,
- the equipment to be applied, buoyancy aids and securing systems (for towing),
- keeping the section afloat during the end-connection operation,
- even distribution of the tension forces,
- the general characteristics of the end-connection device,
- the outputs exepected compared to the conventional ahead laying method.

Two major trials were indispensable: it had to be proved that it was possible to tow a pipeline over a great distance and that the section could be connected up at sea.

4 - TOWING TRIAL IN THE NORTH SEA

The purpose of these trials was to check the possibility of towing a pipeline over great distances and to compare the behaviour of the line during the towing period, against the results obtained during tank tests made on several tube models, with the towing speeds, direction of the currents and sea conditions as variables.

The trials took place from 13th October to 11th December 1975 in the Moray Firth in the North Sea, in accordance with the following phases:

- onshore construction of a 1000 metre section of 16" tube equipped with buoyancy units,
- launching of the line,
- surface towing trial,
- subsurface towing trial (depth: -15 metres).

These tests showed that it was possible to tow a pipeline section even in heavy weather conditions (8 to 10 Beaufort). The manoeuvres proved easy to carry out everytime, even in unfavourable weather conditions.

The measurements made during the trials yielded the following information:

- the towing force is more evenly distributed for subsurface towing than for surface towing,
- the bending moments in the vertical plane are related to the period and amplitude of the wave,
- the bending moments in the horizontal plane depend on the direction of the swell, but not its height.

The behaviour of the pipeline turned out to approach fairly closely that forecast by the tank tests and enabled a method of computing this behaviour to be developed.

It emerges as a result of these studies and trials that not all pipelines have the same propensity for towing and that, depending on the diameter and apparent weight, the stresses will vary with the period and significant amplitude of the wave.

In addition, the North Sea towing trial made on an uncoated concrete tube and the tests performed on a test bench have shown that repeated bending of a coated tube did not impair the integrity of the concrete sheath.

5 - RAT METHOD LAYING TRIALS

The objective of these trials was to check the possibility of laying a pipeline under tension in the open sea, from a dynamically positioned laying vessel.

The test programme covered the towing, end-connection, laying and raising of four sections of 20" diameter pipeline (two 1000 metres long and two 500 metres long) in a depth of water of 250 metres in the North Sea, South West of Karmoy island.

The sections were constructed on a barge anchored in Stavenger fjord. Once built and fitted with their buoyancy units, these sections were moored to buoys and then towed to the testing site.

The dynamic positioning ship "Flexservice N° 1" was equipped on the starboard side of the hull with a test bench consisting of:

- an end-connection machine for connecting up two sections,
- a tensioning machine capable of exerting a pull of 40 tons,
- a flexible stinger 80 metres long.

The trials took place from 9th June to 24th June 1977. The operations consisted of the following in turn: installation of the stinger towed from Stavenger fjord and handling of the first section.

Bringing up the first section and the beginning of the "eating" operation took place satisfactorily. Deterioration of the weather conditions led to failure of the cable holding the first section in line when it was partly engaged onto the laying ship. The pipeline section thus released took up a central position forward and during the clearing operation, a sudden manoeuvre of the ship caused the stinger to be damaged.

However, in order to gain the maximum information from this method, it was decided to carry out an end-connection operation, this being the major phase of this technique. The two wells of this operation, which were made relatively quickly applying homologated welding methods, were subjected to X-ray testing and proved satisfactory.

6 - RANGE OF APPLICATION OF THE RAT METHOD

The incidents that occurred did not call into question the RAT method. However, the sea trials showed firstly the limits encountered in towing the tubes and secondly the need to have a properly adapted support. An economic study was performed with the framework of contract N° 09.07/77 to specify the range of application of the RAT method, and its conclusions can be summarized as follows:

- towing pipeline sections sets a limit to the weight of the tubes that can be towed, leading in turn either to a limitation as to the depth, or to the use of steels with higher characteristics and hence at a higher price,
- an onshore construction yard is economically justified for constructing the pipeline sections, compared to the cost of a laying barge, only above a minimum length of 25 to 50 kilometres of line to be laid,
- on the support used with the RAT method, a stinger is used that has the same overall dimensions as those of conventional laying barges. At the same time, dynamic positioning sets the same constraints and in particular the maximum depth is the same,
- overall, the RAT method yields considerable savings only for the development of satellite fields requiring pipelines of different lengths and diameters, amounting to a total length of 25 to 50 kilometres.

The depth limitation of the RAT method has oriented the study work on laying pipelines in great depths towards the method known as "J-configuration curve" laying. The preliminary studies on this method carried out within the framework of contract 09.07/77 were continued under contract 09.19/80. They are described later in the Conference Proceedings.

7 - REPAIR OF PIPELINES BY THE HYPERBAR WELDING METHOD

Since it was indispensable to develop a method of repairing the pipelines, the following were the objectives of this project:

- in the short-term, to choose from amongst the existing methods those to be developed and rendered operational for depths of water of 150 metres (North Sea),
- in the longer term and for greater depths of water going down to 1000 metres, to seek and define the main avenues along which the development work should be conducted.

The shallow depth repair method selected was that of welding the replacement tubes within an air bubble kept at the bottom pressure, known as "hyperbar welding".

The equipment designed and built consisted of: an aligning device and a welding chamber, itself consisting of:

- the welding chamber proper,
- the inspection and testing chamber,
- the saturation chamber.

A connection test was performed in the Mediterranean in a depth of 150 metres of water on 32" diameter pipelines in X 65 grade steel. The tests started on 19th November 1974 and finished on 22nd November 1974.

A direct application was quickly found for this technique, particularly in the North Sea, where numerous construction and repair connections were made using this method.

8 - REPAIR OF PIPELINES BY ATMOSPHERIC PRESSURE WELDING

Since the hyperbar welding chamber is maintained at the bottom pressure, the limits of use of this method are those of the divers.

To go beyond these limits, a new method has been studied, based on the use of a chamber maintained at the atmospheric pressure prevailing on the surface, offering the advantage of eliminating the compression/decompression periods that have to be spent by the personnel sent to the sea bottom. The main drawback lies in the need to ensure a tight seal both on the inside and outside of the two ends of the tubes to be welded, under the external pressure exerted at the sea bottom.

The WELDAP system consists of two basic installations:

- A spherical welding chamber 3 metres in diameter equipped with two diametrically opposed openings to take the ends of the tubes. This chamber is equipped with an internal system for plugging the tube and contains the "spool" to be welded between the two ends of the tube.
- A service module clamped to the welding chamber, consisting of a sphere 4.2 metres in diameter and fitted with the necessary equipment to be pulled along the bottom. It clamps onto the receptacle and acts as base for all the connection operations. It supplies the power needed for the tube machining operations and maintains the air inside the welding chamber breathable.
- A submarine for shuttling the personnel between the surface and the service module.

The studies started in 1975 and construction of the equipment was finished by the beginning of 1976.

Shore trials of the main elements and in particular the sealing gaskets of the welding chamber took place from January to June 1976 in Marseilles, in the COMEX workshops.

The preliminary tests at shallow depths (-30 metres) on all the equipment took place in Marseilles in 1977. During this test, an 80 metre section of 20" OD tube was laid and all the operations of hoisting, cutting and insertion of the tube into the welding chamber performed, after which the spool was connected up by welding.

The final test, performed in February 1978, took place in a depth of water of 265 metres in ERDFJORD, situated 40 miles North of Stavanger, from the TALISMAN.

Manual arc welding was used to make the connection, without preheating on a 20" diameter tube 21 mm thick, in X 65 grade steel.

These operations necessitated 18 transfer dives amounting in all to 112 hours of diving time.

9 - PREPARATION FOR REPAIR

The tests performed on hyperbar welding showed the importance of careful preparation of the ends of the pipelines. This preparatory work is carried out by divers in the case of hyperbar welding. However, divers can not be used at very great depths, in addition to which the WELDAP method requires very clean tube ends in order to achieve the necessary internal and external tightness. An automatic machine is indispensable for preparing the ends of the pipeline. Preliminary study performed within the framework of contract 09.07/77 has shown that this machine was at least as bulky and required just as many implementation facilities as the hyperbar or WELDAP welding methods. This work has been continued under contract 09.19/79 and is developed upon elsewhere in the C and P conference proceedings.

10 - CONNECTION BY MEANS OF MECHANICAL CONNECTORS

The work on the WELDAP method has shown that it has limits, essentially related to the weight of the equipment to be used and its sensitivity to currents on the sea bottom. Beyond these limits, repairs can only be considered by means of mechanical connectors. Many connectors of this type have already been built, but all are equipped with plastic or rubber seals, the behaviour of which with time can not be guaranteed. Preliminary work on a new type of connector ensuring a metal-to-metal seal has been performed. The seal is obtained on a collet that is automatically formed at the sea bottom on the end of the pipeline to be connected. The initial tests carried out within the framework of contract 09.07/77 have proved sufficiently convincing to justify construction and testing of the equipment under contract 10.21/79. This work is described elsewhere in the proceedings of the Symposium.

REMARK

This work was carried out in close collaboration with ALSTHOM-ATLANTIQUE, COMEX, DORIS and ETPM.

1 Onshore construction at yard on sections of specific lengths

2 Launching and storage of sections equipped with buoyancy units

3 Towing to laying site

4 End-connection of sections on surface support

5 Tension laying of sections by "heating" the pipeline and recovering the buoyancy units on the surface support

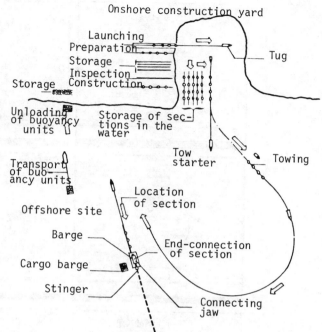

Onshore construction yard

Launching
Preparation
Storage
Inspection
Storage Construction
Tug

Unloading of buoyancy units
Storage of sections in the water

Transport of buoyancy units

Tow starter
Towing

Offshore site
Location of section

Barge

Cargo barge
End-connection of section

Stinger

Connecting jaw

R A T LAYING TRIAL
Principle

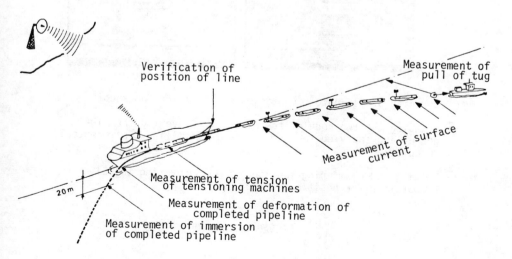

Verification of position of line

Measurement of pull of tug

Measurement of surface current

Measurement of tension of tensioning machines

Measurement of deformation of completed pipeline

Measurement of immersion of completed pipeline

20 m

R A T LAYING TRIAL
Measurements during laying operation - Pipe heating operation

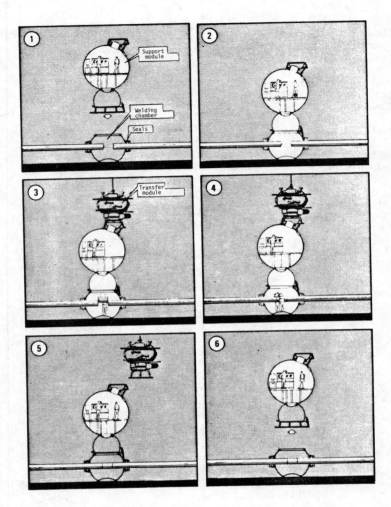

① Lowering of EIS onto receptacle

② Clamping the two modules together and emptying the
receptacle

③ Cutting and chamfering the tubes
Preparation for positioning the
spool

④ Welding the spool

⑤ Recovering the MTPA
Refilling the receptacle

⑥ Recovering the EIS

WELDAP CONNECTION PHASES

(07.13/77)

THE TM 402 DEEP WATER TRENCHING MACHINE

P. VIELMO
Servizio Impianti e Mezzi Marini
Tecnomare SpA

Summary

This report describes the principal features and results of the Tecnomare TM 402 project for the design and construction of a system for the burial, in the seabed, of cables and flexible pipelines. Launched in July 1977, the project has led to the production of a tracked vehicle capable of crossing the seabed and being guided by the cable or pipe to be buried. This device, supplied with electricity from the surface and operating under automatic remote control, can excavate a vertical-walled trench up to 1.5 metres in depth underneath the cable or pipe. In 1980 and subsequent years, operational testing of the system was carried out under a range of trial conditions whereby both cables and bundles of electro-hydraulic lines were buried. Work is currently being carried out on the TM 402 system to optimize its capabilities and so make it suitable for burying certain types of power cable and to enable it to work in a wider range of seabed conditions, especially where the bearing capacity is very low.

1. INTRODUCTION

The TM 402 project has been developed in order to provide an adequate technical solution to the problem of protecting cables and flexible submarine lines from the risk of damage associated with human activity at sea (fishing, anchorage, ...) or with environmental agents such as waves and currents which can cause such phenomena as "scouring" of the sea bottom, the formation of "free spans" and consequent excessive mechanical stresses.

It is well-known that the construction at sea of large hydrocarbon production complexes generally requires the laying, on the seabed, of considerable numbers of submarine lines providing electric power to the platforms or linking (sealines and control lines) minor production units (satellite wells, small clusters of wellheads, etc.). It is felt that the development of such lines will become particularly significant in the case of future production systems based mainly on underwater satellite units. In addition, the existence of intense surface activity (supply vessels, marine construction...) requires the adoption of protective measures for these lines.

Similar problems occur in the case of submarine power and telecommunication cables, particularly where they run through coastal, i.e. shallow, waters.

In the case of large-diameter lines with a substantial wall thickness, adequate protection can be provided by means of extra thickness and coatings. In the case of small-diameter piping or cables, it is accepted that they are best protected by burying them at an appropriate depth in the seabed.

Further, in contrast to large-diameter pipes, flexible pipelines and cables cannot provide adequate mechanical support for positioning and guiding the systems used to excavate the trench; hence the system must be based on active locomotion across the seabed itself. An excavation system based on mechanical tools was adopted because of the need to operate under the vast range of sea and bottom conditions which may be encountered in the course of a single laying operation.

It was on the basis of these main premises that the project in question was launched (in July 1977), developed and completed (in December 1981) with the achievement of all the main objectives.

Fig. 1 shows a perspective drawing of the entire system that was developed.

2. DESCRIPTION OF THE PROJECT

The TM 402 project was carried out over a period of 4 1/2 years, including the operational trial phases, and part of the costs were borne by the EEC and the Istituto Mobiliare Italiano (IMI). As sole sponsor, La Tecnomare has itself supported the project.

There were seven main phases in the project :
a) basic design (4 months)
b) detailed and structural design (12 months)
c) obtaining materials and components, construction (12 months)
d) workshop assembly (6 months)
e) commissioning and land trials (4months)
f) sea trials (4 months)
g) operational trials (12 months)
In total, the project cost 3 280 million lire.

3. DESCRIPTION OF THE TM 402 SYSTEM

The TM 402 system was developed and perfected for the burial of submarine cables and flexible pipelines and essentially comprises :
- a remote-controlled, tracked underwater vehicle, equipped with a chain-driven excavator, guided by the line to be buried which it straddles as it moves across the sea bed (Figs 2 and 3),
- an electrical umbilical cable to transmit power and command signals to control the device when it is submerged,
- a surface system consisting of a motor generator unit, a control cabin, an auxiliary power generation module and a power winch for paying out and reeling in the umbilical cable.

The underwater vehicle has been designed to cut a trench below the line, already laid on the seabed, which has to be buried. The line is lifted from the seabed and held above the excavator by a system of rollers which, by means of the forward articulated arm, provide the instrument readings needed for guidance. The line being buried is then placed at the bottom of the trench by an appropriate transport device which is also equipped with rollers.

The power supplied is converted in the underwater vehicle into hydraulic energy and then distributed to the various operating systems.

It is possible, using a dedicated computer, to control the underwater vehicle completely automatically from the surface. It can also be controlled in a semi-automatic mode (in which case the computer plays a supervisory role by blocking wrong commands and dealing with the alarms) or, alternatively, in a completely manual mode from the console.

For safety reasons, and solely during the initial positioning phases, the vehicle can be controlled locally by means of the hydraulic valves located on top of it. In this case, it is supplied with power from the surface by means of an auxiliary hydraulic umbilical connection and a dedicated hydraulic panel.

Positioning the device on the sea floor, and inserting the line to be buried, require the assistance of divers, but trenching operations are controlled entirely from the surface.

An instrumentation system has been provided to ensure the safety, throughout all phases of the operations, of the line being buried and to indicate the effective burial depth. The control system can automatically regulate all operating procedures during excavation (the machine encountering and surmounting obstacles etc.) selecting optimal operating parameters (forward speed, depth of trench etc.).

Fig. 4 presents the main features of the system.

As previously mentioned, the excavation system comprises a chain equipped with blades and/or chisel bits (according to the nature of the seabed) supported by means of drop-base rollers on a mobile mounting (Fig. 5).

This mounting is positioned by means of a hydraulic cylinder and its inclination can be varied in order to achieve the various trench depths required by the work on hand.

In addition to providing information about the relative positions of the vehicle and the line being buried, the forward guiding arm (Fig. 6) makes it possible to raise the line itself and to compensate for any slack in the line which may appear as the machine moves forward.

A comprehensive instrumentation system transmits the geometrical parameters of the submerged vehicle (the relative position of the forward and rear arms, the inclination of the excavater) to the surface as well as data concerning the status of the system and of the equipment (forward speed, the forces acting on the tracks, the pressure of the line on the rollers and the conveyor, revolutions of the excavator chain, oil temperature and pressure, etc.).

The submerged vehicle is controlled at the surface from a console which displays the instrument parameters as well as featuring a synoptic video which reproduces the geometry of the system in relation to the cable and on which there is also a display, in clear, of the operating data relating of the system and of any alarms (Fig. 7).

A computer-controlled data acquisition system provides a real time print-out of recorded system and excavation parameters (forward speed, trench depth, system status ...).

A multiplexing system, which uses the power supply line to the auxiliary devices and a carrier current system, is employed for the transmission of signals and commands. Provision has been made for operation with a multi-core control line as a back-up. Fig. 8 illustrates the underwater electronic modules.

The electric motor is connected to the hydraulic pump unit by means of a mechanical coupling (Fig. 9) as well as being directly connected, by means of a dedicated circuit in the umbilical cable, to the power distribution panel located in the surface control module.

4. RESULTS OF THE PROJECT

The TM 402 system was first tested and developed during land trials aimed at checking its overall performance and, in particular, at investigating the excavation system's ability to operate in very compact ground conditions. It successfully operated on a simulated rocky soil with maximum compression resistance of 150 kg/cm^2 (Fig. 10).

Subsequently, during June and July 1980, sea trials were carried out at Toulon (France) in which it was ascertained that the full operating design capacity could be achieved in semi-deep waters. During these trials, carried out on a section of cable laid for this purpose, the trench was cut through soil consisting of compacted clay with layers of calcerous conglomerates.

Fig. 11 shows the equipment line-up in the trials, while Figs 12 and 13 illustrate the trench cut in the sea bed.

In June and July of 1982, an elctro hydraulic control line for a submarine well was buried. Although conditions were sometimes worse than those provided for in the specifications, the line was correctly buried over a total length of approximately four kilometres.

5. CONCLUSIONS

The TM 402 system has proved fully capable of meeting the design specifications and has achieved a level of operating reliability which makes it suitable for the operations envisaged. At the present time, the TM 402 is being modified to make it suitable for the burial of cables which require tensioning during burial.

La Tecnomare has used the TM 402 system as the basis for developing a family of systems for the burial of cables and submarine pipelines and for geophysical prospecting. These systems, some of which are at an advanced stage of development (Fig. 14), are designed with a highly modular construction and a high degree of interchangeability of sub-assemblies and components.

One system in particular, known as the TM 402 B, has been developed for the burial of rigid pipelines of small and medium diameter (sealines, flowlines) even where the seabed has an extremely low bearing capacity. The tracks, which have a very large bearing surface, used in the new system can also be installed on the TM 402.

It is hoped that this systematic approach will make it possible to meet the demands of a wider range of operating conditions created by the seabed or by the trenching requirements concerned.

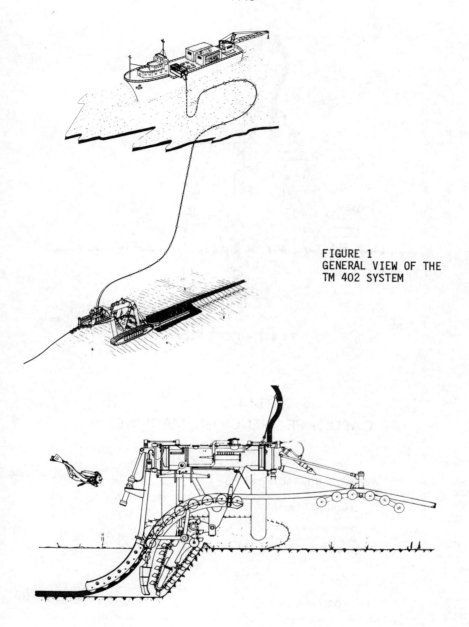

FIGURE 1
GENERAL VIEW OF THE
TM 402 SYSTEM

FIGURE 2 - TM 402 - SIDE VIEW

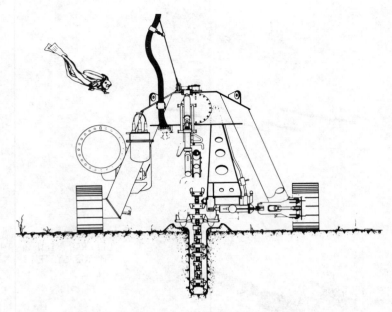

FIGURE 3 - TM 402 - FRONT VIEW

TM 402
CABLE/PIPE TRENCHING MACHINE

VEHICLE CHARACTERISTICS

. LENGTH x WIDTH x HEIGHT	5.6 x 5.6 x 4.0	M
. LENGTH (EXTENDED)	10.0	M
. WEIGHT (IN AIR)	22.0	T
. OPERATING WATER DEPTH	160.0	M
. SHAFT HORSE POWER	200	HP
. CABLE/PIPE DIAMETER	5-30	CM

PERFORMANCES

. TRENCHING SPEED	10-400	M/H
. TRENCH HEIGHT	0-1.5	M
. TRENCH WIDTH	0.25-0.40	M

. TRENCHING ON: ALL SANDY AND CLAYEY SOILS
ROCKY SOIL (UP TO 150 KG/CM^2
COMPRESSIVE STRENGTH)

FIGURE 4 - TM 402 - MAIN FEATURES

← FIGURE 5
TM - 402 - TRENCHING
SYSTEM

FIGURE 6
TM 402 - VIEW OF THE
VEHICLE - GUIDE ARM
↓

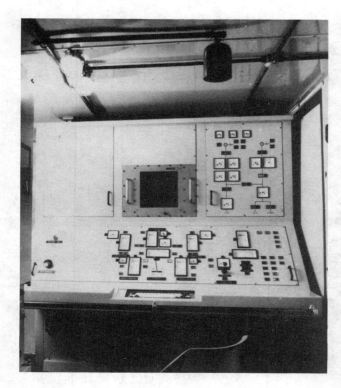

FIGURE 7
CONTROL PANEL

FIGURE 8
TM 402 - ELECTRONIC
MODULES

FIGURE 9
TM 402 - ELECTRIC
MOTOR AND HYDRAULIC
MODULES

FIGURE 10
TM 402 - TRENCHING
TRIALS ON LAND.
TRENCHING IN SIMULATED
ROCKY GROUNDS

FIGURE 11
TM 402 - SEA TRIALS
AT TOULON -
EQUIPMENT LINE-UP

FIGURE 12
TRENCH

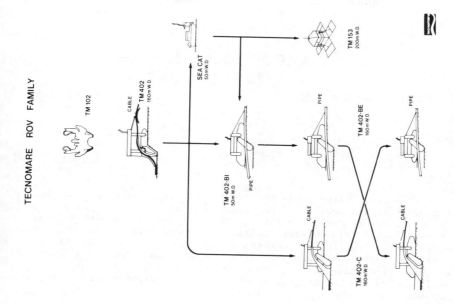

TECNOMARE ROV FAMILY

FIGURE 14 - TECNOMARE DEVELOPMENT PROGRAMME FOR UNDERWATER VEHICLES

FIGURE 13 - BURIED CABLE

(06.05/76)

FEASIBILITY STUDY OF A WAVE DAMPENER

P. FACON, Engineer, Société BERTIN & Cie

Summary

This is a feasibility study the purpose of which is to evaluate the performances and possibilities of application of a submerged wave dampener. The dampener consists of two caissons fitted with flexible check valves opening either inwards or outwards, to ensure that water circulates from one caisson to the other, dissipating the energy by turbulence. Tests in a testing tank on a model has shown that an attenuation of over 60% could be obtained. The submerged position has two advantages: no hindrance to navigation and better behaviour in storms.

1 - INTRODUCTION

Much more than the wind or currents, the waves are a major hindrance to offshore work. They complicate transport, handling berthing of supply ships, drilling work, installation of platforms and the mooring of tankers to loading buoys enormously. Naturally, the efficiency of the equipment can be increased by making it less sensitive to wave action, particularly by increasing its dimensions, as for instance is the case for semi-submersible platforms. Another possibility is attacking the problem at its origin by setting up around the equipment a calm sea zone by means of wave dampening devices. The subject of this study is to verify the feasibility of a submerged wave dampener to protect offshore equipment. The interest of a submerged device compared to a surface dampener is twofold:
- it represents no hindrance to navigation,
- since it does not undergo the impact of the waves, its structure can be lighter and less expensive than a surface dampener.

The latter point is important, considering that the breaking of a 4 metre wave represents about 200 to 800 tons per linear metre and can attain 20,000 to 60,000 tons for a wave with an amplitude of 30 metres.

2 - OPERATING PRINCIPLE

The mechanism of the wave is well known. As an initial approximation, the wave is an orbital phenomenon, the liquid molecules following closed trajectories that can be assimilated to circles. This movement brings about periodic variations in pressure proportional to the amplitude of the wave. Elsewhere, this relation is applied in pressure wavegraphs.

In the wave dampener studied, use will be made of these pressure variations to create a permanent one-way flow, the energy of which can for instance be dissipated by turbulence.

Figure 1 shows a block diagram of the dampener. It consists of two submerged caissons communicating with the sea through a series of check valves distributed throughout their length. In the first caisson, the check valves open under the action of the excess external pressure, i.e. below the crest of the wave. When the wave advances, the check valves open and then close following its movement. In the second caisson, the check valves are arranged in the opposite fashion, that is to say they open when the pressure is a minimum, i.e. below the hollow of the wave. If the length of the caissons is greater than the length of the wave, there is always at least one wave summit above the first compartment and one wave hollow on the second. The result is an appreciably constant pressure difference between the two chambers as the wave propagates.

If the two compartments are placed in communication through orifices with the appropriate cross-section, the liquid flows from the first into the second, dissipating the energy by turbulence. Should one wish to recover this energy, it suffices to install a turbine driven by the flow as it passes from one compartment to the other. As an illustration, one can say that the dampener produces levelling of the surface by taking the water from beneath the crest of the waves and restoring it to a position beneath the hollows.

It is well known that the amplitude of the orbital movement decreases exponentially with the depth. This explains that in practice 70% of the energy of the wave is concentrated in the first 15 metres. Accordingly, one can expect that the efficiency of the dampener will drop as its depth increases.

3 - DESCRIPTION OF THE EXPERIMENTAL MODELS

In the arrangement shown in figure 1, where the two compartments are mounted end-to-end, the length of the dampener in the direction of propagation of the wave reaches at least twice the length of the wave, which may be prohibitive for certain applications.

To reduce the overall dimensions and hence the cost of the dampener, one can "fold" the system back onto itself by mounting the two compartments side by side. The question then arises as to the efficiency of this layout, since the water taken is no longer sent into the hollows of the same section of the wave, but alongside. The amplitude is unquestionably damped less than with the first configuration. To evaluate the performances of this layout, a 1/50 scale model comprising compartments set side by side was built and tested in the wave tank.

The model breaks down into four juxtaposed basic modules enalbing the entire width of the wave tank to be occupied, namely 2.57 metres.

Characteristics of a module (see figure II)
- Length: 2.4 m
- Width: 0.65 m
- Height: 0.13 m
- Diameter of check valve: 5 cm
- Number of check valves: 10 cm
- Number of check valves in each compartment: 75

- Cross-section of intercommunicating passage between compartments:
  adjustable from 0 to 600 cm²,
- Depth of immersion: adjustable from 0 to 1 metre.

The check valves consisted simply of cutout sheets of rubber 1 mm
thick. To enable them to be closed in the rest position, the top check
valves were simply ballasted by a small piece of brass bonded to one of the
faces, whilst a piece of wood acted as float for the bottom check valves.
This being so, the check valves are closed at rest and open as soon as the
pressure difference exceeds a water column pressure of 1 mm.

Characteristics of wave tank
- Length: 36 m
- Width : 2.7 m
- Depth : 1.7 m
- Sinusoidal wave created by a single beater:
  . period: 0.4 to 2 seconds,
  . maximum crest-to-crest amplitude: 15 cm.
- Measurement of wave amplitude by resistive wire gauge mounted in a
  bridge circuit.

4 - EXPERIMENTAL RESULTS

The wave length of the wave depends on its period T

$$= \frac{g}{2} T^2 = 1.56 \, T^2$$

Figure III shows the evolution of the wavelength for periods of from
0.8 to 0.4 seconds. The transposition to full-scale would give 5.6 to
10 seconds for periods of 50 to 150 m wavelength, corresponding to the
usual North Sea values.

The results obtained with a monochromatic wave show that the dampening
effect varies with depth, as was to be expected. Corrected to bring it to
scale 1, the following are the results:

Depth (m)	Amplitude attenuation (%)	Power attenuation (%)
13	64	87
19	40	64
29	22	40

It is interesting to assess the interest of the check valve device to
note the difference in the attenuation obtained when communication between
the two chambers is closed. For instance, the attenuation obtained drops
from 64% to only 28% when the passages are closed. Accordingly, one can
see that the check valves and one-way circulation of water that results
from them enable the attenuation of a single sealed chamber to be more
than doubled.

Figure IV shows how the attenuation varies with the wavelength.

The measurements made on water flows through the caissons and the
resultant head losses revealed that part of the water crossing the
chambers comes from the zone over the second compartment.

Following these tests, a new model dampener was conceived (figure V).
It consists of a submerged structure in the shape of a semi-cylinder with
a horizontal axis, open towards the bottom. This structure, similar to an

overturned hull of a ship is sealed and traps air. The cross partitions set up at regular intervals separate the structure into about ten compartments that are open towards the bottom and in direct contact with the water. When the wave passes, a variation in the hydrostatic pressure results in an equivalent variation in the pressure of the air trapped in each compartment. If the axis of the structure lies parallel to the direction of advance of the main wave, one observes an increase followed by a decrease in the pressure in each compartment in frequency with the wave. Two parallel pipes A and B cross through all the compartments.

Pipe A has openings in each compartment that are closed off by the check valve enabling the air to penetrate from the compartment towards the pipe should excess pressure be generated. Conversely, channel B is fitted with check valves opening in the opposite direction enabling the air to pass from the pipe into the compartment if the pressure in the compartment is below that in the pipe.

When the wave passes, the check valves in pipe A open in turn, whilst the check valves in pipe B open half a phase cycle later. If the length of the structure is at least equal to the length of the wave, there is always at least one compartment under maximum pressure, i.e. situated beneath the crest of the wave, and another at minimum pressure, beneath the hollow of the wave. Consequently, the difference in pressure between pipes A and B remains appreciably constant for a regular wave. By connecting pipes A and B together by orifices with suitable dimensions, air circulates continuously and the energy is dissipated by turbulence. By replacing the orifices by a turbine, a power generator can be driven to produce electricity. This solution offers the following advantages:

- the turbine fed by the air at pressure is much smaller in dimensions than a turbine with the same power that is driven by water. A corollary is that its rotary speed is high, avoiding the use of up-gearing, thus reducing the size of the generator,
- the modules recover the power present in the wave over a much greater width than that of the module. Accordingly, one can set the modules apart so as to cut down the capital cost and the cost per kW installed,
- immersion of the modules below or in the vicinity of the surface improves behaviour in the presence of storms.

5 - APPLICATIONS

Various applications are envisaged:

- Protection of installation sites

Construction and installation of platforms involves delicate phases requiring a calm sea. One can consider protecting the erection site by a wave dampener. By the same token, the dampener could create a calm zone near the coast enabling either a ship to berth where there are no harbour facilities, or work such as excavations for laying pipelines to be carried out, which are operations rendered difficult by the waves.

- Protection of loading buoys

On most simple loading buoys, the tanker can tie up only when the wave amplitudes are less than 8 feet, and beyond an amplitude of 15 feet, the tanker must leave the mooring. The result is that if one refers to the meteorological records of the North Sea, the use of a dampener would enable about 15 days of production to be gained every year, i.e. an increase of about 3 million barrels. Tankers tying up to these buoys often have a displacement of 50,000 tons and hence their shallow draft would be no obstacle to a dampener situated 15 or 20 metres below the surface in a zone where it is efficient.

- Stabilization and protection of semisubmersible platforms

Again, one can envisage the protection and stabilization of semi-submersible production platforms such as those of the Catamaran type, by means of a wave dampener. In this case, the dampener could form an integral part of the structure, thus eliminating the objections linked to the drift of a gravity body and anchorages. Furthermore, since the platform swings to the direction of the waves, the dampener could be designed so as to have to protect only a relatively limited zone.

6 - CONCLUSIONS

The submerged wave dampener represents an interesting solution, since it creates only little hindrance to navigation, offers a good guarantee of withstanding storms and considerably attenuates the waves. The version incorporating air circulation is also interesting with a view to generating electricity.

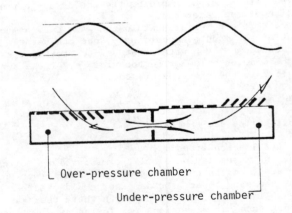

Over-pressure chamber

Under-pressure chamber

FIGURE 1

BLOCK DIAGRAM OP OPERATING PRINCIPLE OF DAMPENER

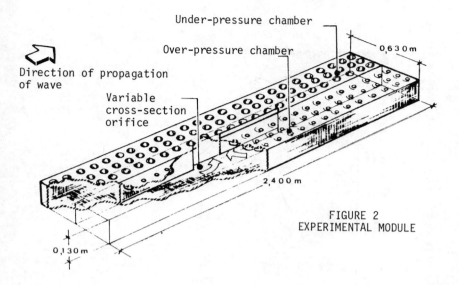

Under-pressure chamber

Over-pressure chamber

Direction of propagation
of wave

Variable
cross-section
orifice

0,630 m

2,400 m

0,130 m

FIGURE 2
EXPERIMENTAL MODULE

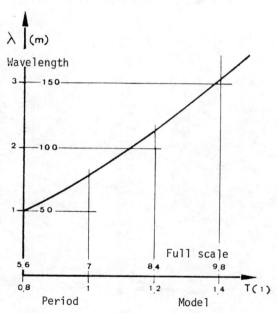

FIGURE 3 - WAVELENGTH IN TERMS OF PERIOD

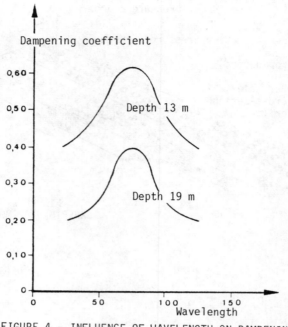

FIGURE 4 - INFLUENCE OF WAVELENGTH ON DAMPENING

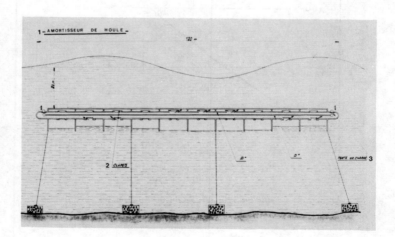

1. Wave dampener - 2. Check valves - 3. Loss of loads
FIGURE 5 - DIAGRAM OF DAMPENER COMPRISING AIR CIRCULATION

PHOTOGRAPHS OF 4 MODULES IN POSITION IN WAVE TANK

SIDE VIEW SHOWING OPENING OF FLEXIBLE VALVES BENEATH HOLLOW OF WAVE

PHOTOGRAPH OF AIR
CIRCULATION DAMPENER

AIR CIRCULATION DAMPENER - DETAIL VIEW OF CHECK VALVES

(07.36/80)

PERSONNEL TRANSFER DEVICE

J.P. MANESSE, Engineer

Summary

The subject of this project is a device for transferring personnel
between a ship and a platform. Its aim is to enable this operation
to take place in complete safety ; thanks to a cabin comprising a
constant tension device, the system enables the cabin to be lifted
from a ship without impact.
This system was built and tested in our workshop. These tests did not
turn out to be satisfactory and the project was abandoned.

1 - INTRODUCTION

The development of the offshore petroleum industry has given rise to
a particular problem, namely "transfer of personnel".
In general, the personnel is transferred between the offshore
platforms and the coast by helicopter. The problem generally arises for
transfers between platforms and ships, particularly where helicopters can
no longer operate (in most cases, owing to fog) and personnel must
rejoin the platform or the ship for urgent reasons.
In this case, the basket is used. This dangerous method is tolerated,
lacking any alternatives today. The disadvantage of the system is its lack
of safety. Another difficulty inherent to the basket is setting it down on
the deck of the ship, the result of which depends on the state of the sea
and the dexterity of the crane operator.
Since the need was becoming felt in this area, ACB decided to design,
build and develop a device for eliminating these disadvantages, namely:
- to install the passengers in a cabin,
- to enable the cabin to be laid on the deck at a definite point on the
  ship,
- to avoid bumps when setting down on the deck and picking up.
Such a device was to bring about considerably better safety during
personnel transfer operations.

2 - DESCRIPTION OF THE PROJECT

The system consists of a cabin (1) comprising a "constant tension winch" at its centre, fitted with two independent drums.

The top drum (2) winds the cable carrying the weight of the cabin, with the end of this cable hooked to a crane (3).

The bottom drum (4) winds the wire hanging beneath the cabin. This wire, hooked to a specific point of the ship (5) enables the cabin to be guided during the setting down and picking up operations. On completion of the picking up phase, an automatically jettisonable hooking system unhooks the deck wire.

The heave of the ship is compensated by one of the drums of the constant tension winch, with the other locked by a pawl system.

This project comprised the following three development stages:
- Phase 1: Design,
- Phase 2: Construction and testing in the workshop,
- Phase 3: Sea trials.

The workshop tests were to provide the solution to the final de-bugging problems so as to achieve a satisfactory level of safety and reliability. The main component to be tested is the top winch locking pawl, the correct operation of which is to ensure the safety of the system.

Static tests to 150% of nominal load (i.e. 1000 kg) proved fully satisfactory. No particular incident occurred and the dynamic tests to 125% of nominal load (i.e. 850 kg) were taking place. They consisted in simulating transfer operations, comprising the following phases:
- pick-up of the device from the platform,
- unwinding of the wire,
- hooking on of the jettisonable hook to a deck ring,
- winding up of the wire,
- operation of the wire compensation system,
- setting down on the deck,
- compensation on the top drum,
- picking up the device and automatically unhooking the jettisonable hook.

For these tests, the device was hung from an overhead travelling crane, the hook movements of which simulated the heave of the vessel.

3 - STATUS OF THE PROJECT

During simulation of a decking operation, a length of wire had just been unwound and the quick fastener secured to the ring anchored to the deck. The following "wire wind-up" phase was then commanded, ths purpose of which was to take up the slack and set the compensation system into action on the bottom drum. At the end of this phase, the device suddenly dropped to the deck when a pawl of the top drum opened. After complete expert examination of the various parts (mechanical, electrical), no precise explanation of the phenomenon could be found.

Following this fall, we were led to call into question the pawl system, since the entire safety of the device resides on its correct operation. This being so, we decided to interrupt the programme in its present concept, since reliability appears inadequate.

— 1163 —

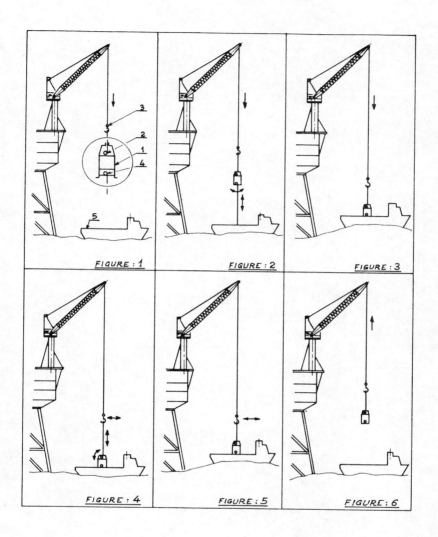

NATURAL GAS :

PRODUCTION – PROCESSING – LIQUEFACTION – STORAGE –

TRANSFER

Development of a system for the production of methanol offshore (03.87/80)

Gas disposal systems for deepwater (03.119/81)

Influence of irregular motion of a floating structure on absorption and distillation (03.70/78; 03.96/80)

Offshore liquefaction of natural gas from deep water deposits (03.29/76)

Development and model testing of a floating natural gas liquefaction plant (03.28/76)

Floating natural gas liquefaction plant for offshore liquefaction and loading of associated gas (03.48/77)

Development of a tension leg platform as supporting structure for a natural gas liquefaction plant (03.58/78)

Cryogenic fuel lines (10.14/78)

Offshore loading of liquefied gases (10.35/82)

Unlined concrete storage facilities for liquefied natural gas (14.06/78)

Construction techniques in limestone for cryogenic storage (14.13/82)

Development of a new technology for LPG storage (14.15/82)

The Schelle cryogenic pilot project cavity (14.10/80)

Constant motion cryogenic swivel joint (10.28/81)

Insulation and barrier system for marine transport and storage of LNG (12.05/78)

(03.87/80)

DEVELOPMENT OF A SYSTEM FOR THE PRODUCTION OF METHANOL OFFSHORE

E. EMERSON, Dr. D.J. BROWN and A. MIDDLETON
Stone & Webster Engineering Limited
Milton Keynes, England

Summary

When Stone & Webster submitted a proposal to the European Economic Community 5 years ago for a system to produce methanol offshore, the flaring of surplus associated gas from North Sea oil production was considerable. Rather than face the cost of pipelining this gas, the on-site production of methanol appeared to be a possible alternative. Whereas numerous schemes have been proposed for methanol and ammonia production on floating facilities these have been primarily barge type designs for shallow, sheltered waters, or for limited use at any location.

The Stone & Webster study for offshore methanol production has focussed on the use of semi-submersible structures with the view of meeting extended operation in hostile environment of the Northern North Sea.

To reduce weight, space and improve safety, the Stone & Webster power reformer concept based on gas turbine and reformer integration is proposed.

The capital cost of the concept is high mainly due to the marine aspects which far out-weigh the process plant costs.

Without pre-supposing possible project applications, the feasibility of a methanol plant offshore is cited as a proposition which may under some circumstances be attractive in remote field developments and subsequent operations.

A significant factor in the economics is the value of methanol. Under the present conditions of oil surplus, the value may be low but this condition may not prevail for long.

1. INTRODUCTION

1.1 Natural Gas Recovery

A major problem with all hydrocarbon resources in offshore locations is their economic extraction and transportation to onshore consumers. Crude oil production and transportation from the North Sea has been successfully handled by means of offshore platforms, subsea pipelines and single anchor leg mooring (SALM) loaded tankers. The gas produced in association with the crude is used as platform fuel and the surplus flared as an unwanted by-product. In 1981 4.2 thousand million (4.2 x 10^9) cubic metres of associated gas were flared in the U.K. sector of the North Sea. The energy environment of the late twentieth century has rendered this view outmoded. As a result, gas reinjection schemes have been commissioned to preserve the associated gas inside the producing formations but this can only be a temporary measure. In addition some extensive (and expensive) sub-sea gas gathering pipelines are now in use (or under construction) but a substantial amount of gas is still going to flare.

One solution to 'recovery' of associated gas especially in small fields in remote locations is its conversion to methanol.

Although methanol production involves a degrading of the energy value of the fuel, as do most other conversion operations, this does not detract from the advantages, and the 'fuel' produced can also be used as a chemical intermediate. As liquid, it is easily handled, clean and relatively safe.

1.2 European Economic Community Loan

In November 1979 Stone & Webster submitted a proposal to the European Economic Community (EEC) entitled 'Development of a System for the Production of Methanol from Associated Gas Offshore', the aim of which was to secure funding of part of the project costs for developing a system using the Stone & Webster power reformer concept. The power reformer replaces the conventional atmospheric firebox reformer used for producing synthesis gas and has the advantage of substantially reducing the size and weight of the plant - vital in any offshore design.

Work formally commenced on the project in March 1981 with the EEC granting the maximum loan of 40 percent of the estimated £1.9 million project cost. Sufficient work has now been completed to enable some conclusions to be drawn.

1.3 Previous Investigations into offshore methanol plants

In recent years many investigations have been made into the idea of constructing offshore methanol plants. Several of these investigations are given in the references. The majority of these studies involve flat-bottomed barge designs for use in sheltered waters. The S&W study involves floating platforms to maintain operations in deep water and hostile North Sea conditions.

The following conclusions can be drawn from the above studies:
(1) The processing and marine problems associated with a hostile environment can be overcome.
(2) The economic incentive for developing offshore methanol plants will improve as the world's oil reserves are depleted and substitute fuels must be found.

2. BASIS FOR STONE & WEBSTER STUDY

2.1 Plant Feedstock

The design feedstock, as shown in Figure 1, has typical natural gas composition.

Natural gas has replaced associated gas as the plant feed-stock because government legislature is forcing operating companies into recovering associated gas from North Sea operations.

2.2 Plant Capacity

The plant capacity is 1,000 tonnes/day of fuel-grade methanol. This requires a feedstock rate of approximately 10^6 standard cubic metres/day of natural gas, roughly equivalent to the production rate from what may today be regarded as a marginal gas reservoir.

2.3 Methanol Synthesis Loop Design

For the purpose of this study the equipment sizing for the methanol synthesis loop is based on the Lurgi design, but the conclusions are not changed by using other processes such as the ICI Low Pressure System.

2.4 Floating Structure

For remote locations, typified by deep water (greater than 200 metres), violent wave action and ice flows, fixed processing structures and subsea pipelines are unsuitable from both technical and economic viewpoints.

A practical solution in the above conditions is a floating production system. For the basic study, the processing facilities are located on an existing design of semisubmersible platform which has a proven stability performance under prevailing worst weather conditions. An alternative study using a floating hull shaped platform offering potential economic advantages has also been undertaken.

3. METHANOL PRODUCTION

3.1 Feedstock Desulphurisation

As sulphur compounds are poisons to the nickel catalyst in the reformer the natural gas feedstock must be desulphurised before reforming.

Much of the current North Sea Gas is essentially sulphur free but facilities are provided to protect the catalyst from trace amounts up to 20 ppm. Sulphur concentration is reduced to 0.2 ppm by hydrodesulphurisation and desulphurisation operations at approximately 400°C. Hydrodesulphurisation involves converting all sulphur containing compounds to hydrogen sulphide by the addition of hydrogen from the synthesis reactor, over a catalyst. The hydrogen sulphide is then absorbed on zinc oxide in one or two absorbers. As only a small quantity of hydrogen sulphide is absorbed, no regeneration facilities are provided and the zinc oxide is replaced when spent after approximately six months.

3.2 Power Reformer System

The power reformer system (Figure 2) consists of a pressurised reformer vessel where the shell side combustion gases have the dual function of (1) providing the heat to produce synthesis gas from methane feedstock, and (2) on exiting the reformer vessel of driving a gas turbine which provides power for the methanol plant. Compared to a conventional system, the power reformer is highly energy-efficient and takes up less space and weight.

Air enters the gas turbine compression stage where it is compressed to approximately 8.0 kgf/cm²g. In normal gas turbine designs up to 70 percent of the compressed air can be diverted to the convective reformer, the remaining air being used for combustion in the gas turbine.

After cooling, the compressed air for the convective reformer is further compressed to approximately 9.0 kgf/cm²g in the air booster compressor.

Fuel is fired stoichiometrically and then moderated with recycle flue gas to obtain a 1090°C flue gas for use as the heating medium. This recycle flue gas is provided by cooling and recycling a portion of the combustion gas via the flue gas circulator. The flue gas mixture has low oxygen content of two to three percent.

The purge from the recycle circuit is routed to the gas turbine for recovery of the compression energy.

FEEDSTOCK AND PRODUCT SPECIFICATION

FEEDSTOCK

COMPONENT	MOL. %	COMPONENT	MOL %
METHANE	95.85	HEXANE	0.12
ETHANE	2.67	C_7+	0.42
PROPANE	0.34	N_2	—
BUTANE	0.52	CO_2	—
PENTANE	0.08	M. WT. OF C_7+	(157)

PRODUCT

COMPONENT	WT. %	
METHANOL	99.85	MIN.
WATER	0.15	MAX.

FIGURE 1

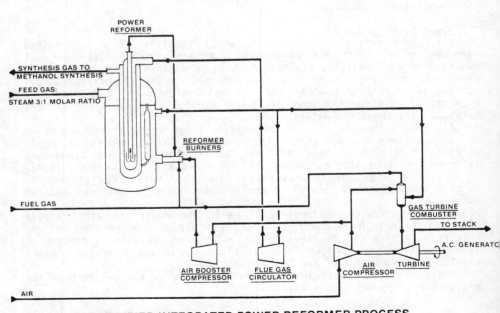

SIMPLIFIED INTEGRATED POWER REFORMER PROCESS

FIGURE 2

The process feed to the reformer is steam and natural gas in a ratio of three moles of steam per mole of carbon in the natural gas. The feed enters the top of the reformer at about 450°C and is distributed to the catalyst tubes for reforming at reaction temperatures up to 850°C. The hot synthesis gas effluent is then cooled to 22°C to remove the water vapour and provide a low suction temperature for the synthesis gas compressor.

3.3 The Power Reformer Vessel

The power reformer is a counter-current tubular heat exchanger. Hot pressurised flue gases circulate on the exchanger shell side while the process gas passes through the catalyst packed tubes. Hot pressurised flue gas is generated in burners which are integrated in the shell of the reformer.

At the top of the tube bundle the 76 mm tubes are necked down to 51 mm diameter to provide increased space for proper combustion gas disengagement and allow for sufficient tube sheet support.

The bottom portion of the tube bundle is reduced to 25 mm transition tubes which pass through external shield tubes. The shield tubes protect the process tubes from burner radiative heat fluxes. The shield tubes are supported on a plate which separates the hot heating zone from the cooled header zone.

The central return tube is utilised to recover sensible heat from the hot process synthesis gas. An exchanger is inserted in the return tube for heat exchange with the recycle combustion gas. In addition to heat recovery the synthesis gas cooling facilities the design of the upper tube sheet by lowering the synthesis gas exit temperature.

3.4 Power Reformer Auxiliaries
Gas Turbine

A minimum of 30 percent of the air compressed in the gas turbine air compressor is required for combustion in the gas turbine. This corresponds to operating the gas turbine with an excess air of approximately 50 percent.

The quantity of excess air can be reduced from the more conventional 500 percent as hot pressurised gas from the reformer is let down through the gas turbine, tempering the combustion products' temperature within metallurgical constraints. The replacement of cold excess air by hot inert gas increases the efficiency of the gas turbine considerably.

The power output from the gas turbine is sufficient to power all rotating equipment on the methanol plant.

3.5 Heat Recovery Systems

The heat recovery systems are fully integrated throughout the feed desulphurisation, the power reformer, the synthesis reactor and distillation units of the methanol plant. Sufficient medium pressure steam is generated to supply the process demand of the power reformer.

The quantity and type of heat exchangers to be installed is optimised on a heat exchanger size/duty basis in order to minimise the weight of equipment.

3.6 Reformer Burners

A key feature of gas turbine technology is used for the reformer burners, namely the burning of fuel at stoicometric conditions of

1600/1700°C, followed immediately by quenching to 1000°C, in this case by almost inert flue gas rather than extra air as for a normal gas turbine.

The combustion and recycle gas mixing takes place in six high intensity burners which are fitted tangentially into a torus-shaped firing chamber encircling the reformer vessel. Tangential firing into the torus chamber improves mixing of the hot burner exhaust gases, which are maintained at 1090°C by recycling cooled flue gas.

3.7 Power Reformer vs Conventional Atmospheric Firebox Reformer

The conventional land-based reformer is of the firebox type, burning natural gas at atmospheric pressure to maintain the reformer reaction temperature by radiative heat transfer to the catalyst-filled tubes.

The power reformer has advantages over the conventional atmospheric firebox reformer. Heat transfer is convective, thereby avoiding the need for a radiant chamber. This gives:-
- Weight reduction factor 2.5.
- Space reduction factor of 4.
- Safer and easier operation in an offshore location.

3.8 Methanol Synthesis

Synthesis gas is initially compressed to about 80 kgf/cm^{2}g and heated to 260°C before feeding to the tubular methanol synthesis reactor. The methanol synthesis reaction takes place in contact with the proprietary catalyst within the tubes. An almost constant reaction temperature is maintained by generating medium pressure steam on the shell side of the reactor, removing the exothermic heat of reaction. Medium pressure steam generated is consumed in the process or used as process heat.

The condensate product from the synthesis reactor system, consisting of 24 percent water and 76 percent methanol is fed to a distillation column for methanol recovery.

3.9 Methanol Distillation

The crude methanol from the synthesis reactor must be purified to a fuel-grade (grade A) methanol specification of 99.85 weight percent.

This purification may be easily achieved onshore using a conventional distillation column. However, in an offshore location where the process equipment is subject to movement, separation efficiency is reduced. Studies have shown that by using a semi-submersible which exhibits excellent stability performance, the distillation can be achieved in packed columns with only a small loss in efficiency.

Steps have been taken to minimise this loss by limiting the height of the columns and using smaller depth of packed beds and more redistributors.

4. SEMI-SUBMERSIBLE STRUCTURE

The process plant is designed to fit on existing semisubmersible structure designs, of the largest available in the industry, to operate efficiently throughout the year in deep water in severe environmental locations.

Equipment sizing, steelwork and piping estimates for the 1000 tonne/day methanol plant shows that the calculated deck load can be accommodated. Motion analysis is carried out to establish the performance of the vessel to determine its on-stream factor and to prove its ability to meet marine certifying authority requirements.

5. METHANOL PRODUCT STORAGE

As the semisubmersible vessel has a very tight weight limitation, no product storage facilities are provided on the structure.

Storage facilities for the methanol production are therefore provided by either a permanently moored tanker or by a purpose built SPAR facility. Operating requirements for a tanker would result in a tanker size of about 150,000 tonne capacity giving over four months production capacity. A SPAR system which comprises a catenary anchored cylindrical loading/storage platform could economically be sized for 30-40,000 tonne of methanol, i.e. about 1 months storage.

6. ECONOMICS

The capital cost of the semisubmersible vessel is considerable and for most existing structures is roughly double the cost of the process plant.

With present day values of the methanol produced the return on capital investment is low.

Methanol production from remote gas only wells is not likely to prove economical if the whole cost of field development has to be taken into account.

Associated gas in conjunction with essentially crude oil production may well change the economic factors, especially if government restrictions are enforced on the amount of gas flaring limit the oil production.

Without taking into account the complex financing considerations of taxation, regulatory requirements and other international factors, it can be said that the offshore methanol production for the North Sea conditions may be a way of making possible, overall field operations in remote areas which otherwise may have been prohibitive.

7. STATE OF ART

The use of an integrated gas turbine/process system is not novel.

For some time gas turbines in combined cycle power plants have been in operation; additionally gas turbine integration with process units have been proposed/used for research projects especially in regard to coal gasification.

The use of the Stone & Webster power reformer concept as proposed for the methanol plant requires some validation of the basic control functions especially the maintenance of gas temperatures under the varying operating modes. To verify heat transfer, flow conditions, gas temperatures, and the control of these parameters, a pilot plant design is proposed.

The operation of the pilot plant is essentially in two phases. One concerns nitrogen in the process system, thereby proving basic control parameters. The second involves hydrocarbons in the reformer circuit.

An ideal location for the pilot unit which is one tenth scale is alongside an existing plant where feedstock is available and which can accept the synthesis gas produced.

CONCLUSIONS

It is expected that the supply/demand outlook for methanol will change within the next ten years, mopping up surplus capacity and creating a demand for methanol.

It is also expected that as the more easily recoverable hydrocarbon reserves become depleted, uneconomic discoveries will move into the marginal field category, and marginal fields will become worthwhile exploiting.

The technology for methanol production is proven; the improvements in weight, space, and safety engineered by Stone & Webster bring the prospect of methanol production offshore from the conceptual into the practicable sphere.

The economics of the production of methanol are heavily influenced by the choice of carrier and field development costs; the use of any carrier will be viable if it can be fully utilised, for example by moving from field to field as they became exhausted. Again for the process plant, savings in weight and space will improve the viability of the production of methanol offshore.

REFERENCES

1. U.K. Government Department of Energy Statistics, 1982.
2. Swedyards Development Corporation - "Barge Mounted Plants have Canadian Potentials" - Canadian Petroleum, February 1983.
3. "Offshore Processing - Methanol Plant on a Submersible Platform" - Dr. G. Meinhold, Veba Oel AG/Chemische and G. Laading, Norwegian Petroleum Constants AS.
4. "The development of Waterborne Alcohol Fuel Plants" - International Maritime Associates Inc, Chem Systems Inc, Avondale Shipyards, U.S. Dept. of Commerce
5. "Schemes for Processing and Loading Gas Offshore" - Offshore Services and Technology, July 1981.
6. "A Shipyard-Built Methanol Plant" - R.G. Jackson, Continental Oil Co, C.E.P. October 1979.
7. "Barge-Mounted Methanol Conversion Plants" - Dr. A.L. Blaxley of Litton Energy Systems.
8. "How to economically recover Marginal Gas Offshore" - R.S. Hollyer & D.W. Fowler, World Oil, July 1980.
9. "Offshore Methanol Production - Selection of Carriers for Different Plant Capacities" - G.C. Widbom & C.O. Thorsson, OTC 4786, Houston 1984.
10. "Influence of Irregular Motion of a Floating Struction on Absorption and Distillation Processes" - B.K. Hoerner, F.G. Wiessner & E.A. Berger, OTC 4276, 1982.

(03.119/81)

GAS DISPOSAL SYSTEMS FOR DEEPWATER

M.A. BROOKES
The British Petroleum Company p.l.c.

Summary

The exploitation of hydrocarbon resources from remote deep offshore locations such as may be found off northern Norway presents many technical and economic problems, one of which is the disposal of gas during production operations. In this paper, the investigations carried out jointly by BP and Statoil into this problem are described. The study comprised an evaluation of the components, technological problems and alternative systems which could be used for the disposal of gas under both normal and emergency production from large offshore fields. Three systems were identified and assessments made of the feasibility of each, with conceptual designs being prepared for two of the systems. It was concluded that on platform systems presented better economic and technically feasible solutions for the scenarios considered than systems located remote from a production facility.

1. INTRODUCTION

The Gas Offshore Northern Norway (GONN) project comprised six studies jointly undertaken by BP and Statoil to investigate the feasibility of hydrocarbon production from deep water locations off the coast of northern Norway. The six studies were:-

S1 - Deep Water Field Development
S2 - Vertically Taut Moored Platform
S3 - Safety of Vertically Taut Moored Platforms
S4 - Subsea Systems
S5 - Safety of Subsea Systems
S6 - Deep Water Gas Disposal and Flaring.

This paper describes the work carried out in the S6 - Deep Water Gas Disposal Systems Study, the overall objectives of which were to investigate, define and develop possible gas disposal systems for both emergency and normal production conditions.

2. BACKGROUND

The study took as its basis two hypothetical scenarios, Scenario 1 being a dry gas field and Scenario 2 an oilfield with large amounts of associated gas, as shown in the following table.

	Gas/Oil Ratio (Sm^3/Sm^3)	Gas Production (Sm^3/D)	Oil Production (Sm^3/D)
Scenario 1	–	32×10^6	–
Scenario 2	181	12.5×10^6	69,000

The fields were assumed to be located in 350m of water in the Tromsoflaket area, about 150 km from shore.

In the S1 - Deep Water Field Development Study, a range of alternative field development solutions for the two scenarios were identified and it was concluded that a single platform production scheme was appropriate for Scenario 1, whilst for Scenario 2 two production platforms would be required. In both cases, the production schemes envisaged tension leg platforms (TLP) with associated subsea systems.

The process systems adopted in the field development solutions determined the rates at which gas would have to be disposed, and whilst a range of process conditions were investigated, it was the emergency situation which set the main design parameters for this study. Based on a 20 minute controlled blowdown of the process trains, gas disposal rates of 17×10^6 Sm^3/D and $8.6 \ 10^6$ Sm^3/D were established for Scenarios 1 and 2 respectively.

3.   STUDY METHOD

A systems approach was taken in the study whereby the various components required for a gas disposal system were identified and evaluated individually and then combined into feasible gas disposal system alternatives.

The study was carried out in two phases, Phase 1 being an initial review and assessment of alternative systems, with Phase 2 devoted to a more detail study of two systems.

In Phase I, alternative components and systems were evaluated and the limits of technology in key areas such as gas dispersion, combustion and radiation, noise, low temperatures, venting and flaring, structures, subsea flare pipelines and risers were explored. At the conclusion of Phase 1, three systems were identified namely:
-    an on-platform vent or flare,
-    a vent or flare bridge linked to the platform,
-    a remote vent or flare, linked to the platform by subsea lines.

In Phase 2, two of the systems were selected for more detail study. These comprised the on-platform and remote systems, both of which were developed in some depth to give recommended systems to satisfy the gas disposal requirements of the two scenarios. Areas of work of both general and specific application where more work is necessary were identified for future development.

4. TECHNOLOGICAL CONSIDERATIONS

4.1 Blowdown Techniques

The manner in which the platform facilities are blown down in the event of an emergency is fundamental to a gas disposal system design. A key premise of the study was that gas would be relieved from the production facilities using a controlled blowdown technique. The use of controlled blowdown techniques enables the peak flowrate to be reduced to approximately the average rate over the 20 minute relieving period, thereby reducing the maximum radiation levels from the flare, with consequent cost savings. This is particuarly important if a remote flare can be then dispensed with by locating the gas disposal system on the production facility.

4.2 Venting and Flaring

Venting and flaring generally present similar problem; a cold vent may be ignited, and a flare may turn into a vent. As a consequence evaluations of gas concentration envelopes as well as flame hot gas envelopes and radiation loads were considered.

In addition to the gas concentration and radiation levels, it is important to consider the hot gas areas being formed by the flame. Specially designed proprietary flare tips can lead to the production of a flame with reduced soot concentration, reduced heat radiation and a more stable flame compared to pipe flares.

The probability of liquid droplet carryover generally favours flares in preference to cold vents. However, this problem can be minimised by the correct design of the liquid knock-out system.

Helicopter operations may be restricted under both emergency flaring and venting conditions. However, in flaring the flames will be visible to pilots as opposed to venting when the disposed gases cannot be seen.

4.3 Low Temperatures

To enable the correct selection of materials for vessels and piping, the calculation of low temperatures arising from emergency blowdown is of importance, particularly for remote systems where risers and flare gas pipelines may be exposed to the low temperatures.

The potential for and consequences of low temperatures arising from emergency blowdown were identified in the study, but more importantly highlighted the need to eliminate present-day uncertainties in calculation assumptions.

4.4 Condensate Removal

The removal of condensate must be considered in any gas disposal system and whilst this is conventionally solved by the provision of knock-out drums in an on-platform system, particular problems arise in remote systems.

Three alternative methods of liquid removal were investigated for remote system, namely:
- pigging,
- vapourisation,
- pumping.

Pigging by conventional means requires the installation of a dual subsea pipeline system. This provides a free gas disposal path during pigging operations.

With a single subsea flowline configuration, a polymer pig would be required of a formulation that will disintegrate when subjected to the shock of an emergency blowdown. In this way, pigging operations could proceed safely without the necessity of a platform shutdown. However, the suitability of existing polymer pig formulations to this mode of operation is uncertain, and further development is required.

The removal of liquids by vapourisation can be achieved by heating and dry gas purging. These are not considered to be effective methods of condensate removal since impractically high temperatures or purge velocities would be necessary to remove water and the heavier hydrocarbons. Similarly, purging would require the installation of subsea collection vessels and pumping equipment which would pose significant maintenance difficulties for this deep water application.

4.5   Purging

A gas purge through the flare/vent headers and stacks is recommended to prevent damage arising from the ignition of explosive hydrocarbon/air mixtures inside the flare or vent stack. For Scenario 1, nitrogen purging was found to be compatible with gas venting, since the density of nitrogen is only marginally less than that of air and the possibility of oxygen ingress due to buoyancy effects is significantly reduced. For Scenario 2, nitrogen purging was found to be unsuitable because of the possible ignition difficulties that may occur at the commencement of an emergency blowdown. Fuel gas purging was therefore preferred.

5.   COMPONENTS

A variety of components are required to make up a gas disposal system, including:
-   support structures,
-   pipelines and riser systems (for remote systems only),
-   flare/vent tips.

5.1   Support Structures
On-Platform

On-platform gas disposal can be achieved by means of a vertical stack or a boom. The main advantages of such on-platform structures are that they can be accessed directly from the production platform and that their cost is far less than bridge-linked or remote structures. A disadvantage is that stack heights and boom lengths will be significant in order to maintain platform radiation levels during flaring within acceptable limits. The possibilty of hydrocarbon liquid droplets being carried over during flaring or venting operations and their effect on personnel and equipment in open deck areas must also be considered.

A single boom offers the advantage of reducing the likelihood of liquid droplet carryover arriving on the platform deck areas under low wind velocity conditions. A disadvantage is that a single boom must be sufficiently high above the platform to prevent the flare flame from impinging on the sides of the platform under prevailing wind conditions. A dual boom configuration will reduce this possibility by enabling the gas release to be diverted to the down wind boom.

A further consideration in this study was the effect an on-platform mounted flare or vent structure would have on the size and stability of the TLP production facility. It was considered that a vertical stack was likely to have less effect on the platform than either the single or dual boom system.

Bridge Linked

Three bridge-linked flare or vent structure concepts were considered. These were a free floating structure, a small TLP and an articulated column each linked to the production TLP by means of a bridge. This type of structure offers the advantage of allowing manned access to the flare. However, the dynamic interaction between the TLP and the flare or vent support structure resulting from the independent motion response characteristics of the two structures would lead to the imposition of high loads on the bridge connection points.

Remote

Three general structural concepts for the support of the remote flare or vent stack were investigated. These were:
- fixed,
- buoyant,
- articulated.

The fixed steel structure alternatives are generally proven concepts even in deepwater applications. Some types, however, are still at the early stages of development. Nevertheless, the fabrication and installation costs will be high compared with the alternative buoyant or articulated concepts.

The buoyant structure alternatives are generally deep water applications of existing tanker mooring and loading bouy designs and as such the technology is relatively well proven. These structures have the advantage of being simple and low in cost compared with fixed structures. However, special consideration is necessary in the design of the associated risers and mooring systems to take account of the buoy motions.

Single articulated concepts are applications of existing designs for a flare tower and an offshore loading point. The multi-articulated and semi-articulated concepts are in the early stages of development. The major disadvantage is that maintenance and repair of the universal joint and the design of the riser crossover points will be problematic.

Following a preliminary screening in Phase 1, the buoy type appeared to be an attractive solution and both a taut-moored buoy (TMB) and a slack-moored buoy (SMB) design were developed in Phase 2. It was found that the tethered buoy design appeared the most suitable taking account of the structural response, installation and maintenance requirements and costs. Further analysis of the tethered buoy design was carried out leading to the establishment of a preliminary basic design. The SMB design was also studied in more detail and was found to be a viable alternative.

5.2 Pipelines and Risers

A number of alternative subsea pipeline configurations were considered for use with the remote gas disposal system taking account of various operating philosophies for condensate removal, namely,
- no pigging with condensate being disposed or by vapourisation or subsea collection,
- pigging from main platform with pigs and condensate being returned there,
- pigging from the main platform with pig receiving and condensate disposal facilities on the remote structure,
- pigging from the remote structure to the main platform.

In the latter case, the use of both conventional and polymer pigs was studied and this resulted in the consideration of both single and dual pipeline and riser configurations. The development of the riser system design was based on the use of a rigid riser with the TMB and a flexible riser with the SMB.

The tension in the risers can have significant effect on the dynamic characteristic of the buoy and on the installation procedures. Three cases were investigated:
- full load carrying riser,
- load distributed between riser and tethers,
- zero load carrying riser,

The extreme low gas temperatures calculated for the emergency depressurisation of the production platform must be considered not only in the selection of the subsea pipeline and riser materials but in the design of the riser flexible joints and couplings. In this respect, the existing proprietary designs would not be capable of effective operation under these anticipated low gas temperatures.

The single rigid riser system was found to be compatible with the TMB without the requirement for heave compensation equipment. However, in the case of the dual riser system, the differential movement of the buoy would necessitate heave compensation equipment on the buoy.

The use of flexible risers with the SMB has the advantages that heave compensation equipment and flexible joints are not required. However, existing designs of flexible pipe would require development to enable their use for this low temperature, deepwater application.

5.3 Flare/Vent Tips

Both pipe tips and proprietary tips were studied in some detail for application to venting and flaring.

For gas flaring, the use of a proprietary flare tip of the Coanda type has significant advantages over a pipe tip in terms of its reduced radiation emission capability during flaring and the consequent effect in minimising the stack height.

The use of a pipe tip for venting represents significant weight and cost savings over a Coanda tip. These cost savings must however be traded-off with the anticipated saving in material and fabrication costs associated with the reduction in stack height possible with the use of a Coanda tip. More significantly, the gas dispersion characteristics of Coanda tips in venting applications are to date unknown.

Among the known methods of igniting the flare pilots, high tension piezo-electric and high energy ignition were considered. The components of the flame-front generator requiring the most maintenance are usually the pilot thermocouples and the inspirators. Flare vendor experience has suggested that in particular, the inspirators are mounted at the base of the stack to allow maintenance access. The most reliable method of ignition is considered to be high energy ignition.

The provision of snuffing facilities is essential to enable the extinction of both an accidentally ignited vent and a flare when the presence of gas leaks results in a safety hazard. Two snuffing agents were considered:

-   inert gas,
-   chemical powder or Halon.

The use of an inert gas such as nitrogen or carbon dioxide as a snuffing agent has the major disadvantage that large quantities of the gas are required for safe snuffing. Halon is considered to be the most practicable snuffing medium.

6.   SYSTEMS

6.1   On-Platform

On-platform flaring or venting can be achieved by means of either a vertically mounted stack or twin booms. The selection of a vertical stack for this application was determined by the minimal impact of these structures on the dynamic stability characteristics of the TLP production facility and the operational capability of the boom system.

The on-platform alternative offers a number of advantages in that condensate disposal does not present a problem, fabrication and installation costs are low, and flare/vent tip maintenance is relatively easily carried out from the platform. A critical aspect of the on-platform design is the flare/vent stack sizing. Acceptable flare radiation intensity levels at platform deck locations must be established taking account of the most probable wind velocities and directions, so that the stack height and consequently weight can be optimised.

6.2   Bridge-Linked

The bridge-linked system also offers the advantage that manned access can be easily achieved for maintenance or operational purposes. The principal problems with this system are associated with the dynamic interactions of the flare structure and the production TLP, and its effect on the bridge connection points. Furthermore, it was found that to achieve acceptable radiation levels on the TLP from the bridge-linked flare, the stack height would be comparable with the height of a platform mounted vertical stack. It was therefore concluded that a bridge-linked system offered no advantages over an on-platform system.

6.3   Remote

The remote system comprises a number of relatively complex components, including risers, subsea lines and a flare or vent support structure. Knock out drums, pigging facilities, flare ignition and snuffing equipment will all be required on the support structure, and if the structure is periodically manned, additional access and safety

features may be required. Whilst the system is considered technically
feasible, a number of areas would require development before
application. These all contribute to an expensive gas disposal
system, costing at least ten times that of a comparable on-platform
system.

7. CONCLUSIONS AND RECOMMENDATIONS

In the foregoing discussion, the various factors influencing the
selection of suitable gas disposal systems to satisfy the requirements
of the two scenarios have been discussed. As a result, it was
concluded that on-platform systems represented the most technically
feasible and economic solutions for both scenarios.

For certain applications, where larger quantities of gas have to
be vented or flared, it may be necessary to use remote gas disposal
systems and these were found to be feasible although development work
is required in several areas. Nevertheless, the cost of such systems
suggests that every effort should be made, by careful facilities
design, to locate gas disposal systems on the production platform.

ACKNOWLEDGEMENTS

The permission of The British Petroleum Company p.l.c. to
publish this paper is acknowledged.

(03.70/78; 03.96/80)

INFLUENCE OF IRREGULAR MOTION OF A FLOATING STRUCTURE ON ABSORPTION AND DISTILLATION

B. K. Hoerner, F. G. Wiessner, E. A. Berger
Linde AG, Division TVT Munich, FRG

Summary

Performance of absorption and distillation columns for integration, for instance, in offshore natural gas liquefaction plants has been successfully tested. The efficiency of moving packed and tray columns was determined during pilot-scale experiments on the motion simulator at DET NORSKE VERITAS (DNV). A distillation column (height 5 m, diameter 0.7 m) was mounted on a random motion simulator that produced movement in all six degrees of freedom. The simulator was controlled by computer programs containing statistical data of sea conditions, combined with the motion response of the structure. The results are used to predict the efficiency of offshore separation units, and column design can be based on these. The effectiveness of measures taken to guarantee reliable process operation are also shown.

1 INTRODUCTION

Offshore processing of natural or oil-associated gas is becoming a more and more attractive alternative to conventional pipeline transport. Marginal gas fields as found in offshore shelf areas can be economically worked by in-situ processing of the gas giving several products. These offshore process plants on floating structures could be production units for LNG (liquefied natural gas), methanol or ammonia.

The plants have to be designed for operation under sea conditions with irregular motion due to wave, wind and current forces. Absorption and distillation columns are considered to be critical components with regard to motion. For these units the uniform countercurrent gas and liquid flow is disturbed by motion leading to reduced heat and mass transfer. Column efficiency is often influenced by the deviations from the vertical position. For example, within a LNG plant the effect of a large efficiency drop in the CO_2 absorption tower or in the fractionation unit could lead to a blockage due to freezing of, for instance, CO_2, H_2S or heavy hydrocarbons in the low-temperature heat exchangers.

For this reason both absorption and distillation units have been investigated in this study.

2 LITERATURE SURVEY

Until now, there has been little information concerning irregular column movement due to sea conditions. The few publications that refer to this problem are based mainly on theoretical estimates which, however, often lack experimental proof.

A permanent inclination is considered to be the most critical situation for a separation column. Several authors have described the efficiency drop of plate and packed columns in inclined positions.

For a distillation column (50 mm Ø, packing, rectification of liquid air), Weedman (1), has shown that, beginning at about 1° inclination, efficiency drastically drops to about 50 % at 2.5°.

Mohr (2) calculated an efficiency drop of about 16 % at an inclination of 1.2° for a one- or two-path tray column of 2 meters diameter due to maldistribution, while Stichlmair (3) reported a maldistribution on a one-path tray of 330 mm width which leads to an efficiency drop of about 31 % at an inclination angle of 1° using absorption of ammonia in water.

From this it is seen that drastic efficiency drops are reported beginning at relatively small inclination angles.

Concerning moving columns, Sasaki et al (4) have investigated the absorption of ammonia in water for a one-path tray (1200 mm Ø) under harmonic motion. These measurements show:
- considerable efficiency drop at angular motion perpendicular to the flow direction
- stabilizing effects due to higher gas throughput
- improved efficiency due to motions parallel to the flow direction.

Our own experiments (5) of ammonia absorption in water (irregular packings, column diameter 400 mm, harmonic motion) showed a slight improvement in operation for oscillation periods of 8 seconds and less.

Due to the wide range of partially contradictory statements based on differing experimental conditions, the previous results seem to be insufficient for effective design of offshore columns.

3 THEORETICAL CONSIDERATIONS

The overall efficiency drop of packed as well as tray columns can be defined in terms of theoretical stages:
$$\Delta E = 1 - \frac{n_{th}}{n_{tho}} \dots\dots\dots\dots(1)$$
where n_{tho} is related to vertical position (nomenclature included at end of text). ΔE is influenced by inclination or motion in several ways:
- increased radial and longitudinal mixing of the liquid due to the disturbance in uniform flow
- the by-pass behaviour of gases and liquids. Parts of a tray run dry while weeping may occur in other regions due to an increased liquid level. For packed columns this effect is caused by a poor local liquid/vapour ratio.
- with trays, blowing through the downcomer may be observed due to a lower liquid level at the inlet weir.

From here it follows that any maldistribution of gas and liquid strongly influences mass and heat transfer. For steady state - ie stationary inclination - maldistribution of gas or liquid can be defined by dividing the cross sectional area of a packed column or the active area on a tray into N sections:
$$u_{G,L} = \frac{1}{N} \sum_{i=1}^{N} \left| 1 - \frac{\dot{V}_{G,L,i}}{\overline{V}_{G,L,i}} \right| \qquad \dots(2)$$

As the stationary inclination is considered to be the most critical "offshore" condition for a separation column, hydrodynamic analysis at stationary conditions can assist in analysing the dynamic behaviour of a moving column.

On the trays, the maldistribuiton of gas and liquid can be calculated as a function of the inclination angle, taking into account the hydrodynamic resistances of weirs and using correlations for the liquid holdup in the froth layer:
$$u_G = f \text{ (geometry}, \dot{V}_G, \dot{V}_L, \vartheta_G, \vartheta_L, \eta_L, \sigma_L, \varphi) \dots(3)$$

The gas maldistribution u_G is induced by the liquid maldistribution u_L, both are considered to influence efficiency. Therefore, efficiency drop in both column types is related to these hydrodynamic parameters.

For packed columns there exist correlations for the maldistribution of the liquid in a vertical position (6). At stationary inclinations, however, this can be measured and further used as a reference for calculating efficiency drop.

4 PROPHYLACTIC MEASURES

One aim of our study was to find and test column internals less sensitive to offshore conditions relative to the above-mentioned conventional onshore installations. A very important criterion was to minimize maldistribution which led to the development of a special radial flow tray with slot weirs.

Because of its axisymmetric structure, efficiency drop is virtually independent of the direction of inclination and additional slot weirs prevent sloshing and provide a better liquid distribution. The hydrodynamic behaviour of this tray is shown in fig. 1. Here it can be seen that gas maldistribution is reduced by:
- increased specific gas load
- decreased column diameter.

Additional calculations show that the maldistribution is also slightly reduced by increased liquid load at constant specific gas loads.

For packed columns, regular packings have been shown to provide a better liquid distribution. Fig. 2 shows the measured liquid maldistribution at several inclination angles for two different packings. Packing R shows a better distribution of the liquid at permanent inclination and, therefore, is considered to produce also an improved hydrodynamic behaviour during irregular sea-induced motion.

Having selected the most suitable trays and packings for offshore operations, we performed experiments in order to test these column internals under sea conditions using pilot-scale equipment.

5 EXPERIMENTAL

Tests on the motion simulator were performed in order to prove the relevance of our prophylactic measures and to check the correlations between maldistribution and separation efficiency and their application to random motion.

5.1 Selection of the Process

As mentioned above, we could have investigated chemical absorption, or distillation for prediction of efficiency drop of offshore columns.

Laboratory-scale absorption tests in packed and plate columns showed that chemical absorption processes are rather insensitive to motion. This is most probable due to the enhanced absorption capacity of the liquid for absorption with chemical reaction. For this reason even a liquid volume which is not renewed for a period, due to maldistribution, is able to absorb reacting gaseous compounds without saturation.

A distillation process, however, is more sensitive to disturbances of uniform flow because equilibrium between the two phases is reached much faster than with chemical absorption.

This was proved for packed columns during earlier tests in the LINDE laboratory. Differing behaviour in physical and chemical absorption was observed for equal column diameter and the same irregular packing. From

this it could be concluded that the results from offshore tests of chemical absorption cannot easily be applied to physical absorption and distillation.
Because of these considerations the more sensitive process was chosen - a distillation unit with total reflux - for the simulation tests.

5.2 Selection of the Test Fluids

As low-temperature test processes would have been impractical, test fluids had to be selected to meet the following requirements:
- a well-tested binary mixture
- a suitable equilibrium curve in order to show significant differences in gas and liquid compositions
- physical properties similar to those of LNG
- a low boiling temperature at ambient pressure and a low heat of vaporization

The mixture of methylcyclohexane and toluene was found most suitable. Table I shows a comparison of its physical properties with those of LNG.

Table I: Comparison of physical properties of methylcyclohexane/toluene (test facility) and LNG (plant)

Plant Test Facility	$\dfrac{\rho_p}{\rho_T}$	$\dfrac{\sigma_p}{\sigma_T}$	$\dfrac{\nu_p}{\nu_T}$	$\dfrac{\eta_p}{\eta_T}$
Range	0.55 - 0.62	0.76 - 0.95	0.65 - 0.9	0.4 - 0.5

5.3 Test Plant

Fig. 3 shows the flow diagram of the test plant. The column (T1) (0.7 m $\emptyset$), which could take 6 special radial flow trays or a regular packing, was mounted on the motion simulator and connected to the other units by flexible hoses; flow was countercurrent. The liquid mixture was vaporized in the reboiler (E1) and fed into the column where the more volatile component, methylcyclohexane, was concentrated in the vapour while the liquid was enriched with toluene. The vapour was condensed (E2) and collected in a tank (D2) before the reflux was fed back to the top of the column, being preheated in the heat exchanger (E3). At the bottom the sump liquid was collected in a tank (D3) before being pumped to the reboiler.

One tank (D1) was used as a feed and a buffer tank. A blow-down tank (D4) was installed in a safe area, outside the laboratory, which could take all the liquid from the plant in an emergency. This tank was also used for changing the test liquid.
A considerable number of safety devices were installed in cooperation with DNV:
- shut-down valves
- blow-down equipment
- a gas detection system
- numerous alarm devices
- separation walls
- ventilators
- fire-fighting equipment
- ex-proof electrical installations.

5.4 Motion Simulator

As mentioned above, the tests were performed on the motion simulator provided by DNV. It was decided to simulate heavy sea and strong storm conditions in the northern North Sea where LNG plants on floating structures could be located. The semisubmersible used was selected from an earlier study by ARGE 76 (7). Its behaviour in regular and irregular waves has been calculated and model tested.

From frequency distributions of maximum roll angles of the semisubmersibles, suitable sea spectra were selected by DNV for the northern North Sea which produced the chosen maximum angles. From these spectra the motion of the semisubmersible in all 6 degrees of freedom was recalculated using the transfer functions of the structure. This random motion was given in digital form on a tape which controlled the motion of the seven hydraulic cylinders of the simulator. Table II shows the random sea motion and the roll and pitch response of the semisubmersible which were considered the most influencial to the separation process. Roll and pitch motion were applied in full scale, while the other motions had to be scaled down due to limitations of the simulator.

Table II: Maximum roll and pitch angles of the simulator performing random motion of the semisubmersible in northern North Sea

Tape	H 1/3 (m)	$\bar{T}$ (s)	Roll φ_{max} (deg)	Pitch φ_{max} (deg)
1	9.6	10	3	1.85
2	18.4	12	6	3.66
3	20.6	16	6	3.65

Besides this random motion, regular sine motion of sway and roll were performed. This was done in order to compare the behaviour of the column during random motion with that during regular motion which is always easier to achieve. Besides vertical position, additional permanent inclinations of the column were tested for comparison purposes.

Fig. 4 shows a view of the plant in the process laboratory of DNV in Oslo.

5.5 Test Runs

About 60 runs were performed. The overall efficiencies of the test columns were determined after the process had reached steady state each time. The loading of the columns was varied over a range of 60 - 120 % of the design case according to a turn-down ratio of 2 : 1.

The pressure in the column was maintained at 1.2 - 1.5 bar corresponding to the temperatures 115 - 125 °C. The initial composition of the test mixture was 60 % toluene and 40 % methylcyclohexane by volume.

Liquid composition was determined by means of a density analyzer. Temperatures, pressures and flow rates were measured at several locations in order to perform mass and energy balances and to calculate efficiencies.

During the experiments, approximately 2 m³ of test mixture were used and hence a variety of safety procedures for start-up, runs and run-down had to be met.

6 RESULTS

Initially, the behaviour of the column was investigated at permanent inclination providing an additional comparison between the distillation and chemical absorption processes. For plate columns this is shown in fig. 5. It should be noted that experimental results are shown in both columns at varying gas and liquid loads, the different diameters having already been taken into account in the calculation of the gas maldistribution. In this generalized presentation the different effects of maldistribution of chemical absorption and distillation processes can be seen.

A comparison of the efficiency drop at permanent inclination of both packed and plate columns is illustrated in fig. 6. Similar behaviour of the two columns is observed. The special plate column is slightly better than the regular packing.

Considering the efficiency drop during irregular motion, our prophylactic measures made the columns almost insensitive to motion at moderate sea states. During heavy sea simulation, especially for long periods, we observed efficiency drops in the range of 10 %. Besides our prophylactic column design, this relatively low efficiency drop is also due to:
- At random motion, the maximum inclination angles were observed to be reached less frequently than during comparable regular roll motion.
- Relevant disturbance of distillation conditions caused by inclination, occurs only at longer mean motion periods.

Nevertheless, the efficiency drop in the pilot-scale column was not negligible. As tests with roll motion were performed at DNV and in the LINDE laboratory with columns of different diameters in order to simulate the behaviour at random motion, a scale-up could be performed, showing that in full-scale columns lower efficiencies will be achieved than in our random motion tests.

During offshore operation, random motion around the vertical position due to the sea alone cannot be expected because of the varying wind forces. Since the ballasting system of a floating structure has a delay time of up to 3 hours the mean position of the column is not vertical. This angle was ascertained and random motion tests with a non-vertical mean position were performed. The results show that in these cases efficiency drop is close to that of permanent inclination in the non-vertical mean position.

7 CONCLUSIONS

1)   Laboratory tests of physical and chemical absorption systems show a differing sensitivity of these processes to flow disturbance.
2)   Pilot-scale tests of our distillation system in a special 0.7 m diameter plate column and a packed column during random motion, show a relatively low efficiency drop compared with permanent inclination.
3)   The effect of inclination of a separation column on efficiency can be described by the maldistribution of gas and liquid for chemical absorption and distillation processes. This approach makes it possible to predict efficiency drop at different column diameters, gas and liquid loadings and physical properties. This is also applicable for random motion.
4)   In our opinion, the design of offshore columns can now be performed more reliably as the calculation of efficiency drop enables us to predict the conditions for proper plant performance. In order to minimize efficiency drop, prophylactic measures in our column design have proved very successful.

NOMENCLATURE

B   liquid load (m3/m2h)
ΔE efficiency drop, defined in eq. (1) (%)
F   specific gas load (m/s   kg/m3)
n_{th} number of theoretical stages (-)
N   number of sections of the cross
       sectional area (-)
u   maldistribution, defined in eq. (2) (-)
Δu $u - u_o$ (-)
$\dot{V}$ flow rate (m3/h)
$\overline{V}$ mean flow rate (m3/h)
$\mathcal{S}$ density (kg/m3)
η dynamic viscosity (kg/m s)
ν kinematic viscosity (m2/s)
σ surface tension (kg/s2)
φ inclination angle (deg)

Subscripts:
o   vertical position
G   gas
L   liquid

ACKNOWLEDGEMENT

The authors wish to thank the BMFT (Bundesminister für Forschung und Technologie) and the European Commission for the financial support of the experiments described.
They also wish to extend their thanks to DET NORSKE VERITAS for their technical support.

REFERENCES

(1) WEEDMAN, J. A.; Dodge, B. F. : "Rectification of Liquid Air in a Packed Column", Ind. and Engng. Chemistry 39 (1947) 732.
(2) MOHR, V.: "Einfluß von Fertigungstoleranzen auf den Wirkungsgrad von Kolonnenböden", Chemie-Ing.-Technik 51 (1979) 139.
(3) STICHLMAIR, J.: "Untersuchungen zum stationären und dynamischen Verhalten einer adiabat betriebenen Absorptionskolonne", Diss. TU München 1971.
(4) SASAKI, A. et al: "Distillation Column Performance for Floating Offshore LNG Plant under Rolling and Surging Motions in Waves", paper no. 3641, 11th Annual OTC, April 1979, Houston.
(5) BERGER, E.: "LNG Production on Marginal Offshore Gas Sources - a Possibility of Short Term Recapitalization of Exploration Expenditures", paper presented at CHEMASIA '81, Oct. 1981, Singapore.
(6) BILLET, R.: "Die industrielle Destillation", Verlag Chemie, Weinheim, 1973.
(7) ARGE 76: "Konstruktive, rechnerische und experimentelle Entwicklung von mobilen, schwimmenden Offshore-Erdgasverflüssigungsanlagen", Final Report, Feb. 1979.

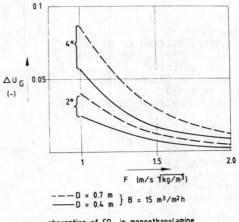

FIG. 1

MALDISTRIBUTION OF GAS ON BUBBLE CAP TRAYS

OVER SPECIFIC GAS LOADING (CALCULATED)

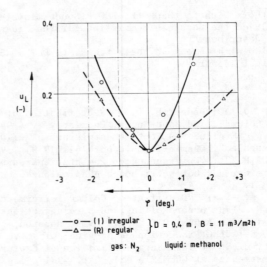

FIG. 2

MALDISTRIBUTION OF LIQUID IN PACKED COLUMNS

OVER INCLINATION ANGLES (MEASURED)

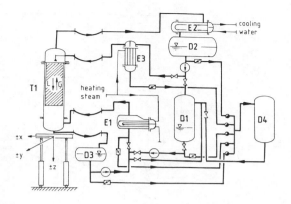

FIG. 3   FLOW DIAGRAM OF THE TEST PROCESS

FIG. 4   VIEW OF THE TEST PLANT

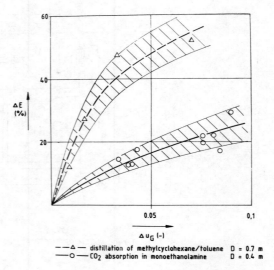

--△-- distillation of methylcyclohexane/toluene D = 0.7 m
--○-- CO_2 absorption in monoethanolamine D = 0.4 m

FIG. 5

MEASURED EFFICIENCY DROP OF PLATE COLUMNS AS
FUNCTION OF THE MALDISTRIBUTION OF GAS AT
PERMANENT INCLINATIONS

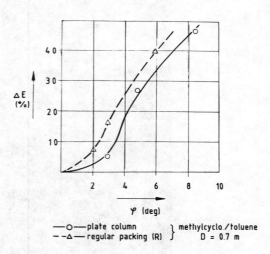

--○-- plate column   } methylcyclo./toluene
--△-- regular packing (R) }   D = 0.7 m

FIG. 6

COMPARISON OF EFFICIENCY DROP IN BOTH
TYPES OF COLUMNS AT DESIGN LOAD AND
PERMANENT INCLINATIONS

(03.29/76)

OFFSHORE LIQUEFACTION OF NATURAL GAS FROM DEEP WATER DEPOSITS

H. BACKHAUS
LGA Gastechnik GmbH
Federal Republic of Germany

Summary

The offshore exploration business is moving into ever deeper areas of the oceans in the search for petroleum and natural gas, and energy policies dictate an ever increasing demand for production, even from "uneconomical" deposits. Tension leg platforms mean that it can now be profitable to work the smaller hydrocarbon deposits lying in deeper waters. The first tension leg platform for petroleum production in the North Sea Hutton Field is under construction. Openings will also have to be found for offshore natural gas from geographically exposed and geologically "difficult" fields to be marketed profitably.

The system described in the following pages for the liquefaction of natural gas on a tension leg platform, using tanker transport, means that both dry and associated offshore natural gas, which today has to be largely burned off, can now be marketed without the use of pipelines.

1. INTRODUCTION

The hydrocarbon deposits in the shelf areas and on the continental slopes under the oceans are becoming increasingly important in supplying man's energy needs. Output is now so high that they have become a factor in regulating prices and securing supplies.

The production and transport of offshore petroleum have now reached the stage of becoming almost common-place techniques. Many platforms have been working for years and the petroleum produced is transported to the mainland by undersea pipelines. If, at the smaller sites, pipeline transport is not economically viable, the petroleum is placed in off-shore buffer stores, from which it is collected in tankers.

Offshore natural gas takes rather more complicated routes to the markets. Where there are large deposits not too far from the centres of consumption, such as those in the southern and central regions of the North Sea, the gas is brought to the mainland in subsea pipelines.

Where there are smaller natural gas fields increasingly far away from the coasts with their energy-consuming hinterland, subsea pipelines soon become an uneconomic proposition. Realizing this, LGA Gastechnik began at the end of the 60's to develop offshore natural gas liquefaction plants (1). Like crude oil, liquefied natural gas can be kept in buffer stores at the offshore production site and transported economically by LNG tankers over even quite long distances.

During the last ten years, several German firms have been involved in the development of various concepts for offshore natural gas liquefaction. This work, which has been both time-consuming and manpower-intensive, has received financial support from both the Federal German Ministry of Research and Technology and the Commission of the European Communities in Brussels. At its current stage of development, the process has been approved by the large oil companies active in offshore exploration, all the technical problems have been solved and the way is now clear for an offshore natural gas liquefier to be started up very shortly.

As with many offshore activities, the principles on which the supporting structures are built are of the utmost importance. Offshore natural gas liquefaction plants can be set up on floating pontoons, on semi-submersibles, fixed platforms and tension leg platforms.

When platforms are designed to be fixed on the seabed, the liquefaction plant can be operated under virtually the same conditions as on land. The relevant concepts have been described in detail in recent years in various papers and specialist publications.

Fixed platforms are, however, sensitive to water depth from both the technical and economic point of view. As a rule of thumb, investment expenditure increases by the square of the water depth (2). In the search for optimum solutions in deep water, the tension leg platform, TLP, came to the forefront relatively early on (3, 4). Designed first of all for relatively light drilling platforms (5), TLPs were then developed quite quickly for oil production platforms with heavy payloads. The TLP designed for use in the North Sea Hutton Field can take a deck load of 15 000 t (6, 7).

The same technological and economic principles which apply to offshore petroleum production, apply also to the production of liquefied natural gas (LNG) on offshore platforms. Together with the other firms in its group, the Howaldtswerke-Deutsche Werft AG and the LGA Gastechnik GmbH, the Salzgitter AG has been very much involved in the development of a tension leg platform since 1976, starting from the concept of a fixed platform. From the outset, the specific requirements of the construction yard installation and offshore operation of large-scale natural gas liquefaction plants were taken into account.

2. NATURAL GAS LIQUEFACTION PLANT

When the natural gas liquefaction plant (Fig. 1) was developed, a certain number of requirements had to be met - as they did in the case of the supporting structure - in order to guarantee flexible use of the plant in offshore conditions :

-   The liquefaction process should be such that natural gas can be processed in as broad an analytical spectrum as possible. In practice, this means a band width ranging from dry natural gas (with 95 % methane or more) to associated petroleum gases (with high C_3 and C_4 content).
-   Special offshore conditions demand that the mechanical equipment used for the process should be designed so that the plant can be shut down for short periods and started up again. For example, weather conditions can be so bad that the LNG tanker cannot be loaded and the liquefaction plant has to be shut down temporarily if the buffer storage tanks are full. This can happen several times a year, and thus, the main features of the process and the plant include the fact that when the plant is shut down, the refrigeration cycles can either be maintained or, using equipment especially designed for use on platforms, quickly emptied, then

cooled and filled up again, during which process the refrigerants are temporarily stored in safe areas below deck; the fact that the quantities to be refrigerated are reduced; the use of gas turbines to produce energy (instead of boilers or stream turbines); supplying the process heat by hot oil cycles with small amounts of circulating fluid.

- In designing the machinery and other apparatus for installation in the plant, care had to be taken that the individual components were not only light but also relatively insensitive to motion. Even though the model tests at the HSVA* showed that even in the heaviest seas the motion of the platform and acceleration will be relatively slight, it was decided not to use such items as gas separation columns for this process. The requirement that the structure should be as light as possible determined, among other things, the use of gas turbines and aluminium-plate heat exchangers, as well as a reduction in the contents of the liquid sumps.

- When the machinery and other apparatus were designed, the relatively small space available on the platform deck had to be taken into account, together with the regulations laid down by the classification societies with respect to safety areas, escape routes, safety equipment, fire protection, ventilation, noise pollution in the crew's quarters etc.

- The liquefaction process to be operated on the platform had to involve as little machinery and apparatus as possible. The development work carried out by the LGA Gastechnik in 1977/78 led to the concept of liquefying natural gas under pressure (9), the pressure being reduced to approximately 8 bars and the boiling point being only approximately −125°C in the case of associated gas. With conventional large-scale natural gas liquefiers, the gas is expanded to atmospheric pressure and has a boiling point of approximately −160°C. This higher LNG temperature is, in effect, the main reason for the savings achieved in the use of machinery, apparatus and energy.

When associated gas is being liquefied, a dual arrangement cascade consisting of a propane refrigeration cycle and an ethylene refrigeration cycle is sufficient. When dry natural gas is being liquefied, "flash gas recompression" is also necessary, although to a lesser extent than with atmospheric liquefiers.

For example, the associated gas to be liquefied comes from an oil production platform by subsea pipeline. It is compressed to the pressure of approx. 40 bars necessary for liquefaction, whereby the heavy hydrocarbons are condensed out and can be used as unstabilized gasoline.

The H_2S and CO_2 present in the gas stream are scrubbed by contact with an alkanolamine solution. The charged amine solution is then regenerated thermally and used again. Finally, the gas to be liquefied is dried in adsorbers (molecular sieves). The purified, dry gas is then cooled and liquefied in four stages in a counter-current arrangement with vaporizing refrigerants, for which purpose the propane and ethylene refrigeration cycles are divided into a low-pressure stage (roughly

---

* = Hamburgische Schiffbau-Versuchsanstalt, the Hamburg shipbuilding testing station.

atmospheric pressure) and a medium-pressure stage (3 to 4 bar). The gas which is liquefied at 40 bar is cooled by expansion in buffer storage tanks (8 bar) to approx. −125°C, during which process about 12 % of the quantities processed vaporizes out. This flash gas is therefore at a low temperature and this coldness is utilized by being fed back in the heat exchangers of the four liquefaction stages for a third time. The heated flash gas is finally used as regenerating gas for the driers and as "fuel" for the gas turbines.

3. BUFFER STORAGE OF THE LIQUEFIED NATURAL GAS

The technical and economic aspects of a natural gas liquefaction plant are such that the product has to be stored temporarily, regardless of whether the plant is on the mainland or sited on an offshore platform.

Whilst the construction of liquefied natural gas stores on the mainland presents no technical problems nowadays if there is a sufficiently large area available and firm sub-soil, offshore there are serious problems.

There is very little space available on a tension leg platform and it is not possible to accommodate the required storage volumes of up to 100 000 m^3 or more.

There is also the fact that the contents of such a store weigh approximately 50 000 t, and the difference between the full and the empty store would have to be made up by water ballast in the buoyancy body of the tension leg platform.

Even assuming that water is twice as heavy as liquefied natural gas, one would have to reckon with approximately 50 000 m^3 volume of ballast in the buoyancy body.

This ballast water would have to be pumped around continually to coincide with the filling and emptying of the liquefied gas store, which means that there would be a considerable amount of machinery (for pumping) in addition to the problems of space.

In order to avoid the problems caused by integrated storage, engineers came up with the design for external buffer storage shown in Fig. 2.

A storage tank ship is anchored to a loading tower. The LNG pipelines lead from the production platform to the seabed, and from there to the loading tower base-plate. They then continue up to the surface again inside the power structure. Finally, flexible lines lead from the top of the loading tower to the storage tanker.

It would be beyond the scope of this paper to describe the technical details of this storage and transfer system. It should be noted, however, that while the system presents no insurmountable technical difficulties, it would be a relatively expensive solution.

The question to be considered then is whether the product, i.e. liquefied natural gas, can cover the costs of the components required for an offshore at liquefier, i.e. can LNG from offshore fields be sold at market prices ?

4. CONCLUSION

The engineering work on the system described above culminated in the approval in principle (quality assurance) certificate issued by

Germanischer Lloyd, confirming that there are no objections in principle, from the point of view of safety, to the technical design and that the system can be considered classifiable, and thus ready for building. Against this technical background it is obviously significant that, with the progress made in offshore exploration activities, self-sufficient production systems for petroleum and natural gas - i.e. systems which do not require subsea pipelines to the mainland - will be increasingly in demand. With oil, this theory has been put into practice already, with the construction of a TLP for the North Sea Hutton Field. There is little doubt that the technical and economic realities of life and the demands of our energy policies will lead to the self-sufficient production of offshore natural gas on specially designed tension-leg platforms.

The potential advantages are obvious :

- Gas liquefaction plant, which incorporates relatively complicated systems, can be assembled very efficiently in a construction yard.
- The plant can be floated out to its offshore location as a single unit ready for operation, which will obviate the need for the protracted and at times difficult transport of individual components and pre-assembled plant modules.
- The fact that the system is broken down into what is virtually a traditional jack-up-platform and tension leg components means that it can be floated out to sea and set up offshore quickly and safely. The relatively complicated and, at times, risky stage of mooring tension leg platforms is over before the most valuable equipment, the process element, is floated out and placed in position, which should reduce the insurance premiums.
- The pressurized liquefaction of natural gas, specially adapted to take place on the platform, works with a minimum of apparatus and machinery (investment). Its total weight is relatively low and the efficiency of the process relatively high (i.e. internal gas consumption is relatively low).
- The TLP-system described is mobile and LNG can be produced during its 15-20 year operating life at different offshore fields (early production). The offshore site can be vacated relatively easily at the end of its operating life.
- The TLP-system moves relatively little in heavy seas; the horizontal acceleration which can be expected according to the HSVA trials will not impede the operation of machinery and equipment in the liquefaction plant.
- The investment costs of the supporting structure are to a large extent unaffected by the depth of water, and longer or shorter tension systems, as a percentage of the total investment, make a negligible difference.

5. REFERENCES

(1) Backhaus, H. : "Floating Offshore Natural Gas Liquifaction Plant", 2nd Int. Conference on LNG, Paris, Oct. 1970
(2) Baldwin, A.H. : "Technische Entwicklung von Offshore-Anlagen für die Nordsee". Erdöl-Erdgaszeitschrift, 92 Jg., March 1976, p. 91/98
(3) Paulling, J.R. and E.E. Horton : "Analysis of the tension leg stable platform". Offshore Techn. Conf., Houston, April 1970, Paper 1263

(4) Brewe, H.J. and S.J. Shrum : "Tension-leg platform will get at-sea test next year". The Oil and Gas Journal, Oct. 8, 1973, p. 88/91

(5) Cranfield, J. : "New offshore platform designs aim to tame hostile areas", Petroleum International June 74, Vol. 14, No. 6 p. 34-40

(6) Taylor, D.M. : "Conoco's tension leg platform will double water depths capability". Ocean Industry, February 1980, p. 35/39

(7) Mercier, J.A. : "Offshore platform concepts, tension leg platform", ONS 1980, Stavanger, Norway, August 1980

(8) N.N. : "New TLP design tested for northern north sea". Ocean Industry, April 1981

(9) Backhaus, H. und E. Flüggen : "Pressurized LNG - and the utilization of small gasfields". 6th LNG/LPG conference, Monte Carlo (Monaco), 7./10. Nov. 1978

(10) Backhaus, H. : "Verflüssigung von Offshore-Erdgas auf spannungsverankerten Plattformen" mt-Meerestechnik Bd. 13 (1982) Nr. 5, p. 127/134

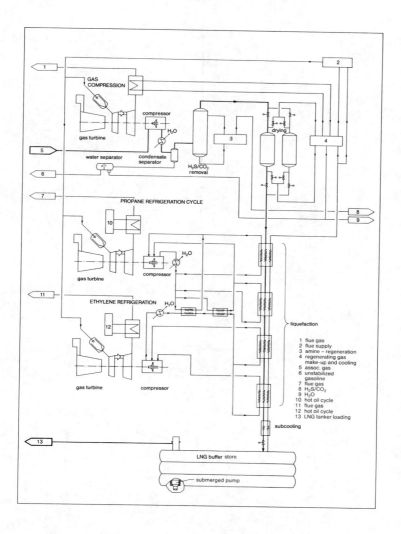

FIGURE 1 : SIMPLIFIED FLOW DIAGRAM, LIQUEFACTION OF ASSOCIATED GASES

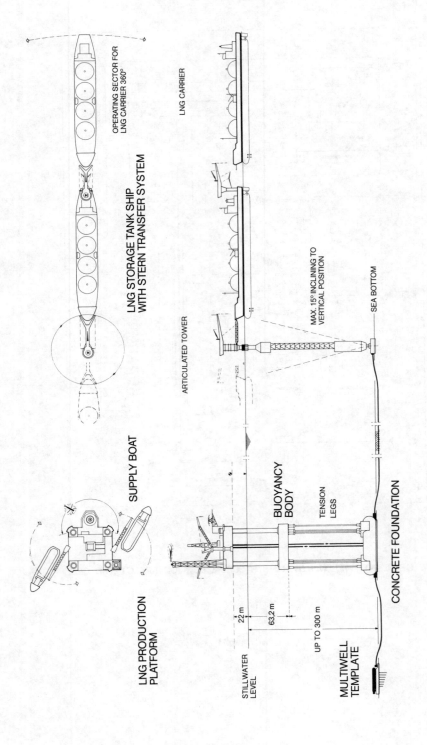

OPERATING SECTOR FOR
LNG CARRIER 360°

LNG STORAGE TANK SHIP
WITH STERN TRANSFER SYSTEM

LNG CARRIER

SUPPLY BOAT

ARTICULATED TOWER

MAX. 15° INCLINING TO
VERTICAL POSITION

SEA BOTTOM

LNG PRODUCTION
PLATFORM

BUOYANCY
BODY

TENSION
LEGS

CONCRETE FOUNDATION

STILLWATER
LEVEL

22 m

63.2 m

UP TO 300 m

MULTIWELL
TEMPLATE

SCALE
0   50   100   150   200

FIGURE 2 : OFFSHORE-LIQUEFIER WITH LOADING-TOWER, BUFFER STORAGE AND LNG-TANKER

(03.28/76)

DEVELOPMENT AND MODEL TESTING
OF A FLOATING NATURAL GAS LIQUEFACTION PLANT

D. MEYER-DETRING
Preussag AG, Hannover, Germany

Summary

Offshore natural gas production by liquefaction on
floating structures, followed by unloading and transportation in LNG carriers, can be advantageous over
conventional methods for conveying natural gas to
consumers in pipelines. This is especially so for
marginal offshore gas fields, or offshore oil fields
with associated gas or even for onshore fields in areas
where there is a lack of infrastructure. The proposed
systems could fill a market gap where fixed platform/
subsea pipeline methods or onshore LNG plants have
reached the limit of economic viability.
The processing of natural gas on floating platforms
presents many challenging problems associated with
the motions of the platform, the need for maximum gas
recovery, minimum space and weight requirements for
the processing plant and maximum safety for men and
equipment.
This paper presents the CONSORTIUM 76 design, a natural
gas liquefaction system integrating LNG plant, LNG
storage tanks and LNG unloading system in one single
platform. The CONSORTIUM 76 design has been developed
by the companies Linde, Preussag, Technigaz, Dyckerhoff
& Widmann, Blohm + Voss and Bilfinger + Berger.

1. INTRODUCTION

The new technology of natural gas liquefaction on
floating platforms offers a technically feasible and, for
certain applications, an economic way to exploit marginal
offshore gas fields as well as offshore oil fields with
associated gas and even onshore gas fields close to the
shore in areas with little or no infrastructure.
The criteria which determine whether an offshore gas
field is to be considered marginal or non-commercial are:
- size and quality of gas deposits,
- distance to shore and/or consumer,
- water depth,
- environmental conditions such as waves, wind, ocean
  current or ice formation,
- sea bottom topography between gas field and shore,
- financial and political concerns such as royalties,
  terms of taxation, level of gas price and political
  stability.

With the advancement of worldwide hydrocarbon exploration activities in areas with more hostile environments, greater water depth and greater distance to the consumer, the number of marginal fields has grown. In view of this, development efforts have been made to replace the expensive method of exploiting offshore fields with fixed production platforms and gas pipelines.

These efforts have consequently led to the design of floating, mobile LNG production and transportation schemes where the natural gas will be pre-treated, liquefied, stored and transported to coastal terminals by LNG tankers.

The new method has the following advantages:
- substantial cost savings,
- a floating plant can exploit successively several smaller fields of similar gas composition without major modifications,
- transport by LNG tanker is more flexible than by pipeline,
- the floating LNG plant can be constructed and equipped completely in industrialized areas and would then be towed to location, thus saving time and money for difficult and expensive offshore work,
- by installing a floating plant close to the shore even onshore gas fields in areas with little or no infrastructure could be exploited at lower cost.

2. GENERAL OUTLINE
   (figures 1,2)

2.1 Design Criteria
   For the development, the following design criteria had to be taken into consideration:
- high efficiency for the natural gas liquefaction plant (low internal gas consumption),
- minimum space and weight requirements for the process and nautical equipment,
- best sea behaviour of the floating structure,
- minimum motion sensitivity of the process plant,
- relative autonomy of operation,
- flexibility regarding processing of different natural gas compositions at successive locations,
- maximum safety.

Furthermore, national and international rules and regulations have been taken into consideration. The design was developed in close cooperation with an experienced classification society and was discussed with the relevant authorities of the Norwegian and British North Sea sectors. The classification society Det Norske Veritas has completed the first phase of an Approval in Principle.

2.2 System Description
   The CONSORTIUM 76 concept offers a compact design, integrating natural gas liquefaction, LNG storage and tanker transfer in one single floating semisubmersible building. This version of an offshore LNG plant was specifically designed for service under heavy environmental conditions.

Through its favourable motion characteristics it can keep up production under North Sea conditions all the year round. The design replaces a fixed production platform in a conventional scheme; the pipeline is replaced by LNG tankers.
The system comprises:
- a circular (buoyant) steel deck providing space for the natural gas liquefaction plant with offsites, accommodation, flare stack and helicopter decks,
- six concrete columns containing sea water ballast tanks and connecting the steel deck with
- six spherical, insulated concrete LNG storage tanks, surrounded by concrete skirts to improve motion characteristics,
- a tanker loading and mooring system which can swing freely around the platform,
- a motion-compensating high pressure gas transfer line for crude gas supply from the ocean floor to the platform.
Main advantages of the system:
- building cost is largely independent of the water depth, no foundation problems,
- excellent motion characteristics guaranteeing a minimum of non-operating time even under northern North Sea conditions,
- the system is mobile so that the capital investment can be used several times (small fields).

2.3 Main Technical Data

Diameter of steel platform	145 m
Height of main deck above baseline	74 m
Operating draft	44 m
Production capacity	640,000 m³/h(Vn)
equivalent to	570 MMSCFD
Storage capacity	125,000 m³
Ballast tank capacity	61,800 m³
Unloading rate to LNG tanker	6,250 m³/h

3. NATURAL GAS LIQUEFACTION PLANT

In order to meet the difficult design requirements on a floating offshore platform, a mixed refrigerant cycle with ethane/propane precooling, based on Linde's CCM-Cycle (C_2/C_3-, Mixed Refrigerant Cycle) has been selected. Four nearly identical lines produce a total of 1100 m³ LNG per hour.

The selection of this natural gas liquefaction process and its associated offsite facilities is dictated by site as well as economic considerations.

The choice is influenced by such factors as the operating conditions on semisubmersible platforms, space and weight restrictions on heat exchangers, machines and other process units, and the desire for as low a power consumption as possible consistent with the use of proven equipment.

The flows and characteristics of tailgas streams can be varied to meet fuel requirements independent of the feed gas composition. This requires little or no alteration to the liquefaction plant.

3.1 Process Description

1 Sour Gas Removal
Natural gas is treated in a MEA wash to remove CO_2.
One regeneration unit serves two liquefaction trains.

2 High Level Ethane/Propane Cooling
Purified natural gas and the mixed refrigerant cycle are
cooled in heat exchangers against high pressure evaporating
ethane/propane.

3 Water Removal
Most of the water in the purified natural gas is removed
by condensation followed by separation. The remainder is
adsorbed in a mole sieve station.

4/5 Medium/Low Level Ethane/Propane Cooling
Dry purified natural gas and the mixed refrigerant cycle
are further cooled and partially condensed against ethane/
propane at two different levels.

6 Fractionation Unit
In this unit the natural gas is fractionated to separate
heavy hydrocarbons which would otherwise freeze out in the
low temperature section. This unit also provides make-up
propane and ethane for the refrigeration cycles.

7 Liquefaction
The fractionated natural gas together with the mixed refrigerant cycle is liquefied and subcooled in a spiral wound
heat exchanger. LNG is sent to storage from which the
resulting flash and boil-off gas are withdrawn for fuel.
The mixed refrigerant stream is expanded to low pressure,
vaporized and recompressed.

8 Compression for Mixed Refrigerant Cycle
Cold cycle gas from exchanger 7 is compressed in two stages
to the final discharge pressure. Cycle gas is inter- and
after-cooled against water.

9 Compression for Ethane/Propane Cycle
The cold ethane/propane streams of the three pressure levels
are compressed in a three stage machine. High pressure
ethane/propane is cooled and liquefied against water in
the following heat exchanger.

10 Ethane/Propane Expansion
The liquefied ethane/propane is expanded in three successive
stages to the three pressure levels. The vapour resulting
from expansion or vaporization at each level is passed to
the compressor for recompression.

3.2 Arrangement of the Plant on the Platform

In order to meet the requirements and safety aspects
laid down by the classification societies and national authorities, only the gas conducting part of the process plant
is arranged on the main deck. All offsites such as the
sea water distribution system, electric power generation,
etc. are installed on the lower deck. The gas and steam
turbines are arranged as far to the edge of the platform
as possible so that air intake and exhaust gas do not affect
the other process area. To reduce the influence of platform
motions to a minimum, the rectification and wash columns
are installed as close to the center of the platform as
possible. For optimum space utilization, some heat exchangers
are erected vertically, while others are installed horizontally above one another. To protect the platform structure
from possible cold liquid spills some components of the
process plant are grouped together and surrounded by special
containment devices.

3.3 Model Tests

The efficiency of moving packed and tray columns was
determined during pilot-scale experiments on the motion
simulator at Det Norske Veritas (DNV). A distillation column
(height 5 m, diameter 0.7 m) was mounted on a random motion
simulator that produced movement in all six degrees of freedom. The simulator was controlled by computer programs
containing statistical data of sea conditions, combined
with the motion response of the structure.

The results are used to predict the efficiency of offshore separation units, and column design can be based on
these.

Initially, the behaviour of the columns was investigated
at permanent inclination. Considering the efficiency drop
during irregular motion, the special design measures taken
made the columns almost insensitive to motion at moderate
sea states. However, during heavy sea simulation, especially
for long periods, efficiency drops in the range of 10% were
observed, taking the motion transfer functions of a semisubmersible steel platform carrier, different from the design
presented here. As tests with roll motion were performed
at DNV and in the Linde laboratory with columns of different
diameters to simulate the behaviour at random motion, a
scale-up could be performed, showing that in full-scale
columns, lower efficiencies can be expected than in the
random motion model tests.

4. PLANT CARRIER - STEEL PLATFORM WITH CONCRETE COLUMNS AND LNG TANKS

4.1 Design Description

The steel platform and the concrete tanks form a mono-
lithic unit. Concrete is used in areas where heavy weights,
resistance against corrosion, cryogenic properties and easy
maintenance are of importance. This concerns the submerged
part of the structure as well as the LNG and ballast tanks.

Steel is used in areas where for reasons of stability and
motion behaviour a low weight in connection with high
strength is needed.

The platform is supported by 6 concrete columns. A
column consists of an LNG tank, the ballast tank positioned
in the cylindrical part above and the skirt, which protects
the tank from exterior influences and improves the motions
of the structure. The storage tanks will be provided with
an internal insulation. The LNG tank is accessible through
a central shaft.

The columns consist of concrete, grade B 45. They
are reinforced and prestressed by cryogenic DYWIDAG thread-
bar St 1325/1470 N/mm², so that the total concrete body
has cryogenic properties.

The total storage capacity of 125,000 m³ LNG is divided
into 6 individual tanks with a capacity of 21,200 m³ each.
The ballast tanks have a capacity of 10,300 m³ each so that
for all loading conditions of the LNG tanks the design draft
of 44.20 m can be maintained.

External forces assumed in the design have been calculated for 100-years conditions in the northern North Sea.
For these calculations, as well as for the extensive finite
element strength calculations and the calculations of the
platform motions, well-proven computer programs of the classification society Det Norske Veritas have been applied.

4.2 Hydrodynamic Tests

The results of these calculations were carefully checked
by a hydrodynamic test program of a scale 1:75 model in
a large towing tank facility.

The hydrodynamic test program comprised the following
tests:
- towing tests in transit and operating conditions at 3
  different speeds,
- tests in waves of 6- 20 seconds period to determine the
  transfer functions of surge, heave and pitch motions,
  anchor forces, moments and forces at the connection plat-
  form/column of two columns of the anchored model for 3
  different loading conditions,
- tests in waves of 6 - 20 seconds period to determine the
  transfer functions of the exciting longitudinal and transversal forces to the total system and of the exciting
  forces and moments to 2 columns for two different skirt
  configurations, measured at the restrained model,
- forced oscillation tests to determine the hydrodynamic
  moments of inertia and the damping forces of the total
  system for two different skirt configurations in a range
  of periods of 6 - 20 seconds in still water, including
  pressure measurements at 3 selected points of the underwater structure.

4.3 Construction and Assembly

The concrete columns are built in a dry dock of a width
of 57 m and a depth of 7 m. Simultaneously the steel deck
is being fabricated in a shipyard and the liquefaction plant

being installed. Upon completion, the columns and the completely equipped steel deck are towed to a sheltered deep
water site with a water depth of approx. 70 m. Each column
is partially submerged by ballasting and floated into position with the central pump shaft fitting into an opening
of the floating steel deck. The structure will then be
lifted by deballasting the LNG tanks. The joints between
steel deck and steel shoe of the columns are completed by
welding. The openings will be closed and the insulation
can be installed after emptying the LNG tanks.

5. COMBINED TANKER MOORING AND LNG UNLOADING SYSTEM

5.1 System Description

The produced LNG has to be transferred in regular
intervals to tankers. Despite the large capacity of the
LNG storage tanks which act as a buffer, the production
continuity of the plant depends to a large extent on the
reliability and efficiency of the unloading system. To
secure a maximum of production continuity and fail safeness,
a combined unloading and mooring system has been developed
which can turn freely about the center of the platform.
The tanker is moored to a carriage running at the outside
edge of the circular steel deck. Should the wind or current
change direction during the loading procedure the tanker
is shifted around the center of the platform. With the
engines pulling back it stays always at the lee side of
the plant. The combined mooring/unloading system has the
following main characteristics:
- an articulated loading arm with a maximum range of 100m,
  carrying the LNG transfer and return-gas pipes and a
  coupling mechanism to be connected to the tanker bow,
- a maximum unloading rate of 6250 m³/h which permits loading
  of a 75,000 m³ LNG tanker in about 12 hours,
- a quick coupling mechanism which allows a very quick and
  entirely automatic decoupling procedure between the ship
  and the loading arm,
- a rotating carriage to which the tanker is moored during
  the loading procedure.

5.2 Tanker Approach and Connection Manoeuvers

To simulate the tanker approach and mooring procedure
in order to find out the best approach under different environmental conditions, tests at the SOGREAH Marine Research
and Training Center near Grenoble, France, were carried
out.

These tests were performed with a loading system somewhat
different from the one described in the previous section.
Its main components are:
- a 90 m long unloading boom,
- a loading arm assembly for LNG and return-gas transfer
  including a motion compensating device.

Also this loading system can be turned around, permitting the tanker to adopt a position of equilibrium with
respect to the resultant force of wind, waves and current.

The loading system was developed on the basis of FMC-components.

The ship-handling operations were successfully tested with the model of a standard 75,000 m³ LNG carrier at scale 1/21. A delicate phase of such a loading operation is when the ship picks up its moorings. For this, two alternative methods have been envisaged and tested:
1. To pick up a small buoy carrying the end of the bow-hawser with the aid of a messenger line of appropriate length ("conventional" method);
2. To pick up and secure an overhead line suspended from the boom, slightly behind the loading arm, and to use it to haul aboard the hawser ("direct boom to tanker" method).

The tanker model was piloted by the captain and an engineer. Hydrodynamic effects were scaled down according to Froude's law.

The main conclusions are the following:
- The ship must be equipped with a bow thruster and stern thruster of adequate power.
- All tests were carried out with a standard propeller, however, a controllable pitch propeller would give better efficiency and security.
- The ship must also be equipped so that the captain can accurately check its position and the remaining distance from the ship's bow to the platform.
- A doppler sonar is necessary to indicate the ship's speed relative to the sea bottom.
- The "direct boom to tanker" method appears to be safer and quicker than the "conventional" method, even with swell up to 4 m high, provided the boom has an adequate length.
- In addition, the floating messenger line seems to be incompatible with the use of a bow thruster, since it is either "swallowed" by it or forced away from the bow when the thruster is working the other way.

6. CRUDE GAS RISER SYSTEM

Six flexible pipes of 6" inner diameter each were chosen for the crude gas transfer riser; five flexible pipes are in service and one on stand by.

The total pipe length between the lower part of the platform and the sea bottom is divided into two parts. These two parts are connected at the level of a submerged buoy which is fixed to the sea bottom by cables and a concrete weight. The buoy is situated at a sufficient water depth to be out of the reach of wave action.

The division into two sections allows a quicker replacement of the flexible pipes in case of a failure of one pipe.

7. ACKNOWLEDGEMENT

The companies participating in this development would like to thank the Commission of the European Communities for the financial support granted.

Special thanks from the author are extended to colleagues of the participating companies for their assistance in the preparation of this paper.

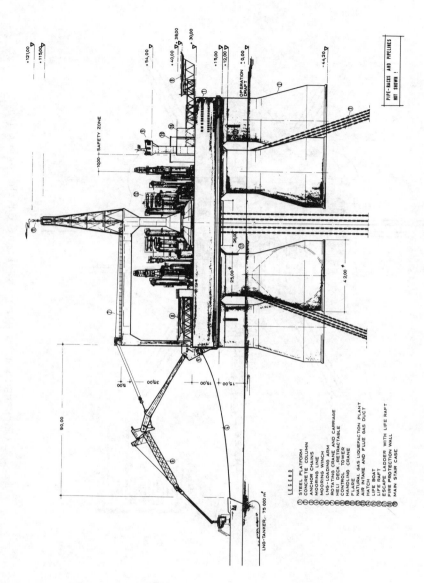

Fig. 1 Side View of the Floating LNG Plant

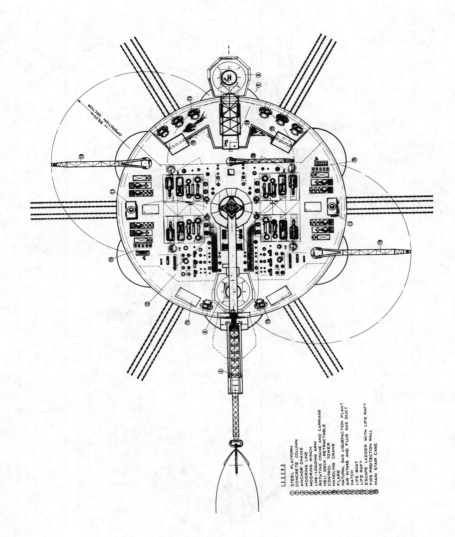

Fig. 2 Arrangement of the LNG Plant on the Upper Deck

(03.48/77)

FLOATING NATURAL GAS LIQUEFACTION PLANT

FOR OFFSHORE LIQUEFACTION AND LOADING OF ASSOCIATED GAS

A. Bath
Preussag AG, Erdöl und Erdgas, Hannover

Summary

Large oil and gas deposits have been found in the North Sea offshore Norway north of 62° latitude and in deep water areas west
of the British Isles. During oil production often large quantities of associated natural gas are produced which presently have
to be flared, reinjected, or brought ashore by pipeline. However,
transport by pipeline from deep water areas and/or to distant
consumers is not always feasible and sometimes uneconomic. Reinjection is only possible during a limited period of time of the
field production. Flaring is more and more prohibited. Consequently the Consortium ALP, consisting of the companies Aker, Linde
and Preussag has been developing an offshore gas liquefaction
and loading system especially suited to associated gas, deep
water and environmental conditions, as are typical for the northern North Sea.

1. General Design Criteria

1.1 Definition of Task

In the last years a number of oil/gas reservoirs have been discovered in the North Sea from which not only oil but also considerable quantities of associated natural gas will be produced. The discovery of additional similar hydrocarbon reservoirs especially in
the northern part of the North Sea is expected. For a number of such
reservoirs due to the insufficient gas amounts transportation of the
gas through a pipeline ashore will not be economical because of great
waterdepth, difficult pipelaying and/or long distance to the coasts
of consumer countries. While the oil can be offloaded by means of
existing offshore loading facilities, the associated natural gas might
have to be flared. To avoid this waste of energy sometimes the production of the oil had to be reduced or to be stopped, on account
of government regulations. To overcome such constraints, we think
a competitive solution is natural gas liquefaction offshore in the
field, intermediate storage and transport of LNG by tankers to the
consumer.

The following general design criteria have been taken into con-
sideration:
- Maximum gas recovery and high efficiency for the natural gas lique-
faction plant
- Minimum space and weight requirements for the process and nautical
equipment
- Favourable sea behaviour of the structures
- Little motion sensitivity of the process plant up to high sea state

- Relative autonomy of operation
- Flexibility regarding processing of different natural gas compositions
- Maximum safety for men and environment

1.2 Gas State

It has been assumed that gas/oil separation and pretreatment of the natural gas would be done on a conventional production platform. The natural gas would then be transferred through a short pipeline to the liquefaction plant. The following table shows the assumed gas composition and state as the gas enters the LNG plant, composition and state of the produced LNG and of the return gas (boil-off and flash gas from the storage tank).

		natural Gas Feed	Gross LNG to Storage Tank	Return Gas
N2	(mol.-%)	1.03	1.06	20.18
CO2	(mol.-%)	0.36	0.01	-----
CH4	(mol.-%)	77.36	77.76	79.80
C2H6	(mol.-%)	11.77	11.72	0.02
C3H8	(mol.-%)	8.29	8.30	-----
C4H10	(mol.-%)	1.07	1.12	-----
C5H12	(mol.-%)	0.06	0.02	-----
C6+	(mol.-%)	0.06	0.01	-----
H2S	(ppm)	3.3	1	-----
S total	(ppm)	35	35	-----
H2O	(ppm)	183	------	-----
Flowrate	(Nm³/h)	419,274	385,550	12,786
Temperature	(K)	283.311	111.00	218.00
Pressure	(bar abs.)	64.70 (inlet LNG plant)	7.0 (ex LNG plant)	6.50 (ex boil-off compressor)

1.3 Environmental Parameters

Location: North Sea, north of 58°, east of 0°
Maximum air temperature   25° C
Minimum air temperature   − 20° C
Maximum water temperature   15° C
Minimum water temperature   4° C
1 min-mean wind speed   50 m/s
Maximum current at the surface   3.0 kn
Design wave height   30 m
Wave period   14 − 20 s
Waterdepth   150 m

2. System Components

2.1 General Arrangement

The ALP system comprises the following main components (Fig. 1):
- Natural gas liquefaction plant
- Floating tension leg anchored steel platform
  (Aker TPP) to carry the LNG plant

- Feed gas pipeline connecting the conventional production platform with the TPP mounted LNG plant
- LNG and return gas transfer system between the liquefaction platform and the intermediate storage
- floating, concrete, semisubmersible storage tank with LNG tanker loading system.

2.2 Natural Gas Liquefaction Plant

For the cooling and liquefaction of the natural gas Linde utilizes its CCM process, essentially consisting of an ethane/propane precooling cycle and a mixed refreigerant cycle for the low temperature section (Fig. 2). Two identical lines produce a total of approx. 690 m³ LNG per hour.

The selection of this natural gas liquefaction process and its associated offsite facilities is dictated by site as well as economic considerations. The choice is influenced by such factors as the operating conditions on Aker's TPP, space and weight restriction of heat exchangers, machines and other process units and the desire for as low a power consumption as possible consistent with the use of proven equipment.

2.3 Tethered Production Platform - TPP

The platform structure, which is carrying the liquefaction plant is an Aker TPP which to some degree is a configuration similar to a semisubmersible drilling platform.

It has four large diameter columns and two rather small pontoons. Horizontal and vertical trusses give the platform the required structural integrity. The reason for the large diameter columns combined with the small pontoons is their reducing effect on vertical wave forces which are taken up by the anchor lines, and improved stability in the free floating mode.

The platform will be moored with vertical mooring cables (Fig. 3). Each mooring cable will be connected to a drilled-in pile, or gravity or other type of anchors, depending upon seabed conditions. A hydraulic jacking system secures equal tension in all mooring cables. The cables, which are designed for a lifetime of 20 years, are protected against the marine environment by a steel/plastic tube filled with a liquid corrosion inhibitor.

2.4 Feed Gas Pipeline

The TPP mounted liquefaction plant is connected to the production platform via a 16-inch pipeline with a length of approximately 400 m and a drilling type steel pipe riser system.

To cut the probability of down-time to a minimum two separate gasrisers, each capable of handling the total gas flow are specified. However, only one gas riser will be in operation at any time, while the other is stored on deck as a stand-by unit.

2.5 LNG Transfersystem

The LNG and return gas transfer system connects the TPP mounted natural gas liquefaction plant with the LNG storage anchored at some

600 m distance (Fig. 4). The LNG transfer lines are designed for appr. 690 m³/hr of LNG, the return gas lines for appr. 13,000 m³ (Vn)/hr of return gas.

The system consists of double corrugated insulated coaxial flexible pipes, arranged in pipe bundles. Additionally nitrogen pipes and high voltage cables are attached to the transfer system.

2.6 Intermediate Storage System

For storage of the liquefied natural gas a weightstable semisubmersible platform has been developed. The storage system consists of a substructure which is mainly made of concrete and a steel superstructure which carries the LNG tanker loading system.

Foundation of the concrete structure is a base plate equipped with pumps and valves for the ballast system. Three spherical storage tanks with a total capacity of 65,000 m³ LNG are mounted on the base plate while a number of cylindrical ballast water tanks provides for constant draft during all loading conditions. A central reinforced concrete column serves as supporting structure for the steel deckhouse and the LNG loading system.

Transport of LNG and return gas between storage-tank and LNG carrier vessel is enabled through the use of a flexible hosepipe. A handling turntable located on the storage-tank connects via the hosepipe to a coupling trestle on the carrier vessels bow. The LNG carrier which is moored to the platform by a bow hawser system can be charged within approx. 6 hours.

The floating storage system is anchored by a conventional catenary spread mooring system. Two LNG carriers with 40000 m³ LNG capacity each are calling at the storage alternately.

3. Main Dimensions and Capacities

3.1 Liquefaction Plant

Gross liquefaction capacity at inlet LNG plant	419,274 m³ (Vn)/h
Production capacity (two trains)	
Hourly gas	385,550 m³ (Vn)
Hourly LNG	690 m³
Daily gas	9.25 x 10⁶ m³
Daily LNG	16,560 m³

3.2 Tethered Production Platform

Overall length	91.00 m
Overall breadth	89.00 m
Operating draft above base line	33.50 m
Displacement at operating draft	54,112 t
Lower deck elevation above base line	57.10 m
Upper deck elevation above base line	66.10 m
Number of mooring cables	16

3.3 Intermediate LNG Storage and Loading Platform

Max. diameter of foundation plate	100.00 m
Operating draft	58.50 m
Helideck elevation above base line	122.50 m
Number of LNG tanks	3
Total LNG tank volume	65,000 m³
Loading rate of LNG loading system	6,250 m³/h

4. Model Tests

Offshore loading of LNG from a floating storage platform to an LNG-carrier has not been practiced so far.

Consequently manoeuvring tests with an LNG-carrier model have been carried out to investigate different methods of calling at the storage platform (Fig. 5).

A critical phase of such a loading operation is when the ship picks up its moorings. For this, two alternative methods have been envisaged and tested: one is to pick up a floating hawser with the aid of a messenger line of appropriate length, and the other is to pick up and secure an overhead line suspended from the boom, slightly behind the loading arm, and to use it to haul aboard the hawser. The first one is the "conventional" method and the latter is called the "direct boom to tanker method".

The tests performed at Port Revel Marine Research and Training Center involved both methods. In either case a very accurate approach to the platform tower is required. This involves bringing the ship's bow to within approx. 70 m from the platform and maintaining the ship in equilibrium in that position until the hawser as been secured.

The human factor is naturally of vital importance in this ship-handling operation, for which the manned model ships at Port Revel Center provide the best available experimental facilities.

The ship-handling operations were simulated with a model of a 75 000 m³ LNG carrier at scale 21, under varying wave, current and wind conditions. The model was piloted alternatively by one captain selected by Preussag and two instructors from Sogreah Port Revel Center.

The main conclusions are the following:
- The "conventional" manoeuvre demands very skillfull ship handling when waves and current are at right angles. Orientation of floating lines depends on current direction, while the ship has to ride the swell with the danger of crossing over the messenger line.
- In addition, the floating messenger lines seem to be incompatible with the use of bow thruster, since it is either "swallowed" by it or forced away from the bow when the thruster is working the other way. Under unfavourable weather conditions this type of manoeuvre can hardly be carried out without the aid of a workboat.
- Even under the most severe weather conditions the participating ship's pilots unanimously favoured the "direct boom to tanker" method. In the direct method manoeuvreability is unrestricted, since there is no need to worry about the floating lines.
- The "direct boom-to-tanker method" appears to be safer and quicker than the conventional method, even with swell more than 4,5 m high.

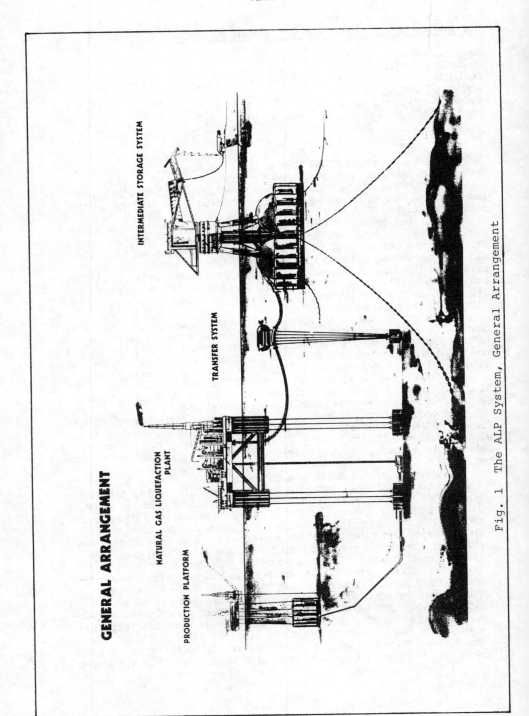

Fig. 1   The ALP System, General Arrangement

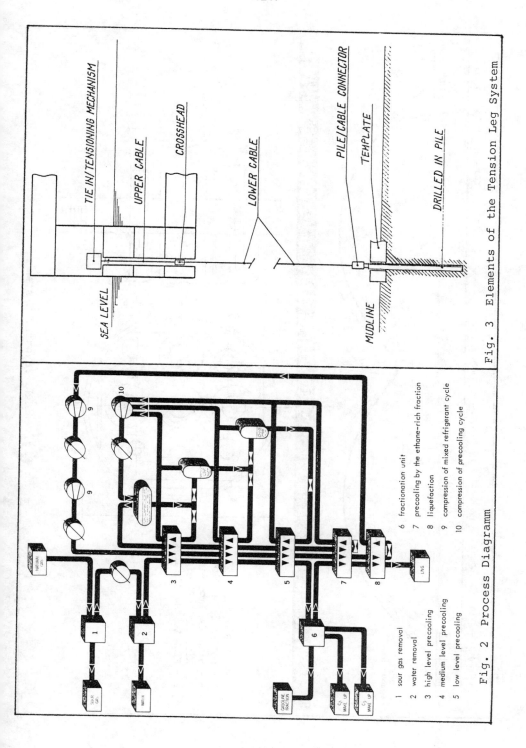

Fig. 3 Elements of the Tension Leg System

Fig. 2 Process Diagramm

1 sour gas removal
2 water removal
3 high level precooling
4 medium level precooling
5 low level precooling

6 fractionation unit
7 precooling by the ethane-rich fraction
8 liquefaction
9 compression of mixed refrigerant cycle
10 compression of precooling cycle

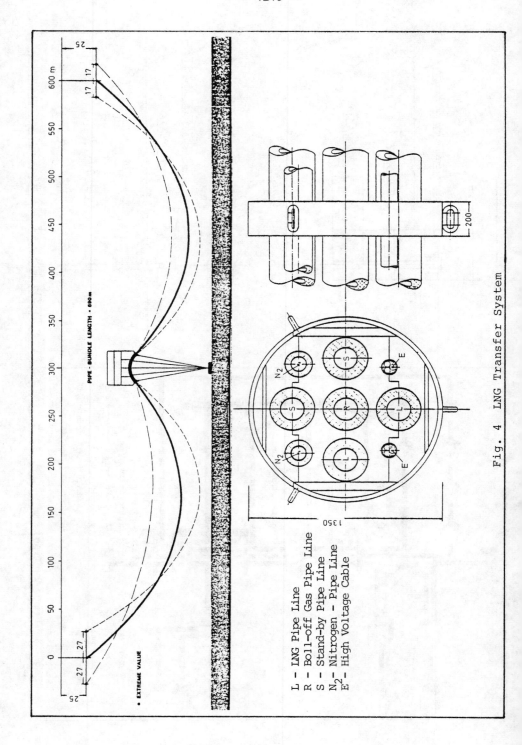

L – LNG Pipe Line
R – Boil-off Gas Pipe Line
S – Stand-by Pipe Line
N_2 – Nitrogen – Pipe Line
E_2 – High Voltage Cable

Fig. 4 LNG Transfer System

Fig. 5 Manoeuvring Tests

(03.58/78)

DEVELOPMENT OF A TENSION LEG PLATFORM AS
SUPPORTING STRUCTURE FOR A NATURAL GAS LIQUEFACTION PLANT

H.-D. BATSCHKO
Howaldtswerke-Deutsche Werft AG
Federal Republic of Germany

Summary

As part of the research project on systems for the production of liquefied natural gas and associated gas in the North Sea, promoted by the Commission of the European Communities, Salzgitter AG, in conjunction with its group companies Howaldtswerke-Deutsche Werft AG and LGA Gastechnik GmbH, has developed a tension leg platform together with a natural gas liquefaction plant. Model tests were successfully carried out in the wave tank at the Hamburgische Schiffbau-Versuchsanstalt on three tension leg platform designs. The results show that the construction principles, with the tension leg system and the liquefaction platform designed as separate units for safe offshore installation, ave sound and reliable even in extreme offshore conditions. The engineering design work of the system described culminated in quality assurance acceptance by Germanischer Lloyd.

1. INTRODUCTION

In connection with a series of research projects supported by the Commission of the European Communities and the Federal Ministry of Research and Technology, technical solutions have been worked out for the exploitation of offshore natural gas deposits, using floating or fixed offshore processing plants, to produce, for example, LNG and LPG.

The platform on which the plant is to be set up must be technologically suitable and economically viable. With floating platforms, there is a risk that some of the equipment in the production plant will function less efficiently if the pontoon moves with the wind and the motion of the sea. A fixed platform is therefore ideal for supporting an offshore natural gas processing plant. However, with fixed platforms, it has been shown that the investment costs for the platform alone rise by roughly the square of the water depth.

The oil companies working offshore are drilling in ever deeper waters (up to 500 metres or more) for petroleum and natural gas.

Experts agree that it will be possible to exploit such deposits economically using such devices as tension leg platforms because the financial outlay for the system depends to only a negligible extent on the depth of water.

A first step in this direction is the tension leg platform for the North Sea HuttonField, which is currently under construction.

This paper describes the design of a tension leg platform for use in up to 300 metres of water as a supporting structure for a natural gas liquefaction plant, with special reference to the model tests and the attendant calculations.

2. THE TENSION LEG PLATFORM (Fig. 1)

A jack-up-platform is used to support the liquefier, ancillary plant and living quarters. The platform and legs, can be lowered and raised by hydraulic lifting gear. The advantages of the jack-up system are :
- it can easily be moved to a different location or water depth;
- the jack-up can be fully fitted out in the yard and then towed out and placed in position;
- it is simple to assemble at sea;
- overall construction time is less.

The buoyancy body is an important component of the TLP system and is used basically for the following purposes :
- To produce the reserve lift required for the initial tensioning of the tension legs. The size is such that a reserve lift of approximately 11 000 t can be produced in the case of a single- strand construction with 9 000 t plant weight and 16 000 t using twin strand construction with 15 000 t plant weight. The buoyancy body is secured to the base with three tension legs at each of its four corners and is thus maintained at a water depth of approximately 60 m.
- To form the platform base on which rest the legs of the jack-up production platform. The legs are seated in an annular structure with packing elements on rubber springs.
- To act as a support for the components which go to make up the adjusting and connecting devices, together with the associated supply and monitoring equipment and quarters for the crew while the system is being set up at sea.

The mooring system of a TLP structure is so designed that the following requirements are met :
- the buoyancy body and the platform set down on it are kept securely in position above the site, as has been shown by extensive calculations and model tests;
- the forces of wind, currents, sea state and gravity, which all act on the platform, its production equipment and the buoyancy body, together with the buoyancy produced by the body itself are transmitted into the foundation system on the seabed;
- when acted upon by external forces, the buoyancy body moves in a horizontal plane around its normal resting place and at the same time is submerged to a slightly lower depth, whilst the tension legs incline;
- the mooring system is constructed in such a way as to be able to cope with changing environmental conditions;
- it is possible in situ to change the length by about 1 m to compensate for any inaccuracies in construction and variations in dimensions after the rig has been assembled at sea;
- the construction is easily moved to a new site and the tension legs can be changed when repairs are needed.

Important design features were a simple and sturdy structure, using tried and tested building components, a favourable weight ratio and distribution pattern with respect to the overall system, compression-proof construction, utilizing compression-proof displacement built in for sea transport and assembly, reliability, minimum maintenance and

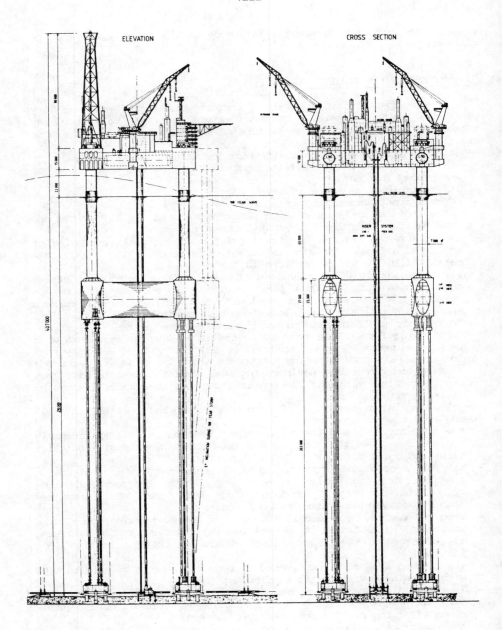

FIGURE 1 — TLP SYSTEM

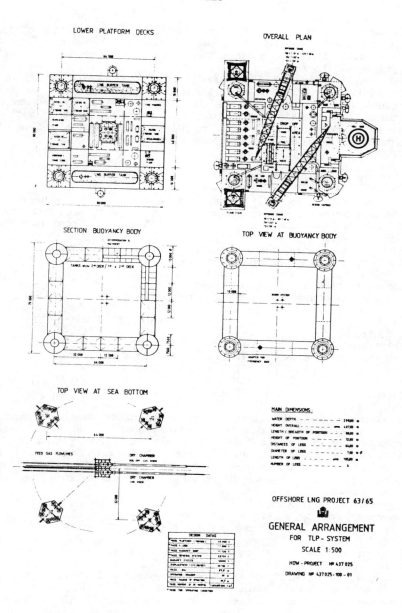

FIGURE 1 (cont.) — TLP SYSTEM

adequate redundancy features. For the TLP platform system, therefore, vertically stressed steel tubes are used between the buoyancy body and the foundations. There are coupling devices at the bottom of the steel tubes and pins at the top.

The coupling devices are connected with the pins which are mounted on universal joints on the foundation base. The pins at the top of the tubes are connected with the coupling devices of the adjusting and tatching mechanism which jutes out from the buoyancy body and is also mounted on universal joints. The coupling devices and joints have a special protective coating against corrosion and sanding up.

3. MODEL TESTS AND CALCULATIONS

Using models of three TLP designs, tests were carried out in regular waves and long-crested, irregular seas, principally with two total initial loads (107873 kN and 158377 kN) in 180° and 135° positions for 250 m depth of water.

The three designs differ in the shape of the buoyancy body :
- square buoyancy body, design I (Fig. 2),
- cylindrical buoyancy body, design II (Fig. 3),
- cylindrical/elliptical buoyancy body, design III (Fig. 4).

The main measurements of the platform examined were as follows :

	Design I	Design II	Design III
Length (platform)	65.92 m	65.92 m	65.92 m
Width (platform)	65.92 m	65.92 m	65.92 m
Height to the platform deck	97.20 m	99.33 m	109.20 m
Draught	63.20 m	65.33 m	75.20 m
Air gap	22.00 m	22.00 m	22.00 m
Displacement	57 197 m^3	55 005 m^3	56 200 m^3

In addition, with the first design model, tests were carried out for the two initial loads and situations mentioned above in wave groups with group periods of 0.85, 1.0 and 1.15 times the natural period of the TLP system in the wave direction. The first design model was also used to simulate a disaster in hundred-year storm conditions by detaching on eof the tension legs. As well as the wave height and period, the following were measured :
- the forces in the mooring,
- the movements of the platform in the wave direction,
- the accelerations in the three principal axes of the model,
- the yaw.

The main results of the tests carried out in regular waves were as follows :
- the forces in the mooring were somewhat smaller in Design II, and considerably smaller in Design III, than in Design I (Fig. 5-7),
- the motions in the wave direction and accelerations in the direction of the three principal axes were approximately the same in all three designs.

The main results of the tests carried out in irregular waves were :
- again substantially smaller forces in the mooring of Design III than in Design I;
- a very stable platform, staying on station in seas of maximum wave heights up to approximately 16 m, with no lifting of the platform legs from the buoyancy body in any of the three designs;

in hundred-year storm conditions, with wave heights up to a maximum of 31.5 metres, average period = 16.9 s, with Designs I and II and an initial load of 107873 kN (11 000 t) the tension legs were several times completely free of tension, there were at times quite strong tilting motions on the platform, in conjunction with an oscillating motion in the buoyancy body. With Design III, no tilting movement of the platform was observed and no oscillating motion was recorded for the buoyancy body. With an initial load of 158377 kN (16 150 t) no oscillating motion of the buoyancy body was observed with Designs I or II either, nor were there any tilting movements of the platform. The platform was stable on the buoyancy body in all sea conditions, even if the platform weight was decreased, as it was in order to increase the initial load. The oscillating motion of the displace-ment body (with Designs I and II) was caused primarily by the effect of hydrodynamic forces directed downwards, which were greater than the initial load in the tension legs. By selecting a hydrodynami-cally more efficient shape for the buoyancy body in Design III, it was possible to avoid oscillating motion of the buoyancy body in the worst seas that could be expected, even with the smaller initial load (107873 kN). It should therefore be possible to achieve the desired goal of full-scale production with only three tension legs per corner.

Together with the TLP on which tests were carried out, a second, larger supporting platform was developed for which no model tests were carried out although extensive calculations were made. The main data for this TLP were :

Length of the pontoon	80.00 m
Width of the pontoon	80.00 m
Height of the pontoon	12.00 m
Distance between legs	64.00 m
Draught	87.00 m
Total system mass	53 720 t
Residual buoyancy	1 600 t
Displacement	69 720 t
Centre of gravity of the total system over the base	87.50 m
Radius of inertia	51.10 m
Mass moment of inertia	140 400 000 t m^2

The following calculations were made :
- hydrodynamic exciting forces;
- motion characteristics in the frequency range (transfer functions);
- calculation of the joint forces in the frequency range (transfer functions);
- significant motion values for selected sea-state spectra
- significant values of forces for selected sea-state spectra.

All calculations were carried out for wave direction angles 0° and 45°.

These design calculations showed that a residual buoyancy of 12 000 t was sufficient to prevent it being neutralized by the hydro-dynamic vertical forces, and was confirmed by the programme calculation for wave direction angle 0°. With a wave direction angle of 45°, it could be seen that the residual buoyancy would have to be increased by 4 000 t to 16 000 t.

These calculations have, however, not yet resulted in optimum platform construction, although the trend would seem to show, that by shortening the platform legs, i.e. siting the buoyancy body nearer the surface, the residual buoyancy could be decreased.

Extensive studies and model tests are still needed here.

An evaluation of these calculation data shows however, that the motion characteristics of the platform system are stable and uncritical and that the forces in the tension legs do not exceed the permitted limits, as can be seen from the following diagrams (Figs. 8 - 11).

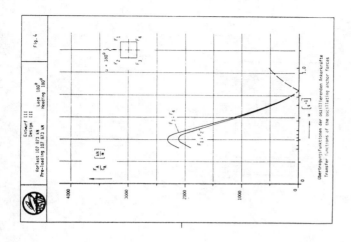

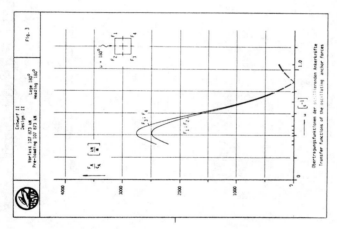

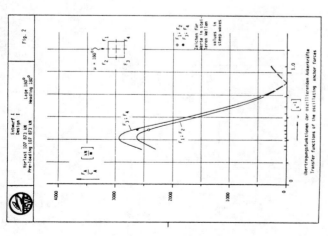

FIGURES 2, 3 and 4 – TRANSFER FUNCTIONS OF THE OSCILLATING ANCHOR FORCES – Design I, II and III

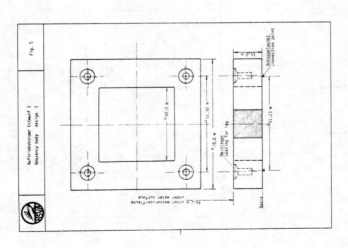

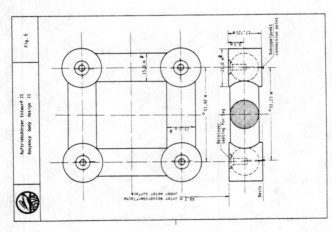

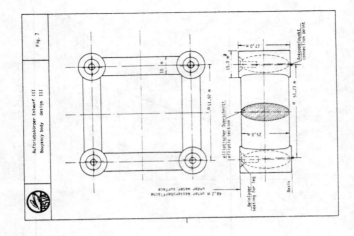

FIGURES 5, 6 and 7 – BUOYANCY BODY – Design I, II and III

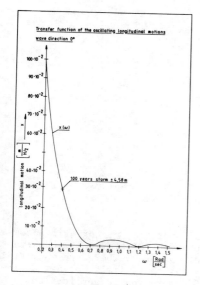

FIGURE 8

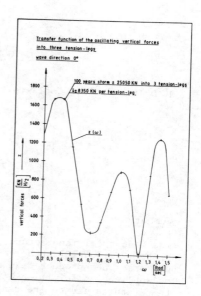

FIGURE 9

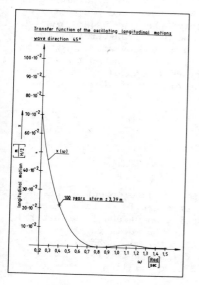

FIGURE 10

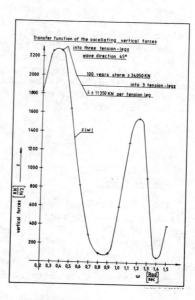

FIGURE 11

(10.14/78)

CRYOGENIC FUEL LINES

H. DRIDI, Société Coflexip

Summary

The industrial manufacture, laying and operation of Coflexip hoses since 1972 justifies the conclusion that the offshore oil industry now has a new product that has proved its worth. There are many fields of application for these hoses, but today their limits of use at low temperatures lie in the range of - 50°C for dynamic applications. Since 1978, Coflexip has launched a research and development programme that has resulted in the manufacture and qualification of a new generation of hoses capable of carrying fluids at very low temperatures and in particular liquefied natural gases (LNGs), namely at temperatures in the range of - 160°C. The internal diameter of these hoses can vary from 203 mm (8") to 406 mm (16") and the unit length can reach 500 metres. Their satisfactory behaviour under dynamic stressing enables them to be used in the main transfer functions, namely: surface links (ship-to-ship, ship-to-buoy with cryogenic swivel joint, storage area aboard ship, etc...) and bottom and bottom-to-surface links (buoy with cryogenic swivel joint - onshore storage facilitie, offshore processing plant - offshore storage facility, etc...).

1 - INTRODUCTION

The liquefied gas industry sprang from the development of the natural gas industry and the need for storage and transfer in certain cases in compact and reduced form (1 m³ of LNG contains about 600 m³ of natural gas). The quantity of natural gas marketed in the world in the form of LNG has risen from 2.7 10^9 m³ in 1970 to 42.7 10^9 m³ (2.8% of world production of natural gas crude) in 1983. Certain specialist bodies consider that this quantity could attain 200,000 million m³ in the year 2000.

This trend requires more flexibility in the LNG exploitation system. One of the links in this system is the transfer function. This is why Coflexip has developed, manufactured and tested hoses for liquefied gases capable in addition to thermal stresses at - 160°C, of standing up to the dynamic stresses encountered in offshore applications. These character-istics, in addition to conventional applications in fixed structures, to be used as transfer lines for future schemes, namely: loading or unloading stations in the open sea comparable to those used for oil and installations for processing and liquefaction of fatal gases (generally burned at the flare or reinjected into the reservoir), or the exploitation of gas fields that were formerly considered "marginal".

2 - STRUCTURE OF CRYOGENIC HOSE

2.1 - Description of structure
The structure of the Coflexip hose for conveying liquefied gas consists of the following, from inside to outside:
- a sealing sheath in 304 L grade steel of corrugated form,
- synthetic fibre reinforcement wires ensuring the stability and strength of the line,
- thermal installation for low temperatures,
- an outer carcass in the form of stainless steel stapled sheeting ensuring strength in the face of external loads,
- an external sealing sheath in plastic material (for example high density polyethylene), obtained by extrusion.

2.1.1 - Undulated internal pipe
The design of the corrugation has been selected to make the pipe flexible. The dimensioning of the corrugation, obtained from different results of calculations, must enable the hose to withstand the internal pressure without causing the corrugations to buckle whilst at the same time withstanding the considerable distortions under dynamic loads for several million cycles, without failing.

Fatigue tests under simple bending were carried out on three corrugated hoses. These hoses were manufactured at the Coflexip works at Le Trait in France. The principle is based on longitudinal undulation of a 304 L grade stainless steel sheet, spiralling of the corrugated sheeting on a mandrel endowing it with the desired internal diameter, continuous welding of the sheet and inspection of the well.

The following were the conditions in which the test was carried out:
- unit length (m) 5
- internal diameter (m) 0.203
- variation in radius of curvature (m) 7 < R < 30
- temperature ambient
- internal pressure (MPa) 1.5

The average lifetime of the three prototypes was about 3.7 million cycles.

These conditions are in fact severe, since the variation in the radius for each cycle is very considerable – from 7 m to 30 m –, and hence the corresponding stress variation is also considerable ; now, in accordance with Goodman's diagram, we know that this parameter is the predominant one in determining the lifetime of a structure. Accordingly, we consider that for the actual conditions of application, where the variation in the curvature is less, the service life could exceed 5 million cycles and more, without allowing for the improvement in the fatigue stress of steels when cold.

2.1.2 - Other layers of the structure
The other two main constituents of the structure are the reinforcement and the heat insulation. Low temperature mechanical characterization has enabled the following elements to be selected: a reinforcement in synthetic fibres and an insulation in EVA foam.

Tests of the compatibility with the final structure have enabled these two materials to be retained for manufacturing prototypes intended for the qualification tests.

2.2 - An example of the main characteristics of a cryogenic hose with an internal diameter of 12"

- internal diameter   300 mm (12")
- external diameter   530 mm (20")
- service temperature   - 163°C
- weight in air, empty   145 kg/m
- service pressure   3 MPa (435 PSI)
- minimum radius of curvature   3.5 m
- heat exchange coefficient   0.6 w/m°C
- loss of head for a flow of 100 m³/h   0.9 10^{-3} bar/m

These characteristics can be modified depending on the various applications (sea-line, riser, surface links, etc...).

3 - QUALIFICATION TESTS ON PROTOTYPES

After selecting and characterizing the various constituent materials of the structure, and of certain subassemblies, for instance the internal core or the end-pieces representing the main elements of this study, three cryogenic hoses with an internal diameter of 203 mm (8") were built in order to subject them to low temperature tests.

3.1 - Static tests

The purpose of these tests is to check and fill out the theoretical approaches to the different overall mechanical characteristics of the finished pipe.

For example, mention may be made of the determination of the stiffness at ambient temperature and at - 160°C, measurement of the variation of the length under pressure and at - 160°C or under external tension, resistance to thermal shock, etc...

3.2 - Dynamic tests

To simulate the behaviour of cryogenic flexible lines at sea, Coflexip has built a cryogenic test bench for bending fatigue tests. This test bench essentially consists of the following (see photograph N° 1):
- a metal frame carrying mobile carriages, occupying an area of 4 x 8 x 20 m,
- a hydraulic power plant actuating 2 cylinders each capable of a maxium thrust of 10 tons,
- a cryogenic plant capable of lowering the temperature inside the pipes to - 160°C at a pressure of 15 bars.

The length of the pipes tested can be up to 20 metres and the internal diameter 500 mm (20"). The radii of curvature can vary from 3 to 40 metres, and the frequency from 3 to 12 cycles per minute.

This test bench is also capable of testing LPG hoses at - 50°C and pressures of 20 bars. Dynamic tests on LPG and LNG hoses are at present taking place in this test bench.

4 - CONCLUSION

At the moment this paper is being written, we can not yet give the final result, since the dynamic, cold and pressure tests on pipes 15 metres in length are still taking place. However, we can already state that the industrial feasibility of cryogenic hoses with an internal diameter of up to 12" has been demonstrated.

For the pipe with an internal diameter of 16", requiring an average diameter of the outer plastic protective sheath of about 25", the appropriate tools will have to be built and extrusion tests carried out.

The future orientation of this project will depend exclusively on the individual specifications of the potential users, since in our view, Coflexip is today capable of satisfying specific orders.

CRYOGENIC TEST BENCH

(10.35/82)

OFFSHORE LOADING OF LIQUEFIED GASES

E. BONJOUR, Compagnie Française des Pétroles
J.M. SIMON, Entreprise d'Equipements Mécaniques et Hydrauliques

Summary

The aim of the "OFFSHORE LOADING OF LIQUEFIED GASES" project is to develop an offshore loading/unloading station for liquefied petroleum gas (LPG, temperature - 48°C) by means of a CALM buoy.
This type of offshore station will be much less expensive and easier to install than are today's harbour structures needed for loading and unloading LPG carriers.
The objective of the studies and tests carried out during the project was to qualify the critical components of such a system, and in particular the following:
- a 16" prototype swivel joint for a CALM buoy was built and qualified by means of low temperature fatigue tests simulating a service life of over 4 years,
- a low temperature test campaign on a prototype sealine section is now taking place,
- development is now going on in the plants of industrial makers of floating and subsea LPG hoses adapted to the requirements of the project. The qualification of these 16" prototypes by low temperature bending fatigue tests will be conducted by TOTAL/EMH on a test bench modified to accommodate these tests.
The results already acquired for the project tests as a whole indicate that an offshore pilot project at industrial scale can be envisage for the near future.

1 - INTRODUCTION

It is accepted that world LPG sea traffic could double by 1990, reaching about 30 million tons per year. This continuous growth that has already been observed for several years, is mainly due to increased production of fatal gas from the Middle East and the commissioning of liquefied natural gas production plants in a number of producer countries.
Most of this sea traffic concerns refrigerated LPG and the number of LPG exporting or importing terminals will increase significantly.
At present, the loading and unloading of refrigerated LPG carriers passes either via a loading arm or overhead transfer hoses operating practically statically.

The design of these links (ship-to-shore) means that the dynamic stresses they undergo in service must be limited as much as possible. Today's solutions consist in carrying out the loading and unloading operations in sheltered water. Accordingly, current terminals operating at low temperature (- 40°C to - 50°C) are of two types:
- mainly, the conventional harbour structure, where the ship is moored alongside a jetty or landing stage. The LPG is transferred by loading arms. The dynamic stresses in these arms are low,
- in a few cases: a process barge anchored permanently to an SPM (single point mooring): the gas is transferred via the SPM at ambient temperature to the barge, where it is refrigerated and stored. Transfer to the ships picking up the refrigerated LPG takes place by ship-to-ship configuration under operating conditions similar to those of a harbour terminal (small relative movements of the two ships).

In many cases of LPG valorization, where no deep water harbour exists near the production plant, one has to envisage building such a loading (or unloading) artificial harbour, which is expensive, representing a major share of the capital investment required for liquefaction and the transport of the liquefied gas.

The object of the "offshore loading of liquefied gases" project is to develop a device permitting a large cut in the investment required to load refrigerated LPG. More precisely, the project involves the study of an LPG offshore loading station of the SPM light type, without intermediate storage, the construction and testing of the critical prototype equipment and the preparation of an offshore loading pilot project.

The development of such a loading station represents an alternative to present techniques, namely an evolution entirely similar with regard to LPG terminals, to that contributed by the development and use of the first single point moorings for petroleum terminals. The main advantages of this progress would be:
- a considerable reduction in investments compared to any harbour structure, all the more so where the coastal site at which the terminal is to be located is inhospitable (unconsolidated seabed soils, shallow depths of water...). This reduction can vary from a factor of 2 to a factor of 5,
- flexibility in choosing the location, enabling construction of an SPM for LPG to be envisaged in sites excluding all possibility of building a harbour,
- in the longer term, the development of a refrigerated LPG offshore loading station represents a concrete initial stage in the face of the problem of the feasibility of transfer of offshore liquefied gas products.

2 - DESCRIPTION (Figure 1)
Conducted by TOTAL/COMPAGNIE FRANCAISE DES PETROLES and Société EMH (Equipements Mécaniques et Hydrauliques) who specialize in offshore terminals, the project has been working since 1982 on the development of a refrigerated (- 48°C) LPG loading station by SPM.

An SPM loading station is designed so that a ship can be moored whilst remaining free to turn completely around the mooring under the influence of winds and currents. Just as in one of today's LPG terminals, the boil-off gas produced whilst the refrigerated LPG is being loaded (heating and evaporation of the fluid along the transfer lines and in the tanks) is returned to shore for processing (reliquefaction...).

Two crucial problems then have to be considered:
- the construction and installation of a rotary joint for refrigerated liquefied petroleum gases on the SPM,

- the construction and installation of the connection between the SPM and the ship, which must be capable of accommodating their differential movements. The difficulty in developing these components, stemming mainly from the operating temperature (- 45°C) combined with offshore environmental conditions (fatigue, dynamic stresses...) is such that no component qualified for long duration low temperature offshore use exists today on the market. Much development work is taking place on these matters. An evaluation of the progress of these various development projects, together with a technical and economical comparison of the various types of SPM that can be conceived for transferring refrigerated LPG resulted during the initial phase of the project in opting for the development of a CALM (catenary anchored leg mooring) buoy for loading the LPG via subsea and floating transfer hoses.

So that it can replace all types of conventional harbour terminals, the loading station has been designed for embarking the contents of a large storage tank situated onshore (60,000 to 120,000 m³) onto a conventional LPG tanker, with only minimum adaptation required to these tankers, within about 15 hours.

Accordingly, the terminal consists in succession of the following subassemblies, starting from the shore storage facility and proceeding towards the ship to be loaded:
- Transfer lines between the storage facility and the pipeline end manifold (PLEM, situated beneath the CALM buoy) ensuring: transfer of the LPG to the ship, transfer of the boil-off gas to the storage facility, return to shore of the purge fluids used after each loading operation... Use of two parallel LPG loading lines enables these lines to be kept cold between each loading operation by closed-loop circulation via a by-pass interconnecting the two lines at the PLEM.
- A single point mooring comprising:
  . a CALM buoy adapted for LPG transfer, completely motor-powered and remote-controlled with a swivel joint for two fluids (LPG and boil-off),
  . subsea transfer fluids between the PLEM and the buoy proper,
  . floating connecting hoses between the buoy and the ship.

3 - GENERAL CHARACTERISTICS

The following operating data were adopted:
- Service temperature: - 48°C (in compliance with the design specifications of the LPG tankers).
- Loading time (including mooring and cast off operations): 20 hours.

During loading, the boil-off produced by the evaporation of the LPG and the displacement of the gaseous volume in the tanks of the ship exceeds the capacity of the reliquefaction units of the tanker. All the transfer lines (sealine, risers, piping, floating hoses) are heat-insulated so as to minimize the boil-off gas to be returned to shore.

By making use of modelization of the thermal and load loss phenomena in the lines, the optimum dimensions were determined: 20" for the sealines, 16" for the hoses and 12" for the boil-off line.

4 - STUDIES AND TESTING OF THE MAIN COMPONENTS OF THE LOADING STATION

4.1 - LPG sea-lines

Aside from careful application, the presence of insulation around the LPG transfer pipes requires a protective sheath and an absolutely tight seal around the insulation material to prevent any penetration of water that would destroy the insulation material by forming ice.

To achieve maximum reliability, it was decided to apply protection by means of an external steel casing. Centering anchorage points set out at regular intervals secure the internal tube and the external casing to prevent any relative motion between the two tubes. The dimensions and choice of the qualities of steel of the tubes and anchor points, together with the density of the polyurethane insulation material, were selected on the basis of the following criteria:
- no formation of ice on the external casing during service,
- selection of a high safety factor with respect to the stresses in the steel and the insulating material, during service.

Société SPIE-CAPAG carried out the research work on laying such a sea-line.

The end-connection of pre-insulated unit lengths takes place in accordance with the following stages:
- welding the internal pipe,
- making up the anti-corrosion lining,
- installing the shells of insulating material,
- welding the two half-shells so as to form a continuous casing.

Mainly because of the time required to make such an assembly, the laying method by pulling a pipeline built onshore turned out to be more economical than conventional laying from a laying barge.

At present, an experimental campaign for low temperature tests on a short length of prototype is now being organized (see figure 2).

These tests will be carried out in 1984 and enable the following main points to be verified:
- the procedures for end-connection of the field joints),
- the thermal and mechanical behaviour of the design,
- propagation of water in the insulating material in the event of failure of the protective casing.

4.2 - CALM buoy for LPG

The specific characteristics of the fluid transferred have resulted in major adaptations to the architecture and structure of a CALM buoy for loading oil and in particular the following aspects:
- choice of the steels of the structure,
- arrangement of the piping,
- choice of the equipment.

. Steels used for the structure

Since there are no regulations specific to offshore LPG terminals, the steels for the buoy have had to be selected by extrapolating the codes applicable to LPG carriers (for example the IMO code).

The grade and toughness at low temperature of the steels were therefore defined in terms of the temperatures at the various parts of the buoy in the event of LPG spreading. For example, the body of the buoy (except for the bottom) will be in manganese-carbon steel with a Charpy V resilience of 41 Joules at - 55°C.

. Arrangement of the piping

The general arrangement of the pipes and valves of the buoy were designed to prevent any possibility of gas gathering in a confined space:

all the valves lie above the surface of the water and the central shaft has been rendered tight by using seals adapted for low temperatures...
. Choice of the equipment
To ensure maxium safety and reliability the following principles have been adapted:
- use of "fail safe" valve on the PLEM,
- remote-control and hydraulic drive of all the SPM valves,
- monitoring of all the operating parameters: statuses of the valves, monitoring of pressure and temperature, gas detection...,
- a high degree of autonomy throughout the system.
The electrical equipment of the buoy was selected in accordance with the definition of the dangerous zones of the buoy, established in agreement with VERITAS.

4.3 - Bi-fluid swivel joint of LPG CALM buoy
A bi-fluid swivel joint is needed on the buoy to ensure transfer of the following from the body of the buoy to the turntable:
- LPG at - 48°C,
- boil-off gas at about 0°C.
A two-stage bi-fluid joint with its bearing was developed. The modular design consists of two coaxial single-fluid joints and offers the following advantages:
- the swivel joint for transfer of the boil-off gas is considered as known and proven, through similitude with swivel joints used for crude,
- the seal linings of each single-fluid joint are easy to change without having to disturb the overall system.
The development of the bi-fluid joint is hence equivalent to that of an LPG single-fluid joint.

4.4 - LPG single fluid swivel joint for CALM buoy
Such swivel joints exist on the loading arms, but mainly work practically statically. Accordingly, TOTAL/EMH has developed a new design for use specifically on a CALM buoy operating at low temperature (see figures 3 and 4).
This swivel joint is equipped with an internal tube and external insulation protected by a metal sheath comprising roller bearings. Here are its main features:
- the existence of a double layer of sealing material with an intermediate counterpressure chamber,
- maintenance of the bearings at ambient temperature during service, thus extending their service life.
EMH carried out the engineering work on a prototype swivel joint for LPG with a diameter of 16", built in 1983. During the first half of 1984, qualification tests for the design were successfully carried out in accordance with the following stages:
- preselection of 16" sealing linings by carrying out static sealing tests at pressure and low temperature,
- verification of the thermal behaviour (temperature in the insulating material, near the bearings...) and the mechanical behaviour (friction torque...) of the prototype joint,
- low temperature dynamic aging tests of the preselected seal linings (see figure 5).
This programme was completed and corroborates the excellent behaviour of the swivel joint and qualifies the seal linings following a fatigue testing programme that simulated over 4 years of service life without any significant leaks occurring in these linings (maximum leak outside

primary sealing barrier of less than 15 cm³ of gas vapour per hour and zero leakage outside the seal barrier).

4.5 - LPG transfer hoses

The technical challenge in developing offshore hoses for transferring LPG mainly lies in their operating temperature. Floating and underwater hoses for LPG must display:
- good mechanical properties at low temperature in order to withstand dynamic stresses,
- good behaviour with respect to the problem of gaseous diffusion in the elastomers.

Whilst LPG hoses are today available on the market, they are mainly intended for onshore applications or short-duration offshore applications. As yet, no large diameter LPG hose has been qualified by dynamic low temperature tests for long duration offshore use.

However, since 1983, several industrial firms are developing these products in collaboration with TOTAL/EMH in order to find a solution to the problem. Two different types of products are now undergoing development:
- rubber hoses derived from buoy hoses for transferring oil (DUNLOP, KLEBER...),
- composite hoses comprising concentrical steel turns ; metal carcass riser (COFLEXIP, HUTCHINSON...) (see figure 6).

Pre-qualifying tests are now being carried out by these firms (fatigue testing by alternating bending at low temperature on 1/2 scale prototypes, static tests on 16" prototypes...). In close collaboration with classification and inspection bodies, TOTAL/EMH have drawn up detailed specifications for acceptance-testing of these prototypes:
- routine tests (pressure, vacuum, bending...) at ambient and low temperatures,
- gaseous diffusion tests,
- bursting tests.

The 16" prototypes selected to these specifications will be subjected during the second half of 1984 to a series of fatigue tests at low temperature. An already existing flexible test bench has been specially modified to enable these tests to be carried out at a temperature of - 50°C, using refrigerated isopentane.

5 - CONCLUSION

The studies and research work carried out during this project, together with the results of the tests already made, make it reasonable to expect that the design feasibility of a loading buoy for refrigerated LPG will be attained.

Already, TOTAL/EMH are studying a prototype project enabling the design studied to be brought to industrial level.

Such a prototype operation will form the substance of a specific 3-year project with a view to:
- simulating and evaluating the operational procedures (in service, emergency...) of the loading station,
- subjecting the elements of the loading station already tested in the workshops to accelerated aging under real-life offshore conditions.

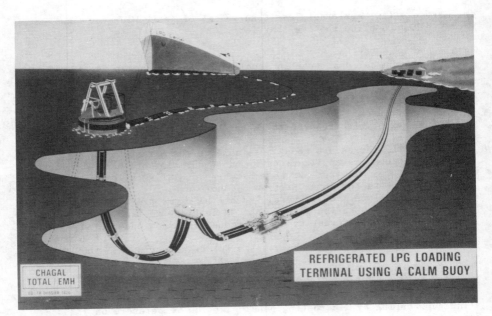

FIGURE 1

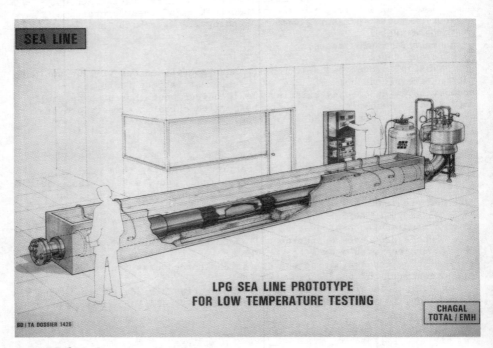

FIGURE 2

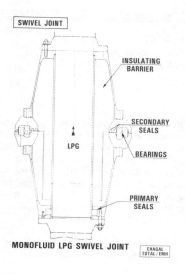

FIGURE 3 - GENERAL DIAGRAMME

FIGURE 4 - PROTOTYPE JOINT FOR TESTS

FIGURE 5 - DYNAMIC TESTING OF LOW
TEMPERATURE SWIVEL JOINT

FIGURE 6 - PROTOTYPE FLOATING LPG
HOSE

(14.06/78)

UNLINED CONCRETE STORAGE FACILITIES FOR LIQUEFIED NATURAL GAS

P.B. BAMFORTH
Assistant Projects Manager
Taylor Woodrow Construction Limited

Summary

A project has been carried out to provide the technical data to
support the design of prestressed concrete facilities for the storage
of liquefied natural gas (LNG). Tests were undertaken to identify
concreting materials and mix proportions most suited to cryogenic
applications. Selected concretes were comprehensively tested at
temperatures down to -165°C to obtain engineering properties required
for design. A range of reinforcing steels and prestressing systems
were also investigated. Finally prestressed concrete elements were
tested under conditions of live and thermal loading.

A general description of the test programme is given, together with
information on the experimental techniques and test results.

1. INTRODUCTION

In October 1979 work began on a programme to develop the potential of
concrete as a primary containment material for the storage of liquefied
natural gas (LNG) at -165°C. Hitherto, when concrete had been used in a
storage system for LNG, some form of impermeable barrier was incorporated
into the system to provide a leak tight barrier.

The principle requirement of containment for LNG is that the structure
shall not leak. To achieve this using unlined prestressed concrete, three
factors require consideration: the inherent permeability of the concrete;
the prevention of cracking for the primary containment; and the limitation
of crack depth for the secondary containment. Provided that the concrete
can be shown to have adequately low permeability, then the prevention or
limitation of cracking is the primary purpose of the design.

To establish likely causes of cracking it is necessary to consider the
state of the structure at each stage of construction and operation,
together with any potential accidents as follows:-
- Construction, prestressing and hydrostatic testing,all at ambient
temperature.
- Cool down or warm up, filling and emptying,giving rise to controlled
transient thermal conditions.
- Accidents giving rise to uncontrolled transient thermal conditions.

It is important to give due consideration to the avoidance of cracking
in the initial conditions at ambient temperature. This can be achieved
using construction practices currently adopted for other high technology
structures such as prestressed concrete pressure vessels for nuclear reactors, and offshore gravity platforms for oil production.

During commissioning and under normal operating conditions of the
primary containment, a knowledge of the thermal and structural properties
of concrete,and the way these change with temperature,is required in order

to establish controlled procedures with safe operational limits of temperature differential. Properties of principle relevance in this respect are the thermal expansion coefficient and tensile strain capacity of the concrete. Thermal diffusivity is also important when time dependence is considered and this is in turn a function of the thermal conductivity, specific heat and density of the concrete.

Under accident conditions, giving rise to thermal overload of the secondary containment, it is not possible to prevent cracking. Attention must be given to the provision of prestress and passive reinforcement to limit the depth of cracking. The relevant properties in this instance are as above, with the addition of the bond strength of concrete to steel, and elastic and creep deformation.

2. TEST PROGRAMME

The test programme, which was completed at the end of 1982, was undertaken in three phases:-
Phase 1 - Literature analysis and preliminary tests.
Phase 2 - Comprehensive property testing on selected materials.
Phase 3 - Component testing and analytical studies.
At the outset of the programme it had been proposed to include a fourth phase comprising a model tank test. However, in view of the downturn in world trade in LNG, and the consequent reduction in demand for new facilities, the programme did not extend beyond Phase 3.
Details of the range of tests carried out in each phase of the programme are given in Table 1, which also shows the number of specimens tested.

2.1 Literature Analysis and Preliminary Tests
The objectives of this initial phase were as follows:-
(1) To avoid duplication of work already undertaken and identify significant gaps in cryogenic technology.
(2) To obtain information on cryogenic testing techniques.
(3) To establish the requirements of materials used in cryogenic storage facilities.
(4) To select materials for comprehensive testing.
To aid in the materials selection, preliminary screening tests were carried out on 19 concretes to establish the influence of factors such as strength grade (w/c ratio), air entrainment, cement type, aggregate type, and water reducing admixtures. Tests included the following:-
(1) Compressive strength at $20^{\circ}C$, $-180^{\circ}C$ and after thermal cycling.
(2) Tensile splitting strength at $20^{\circ}C$, $-180^{\circ}C$ and after thermal cycling.
(3) Load deformation (recorded on test cubes) at $20^{\circ}C$, $-180^{\circ}C$ and after thermal cycling.
(4) Water permeability.

2.2 Phase 1 - Selection of Materials for Phase 2 Testing
Four concretes were selected primarily on the basis of low permeability. Details are as follows:-
Mix 1. Typical structural concrete with OPC, gravel aggregate and air entrainment. This mix was selected as a control against which to assess the modified concretes.
Mix 2. High strength superplasticised concrete with low water-cement ratio for low permeability.
Mix 3. Structural concrete with 30% of OPC replaced by PFA for low permeability.

<u>Mix 4</u>. Lightweight concrete with air entrainment, for low permeability.

Two types of conventional rebar, hot rolled and cold worked, were investigated, together with three special steels:-
- Krybar, a steel developed by Trade Arbed specifically for low temperature application.
- 9% Nickel steel, with known low temperature ductility.
- Stainless steel, also with known low temperature ductility.

In addition,prestressing wire and strand were included, together with tests on the anchorage systems.

2.3   Phase 2 – Concrete Property Tests

The four selected concretes were subjected to a range of property tests at temperatures from 20°C to −165°C. The tests, listed in Table 1, were selected to provide specific information relating to structural leak tightness, i.e. permeability, crack resistance, as well as the data required to establish stress levels for design, i.e. thermal expansion coefficient, elastic modulus and creep.

To ensure that the concretes represented, as closely as possible, the concrete in a structure, all specimens were subjected to sealed curing conditions. This involves storing specimens in moisture tight bags, the only water available for hydration being that which was added to mixing. Having identified the importance of moisture content on the cryogenic properties of concrete, this was felt to be a critical feature of the test programme if the results were to be applied to real structures.

2.4   Phase 2 – Steel Property Tests

The range of tests, listed in Table 1, was on two types of conventional rebar, prestressing wire and strand, and three special steels. A range of bar sizes were tested using procedures which were as close as possible to the normal procedures laid down in the relevant British Standards for quality control at ambient applications

2.5   Phase 2 – Analytical Studies

To support the materials property test programme,analytical studies were carried out. This involved a study of tank design, together with an assessment of various analytical techniques for predicting the performance of prestressed concrete components when subjected to transient thermal conditions at very low temperatures.

2.6   Phase 3 – Component Tests

Having identified concretes and steels with the properties which were most suitable for low temperature applications, tests were carried out on prestressed concrete elements, both to observe the composite performance and to check the analytical methods employed against observed behaviour. Tests were designed to establish the behaviour of a prestressed concrete tank under transient thermal conditions using restrained beams.

In addition, the performance of construction joints has been examined together with the insitu measurement of concrete permeability for comparison with earlier tests on small laboratory specimens.

3.   EXPERIMENTAL PROCEDURES

In general, test rigs,whether for concrete or steel,comprised of a stainless steel loading frame housed within an insulated cabinet. The system was cooled using liquid nitrogen, sprayed into the cabinet at a controlled rate. Experimental features of particular interest were as follows.

3.1 Strain Measurement of Concrete

Vibrating Wire Strain gauges were used to monitor both thermal and load induced strains in the concrete. This type of gauge operates by recording the change in frequency of an electrically plucked wire as the strain changes. Calibration tests were carried out at temperatures down to $-196°C$ to determine changes in gauge factor and thermal response. With only minor modifications, these gauges worked extremely well, and were subsequently used to measure tensile strain capacity, flexural strain capacity, thermal contraction, elastic modulus and creep.

A transducer to measure bond slip of rebar, designed in-house, was also based on the vibrating wire principal. In over 200 tests the failure rate was less than 5%.

3.2 Cooling Cabinets to Increase Rate of Testing

In order to simulate cooldown of an LNG tank, it was necessary to cool the test specimens slowly. A rate of $4°C$ per hour was chosen as a compromise between the more likely rate of $1°C$ per hour in a full size tank, and the limitation of the test programme. However, even at a rate of $4°C$ per hour specimens required a cooling period of about 2 days. A system was therefore adopted whereby a number of specimens, up to 24 at a time, would be cooled in a cabinet adjacent to the test rig, and transferred to the precooled test rig when the required temperature had been achieved. This enabled up to 15 specimens per week to be tested in a single test rig. Cooling a similar number of specimens individually in the test rig would have required in excess of 6 weeks.

3.3 Optical Strain Monitoring of Steel Specimens

In order to monitor strain in rebar and prestressing steel up to failure, an optical extensometer was used. To operate this system it was necessary to build a window into the cryostat to make optical contact onto the specimens. A triple glazed system was used, which was prevented from frosting by passing dry nitrogen between the panes.

3.4 Full Bar Impact Tests

In addition to standard Charpy impact tests, full bar tests were carried out using the British Steel test facility at Motherwell. This has an impact hammer of 9,000 joules capacity, compared with the 300 joules capacity of the standard Charpy impact test machine.

4. CONCRETE PERFORMANCE

The results of the tests undertaken by Taylor Woodrow, together with the considerable volume of published data on the engineering properties of concrete at cryogenic temperatures, clearly illustrates the increased strength and stiffness of concrete. It is therefore safe to assume ambient property data when designing LNG tanks for normal service conditions such as prestress and hydrostatic loading. However, in view of the containment requirements, and the transient thermal conditions which exist during cooldown, filling and emptying and possible accidents, data on concrete permeability and crack resistance is essential for rational design if concrete is to be used as a primary barrier to LNG. A summary of property changes when concrete is cooled to $-165°C$ is given in Table 2, obtained from tests undertaken by Taylor Woodrow on simulated site concrete. The following conclusions may be drawn:-

(1) By carefully selecting concreting materials and mix proportions, low permeability concretes can be achieved which will maintain their performance even after extreme thermal cycling.

(2) The containment properties of concrete, i.e. permeability and strain capacity, are enhanced at low temperature. The permeability reduces by about 50% over the range 20 to -165°C, whilst the resistance to thermal cracking is approximately doubled.

(3) The proportionally higher increase in bond strength at low temperature compared with the tensile strength of concrete, will reduce the size of cracks which occur when the concrete is cold.

(4) The containment properties of lightweight concrete have been found to be markedly better than normal weight concrete. Permeability coefficients have been found to be about 2 orders of magnitude lower, crack resistance is approximately doubled, and if cracking does occur, crack widths will be smaller.

The results clearly indicate that selected concretes can be assumed to have the ability to contain liquefied natural gas. In view of the extreme environmental conditions and the need to ensure safety to both plant operators and the general public, up-to-date materials and construction technology must be applied, however, to ensure that the potential performance of the concrete is fully realised on site.

5. PERFORMANCE OF REBAR AND PRESTRESSING SYSTEMS

In the absence of any detailed specification for rebar and prestressing steels used continually at cryogenic temperatures, their performance at low temperature can, at present, only be compared with that required in specifications for use at ambient temperature. In this respect whilst the yield stress, ultimate tensile stress and elastic modulus were all substantially increased at cryogenic temperature, the only steel tested by TW which was fully compliant was 9% nickel steel, but even this was found to be highly variable in its mechanical properties. Typical property changes are given in Table 3. Cold worked bar and 20mm hot rolled bar failed on the ductility requirement, whilst Krybar and 32mm hot rolled bar failed to meet the minimum uts/ys ratio of 1.15.

Based on current specifications, conventional rebar would not therefore be acceptable for cryogenic use. Cold working results in a severe loss of tensile strain capacity and processing variables means that brittle failure could always be a possibility. Considering the role of rebar in a cryogenic tank, however, indicates that aiming to achieve ambient specification requirements may be unnecessarily stringent. In service, a low level of ductility may be acceptable. High levels of failure strain are only generally required for cold working, e.g. bending. In addition, the Charpy test has been found to be over pessimistic in predicting the impact behaviour of actual bars. A more satisfactory test, such as whole bar impact, is likely to be more relevant to structural applications. A specification based on insitu performance would increase the range of acceptable commercially available reinforcement considerably.

Only two of the seven prestressing steels tested met the ductility requirements of a minimum elongation of 3.5% at maximum load. However, the tests were carried out using standard stress inducing wedges rather than test machine grips, this being a more severe case than intended for compliance testing. Consequently wire and strand with a mean failure strain greater than 3.5 would probably be acceptable, increasing the number of prestressing steels with adequate performance from two to five out of the seven tested.

6. PERFORMANCE OF PRESTRESSED CONCRETE

The following conclusions are based on a study of published data, an analytical study of the performance of prestressed concrete and tests on prestressed concrete beams undertaken by Taylor Woodrow.

6.1 Thermal Shock Loading

Even though the surface of concrete may suddenly come into contact with liquid gas, it will not be cooled instantaneously to cryogenic temperatures. It may take up to several hours for the concrete surface to reach the temperature of the surrounding liquid due to the insulating effect of the gaseous layer which is formed. For accurate thermal analysis the effect of this gas layer must be taken into account, together with the changing properties of the concrete as it becomes colder. Failure to do so may result in errors in predictions of strains and stresses of the order of 25%.

Under shock loading conditions reasonable agreement between experimental and theoretical strains has been obtained using a non-linear dynamic relaxation program. The predicted pattern of cracking is in close agreement with experimental results.

6.2 Controlled Thermal Gradients

During cooldown, the bending moments generated in a restrained section due to temperature gradients can be predicted with reasonable accuracy. However, the changing concrete properties must be taken into account.

With moderate prestress and the selection of a suitable high strain capacity concrete, controlled temperature differentials in excess of 100°C may be imposed on a restrained concrete section without the occurrence of cracking. Even without prestress temperature differentials in excess of 50°C may be tolerated.

6.3 Ultimate Strength

The ultimate strength of a prestressed concrete beam is increased as temperature levels drop to cryogenic regions. Reducing the temperature from ambient to about -165°C increased the ultimate strength by approximately 45 per cent. Other available data would, however, suggest that most strength gain occurs during the early stages of cooling to temperatures of about -70°C.

This increase in ultimate strength at cryogenic temperature can be predicted with reasonable accuracy by using modified material properties in conventional design methods.

6.4 Crack Control

To control cracking, the reinforcement ratio must be such that the yield strength of the steel is greater than the tensile strength of the concrete in the tension zone. The increased tensile strength of concrete at low temperature will therefore necessitate the use of either additional steel or a steel with a higher yield strength.

The relationship between average crack width and the stress in the reinforcement is largely unaffected by the temperature conditions or the nature of loading. This means that, in general, the allowable stresses in the steel need not be changed when members are subjected to low temperatures.

7. GENERAL CONCLUSIONS

To achieve low structural permeability, lightweight aggregate concrete was found to be the most suitable, with low inherent permeability and a high level of crack resistance.

Conventional hot rolled rebar would appear to be acceptable when considered in relation to structural performance, but fails to meet the current specifications. In view of this it is recommended that specifications for low temperature performance be reviewed and new, more relevant compliance tests be adopted. In the meantime, special steels with low temperature ductility should be used.

Conventional prestressing systems with wedge anchors have been found to perform satisfactorily at low temperature and may therefore be considered for use in LNG systems subject to low temperature quality control checks.

Overall, the data obtained should enable the design of tanks to be undertaken on a more rational basis, increasing confidence in the materials performance and the safety factors used.

ACKNOWLEDGEMENTS

The Author wishes to thank the Directors of Taylor Woodrow Construction for permission to publish the paper. The project was funded by the CEC, the UK Department of Energy and Taylor Woodrow Construction Ltd. Technical support was provided by British Gas, Shell, British Petroleum and Britoil. Their advice and comments were greatly appreciated.

TABLE 1. PROPERTIES MEASURED AND NUMBER OF SPECIMENS TESTED

PROPERTY OR PERFORMANCE MEASURED	AMBIENT	CRYOGENIC
Phase 1 Screening Tests – 19 Concretes		
Compressive strength	38	38
Compressive strength after thermal cycling	48	4
Load deformation behaviour	86	42
Tensile splitting strength	38	38
Tensile splitting strength after thermal cycling	44	2
Water Permeability	42	–
Phase 2 Measurement of Engineering Properties		
Concrete		
Gas permeability	12	8
Direct tensile strength/strain capacity	12	48
Flexural strength/strain capacity	12	48
Bond strength	12	48
Elastic modulus	12	12
Creep	12	12
Thermal expansion coefficient	12	12
Specific heat	12	48
Resistance to thermal cycling – rapid cooling (10 cycles)	96	–
– slow cooling (10 cycles)	90	–
Steel		
Charpy impact transition temperature	1260	
Full bar impact resistance	32	
Tensile tests including Young's Rebar	99	100
modulus, yield stress, uniform Prestress	91	31
elongation, ultimate tensile		
strength, failure strain		
Performance of prestress under cyclic loading		3
Chemical analysis	18	–
Microscopic examination	18	–
Phase 3 Component Tests		
Crack resistance under thermal loading	–	2
Ultimate load behaviour	–	2
Leakage through joints	1	1
Structural permeability	1	1

TABLE 2. <u>TYPICAL CHANGES IN THE PROPERTIES OF SEALED</u>
<u>CURED CONCRETE WHEN COOLED FROM 20°C TO −165°C</u>

Gas permeability	— reduced by 50%
Compressive strength	— increased by 100%
Tensile strength	— increased by 150%
Tensile Strain Capacity	— increased by 100%
Flexural strength (modulus of rupture)	— increased by 150%
Flexural strain capacity	— increased by 100%
Elastic modulus	— increased by 60%
Creep	— reduced by 90%
Bond strength to deformed reinforcement	— increased by 200%
Thermal expansion	— reduced by 35%
Specific heat	— reduced by 65%
Thermal diffusivity	— increased by 200%

TABLE 3. <u>TYPICAL CHANGES IN THE PROPERTIES</u>
<u>OF STEEL WHEN COOLED TO −165°C</u>

<u>Reinforcing Steel</u>

Yield stress	— increased by 50%
Ultimate tensile stress	— increased by 35%
Failure strain	— reduced by 40%
Charpy impact energy	
Conventional rebar	— almost eliminated
Special steels	— reduced by 40%
Whole bar impact energy	— increased by 25%

<u>Prestressing Steel</u>

0.1% Proof stress	— increased by 20%
Ultimate tensile stress	— increased by 15%
Failure strain	
5mm wire	— reduced by 25%
12mm, 7 wire strand	— increased by 20%
Charpy impact energy	— reduced by 20%

(14.13/82)

CONSTRUCTION TECHNIQUES IN LIMESTONE FOR CRYOGENIC STORAGE

F.C. BARTER
Cavern Systems Dublin Limited

Summary
    Cavern Systems Dublin Limited is a company formed
in late 1980 to investigate the possibility of
constructing underground LPG storage in the port area
of Dublin. It represents the combined interests of
Calor Gas Ireland Limited (a wholly owned subsidiary of
Calor Gas Limited and a member of the I. C. Gas
Association Group of Companies), and Conor Holdings
Limited (an Irish private venture capital company with
existing energy and natural resource investments).
This project is concerned with the extension and
development of existing underground rock cavern
technology with a view to constructing 100,000 tonnes
of underground refrigerated LPG storage in Dublin Bay
Limestone.
    To date, geological, geotechnical and geophysical
appraisal has established the overall suitability of
Dublin Bay limestone for refrigerated LPG storage, and
various feasibility studies have established the
overall environmental acceptability, the preliminary
design configuration and geometry, and the likely
economics of the project.
    The project has significant advantages to both
Ireland and the Community by providing the necessary
infrastructure to gain from increasing the use of an
available and competitive energy source.

1.   Description of Project

The aim of the project is to extend and develope existing
underground rock cavern technology with a view to
constructing 100,000 tonnes of underground refrigerated LPG
storage in Dublin Bay Limestone and it comprises the
following main elements:-
-    Detailed geological investigation and geological
     appraisal of the port area of Dublin
-    Comprehensive feasibility studies of the environmental,
     hazard, technical design, and economic aspects
-    Construction of an exploratory shaft to an approximate
     depth of 110 metres and performing comprehensive tests
-    Construction of scaled down test cavern and performing
     comprehensive tests to determine whether full scale
     cavern proceeds and the final design configuration.

2. Status of Project

The geological appraisal and geotechnical investigation and the feasibility study are now complete.

The geotechnical appraisal and geotechnical investigation were carried out by performing an overall examination of Dublin Bay geological strata and by an extensive core drilling programme. The cores were subjected to detailed geological and geotechnical analysis and various tests were performed to establish their geological and thermal properties. The results of these tests combined with heat and stress calculations established the overall suitability of Dublin Bay limestone for refrigerated LPG storage.

Having established the geological and geotechnical suitability of the rock it was necessary to establish the environmental, technical and economic feasibility of proceeding with the project. An Environmental Impact Assessment and Hazard Analysis was performed which concluded "that the overall environmental impact was acceptable and that the societal risk was within internationally accepted criteria." Rock cavern design configuration and related geometry has been established.

On the economic feasibility of the project, the overall market for LPG and terminalling requirements in Ireland and Europe has been researched and the likely economics of a Dublin cavern established. Initial findings suggest that the construction of a large refrigerated LPG Cavern located in Dublin is not only a sound commercial proposition but provides essential strategic storage. It also provides social and economic benefits to Ireland and to the EEC.

The current status of the project is that although Planning Permission has been obtained from Dublin Corporation to proceed with actual construction work, third party objections have resulted in a delay due to the holding of a Planning Appeal Hearing, the results of which are not yet known.

3. Geological Appraisal and Geotechnical Investigation

3.1 Examination of Dublin Bay Geological Strata

An examination of the general geological structure of Dublin Bay and its surrounds was undertaken in conjunction with leading Irish geologists. A review of available geological data in conjunction with investigation of limestone outcrops in the Dublin region was carried out. The works which were studied include :-

- Dr. Jackson's Studies of the general geology of Dublin.
- Scientific Proceedings of the Royal Dublin Society 1965 Naylor "Pleistocene and Post-Pleistocene Sediments in Dublin Bay".
- Report on the Clontarf Lead Mine - Weaver, 1884.

The review identified the presence of lower carboniferous limestone close to the surface in the Dublin Bay area and indicated its likely depth to be in excess of 300 metres.

3.2 Core Drilling

A series of rock drilling investigations was carried out to determine the depth of the overburden and the nature and characteristics of the bedrock. The rock cores were recovered and geological observations made on the rock samples. Rock Quality Designation (RQD) values and fracture frequencies were measured. The recovered rock cores were logged and are available for inspection. The results of the core drilling indicate :-

-   Bedrock is present at around 20 to 25 m below datum. The rock is apparently a gray fine to medium grained limestone of Lower Carboniferous age with dark fine grained shaley or slately layers with partings of varying thickness. This limestone is an impure argillaceous limestone locally known as Calp.
-   The intactness of the rock expressed in the form of Rock Quality Designation (RQD) was found to be quite high, generally exceeding 80% at the proposed depth of the Caverns.

3.3 Geological Tests

The following tests were carried out on representative samples of cores extracted :-

Geological Tests
Porosity Tests
Young's Modulus
Poisson's Ratio
Moisture Content
Uniaxial Tests - Compressive strength (unconfined)
Triaxial Tests - Maximum compressive strength (confined)
Brazil Tests   - Tensile strength
Direct Shear Tests
Thin Section Analysis.

In addition to these geological tests, a series of more specialised tests were carried out to establish the thermal properties of the rock. These tests were :-

Thermal Tests
Thermal Conductivity Tests
Temperature Coefficient of Length Test
Immersion Tests in LPG at -40^0C
Dry Sample Cyrogenic Tests
Volumetric Heat Capacity Tests

These tests results were used in the heat and stress calculations which are detailed in 3.4 and 3.5 below.

Water Pressure Tests

Each borehole was water pressure tested using the double packer system. The distance between packers being 1.3 metres. The distance between tests was 1.2 metres thus ensuring that a continuous testing of the full depth of the boreholes. At every test level five seperate tests took place, 71 psig, 100 psig, 128 psig, 100 psig, 71 psig. Each individual test comprised the logging of any water losses under steady state conditions for 20 minutes.

Shell and Auger Drilling

Shell and Auger drilling of the overburden was carried out. The results indicated that there is approximately 5m of fill over the original foreshore despoits which are some 20 to 25m thick and consist of estuarine mud, silts, sands and gravels.

Bedding planes, dips, faults and jointing

The core drilling and analysis carried out indicate :
- The rock was found to be sound there being no sign of any remarkable faults in any of the drill cores.
- Fractures and joints were generally tight and in some cases, in the some open joints (up to 10mm wide), calcite crystals were encountered.

3.4 Heat Flow Calculations

The purpose of the heat flow calculations was to determine the temperature distribution in rock around the LPG storage as a function of time. All three caverns were assumed to be filled with propane at -45^0C, partly in a liquid phase and partly in a gas phase. The rate of heat flow into the caverns was calculated.

The heat calculations considered:-
- the conduction of heat in the rock as a function of time in two dimensions;
- the convection heat transfer between rock surface and adjacent liquid or gas as expressed by Newton's rate equation;
- the terrestrial heat flow at Dublin.

These considerations were presented in their appropriate mathematical formulae and a finite element method of numerical analysis was used to obtain a solution.

The calculations were first of all carried out for the dry state at -45^0C, a temperature which exceeds the proposed product storage temperature. This theoretical limit calculation showed that at no time could the frozen zone reach ground surface level.

The calculations based on the wet state (quantity of ground water present 2%) predicts the frozen zone to extend about 25 metres in the stationary state. This figure concurs with the practical experience of Esso at Stenungsund for similar conditions. The Esso experience is that the stationary state is reached in 15 years at about 20 metres from the cavern walls.

It is concluded that the temperature distribution in
the rock around the LPG caverns will develop a stationary
state over time and result in an ice envelope around the
periphery of the caverns of 20 to 25 m thick for a cavern
temperature at -40°C.

3.5 Stress Calculations

The stress calculations analysed the tensile and
compressive stresses induced by the temperature changes in
the rock during the construction, commissioning and
operational stages. A finite element method of numerical
analysis was applied to the same mesh as was used for the
heat flow calculations. The rock was assumed to be an
elastic isotropic material which satisfies Hook's general
elasticity equations. The calculations were carried out in
a plane strain state. This estimate is considered correct
if both excavation space and rock structure are continuous
and homogenous in the direction perpendicular to the chosen
cross section.

The stress distribution in the rock was also considered
for different conditions of initial horizontal stresses
present in the rock over the range 2.0 MN/m^2 to 20.0 MN/m^2,
although the laboratory results concluded that high
horizontal tectonic stresses were improbable.

The following rock parameters were used in the stress
calculations:-
- Young's modulus for jointed rock 8000 MN/m^2, deduced
  from laboratory tests on the basis of the number of
  fractures.
- Poisson's ratio 0.2, based on laboratory tests (range
  0.14 ... 0.25).
- Coefficient of linear thermal expansion 4.0 x $10^{-6}/^0C$,
  based on laboratory tests.
- Submerged unit weight 17 kN/m^3, based on laboratory
  tests.

The calculations based on the above rock properties
produced an optimised design using computerised techniques.
The calculations indicated the limits and extent of tensile
and compressive zones. The analysis was repeated using
various shapes and sizes of caverns until an optimum design
was established.

When the optimum shape had been finalised a more
detailed analysis was made of the stress strain relationship
corresponding to the various stages of construction,
cooling phase and operating conditions. The following
conclusions were drawn:-
- The stress-strain analysis showed that in the
  construction phase the stress distribution around the
  caverns was advantageous from the stability point of
  view.
- Decreasing temperature in the caverns causes tensile
  stresses to develop in the roof and floor of the

caverns.   The compressive stress zone concentrates in
the sidewalls of the caverns.
- The greatest value of tensile stress corresponds to the
temperature distribution in the stationary state.
However, the calculated compressive and tensile stresses
will not exceed the compressive and tensile strength of
tested rock samples.   The compressive strength of rock
samples, tested in laboratory, is more than 30 times
greater than the maximum calculated compressive stress.
The tensile strength is 3 to 4 times greater than the
maximum calculated tensile stress.   The compressive and
tensile strength of rock samples has been corrected to
correspond to the strength of rock mass.   The
compressive strength of rock mass has been calculated to
be 20 MN/m^2 and the tensile strength 1.0 MN/m^2, which
will be exceeded in the roof of the caverns.
- The thermal calculations show that as the rock
temperature around the caverns is lowered, the water
will freeze and tend to expand.   This freezing process
will fasten the rock and prevent water flow into the
caverns. Ice and water around the caverns will also
prevent the leakage of vaporised gas from the caverns.
- The freezing of water in the rock mass will also
strengthen the rock as determined in the laboratory
tests.

3.6   Further Geophysical Studies
     Prof. Lindblom a Geo-Physicist of Hagconsult, Sweden,
was commissioned to carry out physical cold temperatures
tests at -45°C in the saturated state, both propane and
water, on the extracted rock cores selected at various
depths where the greatest mechanical stresses were projected
to occur. These samples were then subjected to both
comprehensive and tensile forces greatly in excess of the
predicted stresses.   The rock behaviour under these
conditions was unaffected and remained intact.   Prof.
Lindblom has no reservations on the structural strength of
the limestone. His predictions on the ice envelope
propagation were in line with previous predictions.

3.7   Overall Geological Suitability
     The overall geological and rock mechanic properties
were considered and their suitability for refrigerated LPG
storage evaluated with the follow results :-
- The rock mass is suitable for the construction of
underground caverns.
- The compressive and tensile strength of the rock,
Young's Modulus, Poisson's Ratio and Coefficient of
Linear Thermal Expansion were found to be typical of
this type of rock.
- Sufficient drilling and testing of rock had been
carried out in the planned area to conclude that the

chosen location is suited to the construction of the proposed caverns.

4 Feasibility Study

4.1 Environmental Impact Analysis

An Environmental Impact Analysis has been completed in line with the EEC Preliminary Draft Directive EIE/OY/18, CEEC Brussels (1979).

The overall conclusion of all the research is that with proper safety engineering and procedures the risk associated with the project is acceptable.

The Environmental Impact Assesment and Hazard Analysis as above were presented to Dublin Corporation as part of a detailed Planning Permission submission to obtain approval to commence the construction stage of the project. Dublin Corporation, advised by the Institute of Industrial Research and Standards (IIRS), granted planning permission. However, local resident third party objectors contested the grant of planning permission and caused a Public Oral Hearing (enquiry) to be held. This enquiry examined in depth all aspects of the project with particular emphasis on environmental factors. The results of this enquiry are not yet known.

4.2 Technical Design Aspects

The technical design aspects have been studied in depth. The stress calculations which have been performed have resulted in the following design criteria :-
- The shape of the Cavern should be elliptical.
- Due to sedimentary (schistose) structure of bedrock the caverns should be orientated so that the longitudinal axis of the Caverns are perpendicular to the strike of the bedding planes.
- Because of the tensile zone in the roof of the caverns, the nature of the rock, and the effects of the rock blasting during construction, the cavern roof design should incorporate rock bolting and shotcreting with wire mesh reinforcement to form an arch to act as a roof support to the rock.

The above criteria taken together with other technical design and economic criteria have resulted in the following optimised preliminary design for the project :-

		Butane	Propane
Storage Capacity	m^3	80,000	120,000
Number of Caverns	No	1	2
Width	m	17.0	17.0
Height	m	16.5	16.5
Length	m	360.0	260.0
Depth	m	-122.0	-122.0
Cross Sectional Area	m^2	225	225

		Butane	Propane
Circumference	m	58	58
Spacing (wall to wall)	m	60	60
Shape of Roof Arch		Elliptical	Elliptical
Orientation		68^0 East of North	

4.3 Economic Aspects

The conclusion of the economic analysis completed to date can be summarised as follows:-
-       A shortage of European LPG Import/Export terminal capable of handling large ships currently exists
-       The market for LPG will continue to grow throughout Europe as more product becomes available from the Middle East and the North Sea, and European refineries become net importers of LPG
-       The cost of constructing large underground LPG storage is significantly less per tonne than its equivalent overground
-       The project has significant overall financial and social benefits to both Ireland and the EEC
-       The project compliments the Kinsale Natural Gas resource as it provides strategic back-up and peak-shaving capacity as Simulated Natural Gas (SNG) to the Dublin Gas Company
-       Overall the project is a viable and attractive economic and commercial proposition.

5.   Esso Technology

Esso have a number of refrigerated LPG storages at the Chemical Complex at Stennungsund, Sweden. They have developed the techniques for constructing, commissioning and operating refrigerated LPG rock caverns progressively over the past 17 years. This work has culminated in placing Esso in the position of being able to licence its expertise to approved companies. Cavern Systems has applied for and been accepted for such a licence, however as the nature of such licenses is one of confidentiality it is not open to discussion in this paper.

6.   Construction / Commissioning

The main project comprises the construction of three unlined limestone rock caverns at a depth of -120 metres for the storage of 100,000 tonnes of LPG in the refrigerated phase, above ground facilities for maintaining the stored product slightly above one atmosphere,and berthing facilities for loading / unloading fully refrigerated LPG cargoes at up to 1,000 tonnes per hour. The facility will also have the capacity for loading / unloading both semi-refrigerated and hot LPG and berthing facilities for handling fully refrigerated ships up to 75,000 cubic metres capacity and additional berthing for smaller vessels.

Once the full sized caverns have been constructed, sealed off and the above ground facilities fully tested the commissioning programme will begin.

The first stage will be the removal of all air by substituting inert gas with a low freezing point, vis., nitrogen or similar. During this phase water pumps will be operating to remove water ingressing into the cavern.

The chilling process will be affected by spraying chilled propane on the cavern surface at a predetermined rate. At the initial stages an anti-freezing solution will be also added to prevent the ingress water freezing. By monitoring the cavern wall and termperatures it will be possible to determine when to extract the pumps and turn off the anti-freeze solutions. The chilling process continues until the cavern temperature reaches the required levels. Refrigerated LPG will be then introduced and the mechanical refrigeration system brought into operation.

7. Conclusions

The geological structure is suitable for the proposed project and the economic analysis continues to offer an attractive return on the investment.

The importance of the project to Ireland's energy storage infrastructure is considerable and is useful to the EEC as a whole.

(14.15/82)

DEVELOPMENT OF A NEW TECHNOLOGY
FOR LPG STORAGE

P. FUVEL, Compagnie Française des Pétroles
J. CLAUDE, Technigaz

Summary

The "DEVELOPMENT OF A NEW TECHNOLOGY FOR LPG STORAGE" project mainly consists of design, construction and testing of an LPG pilot project tank using a composite sealing membrane in a system known as the GMS (Gas Membrane System). The basic design is similar to that which has long been used for conveyance and storage of LNG. It consists in separating the supporting function from the sealing function. For this tank, the support consists of a prestressed concrete structure and the seal of a composite membrane consisting of aluminium and a glass fabric known as TRIPLEX. The insulator, which lies between the membrane and the concrete, is kept in an inert gas atmosphere. This design ensures a greater safety factor with respect to external incidents (thanks to the concrete) and internal incidents (reliable leak detection system), for a reasonable cost. The project entered the construction phase on 15th August 1983. On 1st July 1984, the civil engineering of the tank was entirely finished and the seal and insulation was 30% complete.

1 - DESCRIPTION OF METHOD

The GMS membrane storage method enables commercial LPGs such as butane, propane, butadiene, propylene or mixtures of these products in various proportions to be stored in a concrete tank at temperatures that can be as low as - 50°C.

The GMS method consists of the following elements, from inside to outside the tank:
- the TRIPLEX cryogenic membrane, tight to gas and liquid ; this membrane is in direct contact with the liquid stored and is simply covered with a protective coating,
- continuous liquid-type insulation transmitting the hydrostatic pressure,
- a self-standing concrete tank capable of absorbing the hydrostatic loads and the steam pressure ; the inner face of this concrete is covered with a steam-barrier lining.

The tank is surmounted by a metal dome that can also be covered with concrete. This dome carries a suspended roof supporting the top insulation of the tank, consisting of a fibre-glass mattress (figure 1).

The load-transmitting insulation consists of a rigid polyurethane or PVC cellular foam formed into prefabricated panels. These panels are covered with a TRIPLEX membrane in the works.

TRIPLEX is a material developed and patented by TECHNIGAZ comprising:
- a sheet of aluminium captive between two layers of fibre-glass fabric: the total thickness of the membrane is 0.7 mm and the thickness of the aluminium sheet 70 μm (guaranteeing that the sheet has no porosity).
. The seal is ensured by means of the aluminium sheet.
. The fibre-glass fabric endows the combination with mechanical strength.

Each panel is bonded in-situ on the inner wall by means of a mechanical application. The panels are joined together directly in the horizontal plane and via a spacer in the vertical plane. The primary barrier is rendered continuous by overlapping strips of TRIPLEX bonded into position where the panels come together (figure 2).

The result achieved is a storage system that is continuous at the primary barrier and insulation

2 - HIGH LEVEL OF SAFETY OF GMS STORAGE METHOD

In similar fashion to the system developed by TECHNIGAZ for LNG (1) (2) (3) (4), the GMS storage method is based on separation of the functions: the seal is achieved by means of the TRIPLEX membrane, whilst the mechanical strength is ensured by the concrete reservoir, via the rigid insulating material.

The level of stresses in the primary barrier remains (these stresses consist exclusively of thermal stresses) and sudden breakage through propagation of cracks is not possible.

The insulation gap and the concrete tank are separated from the gaseous phase of the reservoir, making it possible:
- to maintain the insulation out of contact with the gas, thus improving the safery and facilitating repairs and maintenance,
- permanently to monitor the tightness of the membrane and the integrity of the system, which is not possible with conventional tanks.

The other advantages from the standpoint of safety are contributed by the prestressed concrete pressure vessel.

Fire behaviour

The concrete has exceptional ability to withstand fire, making it an ideal material for a structure that may be subjected to accidental outbreaks of fire. The main incident against which one must guard is the "domino effect", in other words the danger created by radiation emitted by tank in the vicinity of an LPG fire. This case has been analysed in several studies. Assuming that the usual safety gaps are complied with, the average temperature of the concrete will be about 500°C. The concrete pressure vessels of cryogenic tanks can be designed by adding passive projection, for example, to withstand such temperatures.

Resistance to impacts and collisions

The prestressed concrete pressure vessel has excellent ability to withstand external impacts and can easily be dimensioned so as to withstand impacts as severe as those caused by falling aircraft. The two 120,000 m³ LNG tanks built for GAZ DE FRANCE at MONTOIR DE BRETAGNE (FRANCE) in accordance with the TECHNIGAZ design have been designed to withstand a FOKKER 27 crashing onto the dome.

Protection against explosions and attacks by sabotage

The ability to withstand an explosion provided by the concrete envelope is a by no means negligible aspect of the long term safety guaranteed by this type of installation. Indeed, the thickness of the concrete enables the tank to withstand bombardment ten times greater in intensity than can a metal tank of the same capacity.

3 - MAINTENANCE, OPERATION AND CONSTRUCTION

Ease of maintenance

Concrete tanks do not require to be painted externally, other than for aesthetic reasons.

In the GMS method developed by TECHNIGAZ, insulation remains permanently in an inert gas atmosphere at low temperature, protecting it from aging. For this reason, the problems that arise with externally insulated metal self-standing tanks are avoided. The latter require the installation of effective vapour barriers to prevent damp from penetrating, and only rarely succeed in doing so, as a general rule. Damp results in the formation of ice which in the long term reduces the performance of the insulation and increases the boil-off rate.

The GMS method also makes it possible to repair with neither electricity nor flames.

Ease of operation

The low thermal inertia of the system and the absence of thermal stresses in the bearing structure enable the cooling phase required before filling to be eliminated.

Concrete tanks can be designed for high internal pressures (150 to 250 g/cm^2) and for high vacuums (50 g/cm^2). The result is operating flexibility reducing the reliquefaction equipment and increasing the operational safety margins.

Advantages in construction

Concrete tanks can be built in most countries in the world, and, thanks to its high level of works prefabrication, the GMS system can be installed by fairly unskilled labour assisted by a minimum number of supervisors.

In addition, the work takes place inside the reservoir, and is hence unaffected by the weather conditions.

Lastly, it is easy to design and build membrane storage facilities with capacities of up to 160,000 m^3 or more, which is not the case for self-standing metal tanks, the capacities of which are limited by the welding specifications of the thickest plates.

Three LNG membrane tanks are now operating in the world, two of which in France, with capacities of 120,000 m^3 (5).

A 130,000 m^3 tank is now being built, and 300,000 m^3 tanks in South Korea (the first LNG importing terminal in this country). Three 60,000 m^3 tanks are now being studied for Japan.

4 - THE GMS 2000 PROGRAMME

In order to expose all the properties of the GMS membrane storage method, TOTAL-CFP and TECHNIGAZ have undertaken a development programme consisting in building a 2,000 m^3 capacity pilot project tank and operating it under industrial conditions. This tank is now being built at the "Flandres refinery", near DUNKIRK (this refinery is owned by TOTAL-

CFR-COMPAGNIE FRANCAISE DES PETROLES).
The following are the leading characteristics of this pilot project tank:

External diameter (concrete)	16.08 m
Internal diameter (concrete)	15.28 m
Height of skirt (concrete)	12.60 m
Height of liquid	11.20 m
Thickness of concrete skirt	0.40 m
Thickness of concrete slab	0.50 m
Thickness of insulation	100.00 mm
Nominal capacity	2,000.00 m^3
Operational capacity	1,950.00 m^3
Minimum operating temperature	- 50°C
Maximum service pressure (gaseous phase)	300.00 g/cm^2
Maximum product density	0.6

The tanks is equipped with the following main lines, which all pass exclusively through the dome:

One loading line	Ø 3"
One discharge line	Ø 3"
One gas intake line	Ø 6"
One return line for the expanded liquid	Ø 3"
One flare stack line	Ø 6"
One wind line	Ø 6"

One system for injecting nitrogen into the insulation gap and one nitrogen return line.

The tank is equipped with all the instruments usually fitted to this type of tank ; in addition, it is equipped with a system for monitoring the insulation gap.

The pilot project storage facility will be tested with intensive filling/emptying cycles for one year. The total number of cycles will be about 10. The duration of each cycle will be 12 days for butane and 18 days for propane (allowing for the loading/discharge phases). During these tests, the correct operation of the gas detection system in the insulation gap will also be verified ; this system is in fact one of the key safety points ensured by the GMS storage method. Since the insulation gap is kept in an inert gas atmosphere (nitrogen), the tightness of the primary barrier can be monitored during operation and all leaks of any kind that can occur during this operation can be detected.

5 - PRESENT STATUS OF THE PROJECT

The "DEVELOPMENT OF A NEW TECHNOLOGY FOR LPG STORAGE" project is covered by an EEC/GERTH/TECHNIGAZ contract which started on 1st December 1981. This project covers the design, construction and testing of the pilot project tank.

The civil engineering work on the pressure vessel is now complete since 15th February 1984. The civil engineering design work was entrusted to the PX CONSULTANTS design office (France) and the construction contract given to Entreprise LECAT (France), who subcontracted the metal structures to DELATTRE-BEZONS (France).

The materials for building the storage tank were selected on the basis of homologation tests. The following are the main suppliers:

- PVC panels	KRP (France)
- Polyurethane panels	ELF-ISOLATION (France)
- Bonding agent	HENKEL (Federal Germany)

- Supporting and steam barrier sealant — CFPI (France)
- Triplex — HUTCHINSON, CLOUTH (Federal Germany)

The erection and testing procedures were established by TECHNIGAZ on the basis of full-scale tests carried out in the laboratory. The detail studies on the insulation were also carried out by TECHNIGAZ. The insulation work was entrusted to Société ISOTHERMA (France), and on 1st July, this work was 30% completed. The bi-component products are applied by machines supplied by STERMA (France). Prefabrication of the insulating panels, including bonding of the TRIPLEX membrane to the panels and machining of the panels was entrusted to LA RESINE ARMEE (France). On 1st July 1984, 50% of the panels had been completed.

From the standpoint of the equipment needed, in particular to cool the LPG available at the construction site, detail designs are 80% finished. Most of the heavy equipment has already been ordered, and in particular:
- the reliquefaction refrigeration unit, from SULZER FRANCE (France),
- the flooded pump, from CRYOSTAR (France).

REFERENCES

1. PAUTHIER (December 1983). VERITAS Bureau Technical Bulletin
   The TECHNIGAZ project for an ice-breaking methane carrier with a membrane tank.
2. STASI (2nd April 1983). Petrole Information
   Safety and Operation of LNG land storage.
3. Messrs. KOTCHARIAN and SIMON. HAMBURG "GASTECH" Conference (1981)
   Safety of liquefied gases containment system on land and at sea.
4. LEBRIX (1983) 2nd International Conference on Cryogenic concrete.
   Analysis of concrete LNG storage structure using a stainless steel corrugated membrane containment system.
5. Mme RIOU and Mr. ZERMATI. HAMBURG "GASTECH" Conference (1981).
   Commissioning of the 120,000 m^3 storage tanks of the Gaz de France LNG terminal.

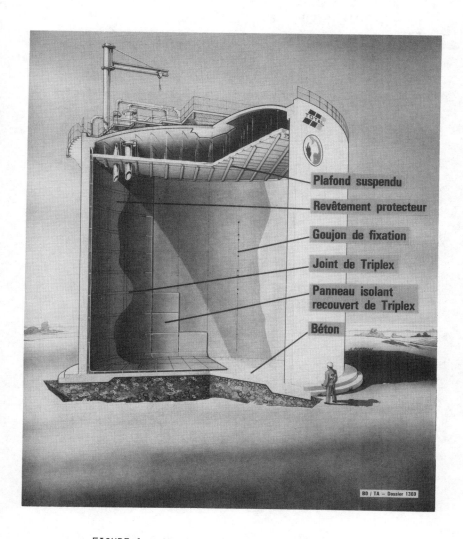

FIGURE 1 - GLP PILOT PROJECT STORAGE TANK

Plafond suspendu = Suspended ceiling
Revêtement protecteur = Protective lining
Goujon de fixation = Securing pin
Joint de triplex = Triplex joint
Panneau isolant recouvert de triplex = Insulation panel covered with triplex
Béton = Concrete

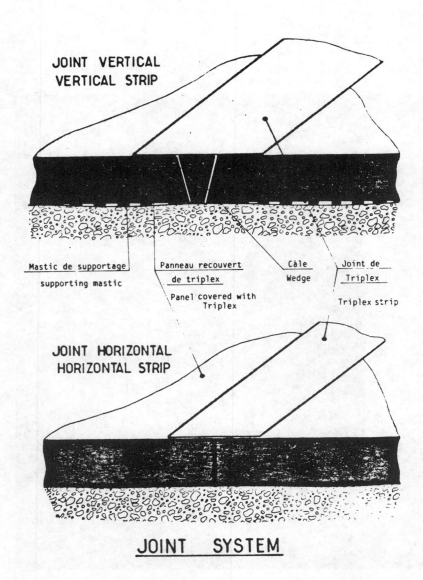

JOINT VERTICAL
VERTICAL STRIP

Mastic de supportage	Panneau recouvert	Câle	Joint de
supporting mastic	de triplex	Wedge	Triplex
	Panel covered with Triplex		Triplex strip

JOINT HORIZONTAL
HORIZONTAL STRIP

JOINT SYSTEM

FIGURE 2 - JOINTING SYSTEM

PILOT PROJECT STORAGE TANK IN NOVEMBER 1983

PILOT PROJECT STORAGE TANK IN JANUARY 1984

FIGURE 3

(14.10/80)

THE SCHELLE CRYOGENIC PILOT PROJECT CAVITY

A. BOULANGER, Société GEOSTOCK, Paris
P.V. de LAGUERIE, Société GEOSTOCK, Paris
W. LUYTEN, Société DISTRIGAZ, Brussels

Summary

Sociétés GEOSTOCCK (Paris) and DISTRIGAZ (Brussels) have built a pilot project demonstration cavity for storing low temperature liquefied gases at Schelle (Belgium), for which they were granted EEC support. In this article, the authors describe in turn the studies and preliminary reconnaissance carried out, the construction and the operation of the pilot project cavity, the tests and measurements performed and lastly the results and conclusions of the experiment. The success of the pilot operation enables the GEOSTOCK-DISTRIGAZ consortium to propose industrial storage for LPG (- 45°C) in all types of rock and for LNG (- 162°C) and ethylene (-,105°C) in clay type rocks.

I - INTRODUCTION

The technical and economic advantages of underground storage of petroleum products have for several years provided a particularly attractive alternative to the solution of conventional above-ground tanks. For liquid hydrocarbons (fuel, motor spirit, naphtha, etc...), for natural gas or again for pressurized LPGs, the variety of today's underground storage facilities in most cases make it possible to adapt to the local conditions (mined cavities, cavities washed out from salt deposits, aquifers, reconversion of abandoned mines, etc...).

However, as regards storage of liquefied natural gas and low temperature LPGs, the underground solution could not so far provide a satisfactory answer, and the only possible solution for the operator was to use above-ground cryogenic tanks.

Aware of the existence of this requirement, GEOSTOCK decided a few years ago to set forth on a major study and research programme in the area of underground cryogenic storage. The outcome of this work was the design and construction of a pilot project demonstration cavity at SCHELLE (Belgium), carried out jointly with the Belgian company DISTRIGAZ. This pilot project operation has been granted the support of the Commission of the European Communities, together with that of ANVAR (Agence Française pour la Valorisation de la Recherche - French Agency for Valorization of Research). The conclusive results that emerge from this pilot storage facility, the construction and operation of which covered the 1980 to 1982 period, enable the GEOSTOCK-DISTRIGAZ consortium to propose a reliable technique for underground storage of LNG in clay formations and of LPG in all types of rock.

II – THE SCHELLE EXPERIMENTAL CRYOGENIC STORAGE FACILITY

1. Preliminary reconnaissance

After systematically seeking out to discover all possible sites, it was decided to carry out this pilot project in a region featuring a major clay outcrop and more particularly in the Boom region of Belgium. The final site chosen was the Schell clay quarry, near Antwerp.

The reconnaissance work studied on the site in May 1980 and revealed the existence of a clay bed 28 metres thick overlaid on an aquifer sandy formation with a pressure of about 3 bars. The Boom clay in this region is overconsolidated and its in-situ Young's modulus is about 300 bars.

The samples taken during reconnaissance borings enabled the thermal and mechanical characteristics of the clay to be determined in the laboratory, together with their evolution with temperature. Amongst other things, the compressive strength, Young's modulus, speed of propagation of sound, thermal conductivity and diffusivity, latent freezing heat, etc...) may be mentioned.

One of the main properties of clay in the presence of low temperatures is its high water content (25% for Boom clay). Freezing of this water content, which moreover does not occur instantaneously at 0°C, but continues down to extremely low temperatures, is responsible for particularly sensitive variations of the thermal and mechanical properties when reducing the temperature. As an example, the compressive strength is multiplied by a factor of 100 between ambient temperature and - 196°C.

The gradual freezing of the water also explains the considerable expansion of the clay, as opposed to the contractions observed in hard rocks of the limestone or granit type as the temperature drops (see figure 1).

To give the general picture, one can say that the tendency of the compression to rise as the water freezes, together with an improvement in the mechanical characteristics, are amongst the key factors contributing towards the success of the project.

The many previous studies performed by GEOSTOCK on cryogenic storage and in particular analysis of the problems and even the failures encountered in the initial cryogenic cavities of the 1960s or in underground storage facilities (of the Arzew or Canvey Island type) enabled the best cooling programme, the shape of the cavities and the depth of storage to be determined.

In particular, several thermal and geotechnical computing programmes were developed, into which the characteristics of the cores extracted at Schelle were entered.

The results firstly made it possible to conclude that the pilot project would be a success, and second to consider the test as representative of an industrial installation.

2. Description of the pilot project

The pilot gallery is cylindrical in shape, excavated from Boom clay at a depth of 23 metres. It has a diameter of 3 metres and a length of 30 metres. It is connected to the surface through a 1 metre diameter shaft (Fig. 2). This shaft contains all the pipes ensuring refrigeration, i.e. a liquid nitrogen line, the gaseous nitrogen extraction line and the gaseous nitrogen recirculation line.

The pilot project is lowered to temperature by liquid nitrogen injected at a flow such that the temperature setting is kept constant within the cavity. To obtain homogeneous temperatures, part of the gaseous nitrogen is recycled in the cavity, the remainder being rejected to atmosphere through pressure regulation enabling the pressure to be kept either at a few grammes above atmospheric pressure, or at a pressure selected between atmosphere and 3 bars.

The liquid nitrogen is injected automatically through a regulation loop controlled by measuring the temperature in the atmosphere within the test cavern.

The behaviour of the cavern and the terrain is trapped by means of 300 measuring points all leading to the control room and comprising in particular:
- temperature probes set up in 19 boreholes and ausculting the terrain through a thickness of 10 metres around the cavity,
- vibrating cord strain gauges on the walls of the cavity and in the clay,
- convergency measurements,
- measurements of the movements of the terrain by strain gauges and clinometers,
- measurements of the speed of sound,
- listening points for "micronoises",
- seismic measurement campaigns.

Excavation of the cavern started from the bottom of the clay quarry, with a machine protected behind a shield. The access gallery (the first 74 metres) was strengthened by conventional reinforced concrete arch segments, whilst the test gallery was reinforced by arch segments capable of withstanding low temperatures. Construction took from March 1981 to October 1981.

The same technique (behind a shield) would be used for industrial caverns, with a diameter which would then be from 7 to 10 metres.

3.   Tests

Refrigeration proper started on 10th November 1981 and took place gradually. This temperature gradient fell in a succession of stages. When brought down to temperature, the gallery was partly filled with liquid nitrogen (figure 3). This filling operation started on 21st June 1982, and the level of liquid was kept constant until 3rd September 1982, when the test was stopped.

On completion of the tests, the ice cocoon was from 5 to 7 metres thick.

Lastly, the behaviour of the cavity during reheating was followed for about 6 months.

The tests as a whole were carried out with the following objectives in view:
- to demonstrate the effectiveness of our cooling programme,
- to analyze the behaviour of the gallery both from the thermal standpoint (determination of heat losses or boil-off) and that of the mechanical stability of the structure, at the following temperatures:
  . − 45°C   liquid propane,
  . − 105°C   liquid ethylene,
  . − 162°C   liquefied natural gas,
  . − 196°C   liquefied nitrogen.

4.   Results

The fundamental result of the Schell experiment is that after 8 months of cooling and 2 1/2 months of operating trials at the temperature of

liquid nitrogen (- 196°C), the cavity remained perfectly tight and surrounded by a cocoon of frozen clay 6 metres thick (see figure 4). In addition:
- the temperature distributions in the massif and the progression of the front of ice are perfectly regular and agree with our predictions,
- the consumption of cooling energy was also very close to what we had estimated, in particular during the phase during which the liquid nitrogen was maintained, when the rate of boil-off within the cavern could be accurately measured. We can hence confirm that industrial storage of propane at - 45°C in clay would have a boil-off rate of about 0.1% per day, whilst for LNG storage at - 162°C in the same clay, the boil-off rate would be about 0.5% per day.
Following a 30-year operating period, the frozen zone would spread about 30 metres for LPG storage and 50 metres for LNG storage.
- the geotechnical behaviour of the cavern was followed particularly carefully. Measurements of the vertical and horizontal movements of the clay (strain gauges and clinometers), measurements of the deformation of the lining and the convergency of the clay enabled us to confirm and refine the results obtained in the laboratory and to test our computing programme. Figure 5 shows the expansion movements of the clay in the centimetre range, around the cavity.

Furthermore, this model is particularly sophisticated, since aside from coupling the thermal and mechanical aspects, it must allow first for the rapid variations in the thermomechanical characteristics of the frozen and unfrozen clay, and second the various modes of behaviour of the clay at low temperatures (plastic, viscoplastic, elastic, expansion and contraction).

As is shown in figure 4, it is also to be observed that the shafts and their installations, in particular the pipes, bring about no appreciable thermal disturbance. This signifies that the technologies and materials used are well adapted to the problem and that we now have available shaft equipment for an industrial installation that is not after all the weak point that was to be apprehended.

Lastly, mention must be made of the use of two seismic techniques during the Schell experiment:
- listening for "micronoises" in the cavern and in the massif, led to the conclusion that there were no cracks,
- determination of the freezing front by seismic refraction, a technique based on the contrast between the speed existing between the frozen clay and the unfrozen clay was developed at Schelle and will be used for the industrial installations.

All these observations hence enable us to conclude that the satisfactory preparation of the cavern observed at - 45°C and which would demonstrate the feasibility of underground storage of propane refrigerated in clays, is maintained at a temperature of - 105°C and - 196°C. Furthermore, the model of geotechnical behaviour the representativity which was confirmed by the test shows that this satisfactory mechanical condition will be maintained throughout the life of an industrial installation.

Because of this, the feasibility of underground storage in clays for propane at - 45°C, ethylene at - 105°C and LNG at - 162°C has now been proven.

CONCLUSIONS

To conclude, this Schelle pilot operation, which demonstrated that a liquid could be stored at - 196°C in a deep underground cavern not lined with insulation fully justifies its cost (18 million French Francs), by the information and the confirmations it has provided.

In addition, this experiment has enabled the following to be developed:
- laboratory tests adapted to determine the characteristics of rock at low temperatures,
- models for computing the behaviour of underground cryogenic storage facilities,
- an excavation, reinforcement and closing technique for the gallery and the shaft adapted to low temperatures,
- trials of the materials adapted for cryogenic storage (shaft equipment, refrigeration installations, processing of boil-off, etc...).

Control of all these technologies will henceforth enable GEOSTOCK and DISTRIGAZ to offer an underground solution for the storage LPG in all types of rocks and for ethylene, and for LNG in clayey materials.

An industrial storage facility in clay will be built in accordance with the same principles as the Schelle pilot project cavity. Access will be made through an access shaft, whilst galleries with a diameter of from 6 to 10 metres will be excavated behind a shield down to a depth of 80 to 100 metres.

The main applications for cryogenic storage involve reception and exporting terminals, and peak-shaving units for LNG. In this respect, it can be pointed out that we have also studied storage of LNG at - 125°C and a pressure of 10 bars, this solution being rendered possible by choosing an appropriate storage depth (minimum hydrostatic head 100 metres), offering the advantage of optimizing the dimensions of the gas liquefaction and boil-off processing unit.

BIBLIOGRAPHY

1. BOULANGER and LUYTEN (1982: Underground Storage of liquefied gas at low temperature. Gastech 82.

2. DE SLOOVERE (1983): Strain gauge measurements in a cavity cooled to a temperature of - 196°C. Symposium on in-situ tests. Paris, May 1983.

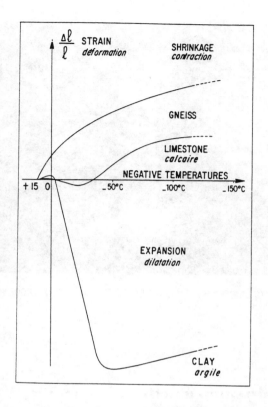

FIGURE 1

REFRIGERATED UNDERGROUND STORAGE OF LIQUEFIED GASES

TEST CAVERN
SCHELLE

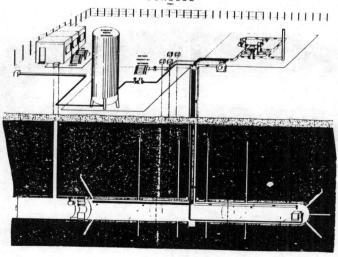

FIGURE 2

FIGURE 3 - VIEW OF THE GALLERY PARTLY FILLED WITH LIQUID NITROGEN

ICE FRONT LOCATION AT THE END OF THE -196° C STEP

Position du front de glace à la fin de l'étape -196 ° C

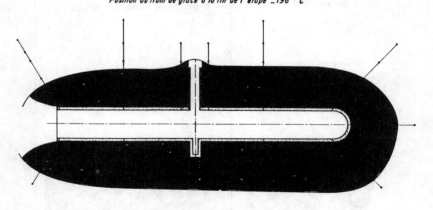

FIGURE 4

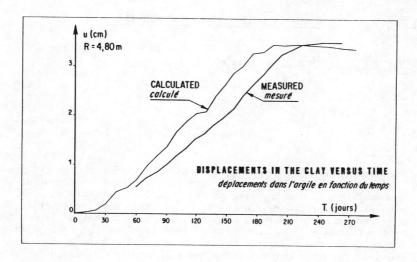

FIGURE 5

(10.28/81)

CONSTANT MOTION CRYOGENIC SWIVEL JOINT

G. GRONEAU)
G. OVIEVE) R & D Department, Société FMC EUROPE S.A.

Summary

The authors have developed a cryogenic swivel joint concept known as "constant motion", intended to form the essential component of a combination of transfer systems for liquefied natural gas or liquefied petroleum gas in the open sea. This novel concept ensures a service life in relation with the practically continuous movements of the system and the considerable importance of the reliability and longevity in these hostile zones, whilst at the same time providing the same performance as existing models mounted in harbour installations. A prototype derived from the concept has been manufactured and is now undergoing tests.

1 - INTRODUCTION

Exploitation of offshore natural gas reservoirs, the use of fatal gases produced in association with crude, the increasing difficulties in setting up new methane terminals in zones with a high population density, the requirements of flexible supplies and depreciation of liquefaction and evaporation installations of fuel gases are all problems facing a number of oil or gas companies.

They are also reasons which have led these companies to turn to liquefaction, transport or regasification of gas in the open sea. All the schemes envisaged assume one or more transfer systems between floating "senders" and/or "receivers" separated by varying distances. The fundamental component common to all these systems is the swivel joint which enables either segments of rigid lines to be used, or the curvatures of flexible segments to be minimized.

Although existing harbour installations already comprise such transfer systems (loading arms), the unagitated state of the water, the size of the buildings and the design of the moorings enable movements related to wave action to be neglected and hence also the oscillations of the swivel joint. This is not the case in the open sea, where waves are of greater amplitude and the possibilities of limiting the movements of the buoy often restricted.

Joints thus undergo in addition to the slow rotations caused by the currents, winds or the diffraction of the waves, the faster oscillations, though of less amplitude, of the movements of the buoy caused by the waves, practically continuously.

2 - TECHNOLOGY OF CRYOGENIC SWIVEL JOINTS

Any swivel joint fulfils a threefold function, namely to provide a degree of freedom, to transmit the mechanical loads and to ensure a tight seal. The degree of rotary freedom is obtained by means of two or three sets of balls running in adjusted ball races enabling an outer female end-piece to rotate with relation to an inner male end-piece. The mechanical loads are transmitted through these balls. The advantage of the fitted ball race is that it can use a cold-worked stainless steel, which is much harder than the basic material of weldable austenitic steel.

The pressure seal is provided by a combination of two lip joint type linings arranged in series so that in the event of a slight leak through the internal lining, the evaporation of the leakage gas generates a sufficient back-pressure to restore the seal of the internal lining. Another lining situated outside the joint protects it from penetration of dust, impurities and especially humidity, which by freezing could jam and deteriorate the ball races. In addition, it renders possible slow circulation of nitrogen gas through the joint, drying any possible condensation and generating an excess pressure thus eliminating all risk of infiltration.

In addition to its extreme simplicity, the cryogenic joint described below offers the advantages of low weight and dimensions and the possibility of withstanding very high thermal gradients without damage.

On the other hand, it has certain limits for applications in the open sea, since a small proportion of the balls absorb the load owing to the play that is inherent in the concept or resulting from thermal gradients, and it is impossible to preload the bearings.

3 - FACTORS AFFECTING THE LIFETIME OF THE JOINTS

The lifetime of a swivel joint, just like that of any mechanical assembly, is determined by a combination of deteriorations that can have mechanical, thermal or chemical origins:
- degradation of mechanical origin: there are two types of such degradation:
  a) fatigue of the constituent material under cyclic loads (ball bearing races and balls, fasteners for rotating parts),
  b) wear of frictional parts (linings and lining facings),
- degradation of thermal origin: whilst there can be no question here of creep, it is not certain that following cycles consisting of cooling and return to ambient pressure by the joints do not cause structural aging in the materials,
- degradation of chemical origin: here, what first comes to the mind is corrosion caused by the marine atmosphere, which indeed is of some importance, even for austenitic steels. However, one should not exclude also the risks of precipitation of chromium carbide during welding, which considerably reduces the oxidation resistance and limits the range of steels that can be selected.

Lastly, mention must be made of all phenomena where the time factor, whilst not directly in question, takes on seriousness that it does not have in the short term. For instance, components that are assumed to be fixed and which move under the effect of thermal gradients. This is also true of high stress concentration points where the material can undergo permanent

deformations harmful to the operation of the components, or conducive to
the birth of cracks.

4 - FACTORS AFFECTING THE SERVICE LIFE OF THE JOINTS

The basic idea is to retain the characteristics making the conventional
joint a success, by modifying those rendering it incompatible with use
offshore.

Tests have revealed that amongst the factors described above, the
most important one for the conventional joint was the mechanical
degradation of the balls and ball race. This ball race is in type AISI 301
cold-worked stainless steel. The balls are generally in type 440 C
martensitic steel. Under the effect of the loads applied, hertzian
stresses hence tend to mark the ball race. Although compatible with the
rules of the art, the print is not uniformly distributed over the
completed circumference of the ball race, since the balls are loaded very
differently, depending on their position, and the heaviest loaded zones are
the same for low amplitude oscillations.

The same also applies to linings, where any impurity or surface
defect will always lie at the same point between two friction surfaces.

The basic idea is hence to modify the two-way and sedentary
character of the movements. To do so, the joint is twinned and comprises
two male end-pieces instead of one, two roller bearings and two sets of
linings. It then remains to ensure that each of the end-pieces, roller
bearings and linings always rotates in the same direction, which is
achieved by two cam free wheels. During an oscillation period, the two
male end-pieces turn one after the other in relation to the female end-
piece, each for a half-cycle. If one of the end-pieces is secured and the
other oscillating, the female end-piece always rotates in the same
direction, and only during a half cycle.

All the dynamic elements of the system work only during half a cycle,
so the life expectation of the joint has been doubled (admittedly, at the
cost of doubling the components), but one can hope to achieve much more
after eliminating local permanent stress concentrations, at least on the
female end-piece. The saving is less obvious on the male end-pieces,
since the effort remains the same on these pieces, and this system does
not hence dispense us with the need to attempt to distribute the loads on
the bearings better.

Another problem linked to the principle adopted stems from the
increased overall size caused by doubling the components. As we have seen,
a conventional joint contained two or three roller races, rarely only one,
owing to the low bending capacity of a bearing, even with a deep groove.
With the principle adopted, each half of the joint takes up all the loads
and hence each half must comprise two or three bearings. To avoid this
solution which would unduly extend the length of the joint, a technology
known as the "four contact points ball" has been adopted, which, while
retaining the advantage of dimensional adaptation of the balls enables the
bending loads to be transmitted with only a single row.

5 - THERMAL PROBLEMS

The thermal problems accounted for a considerable proportion of the
studies prior to manufacturing the joint, the key question being how to
incorporate a case-hardened steel bearing tested for temperatures down to
- 50°C in a joint carrying natural gas at a temperatue of - 160°C. It
proved necessary to insulated the joint internally from the liquid,

together with the adjacent cold parts. The joint was insulated from the adjacent parts by means of an intermediate part in epoxy glass fibre matting also making it possible to transmit the external loads. The fasteners were built in the same material.

The internal insulation was more delicate, since it had to enable the male end-piece to oscillate with relation to the other end-piece. An internal protective part was hence manufactured secured to one of the end-pieces and capable of rotating inside it. This part essentially absorbs the thermal gradients and had to be capable of deforming freely without interfering with the movements of the joint. In this case, it became clear that the seal could not be perfect and that even if one allows for evaporation of the gas in the interstitial annulus, this part should be considered more as a thermal shield than as true insulation.

Parallel study of icing on the external part also showed that one could not count on the natural convection of the external air to evacuate the frigories originating from the pipe.

These considerations led to the conception of an internal heating system, or rather a system for evacuating frigories, by circulation of nitrogen entering at ambient temperature. A screw type heat-exchanger was hence built into the joint, the nitrogen making several rotations around the LNG passage, before circulating in the central void, where it then fulfils the same function as in a conventional joint.

6 - MATERIALS AND HEAT TREATMENT

The steel that is far the most widely used in cryogenics is austenitic stainless steel. It can be considered as the ideal low temperature material, owing in particular to its excellent resilience at temperatures like those of liquid nitrogen (- 196°C).

It is also a metal offering a very wide plastic zone before fracture, and unfortunately for the calculations, a proportionately very low elastic limit. This very high ductility enables it to withstand the thermal shocks of a weld or the sudden arrival of liquefied gas without breaking, though sometimes at the cost of uncontrolled and permanent distortions.

Low carbon grades (below 0.03%) are used for the forgings that have to be welded. This thus reduces the risk of precipitation of carbide, but also the hardness of the seals obtained. Lastly, AISI 316 type grades which contain up to 2.5% molybdenum have displayed better performances as regards the ability to withstand corrosion by sea water, the typical form of which in steels is spot corrosion. Accordingly, one can conceive that the main constructive problems of the joint will involve the deformation of the parts after welding, together with inadequate hardness of the material at places where hardness is desired, namely the bearing faces of the linings.

All welded parts were thus stress-relieved at 400°C after welding and before the finishing machining.

To obtain the necessary hardness for the bearing faces of the linings (55HRC), a carbide powder gun projection method was used. The faces were then ground in a polishing machine to obtain a surface finish of 16 microinch RMC (0.4 microns).

As regards cryogenic linings, much research has been carried out in particular for the aerospace industry, resulting in considerable progress in recent years. Sealing linings comprise all problems that arise in cryogenics: ability to withstand major distortions, capability of withstanding thermal shocks, chemically neutral, tenacity. This is hence the province of synthetic materials or composite materials.

The solution adopted is a lining with a reinforced polyethylene lip fitted with a steel spring endowing it with the necessary elasticity. Alone, the linings justify the high level of machining accuracy required for the joint. Regardless of their quality, correct operation of these linings will always depend on the good execution and dimensional stability of their housing.

7 - TESTS OF THE JOINT

Not knowing the exact conditions in which the subsequent joint will be used, it was difficult to establish a procedure general enough to qualify the joint, regardless of its future application.

A more restricted, though more realistic, objective for this test was rather to seek out the potential difficulties involved in the continuous movement of the joint and to check to what extent the above-described arrangements are effective and viable. Data processing and rheo-electrical models developed in the study phase to predict the thermal status of the joint were of limited scope, lacking the ability to know the following with sufficient precision:
- the internal thermal status of the joint: what happens when the joint is partly filled ? Would this not lead to ovalization distortions ?
- the effect of icing on the external transfers: to what thickness can the icing gather ? What is its density and heat conductivity ?
- the efficiency of the insulating materials: what is the influence of the unavoidable plays ? Are there no interfacial transfer coefficients ?

Even for study of the stresses, the presence of the bearings, free wheels, fasteners and tightening, the fact that the loads applied were not axi-symmetrical, the multiplicity of the flat interface is capable of slip, the variation of the mechanical properties with temperature are all factors rendering modelization by finite elements either extremely complicated and expensive, or not very representative.

All these question marks require as pragmatic an approach as possible, allowing the tests prime place in the project.

These tests will be carried out on a 16" prototype with loads identical to those planned in the specification. The oscillating movements, the amplitude of which is 5 degrees, were accelerated by a factor of 4, i.e. 1.6 seconds. The fluid used is natural gas, with a flow naturally reduced compared to reality (10 m³/h), though nonetheless enabling the fluid to be renewed with sufficient filling of the joint, at least during the transient phases.

Ten million cycles will be carried out, at a rate of 50,000 to 54,000 cycles per day (i.e. about 200 days of tests). Each daily cycle will consist of reduction of the temperature and maintaining cold (about 14 hours), followed by 8 hours of reheating (during which the movements can be continued).

8 - CONCLUSIONS

Even if this new joint is bulkier and more expensive than a conventional joint fulfilling similar functions, an examination of what precedes shows that the simple fact that it can rotate continuously for 2 years, for instance and then be changed simply by unscrewing two flanges, revolutionizes its design.

It is also possible that since the results of the tests effect the very concept of the loading systems, the joint could, which is even probable, operate less well at ambient temperature than at low temperatures, or less well during the transient phases than under steady-state conditions and that allowance will also have to be made for this both during the design and operation phases.

- Figure 1 -

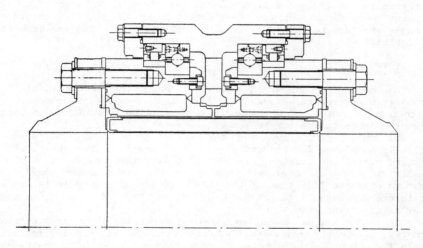

- Figure 2 -

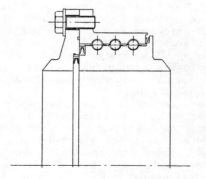

(12.05/78)

INSULATION AND BARRIER SYSTEM FOR
MARINE TRANSPORT AND STORAGE OF LNG

L.R. PREW
SENIOR PROJECT ENGINEER, SHELL INTERNATIONAL MARINE

SUMMARY

The primary purpose of this report is to provide a summary description of the Shell Structural Internal Insulation System for the containment and transportation of LNG in ships. It also serves to document the status of the development when it was terminated in December 1981 due to unfavourable commercial circumstances and technical difficulties.

1. SYSTEM CONCEPT

The system concept is very simple; it is to use the insulation, necessary when storing and handling liquefied gases at low temperatures, to contain as well as insulate the cargo. The insulation therefore forms the inner lining to the tank.

The system design involves the use of a cellular foam material specially developed for structural use at cryogenic temperatures. It is a tough, rigid, non friable, closed cell polyurethane material, with the density selected so that it remains liquid tight under peak cargo pressure over the life-time of the ship. It is sprayed up in layers and reinforced by several woven glass-reinforced epoxy resin laminates interleaved in the foam.

The system can be installed within a structure of any conventional geometry; typically for liquefied gas carriers it may be installed within the inner hull of the ship, or within a series of free-standing tanks inside the ship.

To minimise radial or 'pull-out' stresses at corners on cooldown, large corner radii are employed and for the same reason the system is maintained at constant thickness in both corners and flat walls.

Two densities of foam, of similar but not identical formulation, are employed: 85 kg/m^3 density, designated SP70, and 135 kg/m^3 density, designated SP71. The higher density foam is used in certain areas where greater robustness is required and where its characteristic of improved adhesion to laminates is of value.

The method of application of foam is by spraying in situ in the tank, layer upon layer, using largely mechanised means. The components are mixed in the spray head and are then deposited on the substrate where they rise perpendicularly to the substrate, forming layers on average 10 mm thick. Spray application produces a smaller cell structure for a given density than poured foam, which results in an improvement in mechanical properties.

Foam is deposited by the spray head in bands approximately 1.4 mm wide, and overlaps are arranged between adjacent bands to provide continuity of strength and an even foam thickness. Vertical bands are generally employed for convenience and symmetry - the foam tends to sag slightly if sprayed horizontally - but horizontal bands may be employed if required.

In order to enhance further the performance of the foam, several glassfibre/epoxide resin laminates are interleaved in the foam at predetermined depths as it is sprayed up, and one laminate is applied finally on the inner surface.

When the foam thickness has been built up to a laminate position, the surface skin is removed and surface irregularities smoothed off by a mechanised sanding process.

The laminate is then applied to the sanded surface, using largely mechanised means. Resin is sprayed on to the surface, the glasscloth is rolled into the wet resin and the whole is consolidated to give a 50/50 resin/glass ratio laminate, soundly adhering to the substrate. 130 mm overlaps are arranged between adjacent cloths, to provide continuity of strength. Once the laminate has cured, foam built-up can continue to the next laminate position.

The laminates themselves are made up of an epoxide resin formulation, specially developed for strength and toughness at low temperatures, and reinforced with an open weave glass cloth which provides a controlled degree of through-laminate porosity when laid up on a sanded foam substrate, and so prevents any tendency towards laminate blow-off on warm-up. This applies particularly to the surface laminate. Laminates deeper in the system may have several plies which reduces laminate porosity, but resistance to blow-off is not an essential requirement of deeply buried laminates since LNG will not come into contact with them in the primary barrier condition.

In addition to the conventional laminates, two 'channelled' laminates are also fitted. These are double skinned laminates with a porous core configuration which provide in-plane porosity for gas detection purposes. The double-skinned laminates also fulfil the normal foam reinforcing functions of laminates.

The channelled laminates provide means of checking and monitoring independently the integrity of both the primary and the secondary barriers either continuously or at any required moment, e.g. during construction, in service, before entry into harbour, and at annual survey. They also ensure independent modes of failure of the primary and secondary barriers.

Figure 1 illustrates the system design for the ship.

The system is designed with no penetration throughout the system which could form potential cold bridges or leakage paths. Where it is required to support equipment on the insulation, for example the base of the pipework trellis, pump supports, etc., a circular Invar plate with appropriate seatings attached is bonded both adhesively and with securing laminates to the surface laminate of the structural insulation system. A fibre reinforced epoxy resin mastic has been specially developed for this low temperature application, both to bond the plate to the substrate and to accommodate any variations of the bond line thickness between plate and substrate due to surface undulations.

2. SYSTEM DEVELOPMENT

Early in the development a rational design procedure was adopted where:-

-   service loads and conditions were translated into material design (performance) criteria.

performance criteria were translated into failure mechanisms which were then investigated by theoretical analysis and experimentation to establish the validity of the system.

A crucial element of this procedure was the generation of a comprehensive understanding of the production and performance characteristics of foam that permits its treatment as an engineering material.

The crux of the development was to have been the demonstration of the ability of the system to accommodate the full spectrum of service conditions when installed in a 20 m³ test tank in accordance with commercial application and quality control procedures.

Unfortunately, this test tank failed at the very end of the testing programme after being subjected to almost all the required IMCO service loads and conditions. A very thorough investigation into the causes of this tank failure have been made.

3. CONCLUSIONS

Briefly, the conclusions reached are:-

i) The system as currently conceived is not technically adequate. The failure of the 20 m³ test tank was caused by poor quality control and inadequate design criteria. However, the main problem was caused by a poor understanding of the foam material properties, their variability and the influence of imperfect surfaces (e.g. sanded, cut, overspray, angled) and application conditions on key properties.

ii) There is a suspicion that fatigue deterioration of properties over cut, sanded and overspray surfaces occurs but the evidence is indirect. If this is true then the SP7 series foams may not be able to meet the requirements imposed by the system design stresses.

On the assumption that fatigue deterioration is not significant a technically sound system could be achieved by complete elimination of overspray or the use of SP71 foam throughout the system. However, this will increase the complexity and the cost of the overall system.

iii) The laminates performed well in the 20 m³ test tank where their crack arresting properties were amply demonstrated.

Some laminate deterioration did occur, but whether this could create a problem during a LNG ship lifetime is unknown.

iv) The channelled laminates appear to have all the necessary features for adequate gas detection but a large scale test within the complete system is outstanding as is a commercial method for channelled laminate production. There is a concern that the detection system will not detect a secondary barrier imperfection independent of a primary barrier failure and some Regulatory bodies may not approve of the design for this reason. A combination of gas detection and thermal detection would be appropriate but the system complexity increases as does the installed cost.

v) The pump support rig testing and qualification is outstanding; a smaller version of the rig failed under test. The hatch design is unproven but should not lead to any intractable problems. Detailed design studies are needed to demonstrate how the channelled laminate gas detection system operates and the analysis system is organised.

vi) Studies into the sloshing resistance of the system have shown that with the current tank shape the system has a relatively low resistance to cargo sloshing forces.

While cargo sloshing will remain an inherent weakness of the system it is anticipated that an acceptable tank shape can be designed. It would lead to a more expensive design and some problems gaining acceptance from Regulatory bodies and Classification Societies should be expected.

vii) The necessary application technology to effectively install the system has not been fully developed. Further, the necessary design iterations between automated application technology, design and material properties is outstanding. The rig designed especially for installation of the system in the 500 m^3 pilot proving tank has not been operationally tested. Some design deficiencies exist, and application problems would have arisen. Unfortunately successful operation of the rig equipment does not lead to the necessary equipment for commercial use.

An independent study has investigated possible conceptual application techniques, their associated equipment needs and approximate costs. While this is a sound foundation for further work it should be recognised that considerable development is required.

viii) It was always assumed that the system would be installed in a shipyard. However, because tight environmental controls and appropriate quality control procedures are strictly necessary, shipyard installation may be too difficult to achieve. Factory installed systems may be necessary which would have to be in smaller tanks. Unfortunately the cost of such operations may be prohibitively expensive.

vii) The current commercial climate offers no incentive to further develop this system.

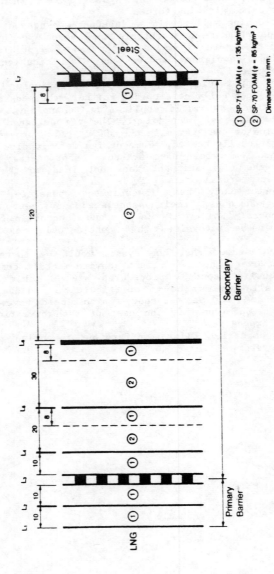

① SP-71 FOAM (ϱ = 135 kg/m³)
② SP-70 FOAM (ϱ = 85 kg/m³)
Dimensions in mm.

Figure 1 Shell internal insulation system

CLOSING SESSION

Summary of discussions and main results

Closing address
G. BRONDEL, Director for Oil and Natural Gas,
Directorate-General Energy, Commission of the
European Communities

SUMMARY OF DISCUSSIONS AND MAIN RESULTS

The main objectives of the symposium were to review the progress made on the projects receiving support under Council Regulation (EEC) No. 3056/73 and, in the process, to take stock of advanced technologies in geophysics, drilling, offshore platforms, production, enhanced oil recovery, pipelines, natural gas and petroleum for underwater operations. Parallel technical sessions provided an opportunity for a more detailed account of 96 of the 340 projects so far supported. Some 400 representatives of the oil and associated industries attended this indepth exchange of views on all the topics listed.

At the closing session the chairmen of each of the working parties submitted their conclusions as regards the technological development still needed in each of the fields covered by the Regulation. All these conclusions are based on the participants' experience, though there is a certain amount of overlapping since it is impossible to draw a clear-cut dividing line between some of these fields. They cover most areas of oil and gas technology, though new ones could perhaps be added to give an even more exhaustive approach.

Part I of this summary sets out the technical objectives and priorities, whilst Part II goes on to suggest ways of improving the Community's work in this field.

Technical objectives and priorities

1. **Exploration**

Geophysical exploration methods are the only reliable means of prospecting for oil and gas reserves in return for a relatively low investment in relation to the total exploration and, even more so, total production costs. But although the direct investment in geophysical exploration is relatively small, the success rate for exploratory wells drilled purely on the basis of geophysical data alone remains extremely modest. Since exploratory wells are still one the leading methods of prospecting, the dramatically high proportion of dry wells leaves prospecting costs high.

Seismic waves generated by signal transmitters are a long-known means of obtaining a more or less clear picture of the general formations underground. Today geophysicists are striving to produce an extremely accurate picture of the nature of the sediments underground and, in the long run, of the fluids accumulated in porous permeable rocks.

The latest trend in this field is that the spectacular progress in electronics and data processing has made it possible to increase the number of separate traces recorded from each shot. This advance has also increased the volume of data generated to such a degree that it would be impossible to process it only thanks to the major advances achieved in software and high storage capacity calculators.

Progress has also been made with the hardware for generating, transmitting and processing seismic waves. These, in turn, have necessitated radical changes in the methods of applying in the field geophysical systems incorporating all these sophisticated electronics and data-processing techniques.

The main priorities are as follows:

1. Surveys of reserves

    Seismic surveys must produce new ways of inpointing the location, nature and fluid contents of oil-bearing strata accurately. To achieve this, two different approaches must be followed, side by side:

    (i) Produce three-dimensional surface profiles, using waves moving in the same direction as, and perpendicular to, the propagation axis (i.e. compression and shear waves);

(ii) Another very promising method relatively little used at the moment because of the difficulty of producing the right equipment, is to transmit a signal into the well and record the results. This is a means of obtaining an extremely dense data network in an area with a large number of wells. It can be used either between wells or between a well and seismic survey equipment on the surface.

2. Survey of difficult cases

When prospecting complex geological strata, the geophysical and geochemical survey data have to be interpreted alongside those derived from the wells. Many geological areas and virtually unprospectable because volcanic or saliferous strata obstruct propagation of the seismic waves underground. No doubt highly sophisticated mathematical processing of the recordings made in the course of the seismic surveys, combined with non-seismic methods such as geomagnetic methods, gravimetry or aeromagnetics, could open up these new geological areas for prospecting.

3. Survey of very deep sediments and boundary layers

Highly sophisticated surface equipment and processing facilities, can now obtain data on geological strata several dozen kilometres underground. But programmes of this type can only cover extremely large areas spanning several different Community countries. They will allow major advances in our knowledge of the structure and development of the continental blocks and in our general knowledge of geological events within them. Geophysical schemes of this type require joint action by the Member States, like the international cooperation on the Glomar Challenger programme which marked such a decisive step forward in our knowledge of ocean substrata.

2. Drilling - anchors - soil prospecting

A. Of all the operations involved in producing oil and gas, drilling is assuming growing importance. After all, the only way to replace the available oil and gas reserves is to strike new ones. This is now even more difficult than in the past and, consequently, requires even more drilling to obtain the same volume of oil or gas.

Besides, drilling is a gruelling job. Consequently, staff, plant and environmental safety must be of constant concern to the operators. After hardly any change over the last fifty years, today drilling technology is undergoing far reaching changes directed towards two essential objectives:

(i) to take as much of the drudgery as possible out of the job. This objective takes priority at sea in particular, where there is the added worry of restricting the number of men on the platforms to the minimum, by using automated methods and robots wherever possible;

(ii) to produce a very marked improvement in the performance of the drilling rigs, which represents such a large share of the investment in exploration and production activities, and to cut the associated costs by approximately 50%.

The following three paths are thus proposed for research on drilling:

1. One priority must be to grant substantial support to projects aiming at boosting the efficiency of drilling systems and at cutting the related costs; for instance, projects to develop automatic drilling control systems with extensive instrumentation fitted to the gear to ensure safety and to monitor the well;

2. Development of the best possible systems for optimizing drilling operations, including any system which would make the data gathered at the bottom of the well available on a real-time basis;

3. Finally, support must be given for horizontal drilling methods and for projects to produce oil and gas from horizontal drains. Methods of drilling horizontal drains with a short radius of curvature must be developed and more powerful bottom-hole engines must be improved.

B. As far as foundation methods and soil investigation are concerned, the development of new production systems for oil and gas at depths ranging from 300 to 1 000 metres has brought with it a need for new types of foundations and anchors to guarantee the safety of the superstructures and for a clearer picture of the soil properties.

Simple extrapolation of conventional methods is not enough to make the switch from conventional foundations subjected to compression stresses to the foundations subjected to the pull exerted by the anchoring of the floating units of the future, for which a more advanced knowledge of soil mechanics is required.

In addition, the experts at the symposium drew attention to the grave shortcomings in the instrumentation for the foundation of the platforms already in place, which should provide a means of acquiring adequate basic data on seabed behaviour.

They recommended that the Commission should take action to centralise the observations and geotechnical data available in the various sector organisations in the Member States.

Accordingly, efforts must be directed towards the following fields:

1. The first priority support must be assigned to projects which develop and test new types of foundations and anchors particularly suited to floating production systems for marginal and deep-sea fields.

2. Support for projects to improve and extend the scope of existing foundation systems;

3. Support for projects on break-testing of foundations to allow better calculation of dimensions in future;

4. Finally, support for extensive programmes on foundation instrumentation with publication of the results.

3. Offshore platforms and structures

Although Shell is the current record holder as regards offshore drilling depths, with its 2 116 metres at Wilmington Canyon, Chesapeake Bay, the production is still only 100 to 150 metres under water. Recently, however, new strikes were made 450 metres under water in the Gulf of Mexico. Similarly, in the North Sea - the strike of Troll at 350 metres, the Tromsoflaket and Haltenbank finds and the gas struck by the British Gas Corporation at 600 metres under water west of the Shetlands, have all once again raised the problem of deep sea production. What is more, despite the failure of the Mukluk well, there is still a great deal of activity in the Arctic, which could be of interest to European industry.

The Working Party felt that in order to exploit these new fields, priority must be given to schemes to improve and make more detailed the existing designs which cover these depths. In general, a major programme is needed in the following fields which are of the utmost importance for the platforms construction.

1. Soil investigation, with the methods of sampling and testing and schemes to devise new types of foundations forming an integral part of the designs developed. Work should also focus on the interaction between the seabed and the foundation.

2. Work on new platform designs should concentrate on:

(a) key component development;

(b) indepth study of methods of constructing and fixing platforms and for maintenance and checking the soundness of the structures;

(c) cutting costs, with particular emphasis on tests on full-size prototypes in realistic environmental conditions.

But it will not be possible to put these methods into practice without extensive detailed studies and developing software compatible with present day data processing hardware.

3. Finally, particular attention must be paid to the technologies needed to produce the resources of the Arctic and in deep-seas such as the Gulf of Mexico or the Mediterranean Sea, and in particular for marginal fields in the Mediterranean Sea.

4. Production

The technology needed to exploit oil fields safely is now perfected. However, to diversify supplies in the long term and to plan ahead for the foreseeable depletion of such major production fields as the North Sea, producers are now looking to regions with severe environmental conditions for new resources, including severe or deep seas and the Arctic region.

Experience over the last ten years has shown the potential of a new concept to take the place of steel or concrete platforms anchored to the seabed and with all the control, monitoring and effluent treatment plant on the part of the rig above the sea surface. This new system would shift the control, monitoring and wellhead equipment to the seabed, leaving only the plant for processing the oil and gas on a floating unit on the surface so that the design would no longer depend on sea depths. The seabed and surface units would be linked by a giant pipe or "riser" for pumping up the effluent, monitoring and operating the wellhead gear and intervening whenever necessary, to tap the field, based on the latest state of the art.

Consequently, the technological production objectives are to produce all the units for this floating unit plus seabed wellhead system for difficult seas and for marginal fields.

Priority will be given to the following types of equipment:

a. single well oil production systems (SWOPS);

b. on-site tanker loading points:

c. risers, which are one major component of floating production units;

d. integration into specific set-ups of maintenance robots and of long-distance remote control systems for underwater wellheads;

e. artificial lift systems;

f. downhole measurement instruments.

5. Enhanced oil recovery

The fact that out of every 100 tonnes of oil in place in the reservoir, on average no more than 30 tonnes are recoverable on the surface shows the need to boost the recovery factor for non-renewable natural resources. But there are a series of economic or technical barriers to using new oil recovery methods which go further than the conventional ones widely used today. It goes without saying that the cost of one tonne of oil recovered by one of the new methods must be economic in relation to the cost of one tonne of oil produced by conventional methods and that this depends on the prevailing crude oil prices. As for the technical problems, although a variety of processes could be conceived and tested in the laboratory the only real way to prove that they are effective is to construct full-size pilot plants in parts of the field. But the danger with tests of this type is not only that they will not bring about any improvement at all but also that they might damage the field.

There are three main types of enhanced oil recovery methods:

(i) chemical processes, entailing the injection of solutions of chemical additives (such as surfactants, polymers and alkaline additves) into the water to wash the liquid hydrocarbons out of the reservoir rock more thoroughly;

(ii) thermal processes, which entail heating the reservoir to help the displacement of the heavy oils;

(iii) processes entailing injecting miscible solvents such as carbonic
gas or hydrocarbon gases into the crude oil in order to reduce
the capilllary force of the effluent.

The review of the EEC sponsored schemes in this field brought out how
very difficult it is to put theoretically effective methods into
practice in full-scale pilot plants which are in the fields themselves.
The reason for this is the poor knowledge of the properties of the
reservoir rocks and of the fluids which they contain.

The following measures were therefore proposed:

(i)   to encourage research into the mechanisms governing the circula-
      tion of the fluids in the rock and also to produce a more
      accurate assessment of the physical and chemical properties of
      the reservoir rocks and of the fluids.   Clearly, this will
      entail fundamental theoretical studies and improvements to
      conventional methods of surveying reservoirs;

(ii) to encourage the construction of pilot units in the field
      following extremely detailed surveys of the reservoirs, in order
      to select cases with the best chance of proving the value of the
      new recovery methods successfully.   A whole panoply of
      complementary methods will have to be mastered since conditions
      in different fields vary widely;

(iii) to encourage schemes to test pilot units in fields outside the
      Member States, i.e. in technical conditions which will add to
      European operators' knowledge.

6. Pipelines

Oil pipelines are another technology of fundamental importance to the
security of the Community's supplies.   Any shortcoming in the pipeline
could cut off the Member States' supplies and, consequently, damage
their economies.   Clearly, underwater pipelines require a most
important technological effort.   Since they carry oil from one
continent to another, they may well have to be laid at far greater
depths than the fields from which the oil was produced.   The Tunisia-
Sicily-Italy pipeline is one good example of the types of technical
problem facing the sponsors and contractors involved in this field.

Accordingly, technical trends for over the next few years will be as
follows:

1. To improve pipe-laying performance, i.e. to speed up the operation at all depths and in all environments, including the Arctic. Welding methods are one of the key components of pipe-laying.

2. The connection to the production systems on the seabed will have to be studied to allow virtually complete automation with remote control from the surface.

3. Development of operational methods of repairing pipelines. Once again every effort must be made to introduce robot and automatic systems, not only for the connections but also for preparing the ends of the pipe to be joined together.

   Several methods have been studied and are now being tested. It should become clear over the next few years which concept is the most efficient.

4. A major effort must also be put into adapting flexible pipelines to a wider range of products (liquefied gases, hot products) and to extreme environmental conditions or pressure levels. Measures must also be taken to encourage research into new materials for flexible or rigid pipelines.

5. Continuation of the effort in the maintenance and inspection field.

7. Technology for underwater operations

For some time to come human beings will still have to explore the ocean floor themselves, whether with or without a vehicle.

Although industrial divers can now reach depths of 300 metres, further advances are still needed to improve diver safety and comfort. Further developments are also needed in the hydrogen mixtures which have to be used if human divers are to go deeper than 300 metres.

Independent manned submarines must be developed so that structures and equipment can be built and placed on the seabed at great depths in seas where it would be difficult to use surface support units. These vehicles will not only be safer but will also cut down the amount of travelling time spent for each hour of work on the seabed.

The participants at the symposium put the accent on two objectives in particular with regard to manned underwater vehicles.

1.Spectacular progress has been made on underwater energy sources: closed circuit diesel engines and Stirling engines. But further developments are needed to develop highly efficient and with great autonomy, independent means of generating energy;

2.The recent progress in data processing and electronic should allow their use in underwater systems. Measures should therefore be taken to promote the use of underwater robots which would give considerably wider possibilities for manned vehicles.

8. Natural gas

It is also of the utmost importance to make better use of natural gas, after the spectacular increase in the available gas resources over recent years. After three years on the decline, in 1984, demand for natural gas rose once again, triggering a further increase in output within the Community and a major surge in imports. At the moment natural gas covers 18% of the Community's primary energy requirements.

At present supply exceeds demand on the natural gas market. This short term phenomenon must not be allowed to lead to any relaxation of the prospecting effort in the Community. On the contrary, it makes more technological development work in the natural gas sector essential. To achieve this, support must be given for action on the following points:

a. cost cutting and safety improvements at all stages of production;

b. surface and underground storage and supply systems technology for natural gas, LPG and other natural gas products, with particular emphasis on safety;

c. improvements to tankers to allow cryogenic loading systems, e.g. microprocessor-controlled manoeuvring aids;

d. installations to exploit secondary gas fields, in the North Sea for example, obtained by reinjecting the associated gases into the reservoirs;

e. exploitation of the Arctic natural gas fields;

f. loading of natural gas products and LNG at sea, including the commissioning of vessels, shipment and handling in extreme environments;

g. space, weight and energy saving methods of processing the gas on the platform, e.g. selective membrane methods;

h. installations to improve the offshore fields exploitation with nitrogen and carbon dioxyde production and distribution systems.

II. General comments on the programme pursuant to Regulation (EEC) No. 3056/73

The experts at the symposium were in favour of continuing the support scheme for Community projects in the hydrocarbons field. They also welcomed the decision to hold this symposium and hoped that it would be repeated every three years in future. They mentioned a number of improvements which could be made to make the programme more effective still.

1. Dissemination of know-how

Some of the working parties were satisfied with the present arrangements for disseminating knowledge. Nonetheless, others felt that more should be done in this direction. Two methods were proposed:

(i) holding meetings on specific themes every year or every two years to present the results of the research;

(ii) publishing a news sheet once or twice a year to keep technical circles informed of the progress made with the projects.

Nonetheless, it still remains the problem that dissemination of know-how must not be made at the detriment of companies that own the projects results of which they ensure the major part of the cost of the projects.

2. Cooperation

Cooperation between promoters in different Member States was also judged highly desirable, subject to the following conditions:

a) It must be spontaneous and voluntary, though the Commission could play a useful advisory role, signposting opportunities for cooperation to the promoters concerned particularly in areas where a certain degree of redeployment is needed following the pursuit of parallel development work.

b) It must be built on the complementary technical skills of the undertakings, with each contributing its own competences, technical, financial and human resources.

c) Cooperation must start at the beginning of the project, since otherwise sharing the industrial property rights and the findings is likely to pose unsolvable problems.

d) Cooperation between different undertakings must not be sine qua non for qualifying for support and must not lead to projects selection to the detriment of the technical quality.

3. Granting of support

The experts regretted that the decision making process to support project took so long, particularly for small companies without the financial resources to cope with this problem.

4. Multiannual programme

In view of the need for a continuous research programme on hydrocarbons, the experts felt that the Commission's multiannual approa was a great plus point.

CLOSING ADDRESS

G. BRONDEL
Director for Oil and Natural Gas
Commission of the European Communities

Ladies and Gentlemen,

Here we are at the end of three days of study and it is now my job to offer a first summing-up.

The reports from parallel sessions and the working parties that their chairmen have presented contain a mass of information and it would be too venturesome of me to try and draw all the conclusions now. We shall study records of the proceedings, which will provide us with an invaluable guide for the future. In particular, they will enable us to establish priorities among the fields of research that remain to be explored and we shall thus be able to set new guidelines to make the programme more effective. Their advice regarding the proper management of the programme will also be most useful.

Nonetheless, certain general conclusions can already be drawn:

In his opening speech, Mr Desprairies showed us how, after 10 years of tension on the oil market and two crises that hit our economies hard, the future looks somewhat more rosy and the chances of obtaining oil at a reasonable price are now stronger.

But, he also warned us of the risk of slackening our efforts now that oil prices have fallen and stressed the need to continue to ensure an adequate diversification of our supplies.

There is thus every reason to continue the support programme for oil and natural gas technology that began 10 years ago, and the Commission, as Mr Audland stated in his introduction, has asked the Council to extend the programme. The Commission's concern is to make the programme more effective in every way.

The funds that can be allocated to a programme of this kind are of necessity limited and are only justified as long as the following conditions are met:

- development and technological innovation must be stimulated; financial support must not be granted by a public authority of any kind unless the risks entailed go beyond the financial possibilities of undertakings;

- the projects supported would not have been possible without this support;

- the maximum multiplier effect must be obtained from the aid granted; a new branch of research should therefore, if successful, have considerable industrial spin-off.

A further point to be considered with regard to the programme's effectiveness is the selection of the areas on which research should be focused in the years to come.

The chairmen's reports that have been presented to you cover the ground very fully and it would be presumptuous of me to try to add anything. They show the numerous areas where progress has been made, illustrate the importance of the programme and give hope for its future.

But I should like to pick up two points:

- Firstly, all the groups stress the need for a multidisciplinary approach to ensure that research projects are mutually consistent; for it is difficult to draw a clear line between one area and another;

- Secondly, I was interested to see in all the reports the concern for concrete plans and the need to concentrate efforts on projects likely to lead to practical applications in the short term. I think it is a very important point in that this is where the programme differs from the research and development programmes put forward by my colleagues in DG XII (Science, Research and Development). But though these programmes may not be geared to immediate results, they are indispensable for the long-term outlook for technical progress.

In its recent Communication to the Council, the Commission proposes improvements to the programme from several aspects which I should now like to outline in brief.

- Adoption of a multiannual programme

The current programme has no time limit. Odd though it may seem, this presents more problems than advantages. The appropriations allocated to the programme, for example, have depended more in the past on the vicissitudes of the annual budget procedure than on rational planning.

The Commission intends to make the programme multiannual and to indicate a total budget for the entire period that the budgetary authorities could use to maintain the requisite continuity. Five years is a reasonable period, with a total allocation for assistance of 200 million ECU.

- <u>Changes to the decision-making process</u>

Currently, the list of projects to be supported must be approved each year by the Council following a proposal from the Commission. The approval decision must be unanimous, which obviously makes the whole programme rather precarious. It is suggested that henceforth the decision should be taken by the Commission after consulting an advisory committee with all the Member States represented. The Commission could thus take the decision quite independently and the advisory committee would be the forum in which main guidelines of grants policy are defined. This new procedure would also cut the time needed to take grant decisions. At the moment, nearly a year elapses between the submission of projects by firms and the Council decision granting aid. If possible, this should be reduced to six months.

The programme has been accused of giving too much support to single projects involving one undertaking only. In techniques as complex as the exploration and production of oil, greater cooperation must be encouraged beween companies belonging to several Member States. It would obviously be unthinkable to make this a binding condition, since individual initiatives may prove to be highly attractive, but joint schemes could lead to more rapid results and widen the potential scope of the techniques developed. The Commission intends in future to look into the ways of encouraging this kind of cooperation.

I was pleased to see that several group chairmen indicated specific areas where this type of cooperation should be developed.

<u>Dissemination of knowledge</u>

Experience over the ten years of running this programme has shown us how difficult it is to disseminate the results obtained. One of the aims of this symposium, like the one in 1979, is to promote better dissemination of information. Some twenty countries have been represented here and we are counting on all the participants to pass on what they have gathered. The objective of our programme is essentially to speed up the development of the techniques needed to exploit oil and natural gas resources in our countries. Although industrial property rights must obviously be respected, the results must be made available to any undertakings that need them. Working on some of the suggestions that have been made, the Commission plans to take three initiatives:

- firstly, as Mr Audland has already told us, the Commission has set up a data bank under the symbolic name of SESAME which will soon be open to wide-scale public access. It will show at once the list of projects supported, their backers, the objectives, the progress of the work and the results obtained when the projects are completed. The publication - the red book - that has been handed out to you gives an example of the type of information that will be found in it. With the aid of companies and firms we hope to be able to go into more detail;

- secondly, we hope to publish the results of the programme in a widely circulated regular bulletin. Information sheets on the most striking projects will be produced like the ones you have seen on the publications stand for both programmes backed by the Commission;

- finally, and probably most important, we hope in future to have more frequent meetings with experts from the various countries. A gap of five years between symposia is too long. We plan to hold annual meetings of specialist groups, which will enable us to keep a closer watch on technological developments. These meetings will also promote cooperation between businesses.

Greater international cooperation

Financial support can obviously only be granted to Community companies. However, projects are often submitted for implementation in conjunction with non-Community firms which, without benefiting from the financial support granted to the project, can usefully learn from the results.

Several projects are implemented in non-Community countries and this practice should be maintained and even expanded.

The Commission feels it would be worth opening up the programme more to contries outside the Community. The development of new technologies can provide a concrete form of cooperation with a large number of countries. Besides Norway, I can think of the ACP countries linked to the Community through the Lomé Convention, the Member States of the OAPEC with which the Commission cooperates in energy matters, as it does with all the developing countries that produce oil. All these countries are equally interested in having the most advanced techniques to enable them to exploit their resources.

I sincerely hope that we can count on the closest collaboration with all these countries. This is my message to the representatives of those countries who are with us here today.

And with that thought I come to the end of the general conclusions I thought should be drawn after these three days of study.

My thanks are extended to the session chairmen and deputy chairmen and to the rapporteurs, whose collaboration has been invaluable and who have performed a particularly difficult task in presenting a summary of this symposium in such a short time.

All that remains for me in closing is to wish this programme on the "technology of oil and natural gas" the same success in the future.

I now close the symposium and wish you all a safe journey home.

LIST OF PARTICIPANTS

ALBERTSEN, M.
Deutsche Gesellschaft für
Mineralölwissenschaft und
Kohlechemie e.V.
Nordkanalstrasse 28
D - 2000 HAMBURG 1

ALESSANDRI, L.
Inspection Maintenance
Sealines Department
SNAM SpA
P.O. Box 12060
I - 20120 MILANO

ALLARD, A.
Société Nationale Elf Aquitaine
7, rue Nelaton
F - 75739 PARIS Cedex 15

ALLITT, M.
Supervising Engineer
Wimpey Offshore Engineers and
Constructors Ltd
Flyover house
Great West Road
GB - BRENTFORD, Middx TW8 9AR

ANDERSON, J.
Head of Engineering Technology
Britoil Plc.
R & D Department
150 St Vincent St.
GB - GLASGOW

APPELL, Y.
Ingénieur
Société Européenne de Propulsion
Division Propulsion à Poudre et
Composites
B.P. 37
F - 33165 ST MEDARD EN JALLES CEDEX

ARIS, R.
Président Directeur Général
Forasol
B.P. 100
F - 78143 VELIZY-VILLACOUBLAY CEDEX

ATLAN, G.
Chargé de Mission
Chambre de Commerce et d'industrie
de Marseille, Maison de
l'Entreprise et de la Formation
Continue
35, rue Sainte-Victoire
F - 13292 MARSEILLE CEDEX 6

AUDLAND, C.
Director-General
Commission of the European
Communities, DG Energy
200, rue de la Loi
B - 1049 BRUXELLES

BACKHAUS, H.
LGA Gastechnik GmbH
Postfach 604
D - 5480 REMAGEN 6

BALLERAUD, P.
Engineer
Single Buoy Moorings Inc.
B.P. 199
MC - 98000 MONACO

BARDON, C.
Ingénieur
Institut Français du Pétrole
1-4, ave de Bois Préau
F - 92506 RUEIL MALMAISON

BARON, G.A.
Ingénieur
Institut Français du Pétrole
1-4, ave de Bois Préau
F - 92506 RUEIL MALMAISON

BARRETT, I.
Technical Manager
B.P. Shipping Limited
Britannic House
Moor Lane
GB - LONDON EC2Y 9BR

BARTER, F.
Consultant
Cavern Systems Dublin Limited
16, Upper Pembroke Street
IRL — DUBLIN 2

BATH, A.
Preussag AG
Erdöl und Erdgas
Arndtstr. 1
D — 3000 HANNOVER 1

BATSCHKO, H.D.
Engineer
Howaldtswerke — Deutsche Werft AG
Department Hok
postfach 11 14 80
D — 2000 HAMBURG 11

BAUDINO, M.
Direttore Studi e Sviluppo
SNAM SpA
P.O. Box 12060
I — 20120 MILANO

BAUDRY, A.
Directeur Recherche & Développement
Syminex S.A.
2, boulevard de l'Océan
F — 13275 MARSEILLE CEDEX 9

BAYAT, M.G.
Petroleum
Britoil Plc.
150 St. Vincent Street
GB — GLASGOW G2 5LJ

BEKKER, G.A.
Chemical Engineer
Shell International Gas Ltd
NGT/4
Shell Centre
GB — LONDON

BELIN DE BALLU, G.
Ing. Civil
Ackermans & van Haaren N.V.
Begijnenvest 113
B — 2000 ANTWERPEN

BELLAMY, D.
Business Development Manager
E.M.H.
196, Bureaux de la Colline
F — 92213 ST CLOUD CEDEX

BENNEHARD, M.
Geologue
Elf Italiana S.p.A.
Via Aurélia 619
I — 00165 ROMA

BERTRAND, G.
Ingénieur
DHYCA Ministère du Redéploiement
Industriel
336, ave Napoléon Bonaparte
F — 92501 RUEIL MALMAISON CEDEX

BJERRUM, A.
Petroleum Engineer
Cowiconsult
Teknikerbyen 45
DK — 2830 VIRUM

BLU, G.
Coordonnateur
Groupement Européen de Recherches
Technologiques sur les
Hydrocarbures, G.E.R.T.H.
4, ave de Bois Préau
F — 92500 RUEIL MALMAISON

BODEN, J.C.
Research and Development
The British Petroleum Company
BP Research Centre
Chertsey Road
GB — SUNBURY-ON-THAMES, Mid. TW16 7L

BOLLEREAU, J.
Attaché de Direction
Souriau et Cie
9-13, rue du Général Galliéni
F — 92103 BOULOGNE BILLANCOURT CEDEX

BOMHARD, H.
Dir., Dipl.-Ing.
Dyckerhoff & Widmann AG
Postfach 810280
D — 8000 MUENCHEN 81

BONJOUR, E.
Ingénieur
Total — Compagnie française des
pétroles
TEP/DP/TA, Tour Chenonceaux
204, rd-point du Pont de Sèvres
F — 92516 BOULOGNE

BONVECCHIATO, G.
Responsabile Ufficio Ricerca e
Sviluppo Tecnologie Gas
SNAM SpA
P.O. Box 12060
I - 20120 MILANO

BOON, C.
Ing. Civil
Ackermans & van Haaren N.V.
Begijnenvest 113
B - 2000 ANTWERPEN

BORMIOLI, L.
Design & Engineering Manager
MIB International LTD
(Subsidiary of MIB Italia SpA)
Via Garibaldi 6
I - 35020 CASALSERUGO, Padova

BORRILL, P.
Chemical Engineer
British Gas Corporation
Midlands Research Station
Wharf Lane
GB - SOLIHULL, West Mid. B91 2JW

BOSIO, J.
Elf Aquitaine
7, rue Nélaton
F - 75739 PARIS CEDEX 15

BOUCKALDER, M.
Journaliste - Directeur
Petrole Informations
142, rue Montmartre
F - 75002 PARIS

BOULANGER, A.
Ingénieur
Geostock
Tour Aurore
Cedex 5
F - 92080 PARIS DEFENSE 2

BOURGEAIS, J.P.
Assistant Manager/Sales Manager
SESAM
132, ave de Villeneuve-Saint-Georges
F - 94600 CHOISY-LE-ROI

BOURGEOIS, A.G.
Ingenieur Direction Operations
Elf Italiana
Largo Lorenzo Mossa, 8
I - 00165 ROMA

BOURGEOIS, T.
Ingénieur de recherche
Société Nationale Elf Aquitaine
Centre Boussens
F - 31360 SAINT MARTORY

BRANCHEREAU, P.
Ingénieur
E M H
196, Bureaux de la Colline
F - 92213 SAINT-CLOUD CEDEX

BRANDS, K.W.
Mechanical Engineer
Shell internationale Petroleum
Maatschappij B.V.
Dep. EP/29
P.O. Box 162
NL - 2501 AN THE HAGUE

BRIN, A.
Chargé de Mission
Mis. Interministérielle de la Mer
9-11, rue G. Pitard
F - 75015 PARIS

BRONDEL, G.
Directeur
Commission of the European
Communities, DG Energy
200, rue de la Loi
B - 1049 BRUXELLES

BRONKHORST, J.W.
Commission of the European
Communities, DG Energy
200, rue de la Loi
B - 1049 BRUXELLES

BROOKES, M.A.
Civil Engineer
British Petroleum plc
Britannic House
Moor Lane
GB - LONDON EC2Y 9BU

BROWN, N.E.
Engineering Consultant
Stone & Webster Engineering Ltd
Stone & Webster House
500 Eldergate
Central Milton Kenes
GB - MILTON KEYNES MK9 1BA

BRUINING, J.
Reservoir Engineer
TH Delft
Dept Petroleum Engineering
Mijnbouwstraat 120
NL - 2628 RX DELFT

BRUMSHAGEN, W.
Geschäftsführer
LGA Gastechnik GmbH
Bonner Strasse 10
Postfach 604
D - 5480 REMAGEN 6

BRYCH, J.
Prof. Dr. Ir.
Université de Mons
Faculté Polytechnique
53, rue du Joncquois
B - 7000 MONS

BUCKLEY, B.
Drilling Engineer
Shell internationale Petroleum
Maatschappij B.V.
Carel vah Bylandtlaan 30
NL - 2501 AN THE HAGUE

BULANG, W.
Dr. rer.nat.
MAN–Neue Technologie
Abt. ENT/V6
Postfach 500620
D - 8000 MUENCHEN 50

BUTT, H.
Head of Offshore Division
Bilfinger & Berger
Bauaktiengesellschaft
P.O. Box 76 02 40
D - 2000 HAMBURG 76

CAHILL, F.
Supply Manager
Irish National Petroleum
Corporation
Warrington House
Mount St. Crescent
IRL - DUBLIN 2

CALVARESE, L.
Funzionario
Ministero Industia Commercio
Artigianato
D.G. Miniere
Via Molise 2
I - 00187 ROMA

CAMPBELL, G.
Chartered Engineer
NEI Peebles Ltd
Peebles Electrical Machines
East Pilton
GB - EDINBURGH EH5 2XT

CARRUTHERS, R.
Study Manager
Taywood Santa Fe Ltd
309 Ruislip Road East
GB - GREENFORD, Middx UB6 9BQ

CARVOUNIS, P.
Commission of the European
Communities, DG Energy
200, rue de la Loi
B - 1049 BRUXELLES

CASTELA, A.
Ingénieur
Institut Français du Pétrole
1-4, ave de Bois Préau
F - 92506 RUEIL MALMAISON

CATHLE, D.
Civil Engineer (Consulting)
S.A.G.E. Ltd, Structural Analysis
and Geotechnical Engineering
2, ave des Tourterelles
B - 1150 BRUXELLES

CAUSIN, E.
E.O.R. Engineer
AGIP SpA S. Donato Milanese
Prav. Dep.
P.O. Box 12069
I - 20120 MILANO

CHAMPLON, D.
Ingénieur
Institut Français du Pétrole
1-4, ave de Bois Préau
F - 92506 RUEIL MALMAISON

CHAPERON, A.
Ingénieur Géophysicien
Compagnie française des Pétroles
Total
39-43, quai André Citroen
F - 75739 PARIS CEDEX 15

CHIERICI, G.L.
Vice president, Petrol. engineering
Agip S.p.A.
P.O. Box 12069
I - 20120 MILANO

CLAUDE, J.
Ingénieur
Technigaz
B.P. 126
F - 78312 MAUREPAS CEDEX

COIRAL, J.C.
Ingénieur
Coflexip
23, ave de Neuilly
F - 75116 PARIS

COLLARD, M.J.
Civil Engineer
McAlpine Offshore Std
40 Bernard St.
GB - LONDON WC1 N1 LG

COLOMBO, A.
Deepwater Repair System Tech. Eng.
SNAM SpA
P.O. Box 12060
I - 20120 MILANO

COLQUHOUN, R.S.
Chartered Engineer, Consultant
Danish hydraulic Institute
Tulstrupgaard
DK - 3230 GRAESTED

CORTEVILLE, J.
Ingénieur
Institut Français du Pétrole
1-4, ave de Bois Préau
F - 92506 RUEIL MALMAISON

COTTIN, R.
Ingénieur
Elf Aquitaine
Centre Micoulau
Avenue Président Angot
F - 64000 PAU

COUVE DE MURVILLE, E.
Ingénieur
Petrorep S.A.
42, ave Raymond Poincarré
F - 75116 PARIS

COX, R.
Commission of the European
Communities, DG Information
200, rue de la Loi
B - 1049 BRUXELLES

COX, S.P.
Reservoir Engineer
Shell U.K. Exploration
and Production
Shell-Mex House
Strand
GB - LONDON WCZR ODX

DARRAGON, J.
Head of Division "Energy"
Commission of the European
Communities, Statistical Office
L - 2920 LUXEMBOURG

DAVIES, D.
Petroleum Engineer
BP Exploration Ltd
Britannic House
Moor Lane
GB - LONDON EC2Y 9BU

DE BAUW, R.
Commission of the European
Communities, DG Energy
200, rue de la Loi
B - 1049 BRUXELLES

DE GRISOGONO, I.
Dipl.-Ing.
Deminex
Dept. S321
P.O. Box 100944
D - 4300 ESSEN 1

DE HAAN, H.J.
Professor Petroleum Engineering
Technical University Delft
SIPM B.V., EP/26
P.O. Box 162
NL - 2501 AN THE HAGUE

DE LAGUERIE, P.
Ingénieur
Geostock
Tour Aurore, Cedex 5
F - 92080 PARIS DEFENSE 2

DE LOMBARES, G.
Ingénieur Géophysicien
Compagnie française des pétroles
Total
39-43, quai André Citroen
F - 75739 PARIS CEDEX 15

DE MARZO, D.
Pipeline Design Leader
AGIP spA
P.O. Box 12069
I - 20120 MILANO

DE RAAD, J.
Manager Development & Systems
Department
Röntgen Technische Dienst B.V.
Delftweg 144
NL - 3046 NC ROTTERDAM

DE SIVRY, B.
Ingénieur
Total, Compagnie française des
pétroles
171, rue de l'Université
F - 75007 PARIS

DE VAULX, C.
R. et D. Manager
Alsthom ACB
Prairie au Duc
F - 44000 NANTES

DELACOUR, J.
Ingénieur
Institut Français du Pétrole
1-4, ave de Bois Préau
F - 92506 RUEIL MALMAISON

DELAPORTE, P.
Project Manager Subsea
E.T.P.M.
33-35, rue d'Alsace
F - 92531 LEVALLOIS PERRET CEDEX

DELAUZE, H.
Président Directeur Général
COMEX
36, bd des Océans
F - 13275 MARSEILLE CEDEX 9

DESPRAIRIES, P.
Président
Institut Français du Pétrole
B.P. 311
F - 92506 RUEIL-MALMAISON CEDEX

DEVILLET, C.
Journaliste
Vers l'Avenir
Salle de Presse du P.E.
97-113, rue Belliard
B - 1040 BRUXELLES

DI MOLFETTA, A.
Professore Universitario
Politecnico di Torino
Dipartimento Georisorse e
Territorio
Corso Duca Degli Abruzzi 24
I - 10129 TORINO

DI TELLA, V.
Technical Manager
Tecnomare SpA
S. Marco 2091
I - 30124 VENEZIA

DIMONT, B.
Vice Président Exécutif
ASTEO
Immeuble Ile-de-France
Cedex 33
F - 92070 PARIS LA DEFENSE

DOREL, M.
Ingénieur
Institut Français du Pétrole
1-4, ave de Bois Préau
F - 92506 RUEIL MALMAISON

DORMIGNY, A.M.
Ingénieur
Elf Italiana
Largo Lorenzo Mossa, 8
I - 00165 ROMA

DOSSI, L.
Chief Eng., Special Research sec.
Tecnomare SpA
S. Marco 2091
I - 30124 VENEZIA

DRIDI, H.
Ingénieur
Coflexip
23, ave de Neuilly
F - 75116 PARIS

DUFAU, F.
Ingénieur
Bertin et Cie
B.P. 3
F - 78373 PLAISIR CEDEX

DUFOND, R.
Ingénieur
Elf Aquitaine
Tour Aquitaine Cedex 4
F - 92080 PARIS LA DEFENSE

DUISBERG, J.
  Dipl.-Ing.
  Hoesch Rohr AG
  Kissinger Weg
  Postfach 1713
  D - 4700 HAMM 1

DUMONT, J.J.
  Souriau & Cie
  9-13, rue Général Gallieni
  F - 92103 BOULOGNE BILLANCOURT

DUPRAT, A.
  Ingénieur à la Dir. Technique
  Compagnie Générale de Géophysique
  1, rue Léon Migaux
  F - 91301 MASSY CEDEX

DURIX, P.
  Ingénieur
  Délégué permanent du Comité
  d'Etudes Pétrolières Marines
  Tour Franklin
  F - 92081 PARIS LA DEFENSE CEDEX 11

DUTRIAU, R.
  Ingénieur
  Institut Français du Pétrole
  B.P. 311
  F - 92506 RUEIL MALMAISON

DYKES, C.
  Senior Drilling Engineer
  BP Petroleum Developement Ltd
  Farburn Industrial Estate
  GB - DYCE, Aberdeen AB2 0PB

ELLIOTT, R.M.
  Engineer
  BSP International Foundations Ltd
  Claydon
  GB - IPSWICH, Suffolk IP6 0GD

EVANS, T.E.
  Chartered Chemist
  Britoil Plc
  R & D Department, Britoil Plc
  150, St. Vincent Street
  GB - GLASGOW

EVANS, S.
  Sales Development Manager
  MIB International LTD
  (Subsidiary of MIB Italia SPA)
  Marketing and Engineering Liaison
  Sun Alliance House
  Little Park Street
  GB - COVENTRY CV1 2JZ

FABIANI, P.
  Adj. au Dir. du Dép. Tech. Avancées
  Total, Compagnie française des
  pétroles
  Tour Chenonceaux
  2d, Rond-Point du Pont de Sèvres
  F - 92516 BOULOGNE BILLANCOURT CEDEX

FAULKNER, P.
  Research and Development
  The British Petroleum
  Company Plc
  B.P. Research Centre
  Chertsey Road
  GB - SUNBURY-ON-THAMES, Mid. TW16 7LN

FAVRE, J.
  Directeur Général, Adjoint
  Institut Français du Pétrole
  1-4, ave de Bois Préau
  F - 92506 RUEIL MALMAISON

FEE, D.
  Commission Official
  Dept. of Energy
  Nassau House
  Nassau St.
  IRL - DUBLIN

FENATI, D.
  Vice-President
  Dept. Geology and Geophysics
  AGIP SpA
  I - 2097 SAN DONATO MILANESE

FERTIG, J.
  Preussag AG
  Erdöl und Erdgas
  Arndtstr. 1
  D - 3000 HANNOVER 1

FEUGERE, G.
  Directeur Commercial
  SBM France
  25, rue d'Astorg
  F - 75008 PARIS

FLOYD, K.
  Manager Drilling Engineering
  BP Exploration Company Limited
  Britannic House
  Moor Lane
  GB - LONDON EC2Y 9BU

FUVEL, P.
Ingénieur
Total CFP
Tour Chenonceaux
204, rond point du Pont de Sèvre
F - 92516 BOULOGNE

GADELLE, C.
Ingénieur
Institut Français du Pétrole
1-4, ave de Bois Préau
F - 92506 RUEIL MALMAISON

GAISFORD, R.W.
General Manager
Research and Development
Britoil Plc
R & D Department
150, St Vincent St.
GB - GLASGOW G2 5LJ

GALANT, S.
Ingénieur Chef de Division
Bertin et Cie
B.P. 3
F - 78373 PLAISIR CEDEX

GARCIA-SINERIZ BUTRAGUENO, B.
General Manager, Dr. Mining Eng.
Hispanica de Petroleos, S.A.
(Hispanoil)
C/ Pez Volador, 2
E - 28007 MADRID

GASPARINI, M.
Ingegnere
AGIP s.p.a.
S. Donato Milanese
I - 20097 MILANO

GEFFRIAUD, J.P.
Directeur Général
Solmarine
6, rue de Watford
F - 92000 NANTERRE

GIANNESINI, J.F.
Directeur Géneral
Horwell S.A.
177, ave Napoléon Bonaparte
F - 92500 RUEIL MALMAISON

GIGER, F.
Ingénieur
Institut Francais du Pétrole
Centre de Documentation
B.P. 311
F - 92506 RUEIL-MALMAISON

GILROY, J.P.
Development Manager
Marine Computation Services Ltd
The Science Park
IRL - NEWCASTLE, Galway

GIRAULT, Y.P.
Direction Développement Promotion
Usinor
Immeuble Ile de France
4, place de la Pyramide, cedex 33
F - 92070 PARIS LA DEFENSE

GLYNN, P.
Commission of the European
Communities; DG Science, Research
and Development
200, rue de la Loi
B - 1049 BRUXELLES

GOLINVAUX, R.
Commission of the European
Communities, Statistical Office
Hydrocarbon
L - 2920 LUXEMBOURG

GRAU, G.
Directeur de recherche
Institut Français du Pétrole
1-4, ave de Bois Préau
F - 92506 RUEIL MALMAISON

GREGORY, J.
Journalist
Offshore Engineer
Telford House
26-34 Old Street P.O. Box 101
GB - LONDON EC1

GRESHAM, J.S.
Mechanical engineer
Shell internationale petroleum
Maatschappij B.V.
dep. EP/25.3
P.O. Box 162
NL - 2501 AN THE HAGUE

GRIFFITHS, A.
Dunlop Limited
Oil and Marine Division
Pyewipe Industrial Estate
Moody Lane
GB - GRIMSBY DN31 2SP

GRIST, D.M.
Petroleum Engineer
BP Exploration
Britannic House
Moor Lane
GB - LONDON EC2Y 9BU

GROSSIN,
Ingénieur
Société Bertin et Cie
B.P. 3
F - 78373 PLAISIR CEDEX

GROUSET, D.
Ingénieur
Socité Bertin et Cie
Zone Ind. de Boucau Tarnos
F - 40220 TARNOS

GUERRIER, G.
Responsable R & D
COMEX
36, bd des Océans
F - 13275 MARSEILLE CEDEX 9

GUESNON, J.
Ingénieur
Institut Français du Pétrole
1-4, ave de Bois Préau
F - 92506 RUEIL MALMAISON

GUILLON, J.H.
Chargé de Mission pour
les Affaires Scientifiques
et Techniques
ASTEO, Groupement Interprof.
pour l'Exploitation
des Océans
B.P. 33
F - 92070 PARIS LA DEFENSE

HADDAD, M.
Directeur de Production
Ent. Tunisienne d'Activités
Pétrolières
11, ave K. Pacha
TN - TUNIS

HAMID, A.
Head of Oil and Gas Division
Pertamina
Jalan Medan Merdeka Ti Mur 1A
Indonesia - JAKARTA, PUSAT

HARBONN, J.
Directeur du département
techniques avancées
Total Compagnie Française des
Pétroles
Tour Chenonceaux
204, rond-point du Pont de Sèvres
F - 92516 BOULOGNE BILLANCOURT CEDEX

HARING, K.
Dipl.-Ing.
Ruhrgas
Postfach 10 32 52
D - 4300 ESSEN 1

HEIERHOFF, F.W.
Managing Director
Wirtschaftsvereinigung industrielle
Meerestechnik e.V.
Association of German Oceanic Ind.
Rossstr. 126/128
D - 4000 DUESSELDORF 30

HENRI, D.
Ingénieur des Mines
Ministère de l'Industrie
Direction des Hydrocarbures
5, rue Barbet de Jouy
F - 75007 PARIS

HEYWOOD, P.T.
Manager
British Petroleum Ltd
Britannic House
Moor Lane
GB - LONDON EC2

HILL, V.
Senior Reservoir Engineer
BP Petroleum Developement Ltd
Farburn Industrial Estate
GB - DYCE, Aberdeen AB2 0PB

HINSTRUP, P.I.
Civil Engineer M.Sc.
Danish Hydraulic Institute
Agern Allé 5
DK - 2970 HORSHOLM

HOLEKAMP, R.
Dipl.-Ing.
Salzgitter AG
Postfach 41 11 29
D - 3320 SALZGITTER 41

HOLSTEIN,
  Ingénieur
  SESAM
  Section technique
  132, ave de Villeneuve St. Georges
  F - 94600 CHOISY-LE-ROI

IMARISIO, G.
  Commission of the European
  Communities; DG Science, Research
  and Development
  200, rue de la Loi
  B - 1049 BRUXELLES

IRVING, R.
  Chemical Engineer
  Monsanto Europe S.A.
  270-272, ave de Tervuren
  B - 1150 BRUXELLES

JACQUEMIN, M.
  Directeur Relations Expérieures
  Compagnie Générale de Géophysique
  1, rue Léon Migaux
  F - 91301 MASSY CEDEX

JORDAN, P.A.
  Director
  Seaforth Maritime Limited
  30 Waterloo Quay
  GB - ABERDEEN AB2 1BS

JOUBERT, P.
  Ingénieur
  Institut Français du Pétrole
  1-4, ave de Bois Préau
  F - 92506 RUEIL MALMAISON

JOULIA, J.P.
  Commission of the European
  Communities, DG Energy
  200, rue de la Loi
  B - 1049 BRUXELLES

JØRGENSEN, H.
  Engineer, Techn. Advisory Service
  European Investment Bank
  L - 2950 LUXEMBOURG

KALINOWSKI, R.
  Ingénieur Conseil
  20, ave Georges-Bizet
  F - 13470 CARNOUX-EN-PROVENCE

KENLEY, R.M.
  Chartered Engineer
  Structural Monitoring Division
  W.A. Fairhurst & Partners
  11 Woodside Terrace
  UK - GLASGOW G3 7XQ

KERMABON, A.
  Président Directeur Général
  Syminex S.A.
  2, boulevard de l'Océan
  F - 13275 MARSEILLE CEDEX 9

KESSEL, D.
  Manager
  Deutsche Texaco AG
  Laboratorium für
  Erdölgewinnung
  Industriestr. 2
  D - 3109 WIETZE

KLAEKE, R.D.
  Dipl.-Ing.
  Ocean Consult GmbH - ArGe SUPRA
  Halbmond 30 d
  D - 2058 LAUENBURG

KLIETZ, R.
  Marine Coordinator
  BEB Gewerkschaften Brigitta und
  Elwerath Betriebsführungs-
  gesellschaft MbH
  Postfach 51 03 60
  D - 3000 HANNOVER 51

KLITGAARD, S.
  Div. Man.
  Aalborg Verft Offshore
  P.O. Box 661
  DK - 9100 AALBORG

KNIGHTS, D.L.
  Research Coordinator
  BP International Limited
  Britannic House
  Moor Lane
  GB - LONDON EC2Y 9BU

KOHRTZ, J.
  Head of Division, Engineering
  Research and Development
  Danish Energy Agency
  Landemaerket 11
  DK - 1119 COPENHAGEN

KOK, P.
Consultant
Shell International Petroleum
mij. N.V.
Trionnaz bte 120
CH - 3941 LENS

KOKKINOWRACHOS, K.
Professor
Techn. University Hamburg &
Techn. Hochschule Aachen
Templergraben 55
D - 51 AACHEN

KRUSENSTJERNA - HAFSTROEM, B.
Head of Section
Ministry of Energy
Slotsholmsgade 1
DK - 1216 KOEBENHAVN K

KVINNSLAND, O.J.
President
Noroil Group
P.O. Box 480
Hillevagsveien 17
N - 4001 STAVANGER

LABABIDI, M.M.
Asst Director
Energy Resources Dept.
Organization of Arab Petroleum
Exporting Countries
P.O. Box 20501
Kuwait - SAFAT

LAEMMERZAHL, D.
Dipl.-Ing.
Salzgitter AG
Postfach 41 11 29
D - 3320 SALZGITTER 1

LARROZE, J.
Ingénieur
Institut Français du Pétrole
Avenue de Bois-Préau
F - 92000 RUEIL-MALMAISON

LE BIHAN, M.
Ingenieur
Elf Italiana
Largo Lorenzo Rossa 8
I - 00165 ROMA

LE GOUELLEC, P.
Chargé de mission
Commissariat à l'Energie Atomique
29-33, rue de la Fédération
F - 75015 PARIS CEDEX 15

LE PAGE, J.F.
Directeur, Coordinateur projets
Institut Français du Pétrole
1-4, ave de Bois Préau
F - 92506 RUEIL MALMAISON

LE XUAN, T.
Ingénieur structure
ETPM
Courcellor II
33/35 rue d'Alsace
F - 92531 LEVALLOIS PERRET

LEBLOND, M.
Président
Groupement Européen de Recherches
Technologiques sur les
Hydrocarbures
1-4, ave de Bois Préau
F - 92500 RUEIL MALMAISON

LEBLOND, D.
Journaliste
Petroleum Economist
Petroleum Information
International
34, rue du Docteur Blanche
F - 75781 PARIS CEDEX 16

LEBOUCQ, V.
Journaliste
AGEFI - SA Le Nouveau Journal
108, rue Richelieu
F - 75002 PARIS

LEDOUX, Y.
Directeur Gén. Adj., Développement
Compagnie Générale de Géophysique
1, rue Léon Migaux
F - 91301 MASSY CEDEX

LEGROS, E.
Journaliste
ENERCOM
65-67, ave des Champs-Elysées
F - 75008 PARIS

LEMOINE, F.
Ingénieur
Alsthom Atlantique
Atelier et Chantiers de Bretagne
Prairie au Duc
F - 44040 NANTES CEDEX

LESTER, T.E.
Research Coordinator
BP International Limited
Britannic House
Moor Lane
GB - LONDON EC2Y 9BU

LIEGEOIS, E.
Rédactrice en Chef
Petrole Informations
142, rue Montmartre
F - 75002 PARIS

LIEKE, N.
Diplom-ingenieur
Preussag AG Hauptverwaltung
Personalabteilung
Leibnizufer 9
D - 3000 HANNOVER 1

LOSECKE, W.
Physicist
Bundesanstalt für
Geowissenschaften und Rohstoffe
Postfach 51 01 53
D - 3000 HANNOVER

LOWENSTEIN-LOM, W.
Asst. Chef de Corps CT
European Investment Bank
L - 2950 LUXEMBOURG

LUECHAU, P.
Assesor
Ferrostaal A.G.
Department FN
Postfach 10 12 65
D - 4300 ESSEN 1

LUKIC, P.
Geophysiker
Preussag AG
Arndtstr. 1
D - HANNOVER

LUNDSTEN, L.C.
Engineer
United Stirling AB
Box 856
S - 20180 MALMÖ

MACLEOD, I.
Deputy Divisional Manager
GEC Mechanical Handling Limited
Marine and Hydraulics Division
Birch Walk
GB - ERITH, Kent DA8 1QH

MADEC, M.
Ingénieur
Institut Français du Pétrole
B.P. 311
F - 92506 RUEIL MALMAISON CEDEX

MAGLOIRE, C.
Ingénieur
ETPM
33-35, rue d'Alsace
F - 92531 LEVALLOIS PERRET

MAKRIS, J.
Geophysicist
Institut für Geophysik
Universität Hamburg
Bundesstr. 55
D - 2000 HAMBURG 13

MANGIALAIO, M.
CISE
C.P. 12081
I - 20134 MILAN

MARI, J.L.
Ingénieur de Recherche
Institut Français du Pétrole
1-4, ave De Bois-Preau
F - 92506 RUEIL MALMAISON

MARIANI, E.
Ingegnere
AGIP
I Palazzo Uffici
Pza Enrico Mattei
I - S. DONATO MILANESE, Milano

MARRIOTT, J.
Development Engineer
British Petroleum Co. PLC
Britannic House
Moor Lane
GB - LONDON EC2Y 9BU

MARSLAND, G.
ZF-HERION-Systemtechnik GmbH
Postbox 21 68
D - 7012 FELLBACH

MARTIN DIAZ, A.
Jefe de Exploracion y Produccion
Instituto Nacional de Hidrocarburos
Paseo de la Castellana, 89
6a Planta
E - 28046 MADRID

MARTIN, P.
Adjoint Financier au Directeur
Département Techniques avancées
Total Compagnie Française des
Pétroles
Tour Chenonceaux
204, rond-point du Pont de Sèvres
F - 92516 BOULOGNE BILLANCOURT CEDEX

MARTIN, J.
Ingénieur en Chef
C.G. DORIS
58A, rue du Dessous des Berges
F - 75013 PARIS

MARTINO, D.
Ingegnere Capo del Corpo Delle
Miniere - Ufficio Idrocarburi
Ministero Industria
Via Molise 2
I - 00187 ROMA

MARTIN, R.S.
Marine Consultant
London Offshore Consultants
1 Fenchurch Buildings
GB - LONDON KC3

MARTISCHIUS, F.D.
BASF AG
ZKM - G201
D - 6700 LUDWIGSHAFEN

MAURY, J.L.
Total Compagnie Française des
Pétroles
39-43, quai André Citroen
F - 75739 PARIS CEDEX 15

MAZZON, M.
Project Manager
Tecnomare
S. Marco 2091
I - 30124 VENEZIA

MELDRUM, I.
Research Chemist
B.P. International Ltd.
B.P. Research Centre
Chertsey Road
GB - SUNBURY ON THAMES TW16 7LN

MENENDEZ, R.
Ingénieur
Société Nationale Elf Aquitaine
Production
Tour Aquitaine
Cedex 04
F - 92080 PARIS LA DEFENSE

MENNEBEUF, A.
Chef du Département Administration
Générale
Compagnie Générale de Géophysique
1, rue Léon Migaux
F - 91301 MASSY CEDEX

MERCIER, B.
Ingénieur
Groupement Européen de Recherches
Technologiques sur les
Hydrocarbures
4, ave de Bois Préau
F - 92500 RUEIL MALMAISON

MEUNIER, J.
Secrétaire Assistante R.H. Cottin
Elf Aquitaine
Centre Micoulau
Ave du Président Angot
F - 64000 PAU

MEYER-DETRING, D.
Dipl.-Ing.
Preussag AG
Postfach 4827
D - 3000 HANNOVER

MILLICH, E.
Commission of the European
Communities, DG Energy
200, rue de la Loi
B - 1049 BRUXELLES

MINEBOIS, J.L.
Ingénieur Production
Société Nationale Elf Aquitaine
Service Production
B.P. 1 Lacq
F - 64170 ARTIX

MONTEBRUSCO, L.
Journaliste
Agence de Presse Novosti
Zeitung vum Letzebuerger Vollek
41, bd de la Petrusse
L - LUXEMBOURG

MORAND, P.
Ingénieur
Technigaz
B.P. 126
F - 78312 MAUREPAS CEDEX

MOREAU, J.P.
R/D Engineer
Chantiers Nord- Méditerranée
B.P. 1 503
F - 59381 DUNKERQUE CEDEX 1

MØLLER ANDERSEN, H.
Hans Møller Andersen ApS
Falkonér Allé 7
DK - 2000 COPENHAGEN F.

MUNK NIELSEN, M.
M. Sc., B.Comm.
Danish Ministry of Energy
Section Research and Development
Slotsholmstrade 1
DK - 1216 COPENHAGEN K.

MURRAY, N.
Joint Research Centre
Ispra Establishment
I - ISPRA (Varese)

MUSCARELLA, G.
President
Dept. Geology and Geophysics
AGIP SpA
I - 2097 SAN DONATO MILANESE

MYRIANTHIS, M.
Deputy Head Geophysiscs dpt
Public petroleum Corporation
of Greece
199 Kifissias Ave
GR - 151 24 MAROUSSI, Athens

McCABE, E.T.
Commission of the European
Communities, DG Energy
200, rue de la Loi
B - 1049 BRUXELLES

McLEISH, A.
Engineer
Taylor Woodrow Construction Ltd
Taywood House
345 Ruislip Road
GB - SOUTHALL, Middx UB1 2QX

NGUYEN, V.T.
Ingénieur
O.T.P.
5, rue Chante Coq
F - 92808 PUTEAUX CEDEX

NICOLAY, D.
Commission of the European
Communities, DG Information
Market and Innovation
L - 2920 LUXEMBOURG

NIEMANN, K.
Dipl.-Chem.
VEBA OEL Entwicklungs-
Gesellschaft mbh
Postfach 45
D - 4650 GELSENKIRCHEN 2

NILSSEN, K.
Industry and Offshore Division
P.O. Box 300
N - 1322 HOEVIK

NISTA, A.
Tecnomare S.p.A.
S. Marco 2091
I - 30124 VENEZIA

OBST, W.
Commission of the European
Communities DG Employment,
social affairs and education
L - 2920 LUXEMBOURG

ODELLO, R.P.
Ingénieur
Total Exploration Production
204, rond point du Pont de Sèvres
F - 92516 BOULOGNE BILLANCOURT

PAGEAUD, D.
Direction Commerciale
Société ECA
17, ave du Château
F - 92194 MEUDON CEDEX

PALLA, E.
Marine Systems Engineer
Tecnomare SpA
San Marco 2091
I - 30124 VENEZIA

PATTEN, T.
Mechanical Engineer
Consultant
15 Frogston Road West
GB - EDINBURGH EH 10 7AB

PEINADO, M.
Ingénieur
Institut Français du Pétrole
1-4, ave de Bois Préau
F - 92506 RUEIL MALMAISON

PEREGO, M.
Reservoir Engineer
AGIP SpA
P.O.B. 12069
I - 20120 MILANO

PEREZ DE GUZMAN, J.
Managing Director
Inidermott, S.A.
Av. Brasil, 17 - 9°C
E - 28020 MADRID

PIEUCHOT, M.
Ingénieur Conseil
Rep. Sercel
rue Claude Tillier
F - 58330 SAINT SAULGE

POMMIER, G.
Directeur Total Exploration
Total Compagnie Française des
Pétroles
39-43, quai André Citroen
F - 75739 PARIS CEDEX 15

PROYER, G.
Senior Reservoir Engineer
Wintershall Aktiengesellschaft
Postfach 10 40 20
D - 3500 KASSEL

PUISAIS, X.
Ingénieur
ETPM
33-35 rue d'Alsace
F - 92531 LEVALLOIS PERRET

PULLAN, P.
Managing Director
Advanced Production Technology Ltd
7, Albert Street
GB - ABERDEEN AB1 1XX

PUTZ, A.
Chef Service Récupération Chimique
Elf Aquitaine
SNEA(P)
Boussens
F - 31360 SAINT-MARTORY

PUYO, A.
Ingénieur
SNEA(P)
26, ave des Lilas
F - 64018 PAU CEDEX

REDERON, C.
Directeur Coord. Planif.
Total Compagnie Française des
Pétroles
39-43, quai André Citroën
F - 75739 PARIS CEDEX 15

REISS, H.
Ingénieur
Société Nationale Elf Aquitaine
Tour Général La Défense 9
5, place de la Pyramide
F - 92088 PARIS CEDEX 22

RENARD, B.
Technical Manager of the East
Frigg Project
Elf Aquitaine Norge A/S
P.O. Box 168
N - 4001 STAVANGER

REY-GRANGE, A.
Partner/Business Development
Manager
SEAMET
61, rue de la Garenne
F - 92310 SEVRES

RICHARDSON, A.
Sales Manager, Chartered Engineer
APV Hall International
P.O. Box 555
GB - HYTHE ST DARTFORD, Kent DA11EP

ROBLIN, J.P.
ACB Nantes
Alsthom Atlantique
F - NANTES

ROEMELING, J.U.
Mech. Eng., Lic. Tech., Head of
Hydrodynamic Dept.
Danish Maritime Institute
Hjortekaersvej 99
DK - 2800 LYNGBY

ROONEY, P.
Engineer
IIRS
Ballymun Road
IRL - DUBLIN 9

SADARE, R.A.
Petroleum Geologist
Organization of Petroleum
Exporting Countries
Obere Donau Strasse 93
A - VIENNA 1020

SAHUQUET, B.
Ingénieur
Elf Aquitaine
Centre Micoulau
Avenue Président Angot
F - 64000 PAU

SANDER, A.
Engineer
Howaldtswerke - Deutsche Werft AG
Department Hok
Postfach 11 14 80
D - 2000 HAMBURG 11

SANTI, G.
Project Engineer
Sub Sea Oil Services S.p.A.
Via della Scafa, 19
I - 00054 FIUMICINO

SAUZADE, D.
Ingénieur
IFREMER Projet SAGA
75, rue Floralie
F - 13008 MARSEILLE

SCHNEIDER, R.
Geophysics
Projektleiting Energieforschung
in der KFA Jülich
KFA/PLE
Box 1913
D - 5170 JUELICH

SCHULZE - GATTERMANN, R.
Sales Manager
Prakla-Seismos GmbH
Buchholzer Strasse 100
P.O.B. 510530
D - 3000 HANNOVER 51

SCOTT, R.
Manager Civil Istructural Engineer
BP International Ltd
Britannic House
Moor Lane
GB - LONDON EC2 9BY

SEDILLOT, F.
Ingénieur en Chef
C.G. Doris
58 A, rue du Dessous des Berges
F - 75013 PARIS

SIMANDOUX, P.
Institut Français du Pétrole
1-4, ave de Bois Préau
F - 92506 RUEIL MALMAISON

SIMONSEN, I.
M. Sc.
National Agency of Technology
Tagensvej 135
DK - 2200 COPENHAGEN N

SIVENAS, P.
Managing Director
Public petroleum Corporation
of Greece
199 Kifissias Ave
GR - 151 24 MAROUSSI, Athens

SNOWDEN, D.
Civil Engineer
Taywood Engineering
Offshore and Marine Engin. Ltd
345 Ruislip Road
GB - SOUTHALL, Middx UB1 2OX

SOEDJONO, N.
Kadin Produksi Migas EP
Pertamina Head Office
Jl Merdeka Timur 1A
17 th Floor
Indonesia - JAKARTA

SOERYOGO,
Pertamina
Jalan Medan Merdeka Ti Mur 1A
Indonesia - JAKARTA, PUSAT

SOILLE, P.
Ingénieur
Distrigaz S.A.
31, ave des Arts
B - 1040 BRUXELLES

SOKOLOV, V.
Journaliste
Agence de Presse Novosti
41, bd de la Petrusse
L - LUXEMBOURG

SOLIER, J.
  Ingénieur
  Total Compagnie Française des
  Pétroles
  39-43, quai André Citroen
  F - 75739 PARIS CEDEX 15

SOUDET, H.J.
  Ingénieur Géologue
  SNEA (8)
  Laboratoire de Géologie
  Rue du Président Angot
  F - 64000 PAU

STASCHEN, D.
  Regierungsdirektor
  Bundesministerium für
  Wirtschaft
  BMWI
  Postfach
  D - 5300 BONN

STRANGE, S.
  Senior Petroleum Engineer
  Danish Energy Agency
  Landemaerket 11
  DK - COPENHAGEN K

TASSINI, P.
  Manager Engineering Technologies
  AGIP SpA
  P.O. Box 12069
  I - 20120 MILANO

TEYSSEDRE, J.
  Ingénieur
  Total - CFP
  204, rond-point du Pont de Sevre
  F - 92516 BOULOGNE BILLANCOURT

THOMSON, M.
  Chartered Mechanical Engineer
  ICI Petroleum Services Ltd
  Wilton headquarters
  P.O. Box 90
  GB - WILTON, MIDDLESBROUGH TS6 8JE

TINDY, R.
  Dir., Adj. au Dir. du Budjet
  Institut Français du Pétrole
  1-4, ave de Bois Préau
  F - 92500 RUEIL MALMAISON

TOPHAM, W.H.
  Control Engineer
  BP International Ltd
  Central Engineering Dept
  Britannic House
  Moor Lane
  GB - LONDON EC2Y 9BU

TOURRE, J.J.
  Ingénieur
  Total CFP
  204, Rd Pt du Pt de Sèvres
  F - 92516 BOULOGNE BILLANCOURT CEDEX

TREWHELLA, E.
  Commercial Manager
  Harwell Laboratory
  Marketing and Sales Dept.
  Atomic Energy
  Research Establishment
  GB - HARWELL OX11 ORA

TURCICH, T.
  Manager, Development Engineering
  Camco, incorporated
  P.O. Box 14484
  USA - HOUSTON, Texas 77221

TUREGANO VALIENTE, J.A.
  Exploration Manager
  CAMPSA, Departamento de
  Investigacion y Production
  de hidrocarburos - Planta 9a
  Capitan Haya 41
  E - 28020 MADRID

UEBEL, H.
  Conseil des Communautés
  Européennes
  170, rue de la Loi
  B - 1048 BRUXELLES

VACHE', M.
  Chef de Projet
  C.G. Doris
  58a, rue du Dessous des Berges
  F - 75013 PARIS

VAN ASSELT, D.
  Commission of the European
  Communities, DG Energy
  200, rue de la Loi
  B - 1049 BRUXELLES

VAN DER BURGH, J.
Head Reservoir Engineering SIPM
Shell Internationale Petroleum
Maatschappij B.V.
EP/22.1
P.O. Box 162
NL - 2501 AN THE HAGUE

VAN ESPEN, M.
Commission of the European
Communities, DG Energy
200, rue de la Loi
B - 1049 BRUXELLES

VAN HERWIJNEN, J.
Mechanical Engineer
Shell Internationale Petroleum
Maatschappij B.V.
EP/23.3
P.O. Box 162
NL - 2501 AN THE HAGUE

VAN LUIPEN, P.
Senior Project Manager
Bomag-Menck GmbH
Werner V. Siemens Str.2
D - 2086 ELLERAU

VENTRE, J.
Ingénieur
SNEA(P)
Division Gisements
26, ave des Lilas
F - 64000 PAU

VIDALINC, D.
Président
Société Rhone Poulenc
Pétrole Services
125 bis, rue Raspail
F - 69150 DECINES

VIGIER, L.
Géophysicien
BEICIP
232, ave Napoléon Bonaparte
F - 92500 RUEIL-MALMAISON

VISSER, W.
Shell Exploration and Production
Shell-Mex House Strand
GB - LONDON WC2R ODX

WALKER, P.A.
Commission of the European
Communities, DG Employment, social
affairs and education
L - LUXEMBOURG

WALKER, R.I.
Chartered Engineer
Britoil Plc
150 St Vincent Street
GB - GLASGOW, Scotland

WEIR, G.
Oil Recovery Projects Division
Atomic Energy Etablishment
Winfrith
Aee Winfrith
GB - DORCHESTER, Dorset DT2 8DH

WEYER, T.
ZF-HERION-Systemtechnik GmbH
Postfach 2520
D - 7990 FRIEDRICHSHAFEN

WIERCZEYKO, E.
Geophysiker
Prakla-Seismos GmbH
Buchholzer Str. 100
Postfach 51 05 30
D - 3000 HANNOVER 51

WIESSNER, F.
Dipl.-Ing.
Linde AG.
Carl von Linde Strasse
D - 8023 HOELLRIEGELSKREUTH

WILKE, K.
Manager Development New Systems
AEG-Schiffbau
Abt. A44 E5
Steinhöft 9
D - 2000 HAMBURG 11

WILLEMSE, C.
Research Engineer
Heerema Engineering Service BV
P.O. Box 9321
NL - 2300 PH LEIDEN

WILLIAMS, D.
Engineering Services, Manager
Offshore Systems Engineering Ltd.
Boundary Road
Harfaers Industrial Estate
GB - GREAT YARMOUTH, Norfolk

WILLM, P.
Directeur Scientifique Génie Mar.
Institut Français du Pétrole
1-4, ave de Bois Préau
F - 92506 RUEIL MALMAISON

WOEBCKE, H.
  Manager Technology Development
  Stone and Webster Engineering
  Corporation
  145 Summer St
  P.O. Box 2325
  USA — BOSTON MA. 02107

WOODHEAD, I.A.
  Production Engineer
  BP Exploration
  Britannic House
  Moorlane
  GB — LONDON EC2Y 9BU

WOOTTON, L.
  Company Director
  Atkins Research and Development
  Woodcote Grove
  Ashley Road
  GB — EPSOM, Surrey KT3 5BW

ZIEMANN, F.
  Germanischer Lloyd
  Vorsetzen 32
  D — 2000 HAMBURG 11

INDEX OF AUTHORS